4th International Conference on Environment and Renewable Energy (ICERE 2018)

IOP Conference Series: Earth and Environmental Science
Volume 159

Da Nang, Vietnam
25 - 27 February 2018

Editor:

Vu Dinh Tuan

ISBN: 978-1-5108-6477-1
ISSN: 1755-1307

Printed from e-media with permission by:

Curran Associates, Inc.
57 Morehouse Lane
Red Hook, NY 12571

Some format issues inherent in the e-media version may also appear in this print version.

Printed by Curran Associates, Inc. (2018)

For permission requests, please contact the Institute of Physics at the address below.

Institute of Physics
Dirac House, Temple Back
Bristol BS1 6BE UK

Phone: 44 1 17 929 7481
Fax: 44 1 17 920 0979

techtracking@iop.org

Additional copies of this publication are available from:

Curran Associates, Inc.
57 Morehouse Lane
Red Hook, NY 12571 USA
Phone: 845-758-0400
Fax: 845-758-2633
Email: curran@proceedings.com
Web: www.proceedings.com

Table of contents

Volume 159

2018 4th International Conference on Environment and Renewable Energy (ICERE 2018)

25–27 February 2018, Da Nang, Vietnam

Accepted papers received: 23 May 2018
Published online: 14 June 2018

Resources and Environmental Engineering

Renewable Energy and Building Energy System

Biomass Pyrolysis and Combustion

Power and Mechanical Engineering

Urban Engineering and Sustainable Development

IOP Conf. Series: Earth and Environmental Science **159** (2018) 011001 doi:10.1088/1755-1315/159/1/011001

PREFACE

It is our great pleasure to welcome you to 2018 4th International Conference on Environment and Renewable Energy (ICERE 2018) which was held in 25-27 February 2018, Da Nang, Vietnam. ICERE 2018 is dedicated to issues related to Environment and Renewable Energy.

ICERE 2018 is supported by Duy Tan University, Vietnam. The major goal and feature of the conference is to bring academic scientists, engineers, and industry researchers together to exchange and share their experiences and research results, and discuss the practical challenges encountered and the solutions adopted.

The rector of Duy Tan University, Vietnam Sir Le Cong Co did opening remark for the conference. Prof. Kondo Akira from Osaka University, Japan have attended the conference as keynote speaker and delivered keynote speech on 25th February. Prof. Wallace C. H. Choy from Hong Kong University; Assoc. Prof. Dam Sao Mai from Industrial university of HCMC, Vietnam have attended the conference as keynote speakers and delivered keynote speeches regarding latest information in their respective expertise areas on 26th February. There were six sub-sessions and one poster session with different topics: Power and Mechanical Engineering, Renewable Energy and Energy Saving, Biomass Pyrolysis and Combustion, Power Generation and Energy System Management, Environmental and Chemical Engineering, Environmental Management and Sustainable Development. It was a golden opportunity for students, researchers and engineers to interact with the experts and specialists to get their advice or consultation on technical matters, sales and marketing strategies. On the third day of the conference, we had academic visit in two campuses of Duy Tan University visiting their laboratories, exchanging ideas with professional staff there.

This proceeding presents a selection from papers submitted to the conference from universities, research institutes and industries. All papers were subjected to peer-review by conference committee members and international reviewers. The papers selected depended on their quality and their relevancy to the conference. The volume tends to present to the readers the recent advances in the field of Environment and Renewable Energy and various related areas, such as Energy-saving Technology, Energy Materials, Environmental Protection, Nuclear Energy Engineering, Power Machinery and Engineering, etc.

We would like to thank all the authors who have contributed to this volume, and also the organizing committee, reviewers, speakers, chairpersons, sponsors and all the conference participants for their support to ICERE 2018.

Dr. Vu Dinh Tuan
Institute of Research & Development, Duy Tan University, 3 Quang Trung, Da Nang, Vietnam
March 14, 2018

Organizing Committee

Conference General Co-Chairs

Prof. Kondo Akira, Graduated School of Engineering, Osaka University, Japan
Prof. Wallace C.H. Choy,Hong Kong University, Hong Kong
Assoc. Prof. DAM SAO MAI, Industrial university of HCMC, Vietnam

Program Co-Chairs

Prof. Tseung-Yuen Tseng, National Chiao Tung University, Taiwan
Prof. Nga P. Le, Vietnam National University, Vietnam
Dr. Thi Minh Phương NGUYEN, Dean of Faculty of Environment and Chemical Engineering, Duy Tan University, Da Nang, Vietnam
Dr. Vu Dinh Tuan, Faculty of Environment and Chemical Engineering, Duy Tan University, Da Nang, Vietnam
Dr. Duy Quang Dao, Duy Tan University, Institute of Research and Development, Da Nang, Vietnam

Contact Chair

Ms. Alice Lin, Hong Kong Chemical, Biological & Environmental Engineering Society

International Technical Committee

Prof. Nga P. Le, Vietnam National University, Vietnam
Prof. Róbert Németh, Univesrity of Sopron, Hungary
Prof. Asher Brenner, Ben-Gurion University of the Negev, Israel
Prof. K.S.K. Rao Patnaik, Andhra University, India
Prof. Piotr Sołowiej, University of Warmia and Mazury in Olsztyn, Poland
Prof. Geum-Bae Cho, Chosun University, Korea
Prof. Asher Brenner, Ben-Gurion University of the Negev, Israel
Prof. ENRIQUE ROMERO-CADAVAL, UNIVERSITY OF EXTREMADURA, Spain
Prof. Ali CHEKNANE, Amar Telidji University of Laghouat, ALGERIA
Prof. Subhransu Sekhar Dash,SRM University, India
Prof. Jian Wang, Kingston University, UK
Prof. Nakorn TIPPAYAWONG， Chiang Mai University, Thailand
Prof. Joong Hee Lee, Chonbuk National University, Korea
Prof. Chan Lee, University of Suwon, Korea
Assoc. Prof. Christia Meidiana,Brawijaya University, Indonesia
Assoc. Prof. Yezi Dai, Soochow University, China
Assoc. Prof. Youn-Suk Son, Pukyong National University, South Korea
Assoc. Prof. Fang-Lin Chao, Chaoyang University of Technolog,Taiwan

IOP Conf. Series: Earth and Environmental Science **159** (2018) 011002 doi:10.1088/1755-1315/159/1/011002

IOP Conf. Series: Earth and Environmental Science **159** (2018) 011003 doi:10.1088/1755-1315/159/1/011003

Peer review statement

All papers published in this volume of *IOP Conference Series: Earth and Environmental Science* have been peer reviewed through processes administered by the proceedings Editors. Reviews were conducted by expert referees to the professional and scientific standards expected of a proceedings journal published by IOP Publishing.

IOP Conf. Series: Earth and Environmental Science **159** (2018) 012001 doi:10.1088/1755-1315/159/1/012001

Hydrochar Generation from Hydrothermal Carbonization of Organic Wastes

P Kantakanit[1], N Tippayawong[1]*, S Koonaphapdeelert[2] and A Pattiya[3]

1 Department of Mechanical Engineering, Chiang Mai University, Chiang Mai, 50200, Thailand
2 Energy Research and Development Institute - Nakornping, Chiang Mai University, Chiang Mai, 50100, Thailand
3 Department of Mechanical Engineering, Mahasarakham University, Mahasarakham, 44150, Thailand
E-mail: n.tippayawong@yahoo.com

Abstract. Organic wastes from leftover food contributes to a large fraction of municipal solid waste. Because of its high moisture content, it is not suitable to be processed in a conventional way of burning. This study is interested in the possibility to apply hydrothermal carbonization (HTC) process to organic wastes. The objective was to study characteristics of HTC products as a function of temperature and residence time (180-250°C and 1-6 h). Samples of organic waste from discarded foods were processed in a HTC reactor. From the result of hydrochar analysis, it was shown that highest heating value and energy densification were 30 MJ/kg and 1.80 respectively, at 250°C and 3 h. The aqueous solution was found to contain minor concentration of fertiliser nutrients. These products may be used as fuel and chemicals as well as applications in agriculture. It was confirmed that valuable products can be generated from organic waste using hydrothermal carbonization.

1. Introduction

Organic wastes from leftover food is one of the most challenging issues society is encountering globally. Large percentages of the food intentionally prepared for consumption is wasted, with significant environmental and socio-economic implications. Food waste represents a significant fraction of municipal solid waste which is largely underutilized. There are more than 200 kg/day of food wastes from Chiang Mai University canteens. It is not only cooked foods but also fruit juices, seasoning or vegetables from cooking and fruits peels. These wastes contain high moisture and become bad with rancid smell quickly. Sustainable food waste management is urgently needed. In this work, processing of these organic wastes by hydrothermal carbonization (HTC) is chosen.

HTC is a thermochemical conversion process, reacting wet organic substrates in hydrothermal conditions or in hot pressurized liquid water [1]. HTC process conditions are usually in mild temperatures range (180-260°C) at 15-25 bar. Typical residence times range between a few to several hours. Hot pressurized water exhibits a higher ion product than at ambient conditions, behaving as an acid/base catalyst precursor and acting as a solvent and a reactant [2]. The main product of HTC process after conversion is hydrochar, a carbonaceous solid organic material. Its chemical composition is similar to low rank coals such as peat or lignite [3]. The interest in HTC is motivated by its suitability to handle high moisture materials and application of hydrochar as solid fuel [4]-[7].

In this work, application of HTC to food waste was carried out. The process has potential to lower the overall cost for waste disposal. The characteristics of the obtained hydrochar and the liquid filtrate product were investigated.

2. Materials and Methods

2.1. Raw materials

Organic wastes were obtained from a local fresh food market in Chiang Mai University campus, Thailand. They were then sorted to remove solid impurities such as bones, small pieces of wood, seeds and stones. The waste sample was subsequently dried at 105°C for 24 h to achieve constant weight and then comminuted and sieved into uniform size. It was ground to particle diameter of less than 2.36 mm and stored in a desiccator for later use. All chemicals used were of analytical grade from Thailand Institute of Scientific and Technological Research in Bangkok, Thailand.

2.2. HTC experiments

The experiments were carried out in a 10 L cylindrical pressure reactor equipped with 14 kW heaters and internal sensors for temperature control, shown in Figure 1. The reactor was designed and constructed by Febix International Co., Ltd (Chiang Mai, Thailand). Dried organic waste (260 g) was dispersed in 5.7 L of distilled water contained in a round-bottomed flask. The mixture was stirred until a homogeneous suspension was created. It was then loaded to the HTC reactor. The heater was switched on. The temperatures set were 180, 215 and 250°C. It was kept at this condition for 1, 3 and 6 h. After which the heater was turned off. When the reactor was cooled down to room temperature and pressure decreased to ambient level, the products (hydrochar and liquid product) were discharged through drain pipe to a temporary container.

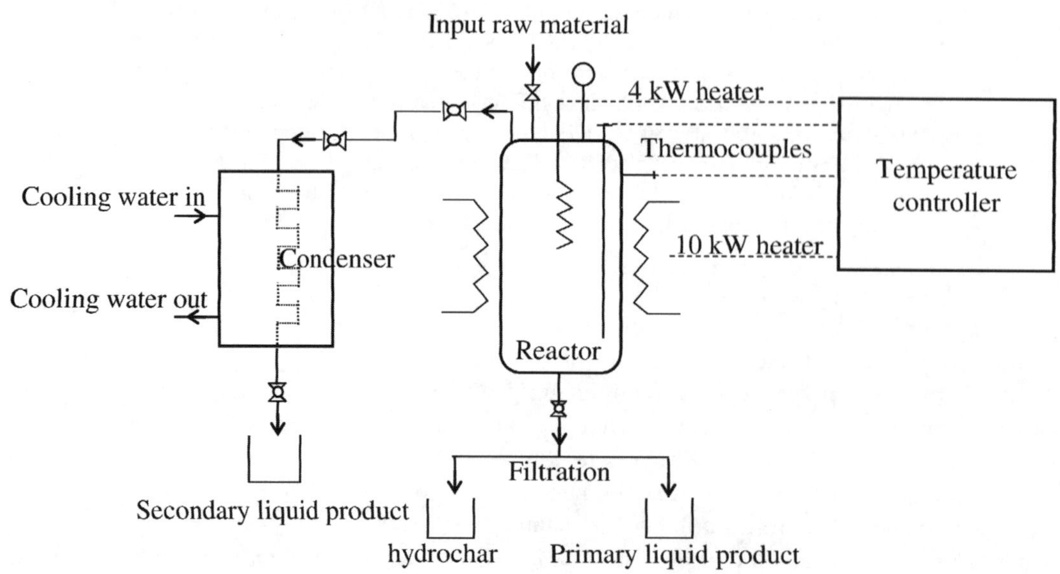

Figure 1. HTC experimental setup.

The solid hydrochar and liquid product were subsequently separated by filtration and all liquid product were collected in a container. The hydrochar was immediately dried in an oven for 24 h at 105°C, then reduced the size and collected in an airtight container for subsequent analyses. The hydrochar samples obtained were denoted according to carbonization conditions as H180-1, H180-3, H180-6, H215-1, H215-3, H215-6, H250-1, H250-3 and H250-6, respectively.

2.3. Characterization of products

The raw organic wastes and hydrochars were analysed for their chemical compositions. Ultimate and proximate analyses were conducted by the Thailand Institute of Scientific and Technological Research (Bangkok, Thailand), according to ASTM D 7582, D 5373, D 4239 and D 5865 standards. Liquid product was analysed by Central Laboratory Co., Ltd (Chiang Mai, Thailand).

3. Results and discussion

3.1. Hydrochars

Characteristics of the food wastes and their hydrochars were evaluated to study the changes of the raw material with respect to physical and chemical components. Figure 2 shows physical appearances for the raw food waste, and its hydrochars at various conditions. It was observed that at more severe carbonization conditions (higher temperature and longer time), hydrochars tended to become darker in colour.

Figure 2. Physical appearances of hydrochars at (a) 180˚C 1h, (b) 180˚C 3h, (c) 180˚C 6h, (d) 215˚C 1h, (e) 180˚C 3h, (f) 180˚C 6h, (g) 250˚C 1h, (h) 250˚C 3h and (i) 250˚C 6h.

Table 1 shows solid yields, ultimate analysis, proximate analysis, and heating values of the hydrochars. The hydrochar yields obtained from the HTC were in the 23-46% range on a dry basis. The values here were lower than the reported by Basso et al. [1] and Li et al. [8], However, regarding energy densification which was defined as a ratio of HHV between the hydrochar and the starting material, this study showed higher values than those reported. It should be noted that treatment conditions were not exactly similar. However, the trend was clear that solid products decreased with increasing residence time. As shown in Figure 3, increasing the carbonization time and temperature led to reduction in yield due to larger overall decomposition of biomass material, or carbon, existing in organic waste as volatile matter, progressively reacted and diffused out.

Increase in fixed carbon and decrease in volatile matter between the raw materials and hydrochars confirmed that carbonization took place. For the first 1 h, only volatile matter at the surface was removed. At longer reaction times, volatile matter in the interior of the particle was also removed, but with more difficulty than at the surface. Figure 4 shows that increasing HTC temperature and time tended to increase fixed carbon and decrease volatile content of the hydrochar. This was in line with change in elemental composition as a result of carbonization. HTC of the organic waste led to an increase in the carbon content from 45% (organic waste) to 58-64% in the hydrochar samples. The high fraction of carbon was retained in the hydrochar products, while the rest mainly remained in the organic compounds that dissolved in the aqueous, similar to those reported in the literature [9, 10]. Transformation of organic material to carbon rich material or coalification process may be visualised using a van Krevelen diagram [11]. Plot of atomic H/C and O/C ratios for hydrochars in Figure 5 allows understanding of the evolution of the solid composition. HTC acts as a process for concentrating the carbon content of biomass, hence H/C and O/C atomic ratios are expected to decrease due to the reaction. The results confirmed such trend. Notably, the raw organic waste had a very high O/C atomic ratio. Increase in severity of HTC conditions (higher temperatures or longer times), the H/C and O/C ratios of hydrochars were expected to approach the values similar to those associated with peat and lignite. Further carbonization was possible to increase the extent of coalification. Reaction temperature selected was 180-250°C in which HTC at this condition was reported to be able to sufficiently generate solid product that was similar to coal quality [9].

Table 1. Chemical characteristics and properties of dried organic waste and hydrochars.

	H180-1	H180-3	H180-6	H215-1	H215-3	H215-6	H250-1	H250-3	H250-6
Yield (% w/w)	46.2	30.8	26.9	36.5	30.8	25.0	34.6	26.9	23.1
Ultimate analysis (%)									
C	59.52	58.61	61.98	57.94	63.37	60.32	60.19	64.26	62.48
H	8.8	8.38	8.52	8.37	8.99	8.48	8.57	8.82	8.47
N	4.32	3.57	3.75	5.54	3.08	3.32	4.38	3.12	3.21
O	23.32	27.44	22.49	24.96	20.98	24.62	23.54	21.27	20.83
Proximate analysis (%)									
Ash	3.79	1.78	3.06	2.86	3.43	3.05	3.08	2.36	4.85
Volatile matter	86.93	84.95	84.53	85.53	86.63	84.97	85.41	84.96	82.16
Fixed carbon	9.28	13.27	12.41	11.61	9.94	11.98	11.51	12.68	12.99
HHV (MJ/kg)	28.28	26.86	28.62	27.36	29.12	28.24	28.03	30.12	28.62
Energy densification	1.69	1.60	1.71	1.63	1.74	1.68	1.67	1.80	1.71

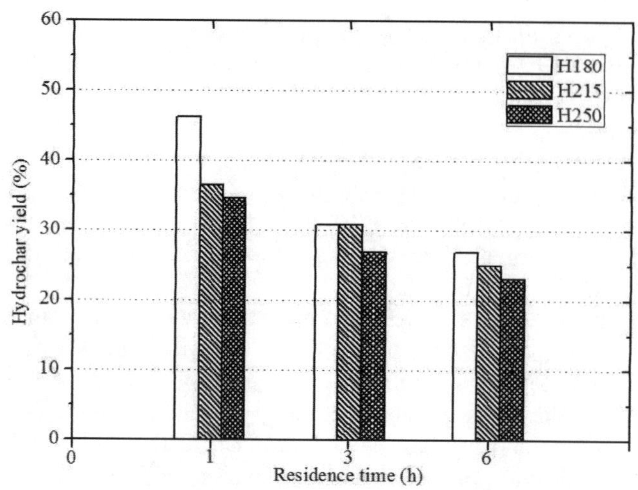

Figure 3. Solid yields of hydrochar.

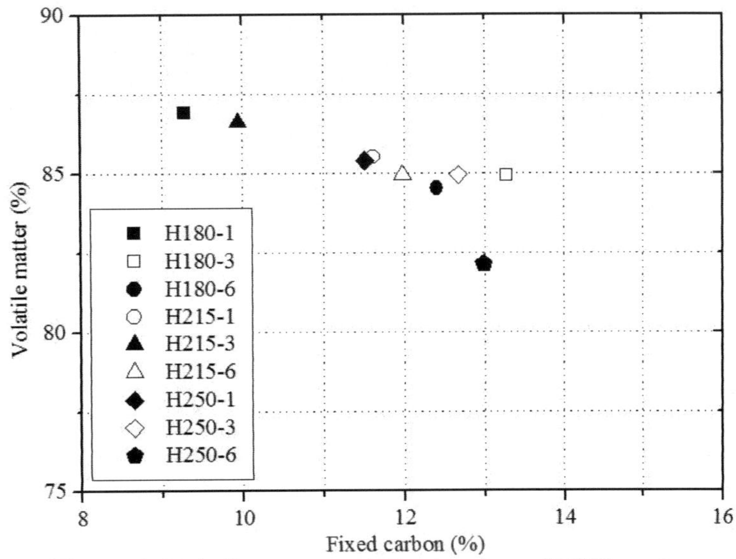

Figure 4. Volatile matter and fixed carbon of HTC products.

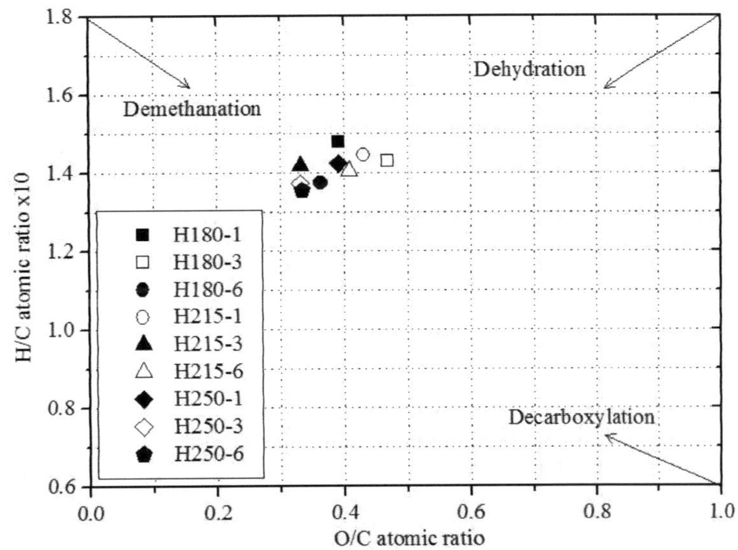

Figure 5. Van Krevelen diagram of HTC products.

3.2. Liquid by-product

It is generally known that water plays a significant role as a solvent and reactant in the HTC process. The liquid phase is expected to contain a high loading of organics and inorganics. It may be recycled as a nutrient solution to agricultural lands. The liquid filtrate obtained was completely sterile after the HTC at 180-250°C. A slight drop in pH of the aqueous phase was observed after HTC reaction, becoming slightly acidic. It is noted here that original pH of deionized water was neutral. The decreasing pH may have some effect to the colour of liquid products. The colours tended to be darker (shown in Figure 6) when the pH was decreased. The liquid phase was also found to contain some nutrient content.

Figure 6. Colour of liquid products of each HTC process conditions.

4. Conclusion

In this work, HTC of food waste was performed at several process conditions. Solid and liquid products were collected and characterized. It was suggested that the hydrochar produced was suitable for energy purposes as its heating value as high as 30 MJ/kg can be achieved. Moreover, its elemental composition was becoming closer to that of peat and lignite, thus the hydrochar may be used in co-combustion with other coals. It was demonstrated that the HTC process offered alternative technique for conversion of organic waste to value-added products. Further works may be performed to gain better understanding of the underlying process, to characterize related properties, and to identify applications for these products.

5. Reference

[1] Basso D, Weiss-Hortala E, Patuzzi F, Castello D, Baratieri M and Fiori L 2015 *Bioresour. Technol.* **182** 217-24
[2] Kruse A and Dinjus E 2007 *J. Supercritical Fluids* **41** 361–79
[3] Schuhmacher J P, Huntjens F J and van Krevelen D W 1969 *Fuel* **39** 223–34
[4] Castello D, Krus A and Fiori L 2014 *Chem. Eng. Res. Des.* **92** 1864–75
[5] Hwang I, Aoyama H, Matsuto T, Nakagishi T and Matsuo T 2012 *Waste Manage.* **32** 410–6
[6] Lu L, Namioka T and Yoshikawa K 2011 *Appl. Energy* **88** 3659-64
[7] Román S, Nabais J M V, Laginhas C, Ledesma B and González, J F 2012 *Fuel Process. Technol.* **103** 78–83
[8] Li L, Diederick R, Flora J R V and Berge N D 2013 *Waste Manag.* **33** 2478-92
[9] Heilmann S M, Davis H T, Jader L R, Lefebvre P A, Sadowsky M J, Schendel F J, von Keitz M G, Valentas K J 2010 *Biomass Bioenergy* **34** 875–82
[10] Heilmann S M, Jader L R, Sadowsky M J, Schendel F J, von Keitz M G, Valentas K J 2011 *Biomass Bioenergy* **35** 2526-33
[11] Sevilla M and Fuertes A B 2009 *Carbon* **47** 2281–89.

Acknowledgments

The authors would like to acknowledge the support from the Fund for Energy Conservation Promotion, Energy Policy and Planning Office, and Chiang Mai University. We also wish to thank the Energy Research and Development Institute-Nakornping staff for technical support.

IOP Conf. Series: Earth and Environmental Science **159** (2018) 012002 doi:10.1088/1755-1315/159/1/012002

Experimental Investigation of hot Water Generation from Small $CaO/Ca(OH)_2$ Thermochemical Energy Storage System

N Takham[1] and N Tippayawong[1]

1 Department of Mechanical Engineering, Chiang Mai University, Chiang Mai, 50200, Thailand
E-mail: n.tippayawong@yahoo.com

Abstract. Thermochemical energy storage (TCS) is a very promising method for energy applications, storage, transport and thermal generation because of its high energy density and negligible heat loss. Several storage materials are being explored, in which $CaO/Ca(OH)_2$ is among the most attractive systems for further investigations. In this work, a small Ca based TCS system was developed, built and tested for hot water generation. Hydration and dehydration reactions involving CaO and $Ca(OH)_2$ powders were performed. Performance in terms of temperature rise in heat transfer fluid and its efficiency during successive discharging/charging cycles was evaluated and discussed.

1. Introduction

Energy storage is crucial in balancing supply and demand of energy. It enables increasing utilization and contribution of renewable energy, helps reduce energy consumption and minimise environmental pollution [1]. One of the established energy storage technologies is thermal energy storage. It is defined as temporary storage of thermal energy in the form of hot or cold substances for later use. There are three main types of thermal energy storage, namely sensible, latent and thermochemical [2]. Thermochemical energy storage (TCS) is considered to be highly attractive as it can provide higher energy storage density than the other types [3]. TCS is mainly based on a reversible chemical reaction: $C + Heat \leftrightarrow A+B$. When material (C) absorbs external heat (solar, geothermal or industrial waste heat) and decomposed into A and B. The materials (A and B) are divided by physical means and stored in different containers. When materials A and B are combined together, the reverse reaction occurs generating C and releases the stored chemical energy. As long as the energy storage system is constituted by a closed loop and appropriately stored, the energy storage period is theoretically unlimited. Generally, a TCS system can be divided into three main processes: charging, storing, discharging [4], shown in Figure 1.

TCS material is a critical component of the storage system. The performance depends on the operating conditions, cycle stability, and operating cost. An appropriate storage medium is extremely important when establishing a storage system. Desirable properties for good thermochemical materials include (i) high energy density; (ii) good heat-transfer performance; (iii) being environment friendly and reasonable cost; (iv) no side reaction under the operating conditions; and (v) excellent cycling stability and high conversion rate. So far, more than 70 kinds of thermochemical reaction have been studied [1-5]. A Ca based $CaO/Ca(OH)_2$ hydration/dehydration reaction couple appear to be among the most suitable TCS system [6].

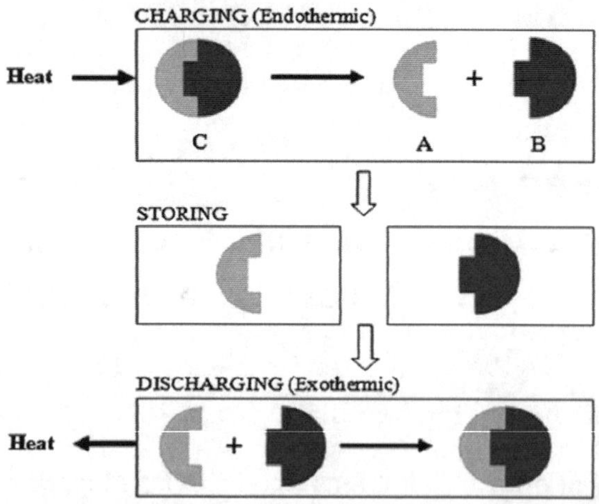

Figure 1. Processes in closed TCS cycle: charging, storing and discharging [4].

At present, the $CaO/Ca(OH)_2$ system has been intensively studied by a number of researchers. Equation (1) shows the basic reaction principle of inorganic hydroxide system. Energy changes during $Ca(OH)_2$ dehydration and rehydration process are shown in Figure 2. It can be seen from that at atmospheric pressure, an input energy 148.6 kJ is needed for decomposing 1 mol $Ca(OH)_2$ at 25°C into CaO and H_2O at 510°C. If CaO is stored as a solid at 25°C, and the product water is condensed into liquid water at 25°C then, 85 kJ heat is released as a form of sensible and latent heat. Thus, the remaining 63.6 kJ is actually stored in the form of chemical energy, which is equivalent to 43% of the total input energy.

$$Ca(OH)_2\ (s) \leftrightarrow CaO\ (s) + H_2O\ (g) \quad \Delta H_R = 104\ \text{kJ/mol} \tag{1}$$

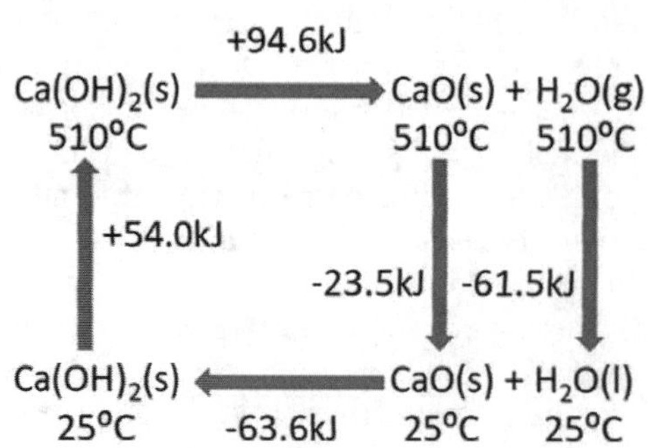

Figure 2. Energy storage/release process in a $Ca(OH)_2/CaO$ system.

$Ca(OH)_2$ dehydrates between 400 and 508°C with the water vapor partial pressure ranging from 0.1 to 1 bar, so it can be used to store high-temperature thermal energy. CaO can be stored at room temperature for a long time. When water vapor is passed into reactors filled with CaO, the reverse exothermic reaction occurs and generates high-grade heat. The $CaO/Ca(OH)_2$ system has many advantages, such as high storage density, fast reaction kinetics, stability and safety, being nontoxic and having low price, and thus has attracted enormous attention especially in concentrated solar power and combine heat and power.

For the $CaO/Ca(OH)_2$ system, one of the dominant disadvantages is related to poor mechanical performance of the material. CaO particles are known to break during hydration/dehydration reactions which can become entrained by the gas and create large pressure drop. A direct-type reactor proposed by Schaube et al. [7] utilized powder reactants was found to perform well but with losses from pressure drop. Research on TCS reactor systems remain to be explored. Further research is still needed to gain better understanding in designing these systems, to improve their efficiencies, and to address other practical issues. In this work, a small TCS unit based on the $CaO/Ca(OH)_2$ system was developed. The powder reactants were used. Its performance was investigated.

2. Materials and Methods

2.1. Material

For the initial reactant, CaO was purchased and obtained from Elago Enterprises Pty Ltd., based on the products data sheet showed in Table 1.

Table 1. Specification of calcium oxide.

	Specification
Assay	90.0% minimum
Description	White powder
Identity (FTIR)	conforms
Insoluble matter in HCI	0.5% maximum
Chloride (Cl)	0.05% maximum
Sulphate (SO_4)	0.5% maximum
Heavy metals (as Pb)	0.005% maximum

2.2. Experimental setup and procedure

Figures 3 and 4 show schematics of the experimental setup and actual reactor and heat exchanger used in the experiment. For each experiment, a total of 28 g CaO or ½ mole was used. It was initially loaded into the cylindrical container acting as the reactor.

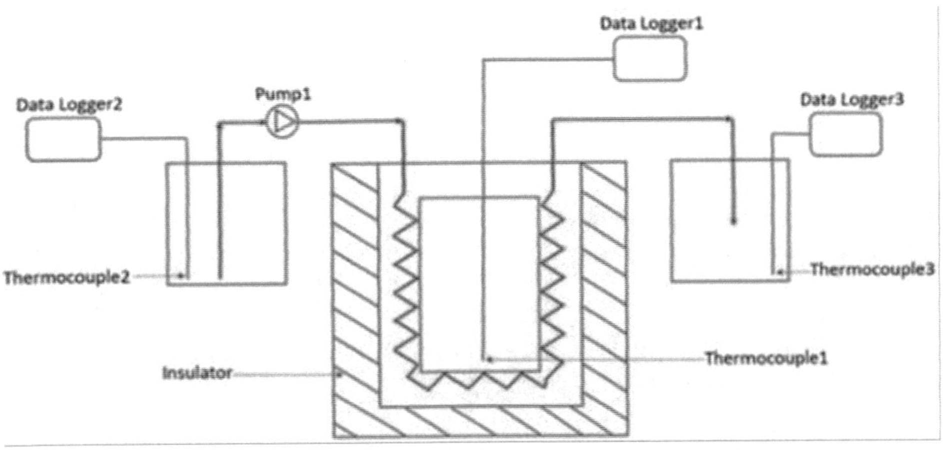

Figure 3. Layout of the reactor and experimental setup.

For discharging process, the initial temperature of the fresh sample of CaO and the reactor was recorded. The water syringe was started to deliver jet spray of liquid water of 9 g into the reactor for hydration reaction. The hydration reaction was started. The temperature in the reactor was increased. Water was used as a heat transfer fluid in with flowrate of 0.1 Lpm. Once the reactor reached the maximum temperature, the pump was kept switched on for circulating the cooling water to receive the energy release from the chemical energy until there was no temperature difference between the reactor

IOP Conf. Series: Earth and Environmental Science **159** (2018) 012002 doi:10.1088/1755-1315/159/1/012002

and heat transfer fluid. The data recording process was started immediately after delivering the sprayed water and monitored continually over the reaction (heating) and cooling period for a further 30 min. Since the specimen was a homogeneous and isotropic material, and housed in an insulated chamber, heat generation was expected to be uniform with the exothermic reaction. The transient temperature response was measured and recorded at the center of the material. The reaction time for each cycle was about 20 min. For charging process, in order to reuse the hydrated specimen, it was dehydrated in a furnace at 600°C for about 60 min and was allowed to completely cool down overnight before being used again.

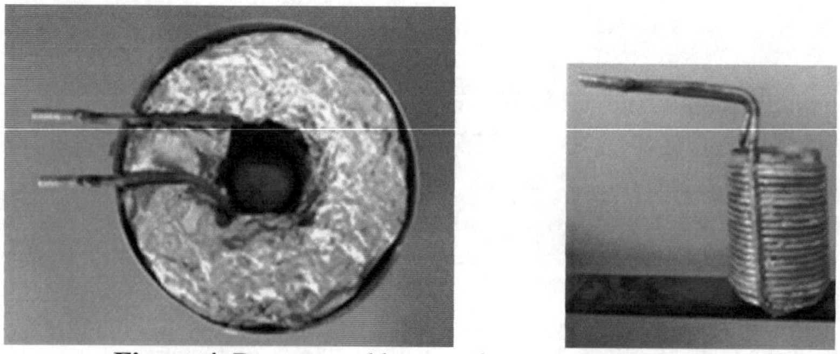

Figure 4. Reactor and heat exchanger arrangement.

2.3. Energy analysis

The net input energy is equal to the energy released by the reaction (enthalpy of formation). An energy balance for this process is as follows [8]

$$-(\text{Energy recovered} + \text{Heat loss}) = \text{Energy accumulation} \tag{2}$$

where the energy accumulation is a negative value equivalent to the energy released during the exothermic reaction. Energy efficiency of this process is evaluated from the energy recovered by the working fluid divided by the heat of reaction as follows:

$$\eta = \frac{m_d C_{pw}(T_2 - T_1)}{\Delta H_R} \tag{3}$$

3. Results and Discussion

During hydration of CaO, rapid rise in temperature at the beginning was observed. Once maximum temperature of the reactor was reached, heat transfer water was operated in which cooling was enabled. Temperature rises from room condition were in the range of 45 – 50°C. Table 2 is the summary of the maximum temperature recorded for the discharging in the reactor ranging between 72 and 82°C (an average maximum temperature of 76.1°C). Figure 5 shows the sequence of repeated cycles 1, 2, 3 and 4 for the reactor temperature and water outlet from heat exchanger of fresh/dehydrated samples of CaO. The energy efficiencies were found to be in the range of 60 – 83%.

Table 2. Summary of maximum recorded temperatures (reaction of ½ mole CaO and ½ mole H_2O)

	Storage material		Mass of H_2O	Maximum reactor temperature (°C)	Maximum water outlet of heat exchanger (°C)
	Condition	Initial temperature (°C)			
1	Fresh CaO	30	9.0	72	43
2	Dehy. CaO	30	9.0	82	44
3	Dehy. CaO	25	9.0	73	41
4	Dehy. CaO	27	9.0	77	41

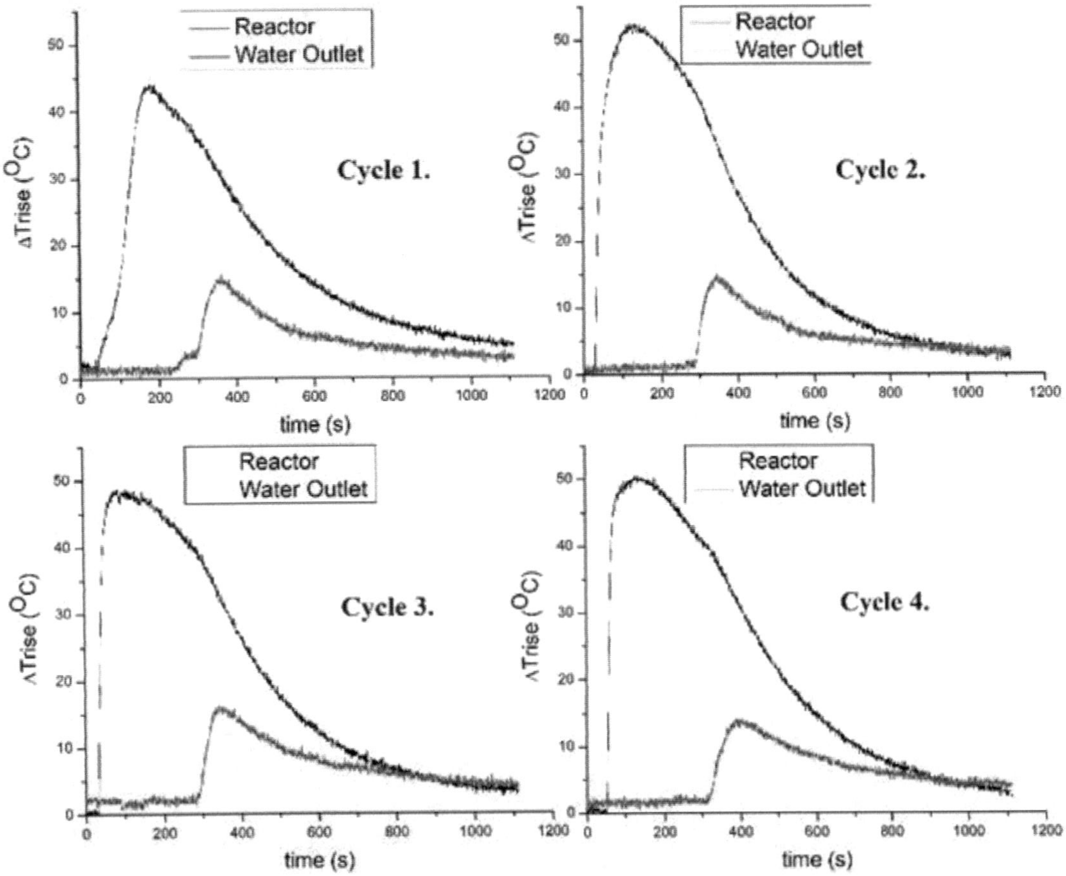

Figure 5. Discharging sequence of cycles 1,2,3,4.

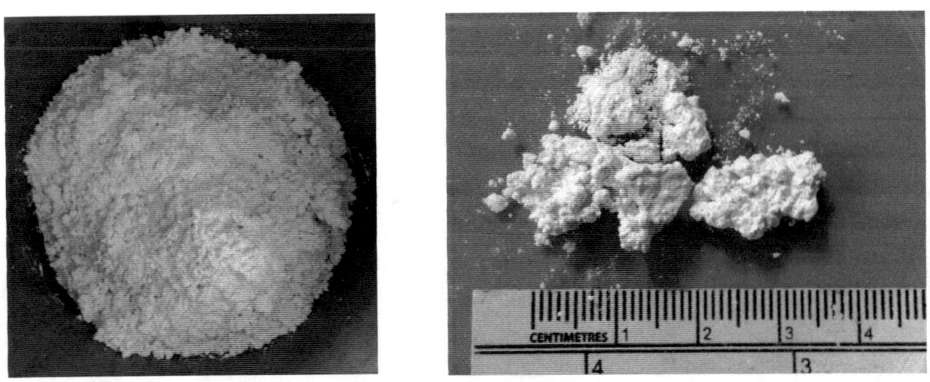

Figure 6. Agglomeration of powders in the reactor after hydration reaction.

It was also observed that agglomeration of powders formed inside the bed, as shown in Figure 6. This might result in local non-uniformity in flow distribution that would affect the reaction. It was suggested that addition of nanoparticles may ease this problem [9].

4. Concluding remark

A small TCS unit based on the $CaO/Ca(OH)_2$ system has been built to generate hot water in this work. Its performance in terms of hot water generation, thermal efficiency and cyclic behaviour was investigated. Hydration reaction rates was immediate with rapid temperature rise. High thermal efficiencies were realized. Nonetheless, greater effort must be exerted to test the reactor setup continuously and to enhance the heat output.

5. Reference

[1] Prieto C, Cooper P, Fernandez A I and Cabeza L F 2016 *Renew. Sustain. Energy Rev.* **60** 909-29
[2] Pan Z H and Zhao C Y 2017 *Energy* **130** 155-73
[3] Pardo P, Deydier A, Anxionnaz-Minvielle Z, Rouge S, Cabassud M and Cognet P 2014 *Renew. Sustain. Energy Rev.* **32** 591-610
[4] Wu J and Long W F 2015 *Int. J. Energy Res.* **39** 869-88
[5] Tatsidjodoung P, Le Pierres N, Luo L 2013 *Renew. Sustain. Energy Rev.* **18** 327-49
[6] Criado Y A, Alonso M and Abanades J C 2016 *Solar Energy* **135** 800-9
[7] Schaube F, Worner A and Tamme R 2011 *J. Solar Energy Eng.* **133** 031006
[8] Abedin A H and Rosen M A 2012 *Appl. Energy* **93** 18-23
[9] Rosskopf C, Haas M, Faik A, Linder M and Worner A 2014 *Energy Convers. Manag.* **86** 93-8

Acknowledgment
The authors wish to acknowledge the financial support from Chiang Mai University.

IOP Conf. Series: Earth and Environmental Science **159** (2018) 012003 doi:10.1088/1755-1315/159/1/012003

Decomposition of Methane Hydrate in the Presence of Sodium Dodecyl Sulfate

A O Dudoladov[1], M S Vlaskin and M V Gaykovich

1 Joint Institute for High Temperatures of the Russian Academy of Sciences, Izhorskaya 13 Bldg 2, Moscow 125412, Russia
E-mail: advert.geom@gmail.com

Abstract. Presented work is devoted to the study of the process of decomposition of methane hydrate obtained in the presence of sodium dodecyl sulfate. It was found that the rate of dissociation increases with temperature and surfactant concentration increasing. It was found that free gas volume inside the reactor plays important role in the process of methane hydrate formation and influences on kinetics of hydrate formation and methane concentration in produced hydrate. It has been suggested that the decrease in the gas content in the resulting hydrate is associated with a decrease in the cooling efficiency with a larger reactor load and a decrease in the area of the effective cooling surface of the reactor walls and the formation of a hydrate on the reactor walls that reduces heat exchange between the remaining liquid phase and the refrigerant.

1. Introduction

Gas hydrates are inclusion compounds in which gas molecules are enclosed in crystalline cells consisting of water molecules joined by hydrogen bonds. Gas hydrates can be formed and stably exist in a wide range of pressures and temperatures (for methane from 2×10^{-8} to 2×10^{3} MPa at a temperature of 70 to 350K) [1].

In nature, hydrates are found mainly near the surface of the earth in those regions where the temperature and pressure correspond to the conditions of their formation. If the temperature and pressure do not match the conditions required for the existence of the hydrate, then the hydrate becomes unstable. For this reason, for example, it is rather difficult to study the deposits of methane hydrate, because when it is delivered to the surface the temperature increases and the pressure decreases that leads to the decomposition of the hydrate.

A great interest in methane in the form of gas hydrate is mainly caused by huge resource of gas hydrates and their wide distribution in the earth's crust (in the cryolithozone) and in the world oceans. The extraction of methane from natural hydrates has a clearly expressed practical meaning: the extraction of natural gas as a fuel. The first estimation for permafrost areas in 1977 gave a value of 37 trillion m^3 — for the USSR and 57 trillion m^3 — for the world in total [2]. To date, several hundreds of gas-hydrate deposits have been identified in the world. Only for the North American continent, according to the US Department of Geological Survey, gas resources in gas hydrate deposits identified by seismic sounding and deep drilling with core sampling and a full range of geophysical studies exceed 9,000 trillion m^3.

In recent years, one of the main areas of research and development related to methane (natural gas) hydrate is the storage of methane (natural gas) in gas hydrate state. Methane hydrate allows storing the gas in compact form and safe conditions. One volume of hydrate contains about 160 volumes of methane [1]. In addition, methane hydrate is interesting from the point of view of methane transportation. Particular interest is bounded with the idea of dividing the delivered hydrate into methane and pure water (without impurities), which can be used for cooking or backed into the hydrate production cycle. Hydrate transport is also important for areas with poorly developed oil and gas infrastructure. Speaking about methane hydrate as an option for storing and transporting energy, it is important to note the effect of self-preservation of gas hydrate — the slow decomposition of the hydrate at negative temperatures due to the coating of its surface with a film of ice [3]. Due to this effect, natural gas hydrate can be in a metastable state for a fairly long time (up to a day) at atmospheric pressure and temperature of –20°C [4], whereas, according to the phase diagram of methane hydrate, at atmospheric pressure for stability of methane hydrate, temperature should be lower than –80°C.

The use of hydrate technology for gas storage and transportation has several advantages. Hydrate technology does not presuppose the use of cryogenic temperatures in contrast to the storage of natural gas in a liquefied state, however, more carriers will be needed to use hydrate technology because of the lower density of natural gas to the same volume, which leads to an increase in transportation costs. Even so, the additional transportation costs will be balanced by the cost of production and storage reduced with respect to the technology of liquefying natural gas, thanks to the low temperature of gas storage in the hydrate state. Thus, the technology of natural gas storage in a hydrate state wins about 20% of the cost of using it in comparison with liquefaction in the case of production in the amount of up to 1.5 million tons per year and a transportation range of 1—6 thousand km [5]. In particular, it says that transportation of natural gas in gas hydrate state may be more preferable for APEC region if Russian Far East is considered as natural gas source.

One of the main problems in natural gas storage in hydrate state is the slow rate of hydrate formation. Scientists from all over the world have made efforts to find the solutions of this problem, such as the development of new designs of reactors that ensure maximum contact of the gas-water phases or the introduction of additives to the system that enhance the kinetics of formation of gas hydrates [6—8]. These additives are commonly known as kinetic promoters and they represent usually surface-active agents. Sodium dodecyl sulfate (SDS) is one of such surface-active agents that can be used for accelerating the process of hydrate formation [9—10]. At the same time the process of dissociation of gas hydrate obtained in the presence of SDS is also important from practical point of view. This work represents the study of the process of decomposition of methane hydrate obtained in the presence of SDS in order to determine the storage and transport conditions for methane in such hydrate state.

2. Materials and methods

Methane was used with a purity of 99.6% (Moscow Gas Processing Plant). To intensify the hydrate formation process, solutions of sodium dodecyl sulfate ($NaC_{12}H_{25}SO_4$) manufactured by PanReacApplyChem with a purity of 95% was used.

In the first series of experiments, a steel cylindrical reactor with internal volume of 300 ml with cooling jacket was used to produce methane hydrate. The pressure in the reactor was measured by an AIR-10N pressure transducer with an accuracy of 0.01 bar, the temperature was measured by thermocouple with an accuracy of 0.1°C. The temperature during hydrate formation experiments was maintained with the help of LOIP FT liquid cryo-thermostat, the accuracy of keeping the temperature was 0.5°C. Data received from the sensors were transmitted to the computer. Scheme of laboratory plant for studying the kinetics of the methane hydrate decomposition process is shown in figure 1.

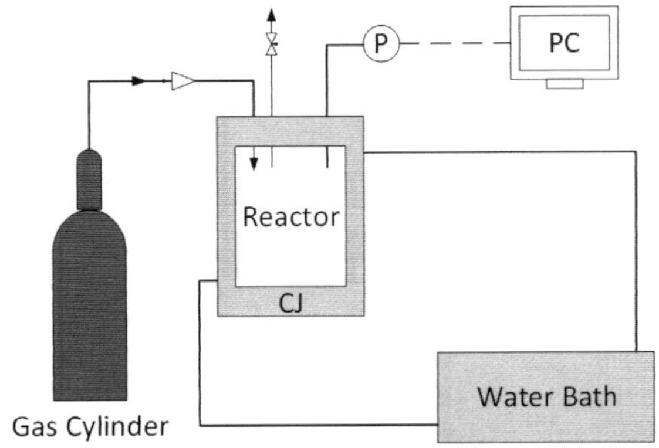

Figure 1. Schematic of the tank reactor setup for hydrate formation and decomposition in the first series of experiments.

The SDS aqueous solutions with concentrations of 0.1, 1 and 10 wt% were prepared and loaded into the reactor. In each experiment 150 cm^3 of distilled water was used. Reactor was sealed, a temperature was adjusted by cryostat in order to attain the desired experimental temperature (270 K). The vessel was flushed with pure methane gas by repeating rapid pressurization (1 MPa) and depressurization cycles. Then the reactor was pressurized with the pure methane gas up to the pre determined experimental pressure of 5.0 MPa thus providing sufficient driving force for hydrate formation. As soon as the desired experimental pressure was reached, the stirrer was turned on at 200 rpm to provide enough agitation of reaction medium to facilitate rapid formation of hydrate. After 3 hours, the reactor was cooled to –10 and kept in this state for 1 hour, then the residual pressure was discharged, the temperature was raised to –5 °C,–3 °C, –1°C and the pressure increase in the reactor was measured.
In the second series of experiments, the methane content in the hydrate was measured as a function of the volume of the liquid. The reaction was carried out in a 500 ml cylindrical reactor with cooling jacket. In this series of experiments the reactor was filled by different amount of SDS aqueous solution: 100, 200, 300 and 400 cm^3.

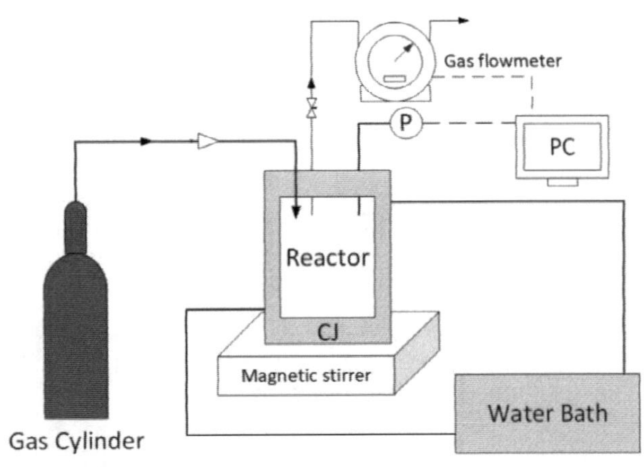

Figure 2. Schematic of the tank reactor setup for hydrate formation and decomposition in the second series of experiments

In all experiments the same solution of 0.1 wt% of SDS was used. Other hydrate formation conditions of these experiments were the same as in the first series of experiments. Hydrate decomposition was studied with the help of gas flow meter Ritter TGO. Scheme of laboratory plant for second series of experiments is shown in figure 2.

3. Results and discussion

3.1 Dependence of the rate of methane hydrate decomposition on temperature and concentration of surfactants

Pressure growth curves in the reactor during the decomposition of methane hydrate at different temperatures and surfactant concentrations are shown in figure 3 (left). It can be seen that the hydrate obtained in the presence of aqueous solution of SDS with concentration of 0.1% at temperatures of –3 ° C and –5 ° C practically does not decompose. The decomposition for this hydrate is only observed at temperature of –1 °C. Methane hydrate obtained in the presence of aqueous solution of SDS with concentration of 1% does not decompose at temperature of –5 ° C, at a temperature of –3 ° C there is slight increase in pressure in the reactor is observed, and at temperature of –1 °C the pressure increased up to 1 MPa after 2 hours. The decomposition of methane hydrate obtained in the presence of a aqueous solution of SDS with concentration of 10% is observed already at temperature of –5 °C as can be seen from figure 3 (right)

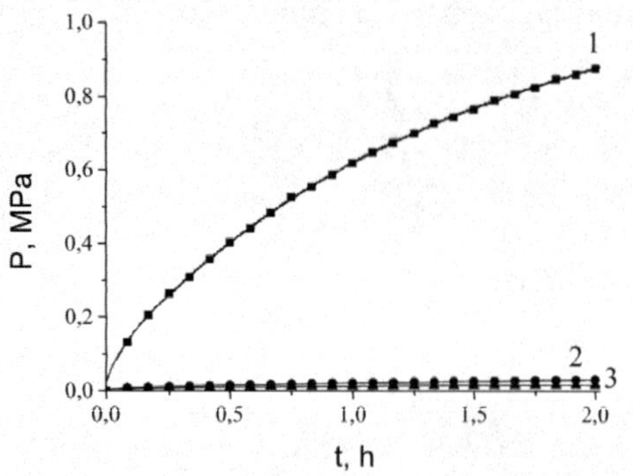

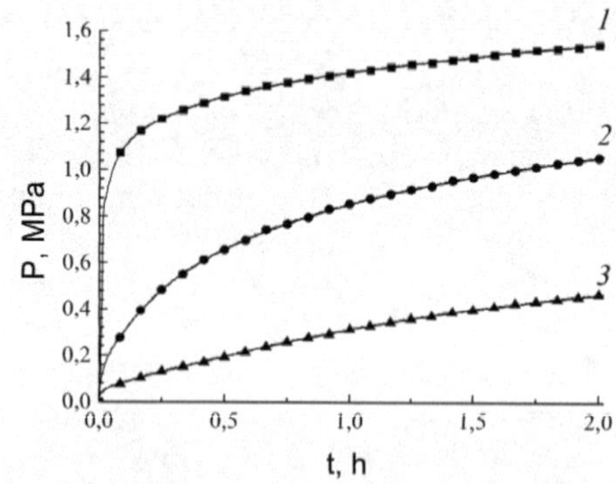

Figure 3. Growth of pressure in the reactor during the decomposition of methane hydrate for a concentration of 0.1 wt% (left) and 10 wt% (right), temperatures: *1* — –1° C, *2* — –3° C, *3* — –5 °C.

The comparison of experimental data says that with SDS concentration increasing the hydrate becomes less stable. This is due to the fact that at a higher SDS concentration, the hydrate contains more methane and so it is less stable. The highest decomposition rate is observed at SDS concentration of 10% and temperature of –1 °C. At temperature of –5 °C and concentrations of 0.1% and 1%, dissociation was almost not observed during 2 hours.

Table 1. The rate of dissociation (cm^3/min) of methane hydrate at various temperatures and SDS concentrations

	–1 °C	–3 °C	–5 °C
0,1 wt%.	40,36	1,84	0
1 wt%.	46,96	3,04	0,61
10 wt%	76,51	53,36	23,82

Table 1 shows the decomposition rates of methane hydrate (cm^3/min) that were determined from pressure grows in experiments with different SDS concentrations and temperatures. It can be seen that with an increase in the SDS concentration and temperature, the rate of dissociation increases substantially.

3.2. Dependence of the methane content in the hydrate on volume of the liquid phase within the reactor

In the second series of experiments, hydrate formation was studied by feeding methane to the reactor and maintaining a constant pressure of 50 ± 2 bar for different initial volumes of the liquid phase. The

SDS concentration in the initial solution was the same (0.1%). Results of this series of experiments are shown in table 2.

Results of experiments show that the methane concentration in hydrate decreases with increase in the amount of the solution loaded into the reactor. This is bounded with insufficient cooling efficiency (that was the same in all experiments) with increased loading. Moreover, the increase of the amount of the solution loaded into the reactor leads to the decreasing of surface of reactor walls that is in gas phase. At the same time, our experimental study showed that surface of reactor walls situated in gas phase is more active than the surface located in liquid phase. This statement is confirmed by figure 4 showing that hydrate inside the reactor produced predominantly on the surface of reactor walls. Even at the lowest loading the all surface inside the reactor was covered by methane hydrate crystals. It says that free gas volume inside the reactor plays important role in the process of methane hydrate formation and influences on kinetics and methane concentration in produced hydrate.

Table 2. The methane content in the hydrate as a function of the volume of the liquid phase inside the reactor (with internal volume of 500 ml)

Volume of liquid (SDS aqueous solution) loaded into the reactor, cm^3	Methane volume obtained after dissociation, cm^3	Concentration of methane in obtained hydrate, cm^3 of methane per 100 cm^3 of solution
0.1	18,5	18,5
0.2	35,1	17,6
0.3	45,3	15,1
0.4	50,3	12,6

4. Conclusion

The study of dissociation of methane hydrate obtained at various temperatures and SDS concentrations showed that the rate of dissociation increases with temperature and SDS concentration increasing. Methane hydrate obtained in the presence of SDS with concentration of 0.1%, begins to dissociate at temperature of –3 °C with a rate of 1.84 cm^3/min. The methane hydrate obtained in the presence of SDS with concentration of 1% dissociates with rates of 46.96 cm^3/min, 3.04 cm^3/min and 0.61 cm^3/min at temperatures of 1 °C, –3 °C and –5 °C respectively.

Figure 4. The image of 500 ml reactor with methane hydrate obtained from 100 cm^3 aqueous SDS solution (0.1%).

IOP Conf. Series: Earth and Environmental Science **159** (2018) 012003 doi:10.1088/1755-1315/159/1/012003

Methane hydrate obtained in the presence of SDS with concentration of 10% dissociates intensively already at a temperature of –5 °C with a rate of 23.82 cm^3/min. Based on the results of hydrate dissociation, it can be concluded that the most optimal concentration of sodium dodecyl sulfate in aqueous solution is the concentration of 0.1%, because at this concentration the intensive hydrate formation is achieved while the methane hydrate is stable at a temperature of –5 °C and lower. During the study of the effect of the initial volume of liquid loaded into the reactor (volume of gas phase) on the methane content in the produced hydrate, it was found that the maximum specific methane content is achieved with the smallest volume of loaded aqueous solution. It was bounded with the fact that surface of reactor walls situated in gas phase is more active in hydrate formation process than the surface located in liquid phase.

5. References

[1] Makogon Y F 2003 *Russian chemistry journal* XLVII **3** pp.70—79.
[2] Trofimuk A A, Makogon Y F, Steclanin Y I 1977 *Geologiya i Geofizika* **9**.
[3] Makogon Y F 1974 *Gydraty prirodnikh gasov* Moscow Nedra p.204
[4] Sloan E D, Koh C 2007 *Clathrate Hydrates of Natural Gases, Third Edition*. CRC Press p. 752.
[5] Gudmundsson J, Borrehaug A 1996 *Proceedings of the Second International Conference on Natural Gas Hydrates*, 2–6 June. — Toulouse, France P. 415—422.
[6] Kumar A, Bhattacharje G, Kulkarni B D, Kumar R 2015 *Ind. Eng. Chem. Res.* **54** pp 12217—12232.
[7] Nakai S 2012 Development of natural gas hydrate supply chain *Proceedings of the World gas conference*, Kuala Lumpur
[8] Veluswamy H P, Chen J Y, Linga P 2015 *Chem. Eng. Sci.* **126** pp 488–499.
[9] Dicharry C, Duchateau C, Asbaï H, Broseta D, Torré J P 2013 *Chem. Eng. Sci.* **98** pp 88—97.
[10] Kumar A, Sakpal T, Bhattacharjee G, Kumar R 2016, *Ind. Eng. Chem. Res.* **55** pp 7964—7972.

IOP Conf. Series: Earth and Environmental Science **159** (2018) 012004 doi:10.1088/1755-1315/159/1/012004

Chemical Composition of Bio-oil Produced by Hydrothermal Liquefaction of Microalgae with Different Lipid Content

M S Vlaskin[1], Y I Kostyukevich[1, 2], G N Vladimirov[1, 2], M V Gaykovich[1*], A O Dudoladov[1], N I Chernova[1, 3], S V Kiseleva[1, 3], A Z Zhuk[1], O S Popel[1] and E N Nikolaev[2]

1 Joint Institute for High Temperatures of the Russian Academy of Sciences, Izhorskaya 13 Bldg 2, Moscow 125412, Russia
2 Skolkovo Institute of Science and Technology Novaya St., 100, Skolkovo 143025, Russia
3 Lomonosov Moscow State University, Leninskiye Gory, 1, Moscow 119991, Russia
E-mail: advert.geom@gmail.com

Abstract. Bio-oil samples were obtained by hydrothermal liquefaction at 300 °C of three different stains of microalgae that differ by the bimolecular composition: *Chlamydomonas*, *Chlamydopodium* and *Arthrospira*. For bio-oil characterization the ultrahigh resolution Fourier transform ion cyclotron resonance mass spectrometry (FT ICR MS) was used. The results of the study showed that the number of reliably identified molecular formulas after filtration and deisotoping were 2794 (*Chlamydomonas*), 1514 (*Chlamydopodium*) and 2110 (*Arthrospira*) and 853 formulas in each sample were the common. At the same time the compounds presented in the bio-oil produced from microalgae with high content of lipids and hydrocarbons (*Chlamydomonas* and *Chlamydopodium*) were more saturated then the compounds in bio-oil produced from *Arthrospira*. In the bio-oil produced from all microalgae the compounds containing 1 and 2 nitrogen atoms dominate, and also ON_2, N_3, ON_2 classes are present with relatively less intensity.

1. Introduction

Microalgae is one of the most promising sources of renewable biofuel [1]. The productivity of microalgae and its oil (lipids) content may be ten times higher than that of terrestrial biomass [2].

In recent decades, the studies devoted to biofuel production from microalgae were focused mainly on cultivation methods. In this context, the main tasks were first of all the increasing of productivity and lipid content in biomass by strains selection [3]-[5], cultivation conditions optimization [6]-[8] and new strains creation by genetic engineering. At the same time, not enough attention was paid to the problem of biomass to biofuels conversion. The rational solution of this problem is one of the key ways to improve the energy efficiency of the process of biofuel production. The traditional conversion method usually includes drying, solvent extraction of lipids and transesterification with the production of fatty acid methyl esters composing the biodiesel fuel [9]. Obvious disadvantages of biodiesel production method are high energy costs and the use of dangerous organic solvents (such as methanol). Moreover, in case of biodiesel production only lipid part of biomass is processed, while great part of biomass including proteins and carbohydrates is not involved in the production of liquid biofuel. At the same time, it is known that lipid-rich strains have relatively low biomass productivity [10], [11].

The problem of biofuel production is connected primarily with high moisture content when leaving the cultivation stage (80-90 % by weight). For the processing of wet biomass into biofuels the hydrothermal technologies seems to be more favorable. Hydrothermal liquefaction (HTL) with the production of crude bio-oil (biocrude oil) as end product is of most interest [12], [13]. One of the main advantages of HTL is that feedstock pre-drying is unnecessary in this case. Biomass raw can be fed into HTL reactor in humid state, e.g. in the form of an aqueous suspension. Another advantage is that by HTL the resulting bio-oil is composed not only from lipids but also from carbohydrates and proteins that increases the overall yield of the product [14]. Additional advantages of HTL are relatively low temperatures of the process and the possibility of organizing the one-step continuous process.

During HTL the biomass is thermally treated under humid conditions at temperatures up to 370 °C and pressures up to 25 MPa. Due to this treatment, the biomass components are passed through hydrolysis and pyrolysis reactions forming a number of liquid hydrocarbons, both soluble and insoluble in water, as well as gaseous and solid reaction products. HTL products are bio-oil, aqueous solution, solid residue and gas product. The target product of HTL is the so-called "crude bio-oil" (biocrude) - liquid hydrocarbons separated from the solid phase and an aqueous solution.

In the present study we report the investigation of the bio-oil obtained by HTL of three different stains: *Chlamydomonas*, *Chlamydopodium* and *Arthrospira*. For bio-oil characterization the ultrahigh resolution Fourier transform ion cyclotron resonance mass spectrometry (FT ICR MS) was used.

2. Materials and methods

Bio-oil samples were obtained by HTL of three different stains of microalgae that differ by the bimolecular composition: *Chlamydomonas*, *Chlamydopodium* and *Arthrospira*. Bimolecular compositions of studied microalgae strains are shown in table 1.

Table 1. Biochemical composition of the investigated microalgae.

Strain	Proteins	Lipids	Hydrocarbons
Arthrospira platensis strain rsemsu 1/02-P	60.7	12.1	7.1
Chlamydopodium starrii strain rsemsu Chcc-14/11	29.1	34.9	31.4
Chlamydomonas globosa strain rsemsu Chlam-15/11	30.1	47.0	17.9

Hydrothermal treatment of microalgae was carried out on a laboratory plant, the scheme of which is shown in figure 1. The reactor represents an autoclave with volume of 500 cm^3, the maximum exploitation temperature and pressure for the reactor are 400 °C and 25 MPa correspondingly. Reactor heating is external ohmic. The heating process (heating rate, maximum temperature and exposure time) is controlled by a PC operator using an automated control system (supplied by National Instruments). Before hydrothermal treatment microalgae was centrifuged and dried at temperature of 80 °C. The reactor was loaded with 150 grams of distilled water and 30 grams of dried microalgae. Then, the reactor was sealed and purged with nitrogen. Then, the reactor was heated up to a temperature of 300 °C. The duration of the heating process to a temperature of 300 °C was about 120 minutes, the exposure at 300 °C was 60 minutes. In the reactor, a pressure close to the pressure of saturated water vapor corresponding to a temperature of 300 °C (closed to 100 bar) was established and maintained with the help of valve. Gaseous products of HTL were passed through gas bubblers where they cooled down and then were passed through gas flow meter. In the end of exposure the ohmic heater was turned off and the reactor cooled to room temperature during about 5 hours. The reactor was then opened and the condensed products (a mixture of an aqueous solution of organic compounds and a solid residue) were removed from the reactor and placed in a glass flask with flat bottom. Flask was placed on magnetic stirrer and the content within flask was stirred by magnetic element during 10 hours at room temperature to take additional organic compounds out from solid residue into aqueous phase. Then the stirring was stopped and the content within flask was left to form sediment on the bottom of the flask.

After sedimentation of solid residue on the bottom of the flask the aqueous solution was extracted from the flask using a syringe for further bio-oil extraction. For bio-oil extraction the dichloromethane was used.

Mass spectrometric analysis was carried out on a LTQ FT Ultra (Thermo Electron Corp., Bremen, Germany) mass-spectrometer equipped with a 7T superconducting magnet [15]. The bio-oil was dissolved in the MeOH to the concentration 1g/l. Ions were generated by an IonMax Electrospray ion source (Thermo Electron Corp., Bremen, Germany) in positive and negative ESI mode. The temperature of the desolvating capillary was set to 300 °C. The infusion rate of the sample was 1 μl /min and the needle voltage was 3000 V. The archived resolving power was 400 000, each spectrum was the averaging of 100 scans. Prior to the analysis the LTQ FT was calibrated using the standard Thermo calibration mixture. For each peak in the peak list following variables were calculated:

$$\text{Kendrick mass } M_{Kendrick} = M_{IUPAC} \frac{\left[m_{IUPAC}^{CH_2}\right]}{m_{IUPAC}^{CH_2}} \quad (1)$$

$$\text{Kendrick mass defect } KMD = round(M_{Kendrick}) - M_{Kendrick} \quad (2)$$

Here [m] means integer part of the mass m. M_{IUPACK} is the IUPAC mass of peak, $m_{IUPAC}^{H_2}$ and $m_{IUPAC}^{CH_2}$ are IUPAC masses of the fragment CH_2 and H_2 correspondingly. Molecules that differ only by the number of repeating CH_2 segments have the same KMD. The homology series formed by C_cH_{2c} for molecules with molecular formula $C_cH_{2c+Z}N_nO_o$ are referred to as ZN_nO_o. Our approach for the spectrum processing is based on the analysis of the weighted Kendrick mass defect histogram. This diagram has the advantage over the conventional Kendrick mass defect plot (kmd vs. m/z) because it allow to see the homology series more distinct [16], [17].

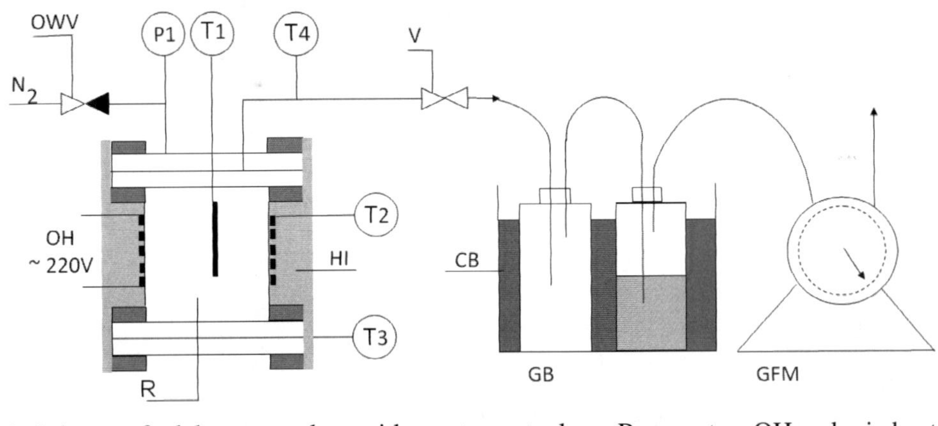

Figure 1. Scheme of a laboratory plant with reactor-autoclave. R – reactor, OH – ohmic heater, HI – heat insulation, CB – cooling bath, GB – gas bubblers, GFM – gas flow meter, OWV – one way valve, V – valve, T and P – temperature and pressure sensors.

3. Results and discussion

Results of the investigation of chemical composition of bio-oil produced from different microalgae are shown in figure 2. The number of reliably identified molecular formulas after filtration and deisotoping were 2794 (*Chlamydomonas*), 1514 (*Chlamydopodium*) and 2110 (*Arthrospira*). We can see that in the bio-oil produced from all microalgae the compounds containing 1 and 2 nitrogen atoms dominate, but also ON_2, N_3, ON_2 classes are present.

It can be seen, that the compounds presented in the bio-oil produced from microalgae with high content of lipids and hydrocarbons are more saturated then the compounds in bio-oil produced from *Arthrospira*. The number of double bonds in molecules spans the region from 1 to 7 for *Arthrospira* (maximum at 2) and the region from 1 to 10 for other microalgae (maximum at 7). Also, the portion of

IOP Conf. Series: Earth and Environmental Science **159** (2018) 012004 doi:10.1088/1755-1315/159/1/012004

oxygen containing compounds in *Chlamydopodium* and *Chlamydopodium* samples is considerably higher. Such results are expectable, because the *Arthrospira* has a considerable content of proteins (~60%) and as a consequence of nitrogen. The high content of nitrogen leads to the formation during HTL of the nitrogen containing compounds which easily ionize in the electrospray ion source.

Figure 3 presents the Venn diagram showing the intersection of the molecular formulas between bio-oil samples. It can be seen that the biggest number of unique molecular formulas (1157) is present in the bio-oil produced from *Chlamydomonas*. For all 3 samples 853 common formulas were found.

In the bio-oil produced from all microalgae the compounds containing 1 and 2 nitrogen atoms dominate, and also ON_2, N_3, ON_2 classes are present with relatively less intensity. Results show that the problem of high content of nitrogen remains for all bio-oil samples: the intensity of compounds containing 1 and 2 nitrogen atoms for bio-oil produced from microalgae with 30 % of proteins was almost the same as for bio-oil produced from microalgae with 60 % of proteins. Combustion of such bio-oil will lead to significant amount of NO_x to the atmosphere. It says that nitrogen reducing treatment of bio-oil is needed irrespectively of microalgae strain.

Our results suggest that the quality of the bio-oil obtained by the HTL depends on the composition of the biomass. High content of proteins leads to the obtaining of low saturated bio-oil with low content of oxygen.

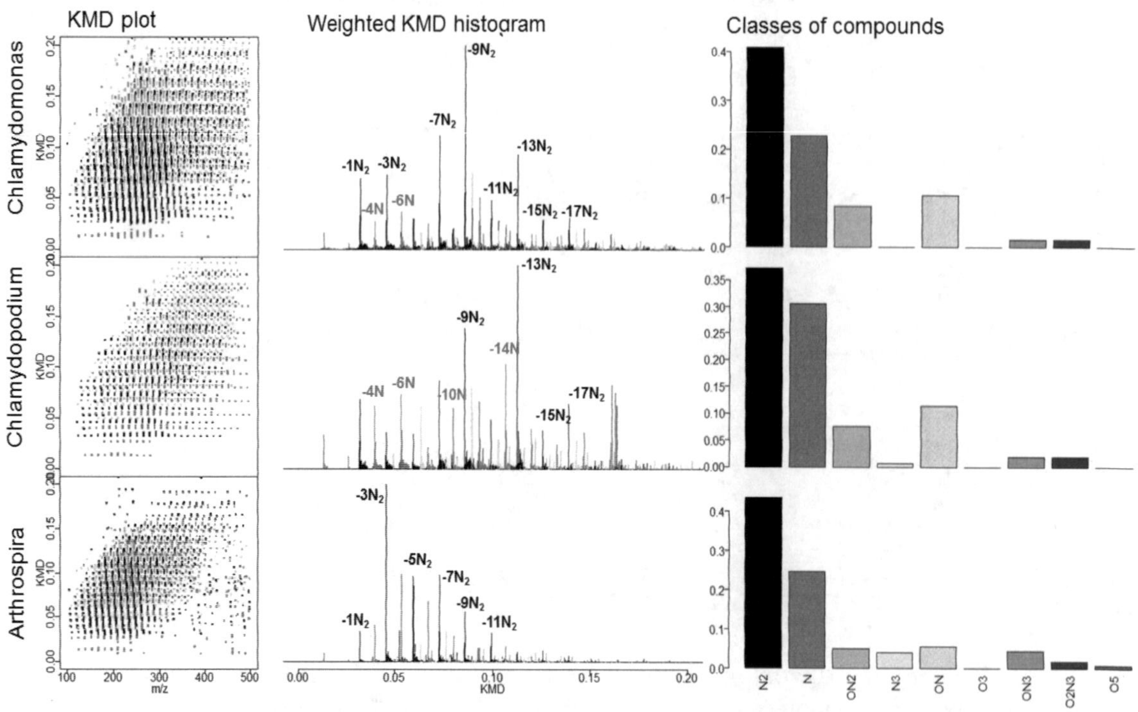

Figure. 2. The visual analysis of the samples: KMD plot, weighted KMD histogram and portions of compound classes. Each class of compounds is labeled with unique color.

4. Conclusion

Bio-oil samples were obtained by HTL of three different stains of microalgae that differ by the bimolecular composition: *Chlamydomonas*, *Chlamydopodium* and *Arthrospira*. The results of the

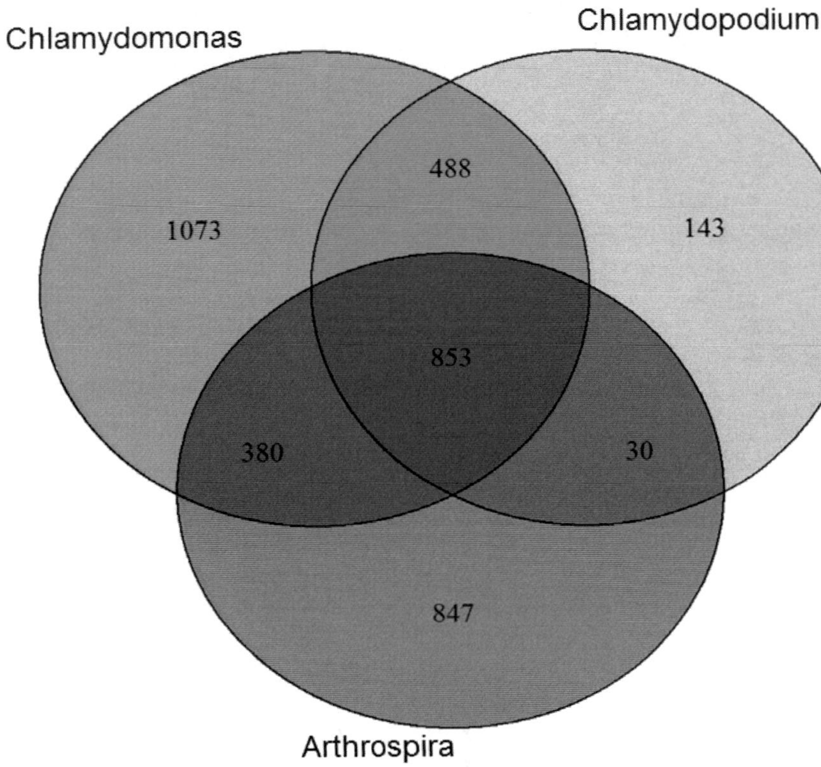

Figure 3. The Venn diagram showing intersection of compounds in different samples.

study showed that bio-oil samples produced from different microalgae differ from each other. The number of reliably identified molecular formulas after filtration and deisotoping were 2794 (Chlamydomonas), 1514 (*Chlamydopodium*) and 2110 (*Arthrospira*) and 853 formulas in each sample were the common.At the same time the compounds presented in the bio-oil produced from microalgae with high content of lipids and hydrocarbons (*Chlamydomonas* and *Chlamydopodium*) are more saturated then the compounds in bio-oil produced from *Arthrospira*. The obtained results can be used in the future for the processing of microalgae into bio-oil.

5. References

[1] Raslavičius L, Semenov V G, Chernova N I, Keršys A and Kopeyka A K 2014 Producing transportation fuels from algae: In search of synergy *Renewable and Sustainable Energy Reviews* **40** 133-42

[2] Chernova N I, Kiseleva S V and Popel' O S 2014 Efficiency of the biodiesel production from microalgae *Thermal Engineering* **61** 399-405

[3] Taleb A, Kandilian R, Touchard R, Montalescot V, Rinaldi T, Taha S, Takache H, Marchal L, Legrand J and Pruvost J 2016 Screening of freshwater and seawater microalgae strains in fully controlled photobioreactors for biodiesel production *Bioresource Technology* **218** 480-90

[4] Hu X, Zhou J, Liu G and Gui B 2016 Selection of microalgae for high CO2 fixation efficiency and lipid accumulation from ten Chlorella strains using municipal wastewater *Journal of Environmental Sciences* **46** 83-91

[5] Neofotis P, Huang A, Sury K, Chang W, Joseph F, Gabr A, Twary S, Qiu W, Holguin O and Polle J E W 2016 Characterization and classification of highly productive microalgae strains discovered for biofuel and bioproduct generation *Algal Research* **15** 164-78

[6] Zheng Y, Chi Z, Lucker B and Chen S 2012 Two-stage heterotrophic and phototrophic culture strategy for algal biomass and lipid production *Bioresource Technology* **103** 484-8

[7] Yee W 2015 Feasibility of various carbon sources and plant materials in enhancing the growth and biomass productivity of the freshwater microalgae Monoraphidium griffithii NS16 *Bioresource Technology* **196** 1-8

[8] Li R, Duan N, Zhang Y, Liu Z, Li B, Zhang D and Dong T 2017 Anaerobic co-digestion of chicken manure and microalgae Chlorella sp.: Methane potential, microbial diversity and synergistic impact evaluation *Waste Management*

[9] Salam K A, Velasquez-Orta S B and Harvey A P 2016 A sustainable integrated in situ transesterification of microalgae for biodiesel production and associated co-product-a review *Renewable and Sustainable Energy Reviews* **65** 1179-98

[10] López Barreiro D, Zamalloa C, Boon N, Vyverman W, Ronsse F, Brilman W and Prins W 2013 Influence of strain-specific parameters on hydrothermal liquefaction of microalgae *Bioresource Technology* **146** 463-71

[11] Rodolfi L, Chini Zittelli G, Bassi N, Padovani G, Biondi N, Bonini G and Tredici M R 2009 Microalgae for oil: Strain selection, induction of lipid synthesis and outdoor mass cultivation in a low-cost photobioreactor *Biotechnology and Bioengineering* **102** 100-12

[12] Chernova N I, Kiseleva S V, Vlaskin M S and Rafikova Y Y 2017 Renewable energy technologies: enlargement of biofuels list and co-products from microalgae *MATEC Web Conf.* **112** 10010

[13] Vlaskin M S, Chernova N I, Kiseleva S V, Popel' O S and Zhuk A Z 2017 Hydrothermal liquefaction of microalgae to produce biofuels: state of the art and future prospects *Thermal Engineering* **64** 627-36

[14] Elliott D C 2016 Review of recent reports on process technology for thermochemical conversion of whole algae to liquid fuels *Algal Research* **13** 255-63

[15] Nikolaev E N, Kostyukevich Y I and Vladimirov G N 2016 Fourier transform ion cyclotron resonance (FT ICR) mass spectrometry: Theory and simulations *Mass Spectrom Rev* **35** 219-58

[16] Kostyukevich Y, Vlaskin M, Vladimirov G, Zherebker A, Kononikhin A, Popov I and Nikolaev E 2017 The investigation of the bio-oil produced by hydrothermal liquefaction of Spirulina platensis using ultrahigh resolution Fourier transform ion cyclotron resonance mass spectrometry *European Journal of Mass Spectrometry* **23** 83-8

[17] Kostyukevich Y, Stavitskaya A, Zherebker A, Konstantinova M, Vlaskin M, Borisova L, Kononikhin A, Popov I and Nikolaev E 2017 Investigation of the ozonation products of natural complex mixtures using Fourier transform ion cyclotron resonance mass spectrometry *European Journal of Mass Spectrometry* **23** 152-5

Acknowledgements
The study was supported through a grant of the Russian Science Foundation (project No. 17-19-01617).

IOP Conf. Series: Earth and Environmental Science **159** (2018) 012005 doi:10.1088/1755-1315/159/1/012005

Effects of Lithium Salt on Electrochemical Reactions at a SiC Electrode in Propylene Carbonate-based Solutions

S Jeong[1]

1 Department of Chemical Engineering, Soonchunhyang University, Asan, Chungnam 31538, Republic of Korea
E-mail: hamin611@sch.ac.kr

Abstract. The effect of electrolyte salts, such as $LiPF_6$, $LiClO_4$, $LiCF_3SO_3$, and $LiBF_4$, on the redox reactions of lithium ions at a SiC electrode was investigated, focusing on the interfacial reaction between the SiC electrode and the electrolyte solution. Lithium salts were found to exert a large influence on the magnitude of resistance of the solid-electrolyte interphase (SEI), and its reversible and irreversible capacities. There was negligible effect on the potential at which the SEI was generated, but the amount of charge required for generation was greatly affected by the type of lithium salt used.

1. Introduction

Secondary batteries convert electrical energy into chemical energy for internal storage, and when electricity is needed, they convert the stored chemical energy back into electrical energy for release. This energy storage and release is accomplished by electrochemical redox reactions at the positive and negative electrodes of the battery. Lithium secondary batteries are widely employed in electronic devices and use redox reactions of lithium ions, which are monovalent cations, as the electrode reactions. This means that any material capable of a redox reaction with lithium ions can be used as an electrode material. However, the amount of energy stored or released by a cell depends on the type of electrode material.

Silicon-based materials have been actively studied as negatively charged electrode materials capable of storing and releasing a large amount of energy. Energy is stored when lithium ions are electrochemically reduced at the silicon-based electrodes and released when the ions are electrochemically oxidized. These electrochemical reactions are known to be highly dependent on electrolyte composition. A film is formed on the surface of the electrode due to electrolyte decomposition accompanied by the redox reaction of lithium ions, mainly during the first charge/discharge cycle [1–5]. This film is called a solid electrolyte interphase (SEI) [6, 7], and its physicochemical properties greatly influence the performance of the battery, affecting reversible capacity, rate capability, and safety. In order for the battery to exhibit excellent performance, an SEI with excellent properties must be formed on the electrode surface, and this is closely related to the electrolyte composition.

SEI formation on SiC electrodes is expected to contribute to their capacity during the first cycle and capacity fading in subsequent cycles. Reversible cycling and long-term stability are closely related to the effectiveness of surface passivation, which depends on the physicochemical properties of the SEI [8]. Therefore, SEI formation is a very important step in the electrochemical reaction in lithium secondary batteries that employ SiC as a negatively charged electrode. In graphite electrodes used in commercial lithium secondary batteries, lithium salt—one of the constituent components of the electrolyte solution—strongly affects the properties of the SEI [9, 10]. Various lithium salts have been

IOP Conf. Series: Earth and Environmental Science **159** (2018) 012005 doi:10.1088/1755-1315/159/1/012005

developed and studied to form high-quality SEIs on the surface of graphite electrodes. Similarly, in this study, the effects of lithium salts on the physicochemical properties of the SEI generated on a SiC electrode were investigated. Four lithium salts ($LiPF_6$, $LiClO_4$, $LiCF_3SO_3$, and $LiBF_4$) were used as electrolyte salts for the SiC electrode. This paper reports the effects of lithium salts based on the results of cyclic voltammetry (CV) and electrochemical impedance spectroscopy (EIS) analyses of a SiC electrode, including its charge/discharge capacity.

2. Experimental

SiC powder (Aldrich, 99%) was used for charge/discharge tests. The test electrode was prepared by coating a mixture of SiC powder (80 wt.%), carbon black (20 wt.%), and poly(vinylidene difluoride) binder (10 wt.%) on copper foil. The base electrolyte solvent was a 1:1 (by volume) mixture of propylene carbonate (PC) and dimethyl carbonate (DMC) (PC + DMC, Panax E-Tec. battery grade). Four types of lithium salts ($LiPF_6$, $LiClO_4$, $LiCF_3SO_3$, and $LiBF_4$) were separately dissolved (1 mol dm^{-3} [M] each) in the base solvent. All the reagents were purchased from Panax E-Tec. and used as received. The water content in each solution was <30 ppm, as confirmed by Karl-Fischer titrations (MKC-210, Kyoto Electronics Manufacturing Co., Ltd.).

CV and EIS were performed using a laboratory-made three-electrode cell, as shown in Fig. 1. The working electrode was mounted on the bottom of the cell and brought into contact with the electrolyte solution using an O-ring. The geometrical surface area was 0.8 cm^2. Lithium foil was used as both the counter and reference electrodes. CV was conducted between 3.0 and 0.0 V at a sweep rate of 0.5 mV s^{-1}. The electrochemical impedance was obtained after the electrode was polarized at a constant potential for a sufficient amount of time (i.e., until the current became negligible [typically <1 μA]). The impedance was measured with a potentiostat (PGZ 402, VoltaLab) over the frequency range 0.1 Hz to 100 kHz. The alternating amplitude was 5 mV. Charge/discharge measurements were performed using a coin-type cell (CR 2032) between 1.5 and 0.0 V at a C-rate of 0.1.

CV and EIS measurements and coin cell assembly were performed in an argon-filled glovebox (Three-Shine, SK-G1200) with a dew point below −60 °C. All potentials are presented as volts vs. Li^+/Li.

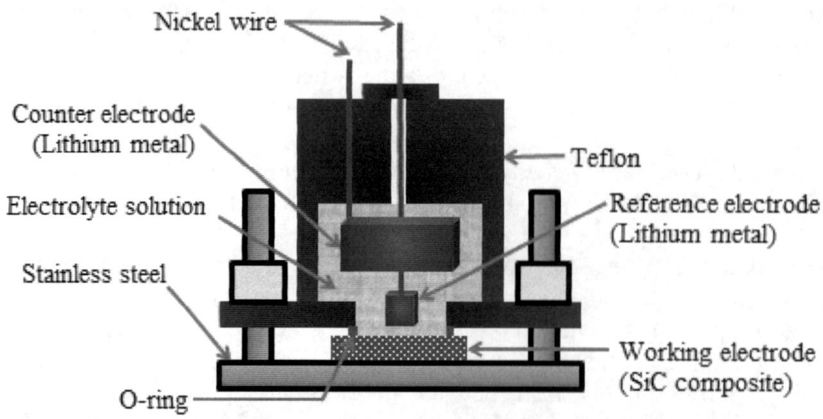

Figure 1. Schematic illustration of the three-electrode cell used for cyclic voltammetry (CV) and electrochemical impedance spectroscopy (EIS) measurements.

3. Results and Discussion

CV was performed to gain a basic understanding of the electrochemical reactions between SiC electrodes and PC-based electrolyte solutions containing different types of lithium salts. The electrochemical behavior of a battery system can be understood by recording the amount of current flowing through the electrode while changing the potential applied to the electrode at a constant rate (Fig. 2). This figure shows the cyclic voltammograms of the SiC electrode in four PC-based solutions. CV is a very useful technique for obtaining information about fairly complicated electrode reactions. During the first cycle, a minor reductive current begins to flow at ~2.4, ~2.3, ~2.5, and 2.2 V in solutions containing $LiPF_6$, $LiClO_4$, $LiCF_3SO_3$, and $LiBF_4$, respectively. This difference is presumably due to the influence of anions present in the electrolyte solution. Three major reductive peaks centered at ~1.9, ~1.1, and ~0.3 V appear for all the solutions. These reductive peaks, including the minor current, are absent from the second cycle, indicating that their presence and subsequent disappearance are the result of irreversible electrolyte decomposition, which might be closely related to the formation of the SEI.

Although scientific correlations between irreversible electrolyte decomposition and SEI formation are unclear in the present work, there is no doubt that SEIs are generated on the SiC electrodes, irrespective of the type of lithium salt, because the stable charge and discharge reactions shown in Fig. 2 do not occur without an SEI [11]. In addition to the irreversible reductive peaks, two reversible reductive peaks are observed at potentials of ~0.7 and ~0 V. The oxidative peaks corresponding to these peaks are observed at potentials of ~0.9 and ~0.2 V. These two reductive and oxidative peaks can be assigned to lithium insertion and extraction events, respectively. However, the reductive peaks are believed to represent a large proportion of the irreversible lithium insertion because the charge consumed in this process is greater than that consumed for oxidation. This is confirmed by the charge/discharge results shown in Fig. 2. Additionally, each lithium insertion and extraction peak varies slightly in shape and size.

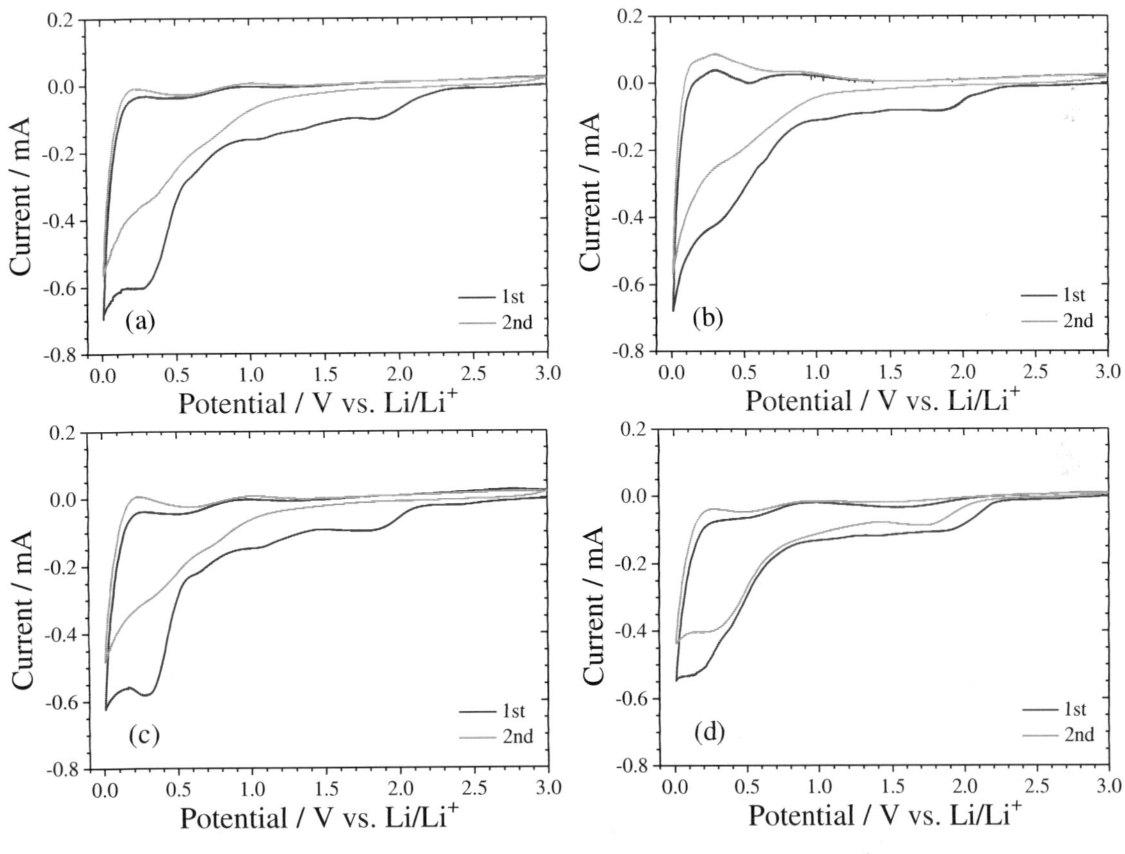

Figure 2. First and second cyclic voltammograms of SiC powder at 0.5 mV s^{-1} in propylene carbonate and dimethyl carbonate (PC + DMC) containing 1 M of (a) $LiPF_6$, (b) $LiClO_4$, (c) $LiCF_3SO_3$, and (d) $LiBF_4$.

The potential profiles of the SiC electrode during initial charging (lithium insertion) and discharging (lithium extraction) in four PC-based solutions are shown in Fig. 3. In the $LiPF_6$ solution, the charge and discharge capacities during the first cycle are ~175 and ~50 mAh g^{-1}, respectively, indicating rather low coulombic efficiency (Fig. 3(a)). Behavior similar to that shown in Fig. 3(a) is observed in the $LiBF_4$ solution (Fig. 3(d)) during the first cycle. In contrast, in the $LiClO_4$ and $LiCF_3SO_3$ solutions, the SiC electrode exhibits large discharge capacities and low coulombic efficiencies (Figs. 3(b and c)). The irreversible capacities observed in all electrolyte solutions are likely consumed during the decomposition of the electrolyte solution to form an SEI on the SiC electrode [8]. The relatively large irreversible capacities in solutions containing $LiClO_4$ and $LiCF_3SO_3$ are observed between 0.75 and 0.25 V, which means that these solutions take a long time to produce a stable SEI. However, SEIs produced in this manner are effective for the redox reaction of lithium ions in terms of reversible capacity.

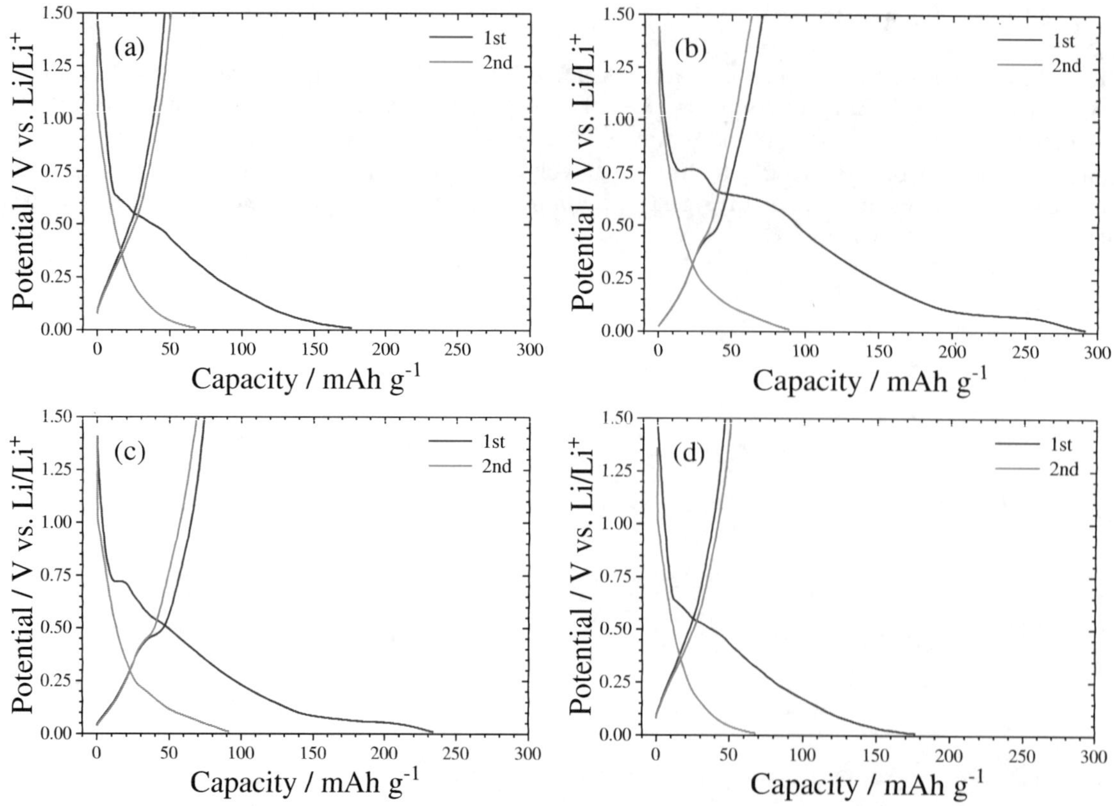

Figure 3. Charge and discharge curves in the first and second cycles of SiC powder in propylene carbonate and dimethyl carbonate (PC + DMC) containing 1 M of (a) $LiPF_6$, (b) $LiClO_4$, (c) $LiCF_3SO_3$, and (d) $LiBF_4$.

The charge and discharge results described above indicate that the nature of the interfacial reactions between the SiC electrode and the electrolyte solutions during cycling vary depending on the type of lithium salt used. Impedance spectra of the SiC electrodes were collected at 1.2 V to understand the interfacial reactions. The spectra obtained in each solution are in the form of a depressed arc (Fig. 4). It is widely accepted that there are two impedance components in the electrode where the redox reaction of lithium ions proceeds. One is impedance due to lithium ion migration through the SEI, and the other is impedance due to charge-transfer reactions [12–14]. The arcs shown in Fig. 4 denote impedance due to the former reaction. The latter reaction hardly proceeds at 1.2 V, where the impedance is obtained. The impedance decreases in the order $LiPF_6 > LiBF_4 > LiClO_4 > LiCF_3SO_3$ corresponding to the SEIs produced in these solutions. The SEI generated in the $LiPF_6$ solution exhibits the largest impedance. In contrast, a larger decrease in magnitude is found in the $LiCF_3SO_3$ solution. This decrease might be closely related to the large reversible capacity shown in Fig. 3(c), indicating that the SEI can have a positive effect on the insertion and extraction of lithium ions into/from the SiC electrode. Structural or compositional differences in the SEI generated in the four

PC-based solutions may lead to differences in the magnitude of impedance; however, direct correlations between the physicochemical properties of the SEI and the magnitude of impedance are unclear. A detailed interpretation of these correlations is beyond the scope of the present work and will not be discussed further here.

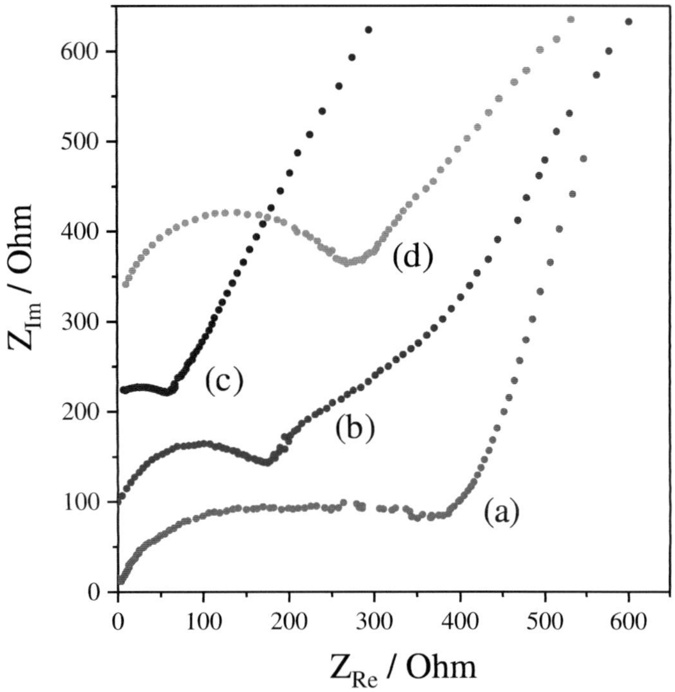

Figure 4. Nyquist plots of the SiC electrode in propylene carbonate and dimethyl carbonate (PC + DMC) containing 1 M (a) $LiPF_6$, (b) $LiClO_4$, (c) $LiCF_3SO_3$, and (d) $LiBF_4$.

4. Conclusions

Formation of a stable and effective SEI on the electrode surface is very important in order to utilize the redox reactions of lithium ions as the electrode reactions. Based on our studies, I suggest that $LiCF_3SO_3$ is a good SEI-forming salt for SiC electrodes employing PC-based solutions as an electrolyte solvent. $LiCF_3SO_3$ plays an important role in reducing the impedance of the SEI and results in improved charge/discharge performance. $LiPF_6$ is known to generate an excellent SEI when graphite is used as an electrode. Therefore, the results obtained in the present work suggest that good performance of an electrolyte in terms of SEI generation depends on the electrode. On the other hand, the specifics of the SEI produced in solutions containing $LiCF_3SO_3$ are still unclear. In the future, we plan to investigate the chemical composition and physical structure of this SEI.

5. References

[1] Aurbach D, Ein-Eli Y 1995 *J. Electrochem. Soc.* **142** 1746.
[2] Winter M, Besenhard J O, Spahr M E, Novak P 1998 *Adv. Mater* **10** 725.
[3] Xu K 2004 *Chem. Rev.* **104** 4303.
[4] Zhang S S 2006 *J. Power Sources* **162** 1379.
[5] Aurbach D 2000 *J. Power Sources* **89** 206.
[6] Peled E 1979 *J. Electrochem. Soc.* **126** 2047.
[7] Verma P, Maire P, Novak P 2010 *Electrochim. Acta* **55** 6332.
[8] Ulldemolinsa M, Cras F L, Pecquenard B, Phan V P, Martin L, Martinez H 2012 *J. Power Sources* **206** 245.

[9] Burns J C, Petibon R, Nelson K J, Sinha N N, Kassam A, Way B M, Dahn J R, 2013 *J. Electrochem. Soc.* **160** 1668.
[10] Agubra V A, Fergus J W 2014 *J. Power Sources* **268** 153.
[11] Li H, Shi L, Wang Q, Chen L, Huang X 2002 *Solid State Ionics* **148** 247.
[12] Lewandowski A, Acznik I. 2005 *Electrochim. Acta* **56** 211.
[13] Wang H Y, Wang F M 2013 *J. Power Sources* **288** 1.
[14] Hoshi Y, Narita Y, Honda K, Ohtaki T, Shitanda I, Itagaki M 2015 *J. Power Sources*, **288** 168.

Acknowledgments
This work (Grants No.C0502212) was supported by Business for Cooperative R&D between Industry, Academy, and Research Institute funded Korea Small and Medium Business Administration in 2017. This research was supported by the Basic Science Research Program through the National Research Foundation of Korea (NRF) funded by the Ministry of Education (No. NRF-2017R1A2B4010544).

Research on Downhole Risk Assessment Method for Production Wells

N T Si[1], Q W Chen[1], J Q Long[1] and Y H Gao[1]

1 Tianjin Branch Company, CNOOC（China）, Ltd., Tianjin, China
E-mail: gewf@cnooc.com.cn

Abstract. At present, because of long offshore oil exploitation time, a large amount of adjusting wells and large offshore oil-gas production, downhole risks are difficult to be evaluated, and safety of offshore oil exploitation faces a great challenge. Aiming at the difficulties, reservoir energy is calculated through PHAST software, and overflow capacity of oil wells is analysed. Based on analysis of factors influencing downhole risks of offshore oilfield wells and analysis of reservoir energy and overflow capacity, a qualitative and quantitative assessment method is proposed for downhole risks of production wells. Downhole risk grades are divided according to the method, and corresponding control measures are proposed. This method is applicable to an offshore oilfield, risk grades of more than one hundred wells are divided, and corresponding control measures are proposed for different risk grades, so that downhole risks are reduced, and safety exploration of the oilfield is guaranteed.

1. Introduction

While oil and gas are continuously exploited, offshore well spacing area becomes dense, downhole oil-gas pressure system is complex and safety of offshore oil exploitation becomes worse, the state continuously updates laws and regulations about offshore oil exploration and development and puts forward higher requirements on safety. Risks on offshore oil and gas exploration mainly come from characteristics of formation reservoirs and properties of downhole fluids. At present, accident risks are often overestimated or underestimated if risk grades are simply judged according to one or two indexes. The assessment method defined according to risks, which assesses risks according to the product of accident frequency and accident severity, has some limitations. Therefore, a method for quantitatively dividing downhole risk grades is needed to be created to divide risk grades of different production wells; well-control equipment and tools shall be rationally configured according to risk grades, so as to avoid a great waste of human and material resources.

2. Downhole risk factors of production wells

After research and analysis, there are many factors influencing downhole risks of production wells, and the factors are divided into four classes according to their types as follows: reservoir geologic conditions, equipment and facilities [1], management system, technology and operations of persons [2]-[5]. At present, factors influencing downhole risks mainly include: geologic conditions of reservoirs, such as formation pressure coefficients of reservoirs, oil-gas ratio, well pattern development methods, reservoir temperatures and reservoir depths; equipment and facilitates, such as function failures of well-head assembly, failures of annulus packers of casings, failures of safety instruments, failures of downhole safety valves, failures of through-cable packers and failures of well-control systems. If downhole risks are assessed from the foundational factors, risk statuses of the reservoir itself are needed to be assessed; therefore, main influence factors [6], such as formation

IOP Conf. Series: Earth and Environmental Science **159** (2018) 012006 doi:10.1088/1755-1315/159/1/012006

pressure, formation pressure coefficient, oil-gas ratio, gas production rate, reservoir temperature and pressure coefficients of wellbore fluids, shall be considered as main influence factors to quantitatively assess downhole risks.

3. Quantitative assessment of downhole risks

Overflow or artesian flow of oil-gas wells essentially depends on the overflow energy [7] of oil wells. Therefore, downhole risks shall be researched from the reservoir energy. In an ascending process of fluids, reservoir gas inevitably escapes due to reduction of pressure, and gas channelling is one of main factors of blowout accidents; therefore, when factors influencing downhole risks are considered, overflow energy of production wells and reservoir gas energy of oil wells shall be considered as main influence factors to comprehensively assess downhole risks.

3.1. Determination of overflow capacity

Overflow energy is the least energy that well bottom fluids overcome their own gravity and resistance in a flow process while fluids flow to well head from well bottom. Main factors influencing overflow energy are pressure coefficients of reservoirs and pressure coefficients of wellbore fluids. Overflow capacity reflects the difference between the formation pressure coefficient and the fluid pressure coefficient. Formation pressure stabilizes in a certain time; the lower the pressure coefficient of the wellbore fluid, the stronger the overflow capacity. Therefore, the key of research is to calculate the pressure coefficients of wellbore fluids.

In a reservoir exploitation process, the wellbore pressure coefficients vary for flow states and composition of fluids, and the main influence factors are water content, gas-oil ratio, exploitation type and formation pressure [8], [9] in the exploitation process. While exploitation continues, formation pressure is reduced, water content rises, and the wellbore pressure gradient rises correspondingly. The wellbore fluid pressure varies for well depths; therefore, the wellbore fluid pressure shall be collected and sorted to seek the pressure change rules.

Overflow energy of well fluids is mainly calculated according to static pressure test reports, productivity test reports and pressure recovery test results reports. The static pressure test reflects static pressure gradient at different depths of the wellbore, and virtually reflects the static pressure coefficient of the fluid at different well depths via static pressure gradient. Data analysis and calculation steps are shown in figure 1.

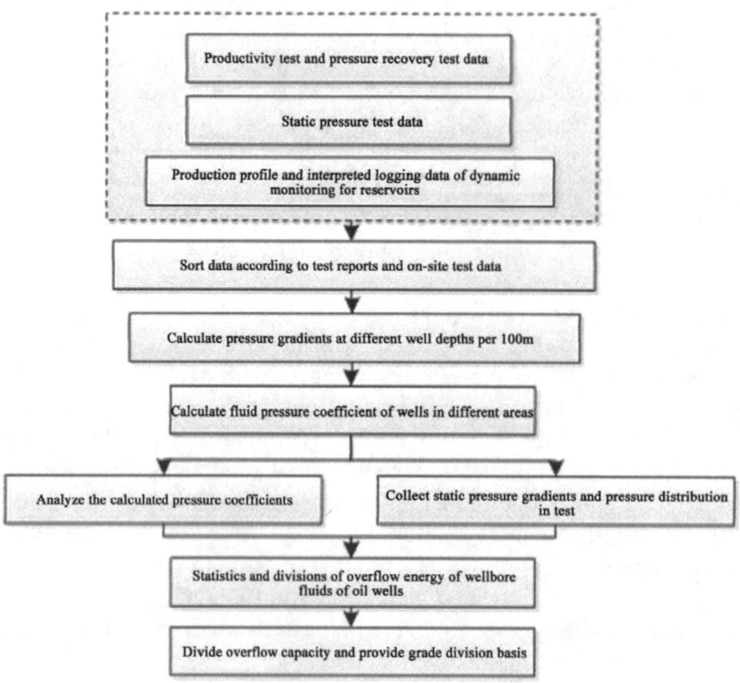

Figure 1. Calculation steps of overflow capacity.

Fluid pressure coefficients of oil wells in different areas can be obtained by calculating the overflow energy of some wells in different areas, and the overflow capacity can be measured according to formation pressure coefficients. Taking a well XX as an example, we introduce how to calculate the overflow capacity. The static pressure test data of oil line I of well XX is shown in Table 1.

Table 1. Static pressure test data of well XX.

Slant depth (m)	Vertical depth (m)	Static pressure (MPa)	Static pressure gradient (MPa/100m)
1651.5	1651.5	13.56	0.97
1551.7	1551.7	12.59	0.97
1451.5	1451.5	11.63	0.97
1351.5	1351.5	10.66	0.96
1252.8	1251.5	9.69	0.963
1152.7	1151.5	9.332	0.950
1052.3	1051.5	8.866	0.942
952.6	951.5	8.352	0.932
852.9	851.5	7.784	0.910
752.3	751.5	7.083	0.873
652.1	651.5	6.184	0.855
551.9	551.5	5.287	0.843
451.8	451.5	4.457	0.841
351.5	351.5	3.748	0.834
251.5	251.5	3.119	0.832
151.5	151.5	2.595	0.832
51.5	51.5	2.159	0.811

Table 1 reflects distribution and characteristics of static pressure fluids in oil sections. From data, only static pressure gradients of oil sections are reflected, but pressure gradients at different depths of the whole wellbore cannot be reflected. In order to understand the variation trend of wellbore static pressure gradient, initial data of tests is needed to be sorted, corresponding pressure values shall be sorted per 100m, and the fluid state data of well section 50-1,650m forms according to the initial data of tests. If linear fitting of pressure gradients at different well depths linearly and corresponding well depth is used, change rules of pressure gradients of wellbore fluids can be reflected through a fitting curve as shown in figure 2.

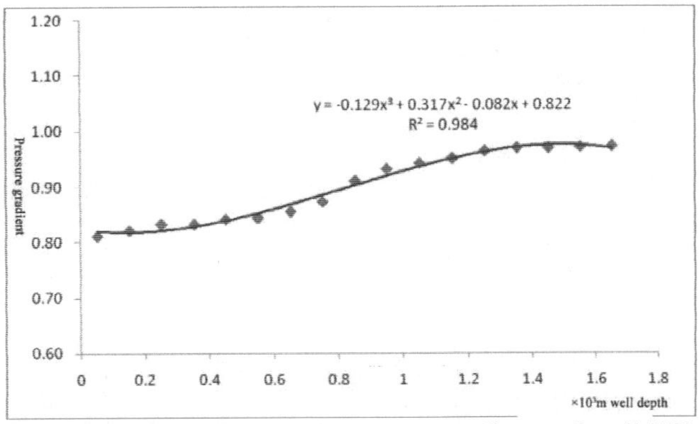

Figure 2. Fitting curve of pressure gradients of well XX.

The formula for fitting curve of pressure gradient and depth is:

$$y = -0.129x^3 + 0.317x^2 - 0.082x + 0.822;\quad R_2 = 0.984 \tag{1}$$

The relative mean pressure coefficient of wellbore fluids can be calculated according to the formula. Because the special well depth corresponds to the relative pressure gradient, the weight pressure coefficient of all wellbore fluids can be obtained by integrating the fitting curve and calculating the ratio of the vertical depth to the pressure of clear water.

$$\rho = \left[\int_{H_0}^{H_1} (-0.129x^3 + 0.317x^2 - 0.082x + 0.822)dx\right] / (H_1 - H_0) \quad (2)$$

Substituting $H_0=51.5\times10^{-3}$ and $H_1=1651.5\times10^{-3}$ into formula (2), we obtain:

$$\rho = \left[\int_{0.0515}^{1.6515} (-0.129x^3 + 0.317x^2 - 0.082x + 0.822)dx\right] / 1.6 \quad (3)$$

$$\rho = \left[-\frac{0.129}{4} \times (H_1^4 - H_0^4) + \frac{0.317}{3} \times (H_1^3 - H_0^3) - \frac{0.082}{2} \times (H_1^2 - H_0^2) + 0.822(H_1 - H_0)\right] / 1.6 \quad (4)$$

$$\rho = 0.893 \quad (5)$$

The pressure coefficient of the wellbore fluid is 0.893, but from reservoir test data, the current formation pressure coefficient is changed between 0.8 and 0.84, and it is lower than the pressure coefficient of the wellbore fluid.

The pressure coefficients of wellbore fluids in a hundred of wells in an oilfield are collected and calculated according to the method, so as to seek the distribution states of pressure coefficients of fluids in the oilfield. After pressure coefficients of wellbore fluids in production wells are calculated and collected, the most wellbore pressure coefficients are within a range of 0.85-0.90, and the lowest wellbore pressure coefficient is 0.8 as shown in figure 3.

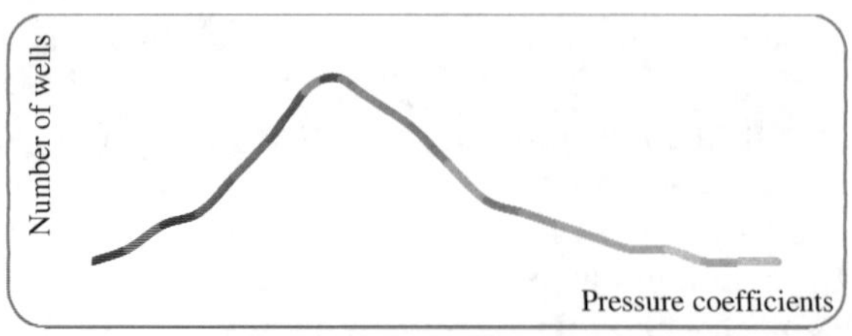

Figure 3. Pressure coefficient chart of production wells in an oilfield.

From calculation results and statistical results of overflow energy of oil wells in the oilfield, the current formation pressure is gradually reduced along with exploitation time, and water content rises gradually in the exploitation process. From analysis of risk principle 20/80, when the formation pressure coefficient of the oil well is lower than 0.80, the oil well has hardly any overflow capacity; when the formation pressure coefficient of the oil well is between 0.8 and 0.905, the oil well has smaller overflow capacity; when the formation pressure coefficient of the oil well is larger than 0.905, the oil well has larger overflow capacity.

3.2. Determination of reservoir gas energy

While fluid pressure is reduced, gas in production wells escapes gradually. Gas itself has compressibility, and has huge destructive power in the exploitation process. Gas is also one of the main factors causing well-control damage. Therefore, we should calculate the reservoir gas energy. From the sorted production data, oil production rate and gas production rate of current oil wells are reduced gradually, and oil-water ratio becomes higher and higher. Therefore, only the maximum data of daily gas production rate in recent years not only represents the maximum amount of gas, but also represents the maximum damage risk of gas in recent years.

Pressure, temperature, oil production rate, gas production rate of the oilfield, sizes of casings and other parameters are entered in software PHAST to calculate the gas energy via a leakage model; kinetic

IOP Conf. Series: Earth and Environmental Science **159** (2018) 012006 doi:10.1088/1755-1315/159/1/012006

energy of gas blowout of the oil well can be calculated according to parameters of various oil wells. The kinetic energy reflects the damage capability of the gas in oil wells.

Energy of gas in oil wells is mainly related to reservoir pressure, gas production rate and temperatures of oil wells, leakage size and other parameters, and the acting energy can be calculated through quantitative calculation software PHAST. If the change section of the acting energy of gas in production wells is (7,956.55, 215,954.57) MJ, according to requirements of risk principle of 20/80 and requirements of risk grade division, the gas energy of oil wells is low if less than 26,000 MJ, medium if between 26,000 MJ and 50,000 MJ, and high if more than 50,000 MJ.

4. Distinction of downhole risks

Downhole risks of oil wells shall be assessed according to intrinsic safety factors of well control, namely formation energy of oil wells. Downhole well-control risks shall be comprehensively assessed by calculating overflow capacity and reservoir gas energy of oil wells. Based on overflow energy, a risk matrix is created according to the acting energy of the reservoir gas to divide the downhole risk grades of production wells. Downhole well-control risk assessment matrix diagram is shown in Figure 4-1. The red area represents high risk, the yellow area represents medium risk, and the green area represents low risk.

		Formation pressure coefficients		
	Risk grades	P<0.80	0.80-0.905	>0.905
Reservoir gas energy	0-26000MJ			
	26000-50000MJ			
	>50000MJ			

Figure 4. Risk matrix diagram.

According to requirements of the risk matrix diagram, formation pressure coefficient and acting energy of reservoir gas of oil wells containing gas can be calculated through the method, and the downhole well-control risk grades of production wells can be comprehensively assessed according to the risk matrix principle.

An offshore oilfield has a hundred of production wells, and downhole risk grades are divided. A statistical analysis shall be made according to high, medium and low grades: 1) oilfield wells with high downhole risks are accounted for 24%; 2) oilfield wells with medium downhole risks are accounted for 47%; 3) oilfield wells with low downhole risks are accounted for 29%, as shown in figure 5.

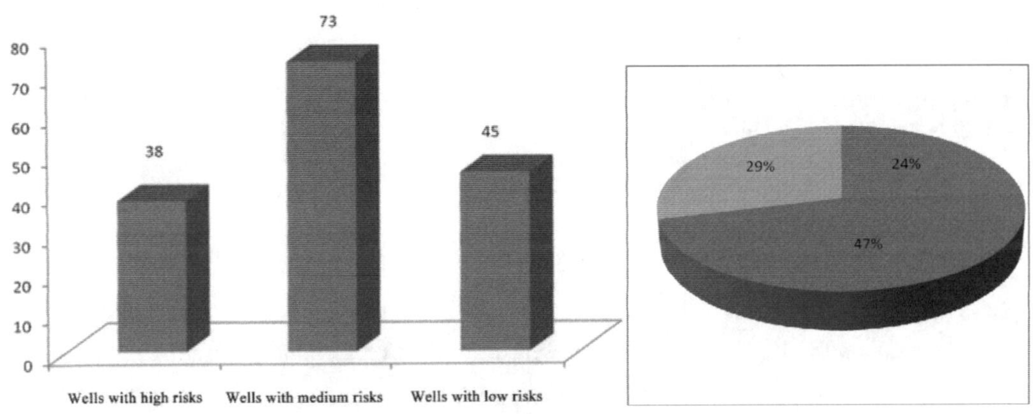

Figure 5. Risk grade scale.

5. Measures and suggestions

(1)For wells with high risks, enterprises shall highly input, list critical focus wells in daily production, regularly monitor and record well bottom pressure, well head pressure, temperature and other production parameters. If parameters fluctuate severely, enterprises shall reduce the monitoring period, and master the well bottom pressure, annulus under-pressure statuses and trend to make sure that the aboveground and downhole well-control equipment is in good conditions.
(2)For wells with medium and low risks, enterprises shall understand well bottom pressure and trend, monitor and record well bottom pressure, well head pressure, temperature and other production parameters. Well head equipment, safety instruments, downhole safety valves and other well-control facilities shall be timely maintained or replaced if having any failure.

6. Conclusion

(1) This paper analyses various factors influencing downhole risks of production wells, and creates an downhole risk analysis and assessment method for production wells based on essential safety;
(2) A downhole risk assessment and grading model is created for production wells according to the proposed assessment method and such factors as formation pressure coefficients, gas-oil ratio of reservoirs, oil production rate and gas production rate.
(3) The acting energy of downhole gas is calculated through the model via quantitative software PHAST. Downhole risk situations are obtained from essential safety according to overflow capacity of oil-gas wells, and the grades of the risks are divided: More than 100 production wells in XX Oil Field are assessed with 24% of high risk wells, 47% of medium risk wells and 29% for low risk wells.
(4) The next step will carry out the well integrity analysis of high risk wells, further quantify the downhole risk, and put forward immediate corrective measures for the high risk wells that have been assessed.

7. References

[1] Gao Y H, Sun B J, Cao S J, Song L S and Shi Z F 2008 Risk assessment on well control in deep-water drilling based on fault tree analysis *Oil Drilling & Production Technology*, pp 23-27
[2] Zhang H, Wang K and Zeng H 2010 Feature analysis and countermeasures of human error in drilling blowout incidents *Natural Gas Industry* pp 98-104
[3] Li Y J, Ren M P, Li X F, Nie M H, Du G and Shi F Q 2012 Risk ranking and well control measures of drilling blowout in Xinjiang oilfield *Journal of safety science and technology* pp 113-117
[4] Ren M P and Li X F 2010 Risk evaluation method of drilling blowout based on fuzzy mathematics *Proceedings of the 3rd World Conference on Safety of Oil and Gas Industry WCOGI, Sept. 26-27, Beijing, China*
[5] Cui W, Luo Y, Zeng Z, Huang H Q, Liu X D, Wang Q S, Li S X and Chen Z H 2014 The study of ranking methods of the typical accidents risks of machinery & electronic special equipment for early-warning and pre-controlling *Industrial Safety and Environmental Protection* pp 59-62
[6] Guo H and Chen G M 2012 The accident-causing analysis and evaluation indexes for offshore drilling platform *Journal of safety science and technology* pp 108-113
[7] Zhu Q Y and Sun B Q 1997 Downhole safety valve for offshore holes *CPM* pp 24-25
[8] Xi Z C, Sun M, Gu X P and Ren M P 2011 The method of workover blowout risk evaluation based on fuzzy mathematics *Science technology and engineering* pp 8334-8338
[9] Ren M P, Li X F, Yin B T, Cui S, Zhang X Q and Xu D R 2012 Study on calculating model of drilling blowout probability based on fuzzy mathematics *China Safety Science Journal* pp 81-86

Acknowledgments
This study was supported by National key research and development program of the Ministry of science and technology (No. 2017YFC0804500).

IOP Conf. Series: Earth and Environmental Science **159** (2018) 012007 doi:10.1088/1755-1315/159/1/012007

A Novel Hybrid Ensemble Classification Approach to Candidate Well Selection for CBM Well Fracturing

Lei Xu[1,2] and Fang Liu[1]

1 Neijiang Normal University, Neijiang, P.R.China
2 Data Recovery Key Laboratory of Sichuan Province
E-mail: 1986_xulei@163.com

Abstract. For coalbed methane commercial development, hydraulic fracturing is the most common stimulation treatment. The selection of wells or layers is the first link. Fracturing with big yield potential and good effect can not only reduce fracturing risk but also offer more technological choices for reservoir stimulation. A method combined with qualitative and quantitative analysis is adopted to construct the index system of candidate wells selection according to coalbed reservoir properties and development characteristics. In order to study the complex uncertainty and unbalanced classification problem in the process of candidate well selection, a fusion of interval type 2 fuzzy logic system (IT2FLS) and extreme learning machine(ELM) based selective ensemble method is proposed in this paper. Illustrative examples are provided to show to the validity of the proposed method.

1. Introduction

Hydraulic fracturing (HF) is one of the most important stimulation treatments for coalbed methane production. There are many factors that have influence on the effectiveness of hydraulic fracturing. In addition to the characteristics of the reservoir itself, the most important thing is the three aspects that human can change, namely to candidate well selection (CWS), fracturing design and field operation.

There have been a large number of methods dealing with CWS [1-3]. Existing approaches to candidate well selection can be roughly divided into two categories: conventional and advanced.

Conventional method is a kind of qualitative research methods based on geology factors including experience criterion method, screening method, reservoir dynamic analysis method and classification. For more details about conventional methods see [3]. Conventional methods have some weakness. For example, the result cannot be obtained when one or more of the key parameters were unavailable. In addition, the lack of data, data conflict, and data measurement error often exists in application process which causes the complexity and uncertainty of CWS.

Advanced method provides a technical framework to solve the problem with complex, uncertain, including artificial neural network (ANN), support vector machine (SVM), pattern recognition, grey correlation analysis, fuzzy logic (FL), etc. Those methods has been paid more attention by scholars and applied successfully including ANN and FL.

However, an accurate ANN model needs a large number of labeled examples, and what's more does not account for uncertainty. FL is an extension of conditional logic which is introduced by Zadeh [4]. It also called type-1 fuzzy logic (T1FL). To some extent, it can overcome the conventional difficult. But the fuzzy membership degree is a precise value, so it can't completely handle the uncertainty.

Type-2 fuzzy logic (T2FL) is introduced by Zadeh as an extension of T1FL. Its membership function itself fuzzy so it has the ability to deal with uncertainties that caused by information loss. At present, it has become an efficient tool for dealing with uncertainty [5].

Uncertainty refers to a kind of generalized error, including the concept of value and error, and also the error of measurable and unmeasurable. At present, it is well known that the petroleum reservoir data are inherently uncertain [6].

In terms of oil and gas, Zoveidavipoor et al [7] first introduced interval type 2 fuzzy set and system to study CWS for oil well. The author makes a contrastive study of the differences between type-1 fuzzy logic system (T1FLS) and Type-2 fuzzy logic system (T2FLS). However, directly use this method to CBM has some important weakness. On one hand, some important parameters such as permeability, skin factor, water cut, and porosity are hard to obtain. On the other hand, vertical well drilling account for the vast majority of well in China, daily average production per well is less than 1000 m^3. The proportion of high and low Wells is not equal to 1. As is well known, the greater the gap, the greater the imbalance. Therefore, the study of how to improve the classification performance of imbalanced data and the generalization ability of classifier has the important value and practical significance. However, the existing methods do not take into account the imbalance classification problem.

Extreme learning machine (ELM), introduced by Huang [8], is a single-hidden layer feed-forward neural networks. Compared with BP Neural Network, ELM does not need to adjust the network weights and bias. It can get the only optimal solution by adjust the hidden nodes. It has advantages of fast learning speed and good generalization. However, ELM suffers from low performance when uncertainties are presumed to be rife [9].

The motivation of this paper is to extend the application of T2FLS and ELM to solver unbalanced classification of well and layer selection for CBM well fracturing. The article is organized as follow. In Section 2, we construct the production division and index system using fuzzy cluster analysis and grey correlation analysis. In Section 3, the frameworks of the hybrid integrated model are obtained. In Section 4, the accuracy of the proposed method is tested through several numerical examples.

2. Production division and index system construction

2.1. Production division

The reservoir of Hancheng area is unsaturated. Most of the production wells need to take a long water drainage stage. Therefore, to analyse the production expediently, classification standard is defined as average daily gas production after two years of production. In general, considering three classifications can be decomposed into three binary classifications, so in this paper we discuss binary classification only. High production gas well is greater than 1500 m^3/d and low production gas well is less than 1500 m^3/d.

2.2. Index system construction

In this paper, the main research is machine learning based candidate well selection which can be treated as a classification problem. For complex problems, a single model's results cannot always be satisfied, so we call each model a weak predictor or weak classifier. When we have thousands of models, we can through fusion of the model to get a better effect, such as Bagging, Boosting and random forests.

The development of coalbed methane is a systematic project. Many factors would affect the production. geological analogy and statistical methods are the main methods to study the production characteristics and factors [10]. The existing research predominantly focused on the areas with a higher level of development, for example Qinshui Basin in china. Though the key factors in different areas vary greatly, yet gas content, coal thickness, permeability, fracture characteristics have the highest frequencies.

In recent years, only a few scholars have analyzed the factors affect the CBM well productivity of Hancheng area, and the key control factors are not clear [11]. However, the factors can be grouped into two main categories generally: those that cannot be controlled by human including gas content, permeability, buried depth, thickness, hydrodynamic system, etc; those can be controlled including drilling and well completion method, perforation program, fracturing design and field operation.

We all know that the quality of data collection and processing is the basis of machine learning. Compared with oil reservoir, the reason that it is difficult to get coal seam information or the obtained information is not accurate. There are many reasons among which two aspects must be mainly mentioned here: on the one hand CBM reservoir has characteristics of heterogeneity and friability which affect the data from laboratory, on the other hand due to its low-cost development strategy many important parameters (permeability, porosity, compressibility, and relative permeability) are not available for development well[12]. Fortunately, scholars have done a lot of research to prediction these parameters based on the conventional well log response and got good effect [13].

To determine the predominant factors, we proposed a qualitative and quantitative evaluation method to determine key factors. We investigate the potential influencing factors extensively firstly, as shown in Table 1, and then exclude the factors which do not meet the principles of relative independentability and operability. For example, sand concentration is equal to the quantity total volume minus pad volume and divided by proppant volume, so sand concentration can be removed.

Fuzzy clustering can set up the description of the sample with the categories of uncertainty. Therefore, it can be used to data with poor separation. The advantage of grey correlation analysis (GCA) is that it requires only a little amount of data and workload, and can largely reduce the losses caused by the information asymmetry. In this paper, we combine fuzzy cluster analysis and grey correlation analysis methods to select appropriate inputs for CBM productivity forecasting. For more details see [12].

Table 1. Primary input parameters and ranges

Parameter	Range	Parameter	Range	Parameter	Range
thickness of coal, m	2-12	burial depth, m	384-1005	well diameter, cm	18-41
natural gamma, API	10.5-47	SP, mv	-80-84	acoustic, us/m	240-467
CNL, v/v	22-60	density, g/cm^3	1-2.51	LLD, Ωm	150-18343
LLS, *Ωm*	56-17380	RMSF, Ωm	36-13507	proppant volume, m^3	7-51
pad volume, *m*^{3}	70-500	total volume, m^3	207-1003	injection rate, m^3/min	4.5-8.5
delivery rate, m/day	-6.8-17.2	distance from fault , m	85-424		

SP: Spontaneous potential; CNL: compensated neutron logging; LLD: deep lateral resistivity; LLS: shallow lateral resistivity; RMSF: microspheric focused resistivity

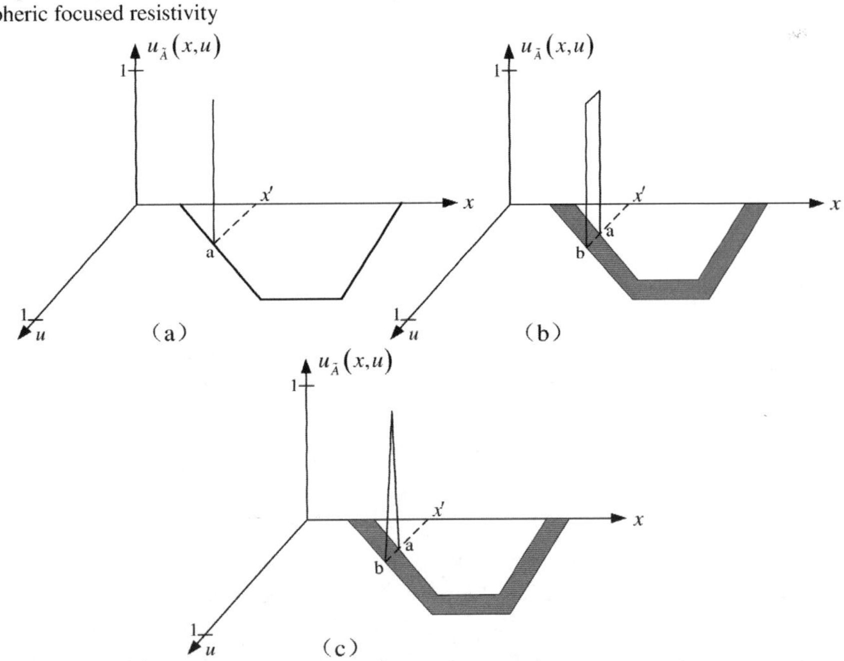

Figure 1. Secondary MFs at x = x′ for (a) T1 FS, (b) IT2 FS, and (c) GT2 FS.

3. The constituent frameworks of the proposed hybrid integrated model

IOP Conf. Series: Earth and Environmental Science **159** (2018) 012007 doi:10.1088/1755-1315/159/1/012007

3.1. Type-2 fuzzy logic system

Zedeh first introduced type-2 fuzzy sets (T2FS) as an extension of ordinary fuzz set which called type-1 fuzzy sets (T1FS). Unlike T1FS, the degrees of general type-2 fuzzy sets (GT2FS) memberships are themselves fuzzy. When the secondary MF equals to 1, then each GT2FS reduces to an IT2FS. IT2FS is the simplest kind of type-2 sets, and there are fast algorithms to compute the output. Therefore, it has practical value.

For a T1FS (Figure. 1a), the primary membership at x = x′ has only one value, so there is no uncertainty. For an IT2FS (Figure. 1b), the primary membership at x = x′ has values within the interval [a, b], so it encompasses a large amount of uncertainty. For a GT2FS (Figure. 1c), each point in the interval have a different secondary membership, so it has a large uncertainty than IT2FS

As shown in Figure 2, a type-2 fuzzy logic system (T2FLS) contains five components: fuzzifier, rules, inference engine, type-reducer, and defuzzifier. For more details see [14].

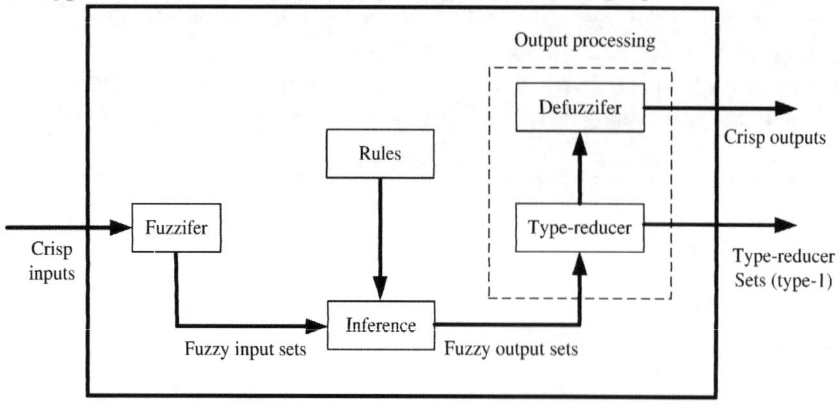

Figure 2. Type-2 fuzzy logic system.

3.2. Extreme learning machine

In general, ELM can be expressed as:

$$t_j = \sum_{i=1}^{N_h} \beta_i g\left(w_i x_j + b_i\right),\ j = 1,\cdots,N_s \tag{1}$$

If we have N_s samples $(\mathbf{x}_i, \mathbf{t}_i)$, N_h hidden node, where $\mathbf{x}_i = [x_{i1}, x_{i2}, \cdots, x_{in}] \in R^n$, $\mathbf{t}_i = [t_{i1}, t_{i2}, \cdots, t_{im}] \in R^m$, transfer function $g(x)$. Or vector form, as: $H\beta = T$, where $H=\{h_{ij}\}$ (i=1, …, N_s and j=1, …, N_h), $h_{ij}=g(w_j x_i+b_j)$. Then, the training procedures for ELM can be summarized as follows.

Step 1: Initialization. Assign random values to the input weight w_i and the bias b_i.

Step 2: compute the hidden layer output matrix H.

Step 3: compute the output weight b as follows: $\beta = H^{\dagger} T$.

where $H^{\dagger}$ is moore-penrose generalized inverse of a matrix of H. See [8] for further details.

Compared with BP network, the learning speed of ELM is faster, generalization performance is better. However, this algorithm has some important weakness. One of the major is its inability to model uncertainties. Therefore, it is necessary to merge unique generalization ability of ELM and the ability of IT2FLS to better model uncertainty, in order to obtain better classification performance.

3.3. Imbalanced classification

Experiments show that: the base classifier with good predict performance cannot guarantee a good performance of the combined classifier. Diversity of the base classifiers is the key of multiple classifier system with high performance and generalization ability. The concept of "selective ensemble ", first proposed by Zhou et al [15], has caused great repercussions. The results reveals that it may be better to ensemble many instead of all of the classifiers at hand. However, there is no unified standard to measure the diversity of the base classifier yet.

At present, there are already many selective ensemble methods in literatures, but due to the different perspectives on which those methods are based, it is difficult to understand them clearly. In this paper, through literature research, we choose margin distance minimization (MDSQ), developed by Martínez-Muñoz G, Suárez A [16]. The method has the characteristics of good prediction performance, fast selection speed, but bigger target integrated classifier size.

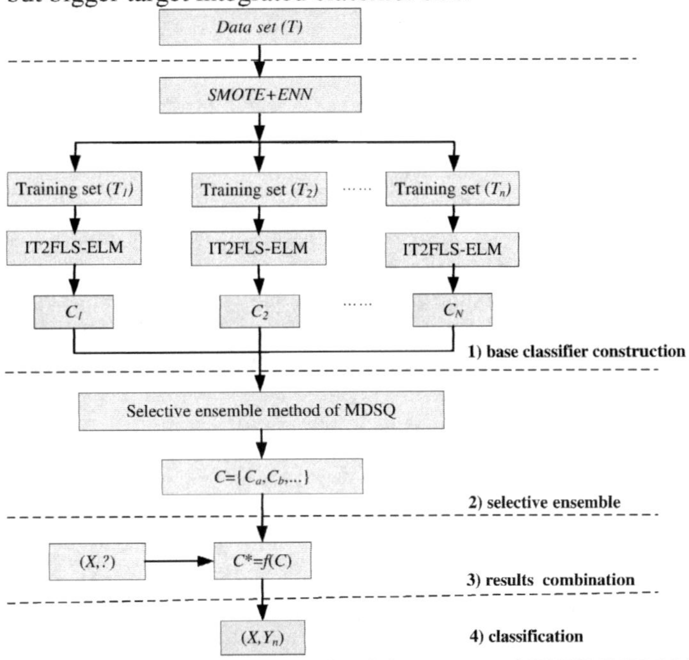

Figure 3. Conceptual integration framework of the proposed IT2FLS-ELM hybrid system.

Figure 3 shows the conceptual integration framework of the proposed IT2FLS-ELM hybrid integrated system. In the first stage, the algorithm of SMOTE-ENN is adopted to get the training sets $\{T_1, T_{2,}...,T_n\}$. The training set of $\{T_i\}$ is passed to the IT2FLS block where all possible forms of uncertainties are adequately handled. Undergoing the five necessary processes, the final outputs from IT2FLS are then passed to the ELM block for training and finally get the base classifier $\{C_i\}$. In the second stage, the MDSQ method is adopted to select M optimal base classifiers based on the validation set. Finally we use the test set to validate the optimal base classifiers, and it is need to note that the weight of the optimal base classifiers is the same in this paper.

4. Experimental results and discussion

4.1. Criteria for performance evaluation

We consider a set of data sample U. $U = P \cup N$, where P represents the positive class and N the negative class. We define $p=|P|$, $n=|N|$, then the imbalance ratio (IR) is defined as $IR=n/p$. The performance of the classifier can be expressed in a confusion matrix, as in Table 2.

For any given classifier, TP is the abbreviation of true positive means a correctly classified positive instance. Similarly, a true negative (TN) means a correctly classified negative instance. In the remaining cases, a false negative (FN) means a positive instance be misclassified as negative, a false positive (FP) means a negative instance be wrongly predicted as positive. G-mean take into account both the classification of positive and negative, the greater the value the better the performance of classifiers, defined as:

$$G-mean=\sqrt{\text{TNR}\times precision} \tag{2}$$

where

Ture negative rate (TNR): TNR= TN/(TN+FP)

$$Precision = \text{TP}/(\text{TP}+\text{FP})$$

Table 2. A two-class dataset of confusion matrix

Acutal/Predicted	positive	negative
positive	TP	FN
negative	FP	TN

4.2. Case 1

In order to compare the propose algorithm with existing methods, we consider 5 data sets from different area to evaluate our proposal. These data sets originate form the UCI machine learning repository [17] and we modified multiple class data sets into two-class. In the data set of Abalone, we take the attribute of no.17 as positive and the remaining as negative. In vehicle, attribute of van as positive and the remaining as negative. In vowel, attribute of hed as positive and the remaining as negative.

The characteristics of the data set can be found in Table 3, showing the number of attributes (Attr), the number of positive instances (PInst), the number of negative instances (NInst) and the imbalance ratio (IR) for each of them. Form Table 3 we can see that there were a broad range in the number of instances and IR. The value of IR in data set Sonar close to 1 which indicate that it not belong to imbalanced data. The choice of the data set is to verify the effectiveness of the proposed algorithm in the balance data set classification problem.

Table 3. Performances of different algorithms on UCI repository（*G*-mean）

Data sets	Attr	Pinst	Ninst	IR	algorithm					
					C4.5	SMOTEBoost-C4.5	EUSBoost-C4.5	RotEasy-C4.5	SMOTE-SVM	Our method
Abalone	8	58	4119	71.02	73.1	0.878	0.933	0.938	0.710	**0.942**
Sonar	60	97	111	1.14	72.8	0.782	0.937	**0.941**	0.905	0.936
Credit-g	21	300	700	2.33	68.9	0.876	0.918	0.923	0.892	**0.932**
vowel	14	90	900	10.00	95.3	0.978	0.972	0.981	0.948	**0.983**
vehicle	19	199	647	3.25	89.4	**0.952**	0.942	0.836	0.902	0.935

Table 4. Input variables for CWS

categories	**parameters**
uncontrollable	thickness of coal, burial depth, GA, LLD, AC, SP, well diameter, distance form fault
controllable	total volume, proppant volume, injection rate, casing pressure, delivery rate

Data in Table 3 is divided into training set, validation set and test set according to the proportion of 6:1:1. Considering the random of division, the experimental results were the mean of 100 times repeated experiments. In order to check the effectiveness of the proposed algorithm, a comparative study with four representative algorithms C4.5 [18], SMOTEBoost [19], EUSBoost [20] and RotEasy [21] were done in this paper. The experimental results are shown in Table 3. From the table it is evident that compared with other algorithms, the proposed algorithm can obtain relatively good results in view of the different data sets.

4.3. Case 2

In order to carry out an empirical study, real-industrial data from Hancheng area were acquired. Among them, positive number is 37, negative number is 135, the proportion of positive and negative classes is 0.215/0.785, *IR* is 3.65. According to input selection algorithm in Section 2.2, the factors are presented in Table 4. The entire data was divided into training set and test set according to the proportion of 6:1. Experimental results were the mean of 100 times repeated experiments. All data were normalized by equation (9) before training.

$$z_{\text{normal}} = \frac{z - z_{\min}}{z_{\max} - z_{\min}} \tag{3}$$

Table 5. Performances of different algorithms evaluated on Hancheng area

Algorithm	Number of base classifiers	TNR	*Precision*	G-mean
Our method	20	**0.873**	**0.914**	**0.893**
ensemble method of T1FLS-ELM	20	0.868	0.891	0.879
ensemble method of ELM	20	0.815	0.893	0.853
ensemble method of IT2FLS	20	0.841	0.902	0.871
IT2FLS-ELM	1	0.756	0.879	0.815
ELM	1	0.681	0.872	0.771
IT2FLS	1	0.733	0.863	0.795
T1FLS	1	0.716	0.854	0.782
BP	1	0.512	0.836	0.654
SVM	1	0.645	0.865	0.747

In the block of IT2FLS, consider a fuzzy system with uncertain mean Gaussian membership functions, singleton fuzzifier, max-product composition, product implication, centre of sums type-reducer, and average defuzzification.

In order to validate the proposed method, ensemble method of T1FLS+ELM, ELM and IT2FLS and advanced approach of BP network and SVM were also being used to predict the classification effect. BP network, T1FLS and SVM were performed by MATLAB toolbox, with the structure of BP network ($13\times18\times1$), T1FLS 18 number of membership function with Gaussian membership function.

A comparative analysis of our study with other methods is shown in Table 5. To our knowledge, no previous studies have been done using the proposed method for CWS.

From the table, it can be concluded that, the hybrid method can make up for the lack of a single method, complement each other's advantages, and improve the effect of the classification. It can also be easily noted that the ensemble method has a better performance than the single method, and IT2FLS-ELM performed outstandingly better in all fronts.

The ELM method performed better than BP and SVM. The essence of the ELM and SVM are problems mapped to high-dimensional space, and then do regression or classification in high dimensional space. In principle, the classification or regression effect is directly affected by the mapping mode, and the more obvious differences in the high-dimensional space the better the classification effect. For ELM, the way map to high-dimensional space is infinite, and the training speed is very fast. However, once the SVM kernel function is settled, the mapping way is uniquely determined.

The overall results therefore indicate that the proposed model is able to consistently deal with the nature of coalbed CWS data due to its ability to cater for different forms of uncertainties, generalization and imbalance.

5. Conclusions

In this paper, we have presented a new hybrid ensemble pruning alorithm, a new algoritnm level solution to two-class imbanlanced classification problems of coalbed candidate-well selection. The solution is based on the combination of IT2FLS and ELM, and MDSQ.

The comparison with four kinds of typical algorithm over a collection of imbalanced datasets from UCI repository verified the feasibility and effectiveness of the proposed algorithm. Experiment results in Hancheng area also show that the proposed method can outperform the classical algorithm such as IT2FLS, T1FLS, ELM, BP, SVM and its ensemble methods.

For future work, we will extend the method to handle multi-class problems and use the method to other reservoir engineering problems and look into its usefulness in other related fields.

6. References

IOP Conf. Series: Earth and Environmental Science **159** (2018) 012007 doi:10.1088/1755-1315/159/1/012007

[1] Barak, S., Mehrgini, B., Maghsoudlou, H., & Branch, Q 2014 Multi-Criteria Decision Making Approach To Candidate Well Selection For Hydraulic Fracturing Treatment *CIE44 & IMSS*. **14** 2092-2106

[2] Xiang-jun, X., & Ting, Y. (2014) Application of TS Fuzzy Model in Candidate-well Selection for Hydraulic Fracturing In *Fuzzy Information & Engineering and Operations Research & Management*. Springer Berlin Heidelberg 563-570

[3] Zoveidavianpoor, M., Samsuri, A., & Shadizadeh, S. R 2012 Fuzzy logic in candidate-well selection for hydraulic fracturing in oil and gas wells: A critical review *International Journal of Physical Sciences*. **7** 4049-4060

[4] Zadeh L A 1965 Fuzzy sets *Information & Control*. **8** 338-353

[5] Zeng J, Liu Z Q 2006 Type-2 Fuzzy hidden Markov models and their application to speech recognition *IEEE Transactions on Fuzzy Systems*. **14** 454-467,

[6] SUBBEY S C M, SAMBRIDGE M 2004 Prediction under uncertainty in reservoir modeling *Journal of Petroleum Science & Engineering*. **44** 143-53

[7] Zoveidavianpoor M, Gharibi A 2016 Applications of type-2 fuzzy logic system: handling the uncertainty associated with candidate-well selection for hydraulic fracturing *Neural Computing &* Applications. **27** 1831-1851

[8] Huang, G. B., Zhu, Q. Y., & Siew, C. K 2006 Extreme learning machine: Theory and applications *Neurocomputing*. **70** 489-501

[9] Olatunji, S. O., Selamat, A., & Abdulraheem, A 2014 A hybrid model through the fusion of type-2 fuzzy logic systems and extreme learning machines for modelling permeability prediction *Information Fusion*. **16** 29-45

[10] Tao, S., Tang, D., Xu, H., Gao, L., & Fang, Y 2014 Factors controlling high-yield coalbed methane vertical wells in the Fanzhuang Block, Southern Qinshui Basin *International Journal of Coal Geology*. **134** 38-45

[11] Zhao, J., Tang, D., Xu, H., Lv, Y., & Tao, S 2015 High production indexes and the key factors in coalbed methane production: a case in the hancheng block, southeastern ordos basin, china. *Journal of Petroleum Science & Engineering* **130** 55-67

[12] Xu, L., Zhu, K., & Yang, X 2016 Production forecasting of coalbed methane wells based on type-2 Fuzzy logic system *The Open Petroleum Engineering Journal*. **9** 268-278

[13] Fu, X., Qin, Y., Wang, G. G., & Rudolph, V 2009 Evaluation of coal structure and permeability with the aid of geophysical logging technology *Fuel*.**88** 2278-2285

[14] Mendel, J. M 2001 *Uncertain rule-based fuzzy logic systems: introduction and new directions* (pp. 131-184). Upper Saddle River: Prentice Hall PTR

[15] Zhou, Z. H., Wu, J., & Tang, W 2002 Ensembling neural networks: many could be better than all *Artificial intelligence*. **137** 239-263

[16] Martínez-Muñoz G, Suárez A 2006 Pruning in ordered bagging ensembles *Proceedings of the 23rd international conference on Machine learning. ACM*: 609-616

[17] Asuncion, A., & Newman, D 2007 *UCI machine learning repository*

[18] Quinlan, J. R 2014 *C4. 5: programs for machine learning* (Elsevier)

[19] Chawla, N. V., Bowyer, K. W., Hall, L. O., & Kegelmeyer, W. P 2002 SMOTE: synthetic minority over-sampling technique *Journal of Artificial Intelligence Research*. **16** 321-357

[20] Wang, S., & Yao, X 2012 Multiclass imbalance problems: Analysis and potential solutions *IEEE Transactions on Systems, Man, and Cybernetics, Part B* (Cybernetics). **42** 1119-1130

[21] Parvin, H., & Minaei-Bidgoli, B 2013 A clustering ensemble framework based on elite selection of weighted clusters *Advances in Data Analysis and Classification*. **7** 181-208

Acknowledgments

This study was supported by SCED of China grant 17ZB0220.

IOP Conf. Series: Earth and Environmental Science **159** (2018) 012008 doi:10.1088/1755-1315/159/1/012008

Using Ion-Exchange to Recovery of Germanium from Waste Optical Fibers by Adding Citric Acid

Wei-Sheng Chen[1], Bi-Cheng Chang[1], and Yen-Jung Chen[1]

1 Department of Resources Engineering, National Cheng Kung University, No.1, University Road, Tainan City, Taiwan (R.O.C)
E-mail: mic830210@gmail.com

Abstract. Germanium is the raw metal of the optical fibers which become more important with the rapid development of optical communication in recent years. Owing to the importance of germanium and the problematic of waste optical fibers, we using the hydrometallurgical process to recovery germanium in this study. The process can be divided into two parts, the first part is pretreatment and leaching, the second is separation and metal recovery. In the first part, we collected the optical fibers and dissolved them by roasting and leaching. Then, the separation process which we used is ion-exchange. At the ion-exchange process, IRA900 is chosen as the resin and citric acid as the new addition agent, and other parameters were also considered. Finally, the metal oxides, GeO_2, was obtained from the separation process through concentration and calcination. With the optimal conditions, the recovery rate of germanium is about 99%. The purity of obtained products is over 99%.

1. Introduction

Germanium is the rare metal in the earth which concentration only has 1-7 ppm [1] and it also is a useful element in optics and electronics. According to the U.S. Geological Survey research, the world reserve of germanium only has 8600 tons and it has none in mineable deposits [2]. With the semiconductor and optical industries increased demand for germanium, the European Union has listed germanium as a critical raw material in 2010 [3]. And germanium is also an indispensable strategic resource in the field of modern military applications [4]. In the global germanium consumer market, the largest amount of germanium used is fiber optics which account for 40% [4]. With the optical fiber technological advances and the demands of the Internet, the global optical fiber market would reach 3.72 billion U.S. dollars in the future [5]. However, germanium which added in the fiber would increase in demand. As the rare and important strategic metal resource, germanium would have the problem of underproducing, so recovering germanium in the future is necessary. And then, 30% of the germanium is obtained from recycled materials which like electronic devices or fly ash [4]. Because the lots of waste optical fibers would be produced and the value of germanium would increase, the process of recycling germanium from the waste optical fibers must be studied.

There are two major recycling ways for germanium recovery which are pyrometallurgy and hydrometallurgy. Pyrometallurgical process is the most used to recover germanium from coal [6]-[9], but it has the disadvantage of the high energy consumption and low purity products. On the other hand, the hydrometallurgical process is more suitable than the pyrometallurgical processes for treating optical fibers. The general hydrometallurgical processes for recycling germanium included chemical precipitation [10], [11], chlorination distillation [12], [13], solvent extraction [14]-[20] and ion-exchange [21]-[26]. Because of the separation efficiency, treating capacity, continuous operation and reusable of agents, the ion-exchange is the better way which was chosen in this study. According

to the literature, the germanium adsorption efficiency of cation resin is lower than anion resin and several anion resins have experimented, such as IRA900[21,22], IRA958[21], PA312[26] and D-403[23]. As mentioned above, it is required which makes germanium ion turn into the anion and it would be formed the anion complexes usually by adding agents, such like catechol[21], [22], [25], [26], 3-methylcatechol[25], [26] or oxalic acid [27]. However, the new agent we used in this study is the citric acid which also can complex with germanium and cheaper than other agents.
In this study, the optical fibers after pre-processing were dissolved in solution by alkali roasting and leaching. The amount of NaOH, pH value and solid-liquid ratio were experimented in this step. To optimize the adsorption efficiency of germanium, we choose IRA900 as the resin and citric acid as the new addition agent. The parameters such as the pH value, citric acid mole ratio and the concentration of eluting agent which is HCl were studied. Finally, we can recover germanium effectively and get high purity GeO2 to reuse.

2. Experimental

2.1. Materials, reagents and instruments
The optical fibers samples were our targets in this study which collected from the waste cable. According to the previous study, the elements analysis was shown in Table 1 which could be found the concentration of germanium in the fiber was 0.1% and other is silica [28]. The sodium hydroxide (Applichem Panreac NaOH 98%) and sulfuric acid (Sigma-Aldrich H2SO4 98%) used as roasting and leaching agents and dissolved in deionized water. In the ion-exchange process, IRA900 (Alfa-Aesar) was used as resin, citric acid as the new addition agent (J.T.Baker 99.5%) and hydrochloric acid (Sigma-Aldrich HCl 37%,) was used as eluting agent. Analyzing the germanium content for roasting, leaching and ion-change efficiency is used by Inductively coupled plasma optical emission spectrometry (ICP-OES, Varian, Vista-MPX). X-ray diffraction (XRD, Dandong, DX-2700) was used to determine crystalline phase of sample and the final product.

Table 1. Elements analysis of optical fiber powder

Elements	Ge	Silica	Fe	Mg	Ca
XRF analysis	0.11%	99.48%	0.23%	0.04%	0.13%
ICP analysis	0.1%	99.9	Not-detected	Not-detected	Not-detected

2.2. Roasting and Leaching
First of all, the optical fibers were collected from cables. Because the fibers contained silica[28], it should dissolve silica and take germanium out of them by alkali roasting. Adding NaOH at 500°C makes silica turn to sodium silicate and leach easily for its water solubility property. The parameters like NaOH/SiO_2 mole ratio, pH value and solid-liquid ratio were investigated. The pH value was set from 1 to 7 and the effect of liquid-solid ratio from 10 to 100 ml/g which were tested to get better leaching efficiency.

2.3. Ion exchange and calcination
IRA900 was strongly basic resin which was used to exchange anion ions efficiently. In this study, we make germanium be the anion complexion by adding the new agent which is citric acid and the formation pH value of complexion is lower than 7[29]. We also compare the germanium adsorption efficiency without citric acid because germanium in high pH value would turn to anion that the pk1 and pk2 of germanic acid is 9.01 and 12.3[30]. At eluting step, we use HCl to strip germanium from resin. The parameters of the ion-exchange experiments were set, the pH value (1-14), addition agents mole ratio of germanium (1-6 mol/mol), bed volume number and eluting molarity(0.1-2M).
The eluting solution which contained lots of germanium was obtained after ion-exchange. The metal oxides can be received through calcination which the products were GeO_2. The purity, composition and crystal phase of the product would be investigated in this study.

3. Results and Discussion

3.1. Roasting and Leaching

The optical fibers were roasting by adding NaOH at 500˚C and considered the $NaOH/SiO_2$ mole ratio effect. The result was shown in figure 1, that amount of NaOH is enough when mole ratio higher than 6. The figure 2 showed that fiber was actually turned to sodium silicate from silicon dioxide by XRD. After roasting, the fibers were leached by dilute H_2SO_4 and showed the pH value and liquid-solid ratio effect in the figure 3 and 4. It was found that germanium and silicon leaching rate decreased significantly at neutral environment in figure 3. The reason of that was that silicon would precipitate and germanium would be precipitated together when pH value at neutral. In figure 4, both of germanium and silicon leaching rate were decreased with the liquid-solid ratio decreasing because of the excessive silicon concentration and insufficient for volume of acid aqueous. The excessive silicon concentration would also make them precipitate that the leaching efficiency decreased. To sum up, adding 6 mole ratio of NaOH at 500 ◦ C to roast and pH value lower than 5, 40ml/g at leaching step was chosen as optimal condition. Under this condition, the leaching efficiency were higher than 99.5%.

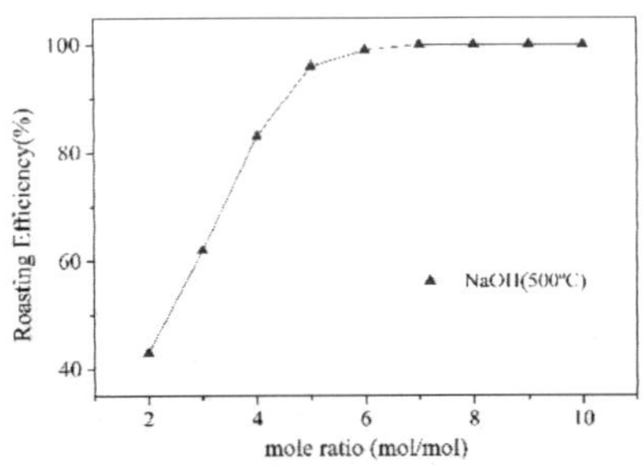

Figure 1. Effect of the roasting efficiency on $NaOH/SiO_2$ mole ratio.

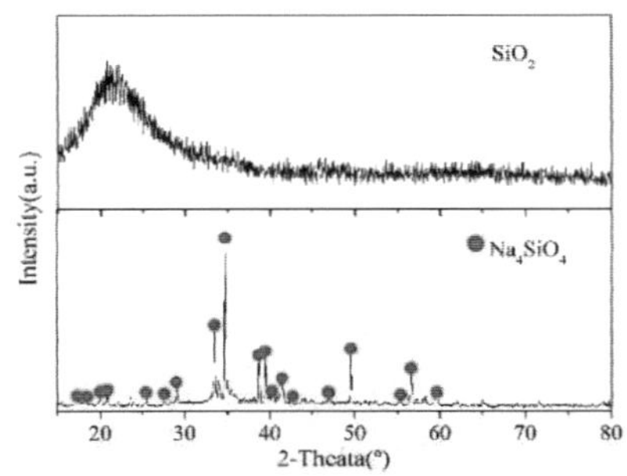

Figure 2. The XRD pattern of optical fibers after roasting.

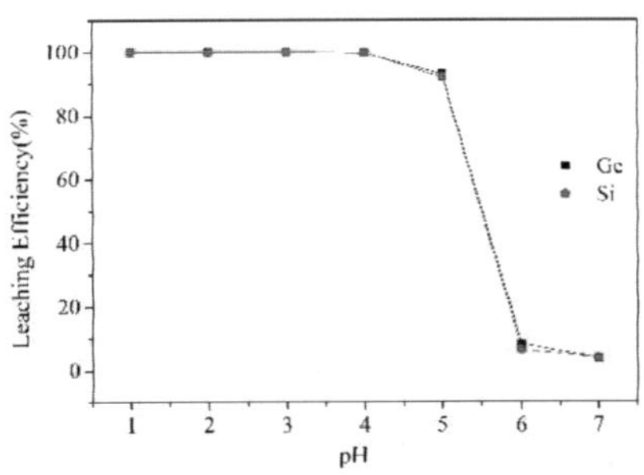

Figure 3. Effect of the leaching efficiency on pH value (L/S=40ml/g, 25˚C).

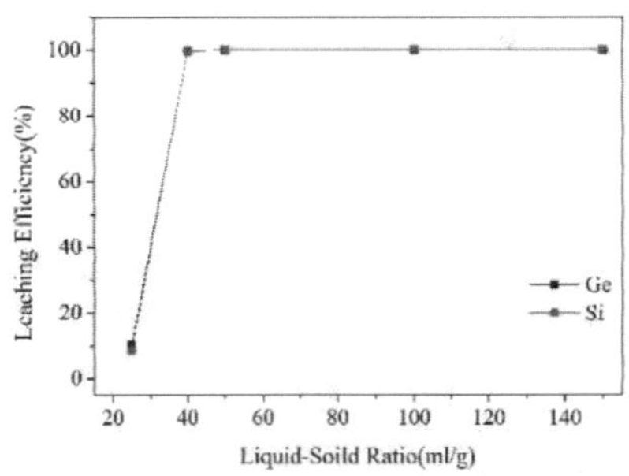

Figure 4. Effect of the leaching efficiency on liquid-solid ratio(pH=3, 25˚C).

3.2. Ion exchange

After leaching process, the elements were almost dissolved in the solution which the major elements were sodium, silicon and germanium. It was used ion-exchange to separate germanium from them in this study. The concentration ratio of germanium, silicon and sodium were set 25、10000、40000 mg/L which was based on the leaching result.

The figure 5a and 5b was shown that the relationship between pH value and adsorption efficiency of germanium. In figure 5a, the adsorption efficiency of germanium without adding citric acid(Cit) was obviously lower than adding Cit. For reasons outlined above, adding Cit is necessary for adsorbing germanium in this study. And then, it was also found that the silicon and sodium were hardly adsorbed and germanium had the highest adsorption efficiency at pH value equaled to 3 in the Figure 5b. By comparing Cit with catechol[21], adsorption capacity of germanium by using Cit is higher than using catechol and the cost is lower than using catechol. It means using Cit as the new addition agent is usable for germanium adsorption. The result of figure 5c showed that Cit mole ratio effect for adsorption efficiency. At the molar ratio equal to 4, it has the highest adsorption efficiency of germanium. After adsorption, it was used HCl to elute germanium and the concentration effect was shown in Figure 5d. To get the best efficiency and benefit, the concentration of HCl to elute germanium is 1 mol/L. Based on the result, we did column experiments at the optimal condition which was chosen the pH value equal to 3, Cit/Ge molar ratio equal to 4, IRA900 as the resin and 1mol/L HCl as the eluting agent at 25 ˚C and its result were shown in figure 6a and 6b. At the optimal condition, the adsorption efficiency of germanium was 92% and silicon, sodium were both lower than 0.001%. After eluting, the recovery of germanium could be recycled to 99% by column in series.

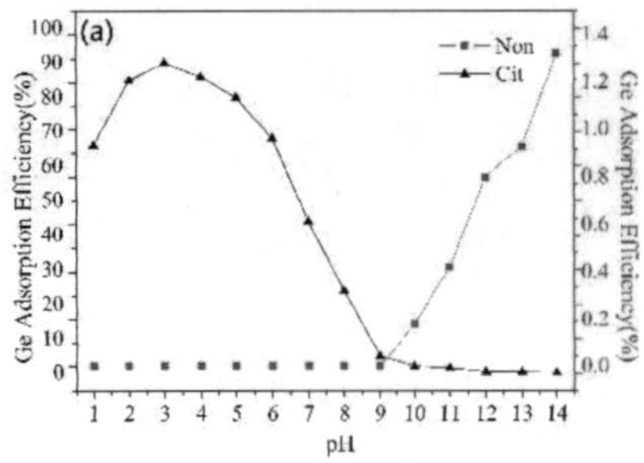

(Cit/Ge=4mol/mol, IRA900, 25 ˚C)

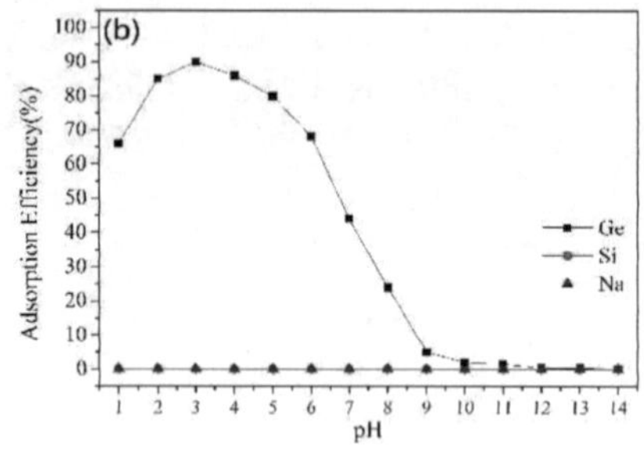

(Cit/Ge=4mol/mol, IRA900, 25 ˚C)

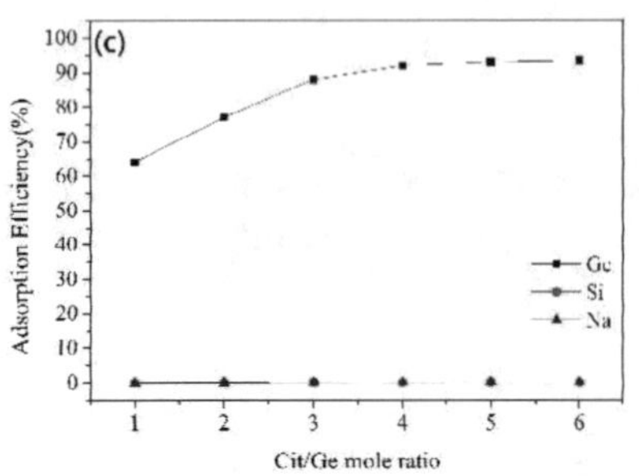

(pH=3, IRA900, 25 ˚C)

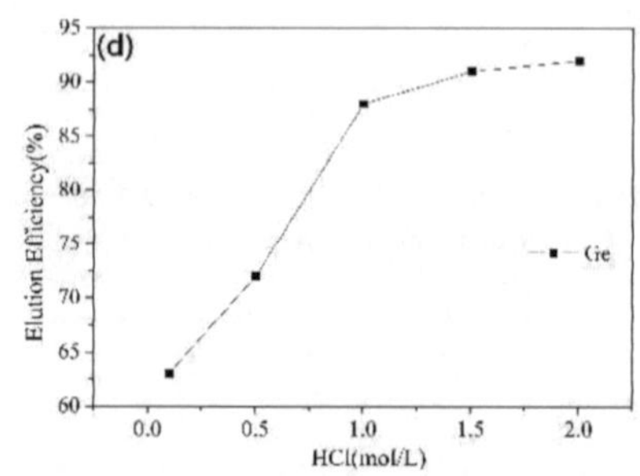

(IRA900, 25 ˚C)

Figure 5. Effect of germanium adsorption or eluting efficiency on (a) addition Cit (b) pH, (c) Cit mole ratio, (d) concentration of HCl.

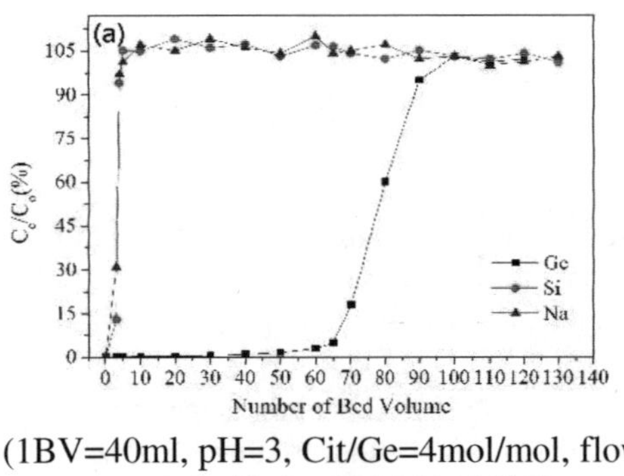

(1BV=40ml, pH=3, Cit/Ge=4mol/mol, flow rate 2.6ml/min, IRA900, 25 ˚C)

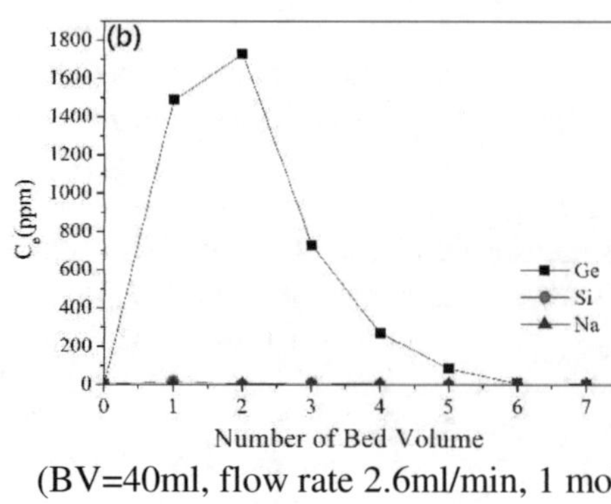

(BV=40ml, flow rate 2.6ml/min, 1 mol/L HCl, 25°C)

Figure 6. Column experiments for germanium (a) adsorption curve, (b) elution curve.

3.3. Calcination

After ion-exchange, germanium was enriched in eluting solution and separated from silicon and sodium. By concentrating and calcining, the TG and XRD analysis of the metal oxides are shown in figure 7 and 8. From the TG diagram, the temperature was set at 500 °C to turn into germanium dioxide. There is no impurity phase is detected and the ICP analysis also showed that the purity of GeO_2 is higher than 99%.

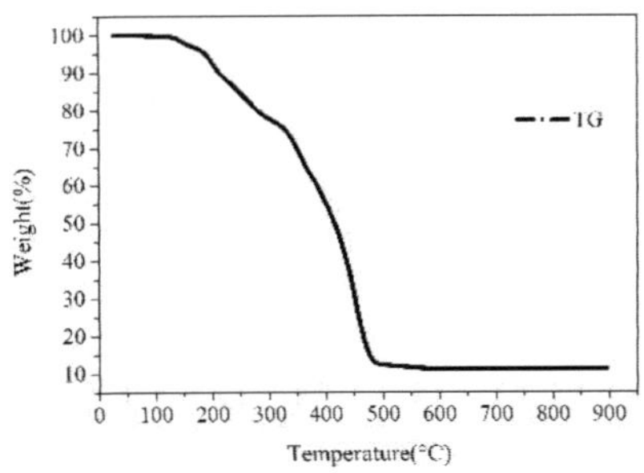

Figure 7. TG analysis of product.

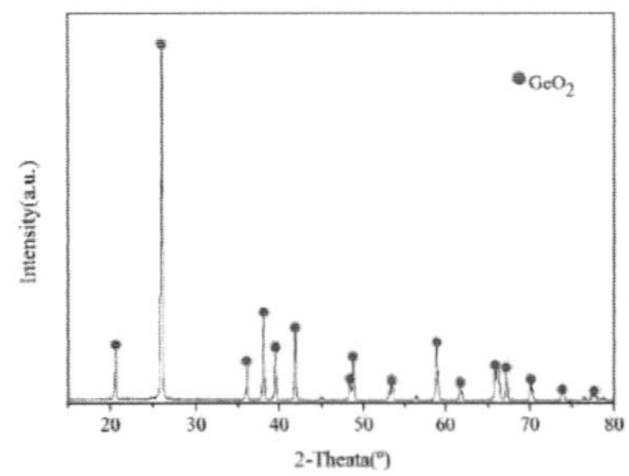

Figure 8. XRD pattern of GeO_2.

4. Conclusions

This study proposed a hydrometallurgical way to recovery germanium effectively from the optical fibers. In roasting step, we add $NaOH/SiO_2$ mole ratio equal to 6 at 500℃ to make SiO_2 be sodium silicate necessarily. And then in leaching step, the leaching efficiency of germanium was higher than 99.5% under the optimal conditions which were pH 6 and 40ml/g. In ion-exchange step, germanium can be separated from silicon and sodium effectively with pH 3, Cit/Ge mole ratio equal to 4, IRA900 as the resin and 1mol/L HCl as the eluting agent at 25 °C. With this condition, germanium adsorption rate was 92% and silicon and sodium was lower than 0.001%. Finally, the GeO_2 product could be get by calcining at 500 °C and its purity was higher than 99%. This process which was shown in this study can provide an effective method to treat the related wastes containing germanium and silicon.

5. References

[1] J. H. Adams, 1990, Germanium and Germanium Compounds. Metals Handbook, 10th ed.; ASM International: Materials Park, OH.
[2] USGS, Kelly, T.D., and Matos, G.R., comps, 2013, Historical statistics for mineral and material commodities in the United States (2013 version): U.S. Geological Survey Data Series 140.
[3] European Commission, 2010, "Critical Raw Materials for the EU: Report of the Ad-Hoc Working Group on Defining Critical Raw Materials," European Commission: Brussels, Belgium.
[4] U.S. Geological Survey, 2017, "Mineral Commodity Summaries 2017."
[5] Zion Market Research, 2017, Fiber Optics Market for Telecom & Broadband, Healthcare, Defense, Private Data Networks, and Other Applications: Global Industry Perspective, Comprehensive Analysis and Forecast, 2016-2022.
[6] Z. Rong-Kun, 1995, Extraction of germanium from lignite, CN1101380 A.
[7] L. Ting et al., 2006, Method for preparing carbocoal and extracting germanium-contained matter from brown coal destructive ditillation, CN1814701 A.
[8] L. Ba-Ke, G. Chi -Ming, 2009, Method for extracting germanium from lignite by pyrogenic process, CN101413063 A.
[9] P. Shih-Kun et al., 2015, Germanium ore washing method capable of improving pyrometallurgy recovery rate, CN105170311 A.
[10] F. Arroyo et al., 2009, "Recovery of Germanium from Coal Fly Ash Leachate by Precipitation," World of Coal Ash (WOCA) Conference.
[11] F. Arroyo et al., 2010, "Precipitation of Germanium from Coal Fly Ash Leachates," *Coal Combustion and Gasification Products*, **2**(1), 28–34.

[12] J. Jandová et al., 2002, "Recovery of germanium from fly ash-waste from coal combustion," In Proceedings of European Metallurgical Conference, Fridrichshafen, BRD; GDMB: Clausthal, 69-75.

[13] J. Jandová, H. Vu, 2001, "Processing of germanium-bearing fly ash. In Proceedings of the Fifth International Conference on Metallurgy," Refractories and Environment, Košice, The Institute of Metallurgy and Materials: 2002, 107-112.

[14] G. Cote, D. Bauer, 1980, "Liquid-liquid extraction of germanium with oxine derivatives," *Hydrometallurgy*, **5**(2-3), 149-160.

[15] H. J. Park, L. L. Tavlarides, 2009, "Germanium(IV) adsorption from aqueous solution using a kelex-100 functional adsorbent," *Industrial and Engineering Chemistry Research*, **48**(8), 4014-4021.

[16] D. D. Harbuck, J. C. Judd, D.V. Behunin, 1991, "Germanium Solvent Extraction From Sulfuric Acid Solutions (and Co-Extraction of Germanium and Gallium)," *Solvent Extraction and Ion Exchange*, **9**(3), 383-401.

[17] B. Gupta, N. Mudhar, 2006, "Extraction and Separation of Germanium Using Cyanex 301/Cyanex 923," Its Recovery from Transistor Waste. *Separation Science and Technology*, **41**(3), 549-572.

[18] J. Liang et al., 2012, "Study on Extracting of Germanium with Trioctylamine," *Energy Procedia*, **17**, 1965-1973.

[19] K. L. Nash, G. R. Choppin, 1977, "The thermodynamics of synergistic solvent extraction of zinc(II)," *Journal of Inorganic and Nuclear Chemistry*, **39**(1), 131-135.

[20] F. Arroyo et al., 2008, "Hydrometallurgical Recovery of Germanium from Coal Gasification Fly Ash. Solvent Extraction Method," *Ind. Eng. Chem. Res.*, **47**(9), 3186-3191.

[21] F. Arroyo et al., 2010, "Recovery of Germanium from Aqueous Solutions by Ion-Exchange Extraction of Its Catechol Complex," *Industrial & Engineering Chemistry Research*, **49**(10), 4817-4823.

[22] F. Torralvo et al., 2011, "Recovery of germanium from real fly ash leachates by ion-exchange extraction," *Minerals Engineering*, **24**(1), 35-41.

[23] D. E. Chirkst et al., 2008, "Sorption of germanium from alkaline solutions on anion-exchange resin," *Russian Journal of Applied Chemistry*, **81**(1), 38-41.

[24] I. Ozawa et al., 2000, "High-speed recovery of germanium in a convection-aided mode using functional porous hollow-fiber membranes," *Journal of Chromatography. A*, **888**(1-2), 43-49.

[25] A. Nozoe et al., 2012, "Germanium Recovery using Catechol Complexation and Permeation through an Anion-Exchange Membrane," *Separation Science and Technology*, **47**(1), 62-65.

[26] H. Takemura et al., 2013, "Germanium recovery by catechol complexation and subsequent flow through membrane and bead-packed bed column," *Journal of Chemical Technology & Biotechnology*, **88**(8), 1468-1472.

[27] D. A. Everest, 1995, "Studies in the chemistry of quadrivalent germanium. Part III. Ion-exchange studies of solutions containing germanium and oxalate," *Journal of the Chemical Society*, **0**, 4415-4418.

[28] W. S. Chen, B. C. Chang, K. L. Chiu, 2017, "Recovery of germanium from waste Optical Fibers by hydrometallurgical method," *Journal of Environmental Chemical Engineering*, **5**(5), 5215–5221.

[29] G. S. Pokrovski, J. Schott, 1998, "Experimental study of the complexation of silicon and germanium with aqueous organic species: implications for germanium and silicon transport and Ge/Si ratio in natural waters," *Geochimica et Cosmochimica Acta*, **62**(21-22), 3413-3428.

[30] B. Pommerrenig et al., 2015, "Metalloido-porins: Essentiality of Nodulin 26-like intrinsic proteins in metalloid transport," *Plant Science*, **238**, 212-227.

Mechanical Extraction of Protein Solution from Microalgae by Ultrasonication

A Kuhavichanan[1,2], P Kusolkumbot[2], S Sirisattha[2] and *C Areeprasert[1]

1 Department of Mechanical Engineering, Faculty of Engineering, Kasetsart University.
50 Ngam Wong Wan Road, Lat Yao, Chatuchak, Bangkok Thailand 10900.
2 Biodiversity Research Centre, Thailand Institute of Scientific and Technological Research.
35 Mu 3 Tambon Khlong Ha, Amphoe Khlong Luang, Pathum Thani 12120 Thailand
E-mail: achinatun@gmail.com and apichay.ku@gmail.com

Abstract. This study investigated the condition on microalgae cell disruption method by the combination of ultrasonication and buffer solution. Various buffer solutions have been utilized and the duration of the sonication was also tested. The quantity in term of protein concentration and the quality of the protein product were evaluated. Results showed that Mixed buffer III and Triton x lysis buffer showed the high concentration of protein but low quality. SDS lysis buffer showed high quality protein product but low the concentration of protein. For the study on the effect of sonication duration, the concentration of protein significantly increased when longer duration of sonication was applied and this result was observed during continuous sonication test as well. The concentration of protein was not only the important aspect, but the quality of the protein needs to be satisfied. Thus, it could be suggested that the optimum condition of protein extraction from microalgae needs the combination of the use of buffer solution to ensure high quality product and an appropriate duration of ultrasonication to maximize the quantity of protein.

1. Introduction

Many researchers have investigated the sustainable and renewable fuels including solar, wind, biomass, geothermal, hydropower, etc. Microalgae have been identified as a potential bioenergy source for fuel production. Advantages of microalgae over terrestrial biodiesel feedstock include easy to cultivate, higher growth rates and productivity than other terrestrial biomass sources, capable of storing solar energy into energy-rich compounds such as lipids and no direct competition for agricultural land [1]. Microalgae contain water around 80-90% [2], so traditional thermochemical processes like pyrolysis and gasification are not economically feasible. However, an innovative thermochemical process, i.e. Hydrothermal Liquefaction (HTL) that uses hot compressed water as a medium, appears to be more suitable for wet microalgae feedstock. The HTL utilizes high pressure and temperature steam at the condition of 250 to 350 °C, 10-20 MPa for conversion of wet biomass. For this result, the high-value intracellular component, such as protein, in microalgae was destroyed by the HTL processes. Protein was denatured easily under high temperature condition; therefore, the high value product must be extracted from the microalgae by other cell disruption methods before the HTL process.
Currently, many literature studies focused on evaluation of cell disruption methods. Cell disruption is important for the extraction of various cell inclusions [3], [4]. Cell disruption methods can be divided

into mechanical, such as ultrasonication [5], and non-mechanical, such as chemical. Ultrasonication is one of the most effective microalgae cell breaking methods. This method used sonic wave with frequencies around 25 kHz for cell disruption. Evaluation on the effectiveness of high-frequency focused ultrasound (HFFU) in microalgae cell disruption was performed [6]. However, only sonication mean was not enough for cell disruption of microalgae; thus, addition of buffer for cooperation on cell breaking was necessary. Buffer solution was able to protect the released protein from disrupt microalgae giving better protein quality. Therefore, this paper aims to optimize the condition on microalgae cell disruption method by the combination of ultrasonication and buffer solution.

2. Material and methods

2.1. Microalgae

Coelastrum sp. microalgae was used in this study. It was cultured with open pond system and was harvested by gravity sedimentation method [7]. The microalgae raw material was supported by the Biodiversity Research Centre (BRC), Thailand Institute of Scientific and Technological Research (TISTR), Thailand. Concentrated microalgae were separated and prepared for pretreatment. For pretreatment method, freeze-thaw was employed as the slurry microalgae were frozen at -40 °C prior to the experiment. Original microalgae (no pretreatment) was also prepared by storing at 4 °C.

(a) (b) (c) (d)

Figure 1. (a): suspension microalgae; (b): slurry microalgae; (c): dewatering by vacuum pump; (d): concentrated microalgae.

2.2. Preparation of microalgae suspension

The freeze microalgae suspensions was prepared at the concentrate (slurry) microalgae (g) and water (ml) ratio of 1:4 and kept at 4 °C. (Figure 1) Similarly, the fresh microalgae suspensions (Figure1a) was prepared at the concentrate (slurry) microalgae (Figure1b) then dewatering by vacuum pump (Figure1c). Finally, The concentrated microalgae (Figure1d) (g) were mixed with other solvents (ml) ratio of 1:4 and kept at 4 °C. Solvents used in this study can be named as follows: Water; Mixed buffer I (pH 7, 100 mM Tris, 10 mM EDTA (Ethylenediaminetetraacetic Acid) and 100 mM NaCl); Mixed buffer II (150 mM NaCl, 1% Triton X (Triton™ X-100) and 50 mM Tris pH 8), Mixed buffer III (150 mM NaCl, 1% Triton X, 50 mM Tris pH 8 and 0.1% SDS (Sodium Dodecyl Sulfate)), SDS lysis buffer (pH7, 100 mM Tris, 10 mM EDTA and 100 mM NaCl added SDS buffer), and Triton-x lysis buffer (pH7, 100 mM Tris, 10 mM EDTA and 100 mM NaCl added Triton x buffer). In this experiment, an evaluation on the effect of the concentration of SDS and Triton x lysis buffers and duration of cell disruption by ultrasonication was performed. Comparison of the concentration of protein and the quality of protein after extracted from the different solvents during ultrasonication was also performed.

2.3. Ultrasonication

IOP Conf. Series: Earth and Environmental Science **159** (2018) 012009 doi:10.1088/1755-1315/159/1/012009

Cells disruption method was done by ultrasonication device VCX 750 by Sonics & Materials Inc. The ultrasonic processors was operated at 750 Watt 20 kHz. The ultrasonic probes are fabricated from high-grade titanium alloy Ti-6Al-4V with 13 mm diameter and 159 mm length. The pulse generator was operated for 20 s and stop for 5 s and worked at 60 % amplitude. In the experiments, the ultrasonic heat, which occurred when probe is working, was absorbed by ice.

2.4. Quality and quantity of the product

After the microalgae suspensions were disrupted by ultrasonication, the product was centrifuged at 10,000 rpm 4 °C for 30 min (Sorvall™ Stratos™ Centrifuge by Thermo Fisher Scientific) for separation liquid and solid phase. A liquid phase (Figure 2) was measured concentration of protein by Bradford assay and quality of protein was determined by Sodium Dodecyl Sulfate polyacrylamide gel electrophoresis (SDS-PAGE). In this work, the test was triplicated and reported in mean value ± SD.

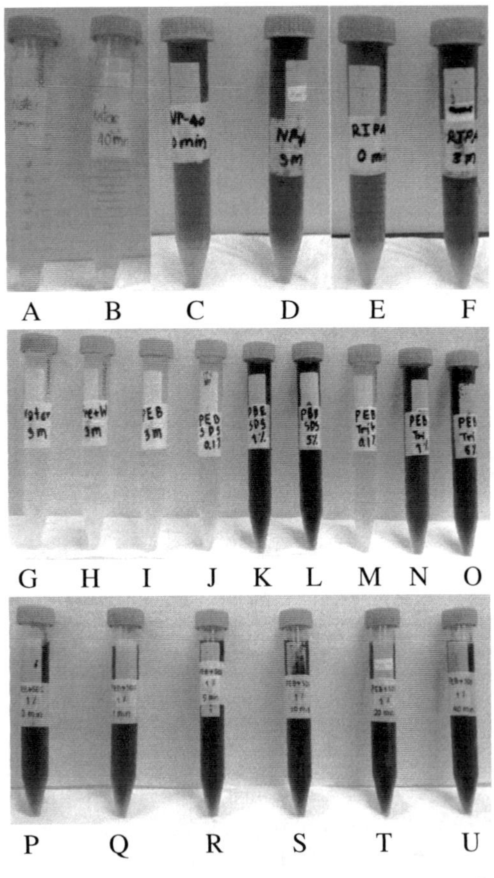

Figure 2. Supernatant solvents at various condition.

In Bradford assay[8], the absorbance was measured at 595 nm by spectrophotometer and the concentration of protein was determined by Monochromator (Infinite M200 PRO, Tecan Trading AG, Switzerland). Bovine Serum Albumin (BSA) was used to a calibration curve. In SDS-PAGE method separated molecules based on the difference of their molecular weight. Gels was stained with Coomassie brillianl blue (CBB). Figure 2 shows supernatant from various condition and solvent. A, G, and B are supernatant from water co-extraction with sonicate at 0, 3, and 40 min. H is supernatant from freeze-thaw co-extraction with sonicate microalgae at 3 min. I is supernatant from Mixed buffer I co-extraction with sonicate at 3 min. C and D are supernatant of Mixed buffer II co-extraction with sonicate at 0 and 3 min. E and F are supernatant of mix buffer III co-extraction with sonicate at 0 and 3 min. J, K and L are supernatant solvents of SDS lysis buffer concentrate 0.1, 1 and 5% co-extraction with sonicate at 3 min. M, N and O are supernatant from Triton x lysis buffer concentrate 0.1, 1 and 5% co-extraction with sonicate at 3 min. P, Q, R, S, T and U are supernatant from SDS lysis buffer 1% co-extraction with sonicate at 0, 1, 5, 10, 20 and 40 min.

2.5. Continuous cell disruption method by ultrasonication
The operation of the continuous ultrasonication cell disruption was performed at three liter/s. The slurry was recirculating until cell disruption was completed. The duration of the test was from 0 to 90 min. Figure 3 shows the schematic diagram of the continuous ultrasonication cell disruption system. The freeze microalgae was used as feedstock for this test. The feedstock was stored and sucked up by centrifugal pump then entering the disruption region from the bottom of the chamber. Sonication was done at 60% amplitude and frequency of 20 kHz. Then, the overflow was guided to the feedstock storage waiting for recirculation and re-disruption. The continuous system was used for cells disruption of microalgae by the combination of buffer solution and ultrasonication.

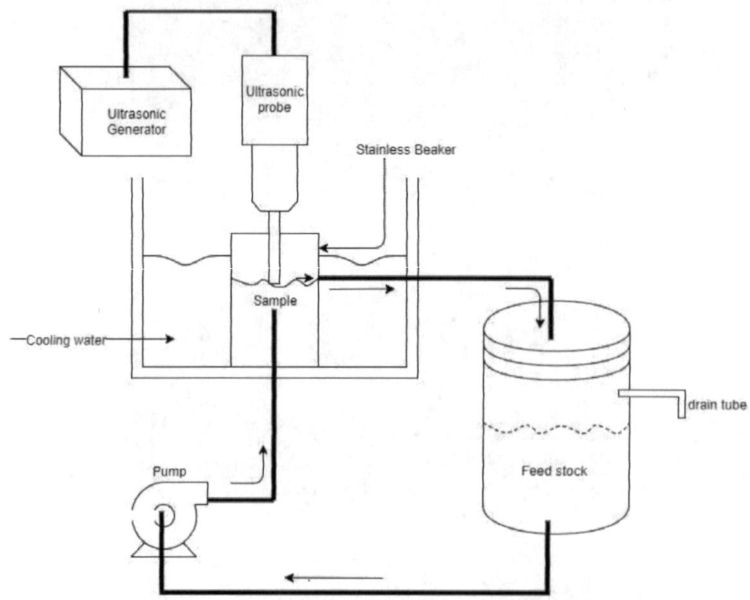

Figure 3. Schematic diagram showing the arrangement for continuous cell disruption method by sonication of the suspension microalgae.

3. Result and discussion
The concentration of protein is shown in Table 1. This method requires comparing the spectrophotometric absorbance of the unknown suspension at wavelength of 595 nm to a calibration curve prepared by standard solution of a protein (BSA). Table 1 presents the comparison of concentration of protein, which was measured from the suspension microalgae after cell disruption by sonication methods. In this Table 1, the highest of concentration of protein was 3.6540 mg/ml by Triton x lysis buffer 5 % 3 min. From the results of the water sonication at 0 min (0.0242 mg/ml), 3 min (0.0303 and 0.0418 mg/ml) and 40 min (0.1324 mg/ml), it revealed that the long duration increased the efficiency of cell disruption giving higher protein concentration. When comparison water, Mixed buffer I, Mixed buffer II and Mixed buffer III, the highest concentration of protein were measured in the case of Mix buffer III at 1.4975 mg/ml. Comparing the water, Mixed buffer I, various concentration of SDS lysis buffer as 0.1%, 1%, and 5%, various concentration of Triton x lysis buffer as 0.1%, 1%, and 5%) at 3 min sonication, result showed that Triton x lysis buffer gave the highest concentration of protein at 3.654 mg/ml. The final test was a comparison of SDS lysis buffer 1% at various sonication duration of 0 to 40 min. It was found that sonication with SDS lysis buffer 1% at 40 min gave the maximum protein concentration value at 0.3505 mg/ml.

IOP Conf. Series: Earth and Environmental Science **159** (2018) 012009 doi:10.1088/1755-1315/159/1/012009

In all cases, the Mixed buffer III and Triton x lysis buffer gave the high concentration of protein; however, the quality of protein determined by the SDS-PAGE shown in Figure 4 was no observed band. It can be suggested that the solvents distorted molecule of protein from long to short chain. This is represented by the poor quality band color shown in Figure 4a and 4b. SDS lysis buffer gave the concentration of protein not higher than that of the other solvents but the quality of protein was the best in all cases (Figure 4b, 4c). Figure 4b shows a comparison of the quality of protein SDS lysis buffer to Triton x with various solvent concentration of 0.1%, 1% and 5%. The H, I, and J were from Triton x lysis buffer 0.1%, 1% and 5%, respectively while the E, F, and G were from SDS lysis buffer 0.1%, 1%, and 5%, respectively. The H, I, and J were intent band but the E, F, and G were no observed band. Therefore, the amount of obtained protein was not only the importance, but also the quality of the protein product need to be consider for increasing the value of the final product.
For the evaluation of sonication duration, the data of the concentration of protein using SDS lysis buffer with various sonication duration as 0, 1, 5, 10, 20 and 40 min are summarized in Table 1. The concentration of protein increased when longer duration of sonication was applied, for example, at 0 min the yield was 0.1730 mg/ml and it was increased to 0.2963 mg/ml at 20 min and 0.3505 mg/ml at 40 min, accounting 71% and 102%, respectively. This was clear that the duration of sonication affected the yield of the protein.

Table 1. Quantification of protein in suspension by Bradford methods.

Solvent	Time (m)	Concentration protein (mg/ml)	SD
Water (1)	3	0.0303	0.0080
Mixed buffer I (1)	3	0.1025	0.0115
Mixed buffer II	0	1.0690	0.0100
Mixed buffer II	3	1.0950	0.0147
Mixed buffer III	0	1.4975	0.0125
Mixed buffer III	3	1.3457	0.0266
Water (2)	3	0.0418	0.0055
Pretreated-water	3	0.1139	0.0141
Mixed buffer I (2)	3	0.1010	0.0103
SDS lysis buffer 0.1 %	3	0.1606	0.0039
SDS lysis buffer 1 %	3	0.3159	0.0080
SDS lysis buffer 5 %	3	0.3346	0.0087
Triton x lysis buffer 0.1 %	3	0.1573	0.0092
Triton x lysis buffer 1 %	3	0.6176	0.0173
Triton x lysis buffer 5 %	3	3.6540	0.0112
Water (3) 0 min	0	0.0242	0.0088
Water 40 min	40	0.1324	0.0257
SDS lysis buffer 1 %	0	0.1730	0.0009
SDS lysis buffer 1 %	1	0.1737	0.0132
SDS lysis buffer 1 %	5	0.1948	0.0151
SDS lysis buffer 1 %	10	0.2270	0.0227
SDS lysis buffer 1 %	20	0.2963	0.0306
SDS lysis buffer 1 %	40	0.3505	0.0458

For continuous cell disruption by sonication, the first test used pretreated freeze-thaw microalgae as feedstock. The test used water to solvents and various duration of sonication. Results are presented in Table. 2. The concentration of protein at 40 min (longest duration in this study) was higher than that of at 20 min (shortest duration) for 12.3%. Additionally, the quality of protein obtained from this process is shown in Figure 4d was not satisfied due to denaturation of obtained protein. Therefore, the solvents need to be consider to improve the quality of the protein. Base on the previous experiment, the protection of extracted protein that can enhance the quality can be the SDS lysis buffer.

Table 2. Quantification of protein in suspension by Bradford methods.

Concentrate of protein from freeze microalgae		
Time(min)	Total protein (mg/ml)	SD
20	0.4232	0.0096
25	0.4300	0.0272
30	0.4488	0.0405
35	0.4584	0.0109
40	0.4754	0.0224

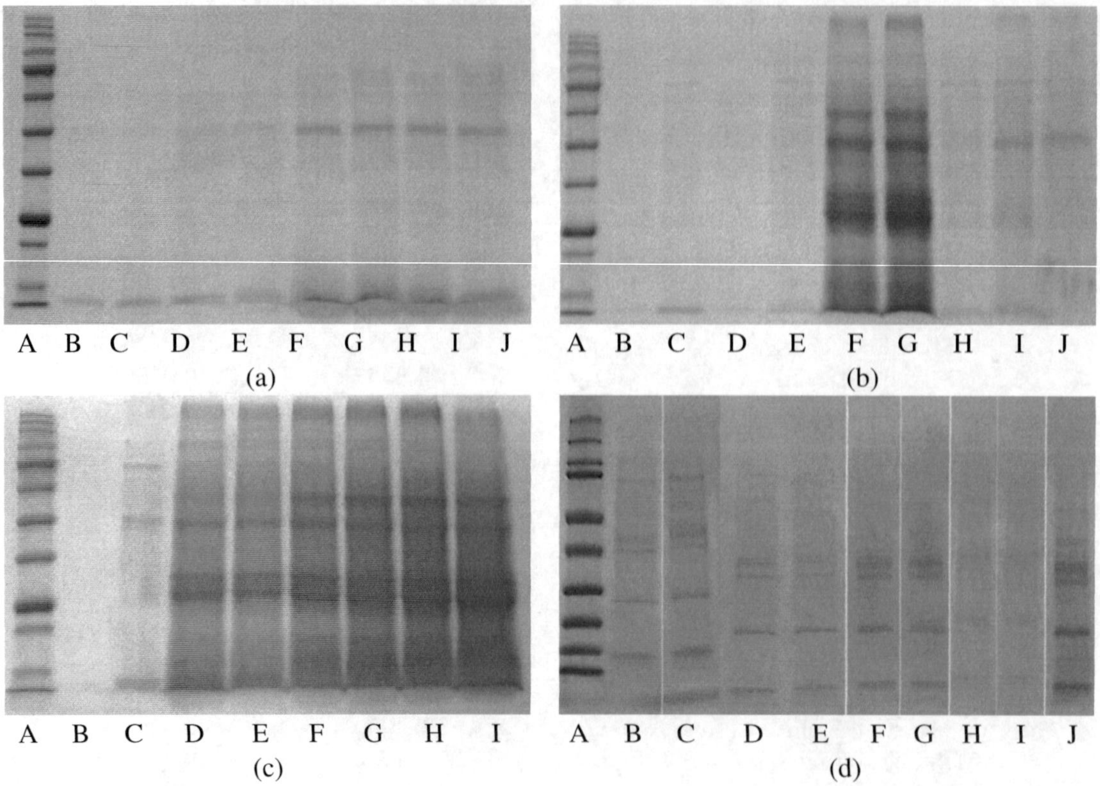

Figure 4. SDS-PAGE gel stained by Coomassie blue (a): A(Maker) B(Water at 0 min) C(Water at 3 min) D(Mixed buffer I at 0 min) E(Mixed buffer I at 3 min) F(Mixed buffer II at 0 min) G(Mixed buffer II at 3 min) H(Mixed buffer III at 0 min) I(Mixed buffer III at 3 min); (b): A(Maker) B(Water) C(Freeze microalgae ,water) D(Mixed buffer I) E(SDS lysis buffer 0.1 %) F(SDS lysis buffer 1 %) G(SDS lysis buffer 5 %) H(Triton x lysis buffer 0.1 %) I(Triton x lysis buffer 1 %) J(Triton x lysis buffer 5 %) at 3 min; (c): A(Maker) B(Water at 0 min) C(Water at 40 min), SDS lysis buffer 1 % at D(0 min) E(1 min) F(5 min) G(10 min) H(20 min) I(40 min); (d): A(Maker), continuous sonication at B(20 min) C(25 min) D(30 min) E(35 min) F(40 min) G(45 min) H(50 min) I(55 min) J(60 min).

Figure 4 shows the quality of total protein concentration which determined by SDS PAGE. A in all gels is protein marker. In Figure 4a as B, C, D and E which had total protein concentration was 0.6821 mg/ml (B and C) and 2.3068 mg/ml (D and E) , in Figure 4b as B, C, D and E which had total protein concentration was 0.9402, 2.5624, 2.2714, and 3.6133 mg/ml, respectively. Figure 4c as B and C, which had total protein concentration was 0.5455 and 2.9795 mg/ml, respectively, show unintent band because the total were low. On the other hand, Figure 4a presents F, G, H and I which had total protein concentration was 24.0516, 24.6383, 33.6934 and 30.2792 mg/ml, respectively. Figure 4b as H, I and J which had total protein concentration was 3.5403, 13.8961, and 82.2149 mg/ml had high concentration of protein. They were not intent band and no observed band. Figure 5 shows a comparison of three processes, water sonication, SDS lysis 1% sonication, and water continuous sonication. The SDS lysis buffer gave the concentration of protein more than water sonication as well as better quality of protein. The highest concentration of protein from this comparison was the continuous sonication. From all curves, it can be summarized that duration of sonication contributed to cells disruption.

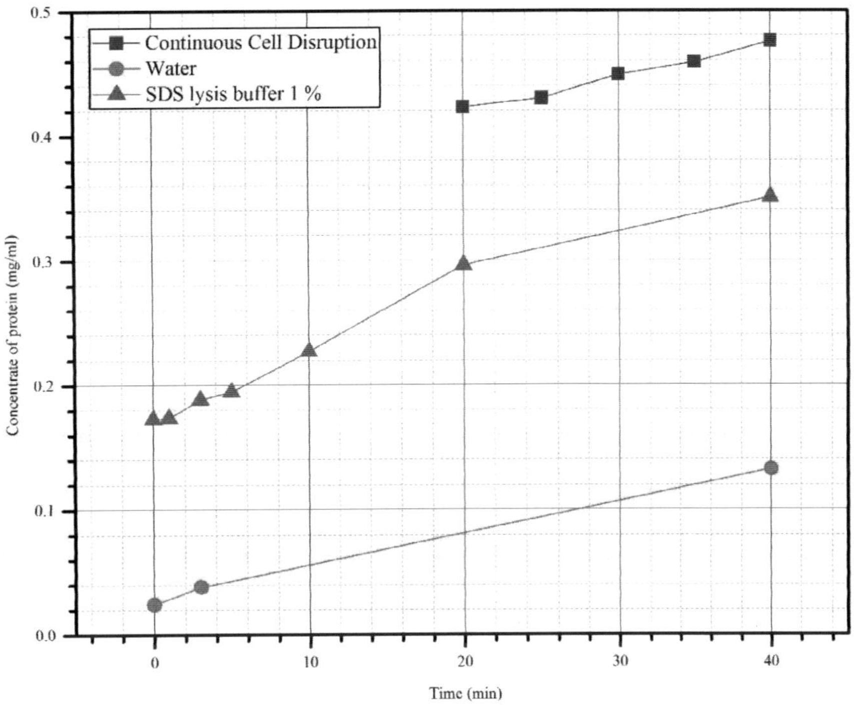

Figure 5. Soluble protein concentration released in each solvents after cell disruption

4. Conclusion

This paper examines the condition on microalgae cell disruption method by the combination of ultrasonication and buffer solution. It can be summarized as follows: (1) Mixed buffer III and Triton x lysis buffer showed the high concentration of protein; however, the quality of protein was not satisfied; (2) SDS lysis buffer showed high quality protein product but low the concentration of protein; (3) When using SDS lysis buffer, the concentration of protein significantly increased when longer duration of sonication was applied and this result was observed during continuous sonication test as well; thus, the duration of sonication affected the yield of the protein; (4) The amount of obtained protein was not only the importance, but also the quality of the protein product need to be consider for increasing the value of the final product; therefore, the optimum condition of protein extraction from microalgae needs the combination of the use of buffer solution to ensure high quality product and an appropriate duration of ultrasonication to maximize the quantity of protein.

6. References

[1]. Alba G L, Torri C, Samorì C, Spek v d J, Fabbri D, Kersten R. A. S and Brilman W. F. (Wim) D
2011 Hydrothermal treatment (HTT) of microalgae: evaluation of the process as conversion

method in an algae biorefinery concept. *Energy & fuels.* **26**(1) 642-57p
[2]. Amin S 2009 Review on biofuel oil and gas production processes from microalgae **50** 1834-40 p
[3]. Lee K A, Lewis M D and Ashman J. P 2012 Disruption of microalgal cells for the extraction of lipids for biofuels: Processes and specific energy requirements *Biomass and Bioenergy.* **46**(Supplement C) 89-101
[4]. Barbarino E and Lourenço O. S 2005 An evaluation of methods for extraction and quantification of protein from marine macro- and microalgae 447-60 p
[5]. Gerde A. J, Lomboy M M, Yao L, Grewell D and Wang T 2012 Evaluation of microalgae cell disruption by ultrasonic treatment. *Bioresource Technology* **125**(Supplement C) 175-81
[6]. Greenly M J and Tester W. J 2014 Ultrasonic cavitation for disruption of microalgae **184**
[7]. Pragya N, Pandey K. K and Sahoo P. K. 2013 A review on harvesting, oil extraction and biofuels production technologies from microalgae. *Renewable and Sustainable Energy Reviews.* **24** (Supplement C) 159-71
[8]. Bradford M. M 1976 A rapid and sensitive method for the quantitation of microgram quantities of protein utilizing the principle of protein-dye binding *Analytical Biochemistry.* **72**(1) 248-54.

Acknowledgement
A Kuhavichanan has received funding from Thailand Institute of Scientific and Technological Research (TISTR) and graduate school of Kasetsart University. This research was under the collaboration between TISTR and Department of Mechanical Engineering, Faculty of Engineering, Kasetsart University.

IOP Conf. Series: Earth and Environmental Science **159** (2018) 012010 doi:10.1088/1755-1315/159/1/012010

The Promotion by Steam on CaO Adsorbing SeO_2 at Medium Temperature

Yu Lou[1], Yaming Fan[1], Chengkai Pang[1], and Yuqun Zhuo[1]

1 Department of Energy and Power Engineering, Tsinghua University, Beijing 100084, China.
E-mail: louy15@mails.tsinghua.edu.cn

Abstract. This study investigated the promotion of steam for CaO adsorbing SeO_2 gas at the medium temperature interval (240℃-800℃), based on the previous studies on CaO adsorbing SeO_2 gas. The experiments revealed both the influences of temperature and steam concentration. The mechanism of the promotion is similar to those of steam in carbonation and is confirmed by a series of designed experiments and characterization of the products of gaseous SeO_2 adsorption.

1. Introduction

Selenium is an important industrial material used in many fields, such as electronics, metallurgy and petrochemical [1]. Although selenium is a necessary trace mineral elements for human beings, excessive intake damages health [2]. The toxicity of selenium compounds varies with the different kinds, both selenite and seleninic acid are highly toxic, which usually come from selenium dioxide [3]. Selenium was generally considered as a dispersive element, whose concentration is only 0.05-0.09 μg/g in lithosphere [4]. However, there is one kind of stone coal in Enshi region, Hubei Province, P. R. China, which contains 80000 μg/g selenium [5]. According to previous researches, coal was adjudged as one important selenium enrichment and the main emission source [6], [7]. In 1995, the selenium emission caused by fossil fuel (mostly coal) combustion reached 4101 tones, more than 89 percent of total anthropogenic emissions [8]. Tian et al. counted 787 tons of the selenium emission resulted from coal-fired power plants in 2007 in China [7].

With the increasing environmental awareness, controlling selenium emission has become more and more important. Since 1998, US EPA has set selenium as one of which needs to be monitored among coal-fired power plant emissions [9], [10]. On Dec. 2011, US EPA announced “Mercury and Air Toxics Standards” (MATS), which contained the emission limits of various power plants’ pollutants, including selenium, and updated in March 2016.

Many researches have focused on adsorption reactions to reduce selenium emission [11]-[15]. A. Ghosh-Dastidar et al. have researched several minerals as adsorbent to gaseous SeO_2 at 500℃ and 900℃, including limestone, kaolin, alumina and calcium carbonate. The results of adsorption reactions proved that CaO was the best adsorbent to deal with SeO_2, and the process was chemical adsorption leading to the product of $CaSeO_3$ [14]. Other studies used the mixture of CaO and selenium-containing substances, which also suggested that CaO was very reactive with selenium [13], [16].

Meanwhile, the reactivity of CaO affected by steam, whether physically or chemically, has been investigated for a long time. As early as 1977, Dobner et al. found that steam could “catalyze” CaO adsorbing CO_2 (carbonation) on 550℃ [17], suggesting that if there was steam at certain flow-rate at 550℃, it would significantly increase for the adsorption rate and extent of carbonation when

IOP Conf. Series: Earth and Environmental Science **159** (2018) 012010 doi:10.1088/1755-1315/159/1/012010

compared with the results of no steam. Some researchers have proved that steam can improve the effect of calcium looping and adsorption efficiency of carbonation [18]-[20]. V. Manovic investigated 8 kinds of limestone at different temperature (350℃-800℃) and found the promotion from steam on carbonation was more obvious under relatively low temperature (below 600℃) [19], similar to other research [21].

The mechanism of promotion was proposed by previous researches that the key step was the dissociative of H_2O molecule on the surface of CaO solid, depending on the oxygen vacancies of the surface of CaO [22], [23], described by (1):

$$H_2O(g) + Vo + Oo = 2OHo \tag{1}$$

The products of this process, i.e. hydroxyl groups on oxygen site have been regarded as the crux of carbonation. The total energy of dissociative adsorption can be expressed by (2). J. Carrasco et al., H. Manzano et al., and D. Ochs et al. have pointed out that the $E_{H2O,\,diss,\,ad}$ is smaller than $E_{H2O,\,ad}$ [24]-[26], which is regarded as the most important part of the proposed mechanism. Although there are some theoretical and experimental implications support this mechanism [22], [23], the direct evidence, especially the evidence from experiments, are generally absent.

$$E_{H_2O,diss,ad} = E_{OH,ad} + E_{H,ad} - E_{H_2O,diss} \tag{2}$$

Due to similar molecule structure with CO_2, it is also very interesting to see if the adsorption of SeO_2 on CaO could be promoted by steam in similar temperature interval (300℃-800℃). CaO, which possesses massive advantages of low cost, wide availability and non-toxic, has been widely used in desulphurization processes. It would be beneficial for power plants to remove multiple contaminants in a single process. Therefore, to understand the influence and mechanism of steam on CaO adsorbing gaseous SeO_2 is of great interests. In this study, both experimental and products characterization analysis have adopted to reveal the interactions among CaO, SeO_2 and H_2O.

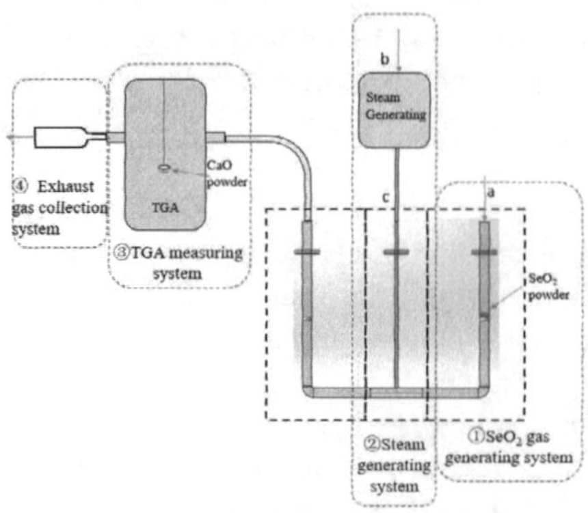

Figure 1. Schematic of SeO_2 adsorption.

2. SeO_2 Adsorption Experiments

2.1. Assemblies

The experimental assembly of SeO_2 adsorption on CaO solid consists of four sub-systems as shown in Figure 1, which are SeO_2 gas generating system, steam generating, TGA(Thermal Gravimetric Analyzer) measuring, and exhaust gas collection. The main reactor includes a three section resistance-heating furnace and TGA (Thermal Gravimetric Analyzer). SeO_2 gas generating system is an improved mimic of the device established by Li et al [11], and Sterling et al [27]. Gaseous SeO_2 is generated in the first section of furnace, in which analytic pure SeO_2 powder is placed on an internal sieve plate. Gaseous SeO_2 is sublimated by controlling the heating temperature, and is carried out by pure nitrogen. Steam generating system located in the second section of furnace consists of an injection pump, a metal furnace. The generated steam is mixed with pure nitrogen and SeO_2 gas at above 250℃ in the second and third sections. Analytic pure CaO powder is placed on a platinum plate, which hangs in a 22mm (inner diameter) quartz tube in TGA. Exhaust gas collection system is a specially designed quartz tube with the purpose to collect all escaped SeO_2 by condensation.

2.2. Material

Analytic pure SeO_2 and CaO solid are used as vapor source and adsorbent respectively in these experiments. The desired SeO_2 concentration in gas can be attained by adjusting the temperature of vapor generator. The balance gas is pure N_2. As calibrated previously, the temperature and N_2 carrier flow rate have been set at 190℃ and 100ml/min, which maintains the SeO_2 concentration at 40ppm in mixed gas [14], [15].

2.3. Steps for without-steam experiments

N_2 purging was used as the step of Pretreatment Ⅰ for all experiments. CaO sample was preheated to 800℃ and held for 5 min. in pure N_2 to remove water and other volatile impurities. Then, the CaO sample was cooled to the set experimental temperature. The analytic pure SeO_2 sample was placed on a sieve plate of the first section in furnace, where the temperature was maintained at 190℃. For without-steam experiments, continuous measurement method was adopted under various temperatures, i.e. 7 temperature stages within 240℃-800℃ range, as shown in Figure 2. The CaO temperature was first set to 800℃ for SeO_2 gas adsorbing as the first stage, and then gradually cooled to 240℃ as the 7th stage. After the last stage (240℃), CaO sample was heated up to 400℃ and 800℃ to see if it maintained the same adsorption activity as the former stages throughout the whole process.

2.4. Steps for with-steam experiments

Pretreatment Ⅰ and similar processes after that were also used. CaO sample was cooled to and maintained at desired temperature stage. Then the steam generating system was turned on and SeO_2 powder was simultaneously placed in. Each with-steam adsorption test was held for 20 min. Due to the simultaneous adsorption of steam and SeO_2 on CaO sample, the adsorbed quantity of selenium has to be determined by ICP-AES. According to previous researches [11], [12], CaO solid could stably adsorb gaseous SeO_2 under experimental conditions for hours. Therefore, the adsorption rate could been considered as constant during the whole adsorption process.

2.5. Reliability

For without-steam experiments, as continuous measurement had been conducted at various temperature, the reliability of experimental data was verified by the deviation of multiple repeated runs. The relative experimental error range has been determined as ±5%. For with-steam experiments, the reliability of experimental data was affirmed by selenium balance across the whole system. The selenium balance was established by: 1) After stopping the SeO_2 generating furnace and taking out the adsorption product from TGA, place fresh CaO powder into reactor; 2) Restart the SeO_2 generating furnace to evaporate the rest SeO_2; 3) Collect the later placed CaO solid and rinse the exhaust gas collection system, by which means all remaining SeO_2 could be collected and later determined by ICP-AES. Some of the selenium balance are shown in Table 1. The relative experimental error range for experimental data from with-steam tests was determined as ±3%.

Table 1 Selenium Balance Measuring (part experiments)

Experiments	Total SeO_2(μg)	SeO_2□ (μg)	SeO_2□ (μg)	Balance
400□-6%H_2O Adsorption	1.23×10^4	3.42×10^2	1.16×10^4	0.983
320□-20%H_2O Adsorption	1.05×10^4	2.55×10^2	9.99×10^3	0.972

2.6. Results and discussion

For without-steam experiments, the SeO_2 adsorption rates are shown in Figure 2. The data from unstable periods due to temperature adjusting had been removed. Only the CaO weight increase within the 5-10 min of temperature-stable segment were presented. The adsorption rates of the last two segments (400℃ and 800℃) of the series agreed well with the previous rates under corresponding

temperature, suggesting the CaO adsorption activity was not affected by either temperature or adsorption itself.
The results of without-steam experiments shows that from 240℃ to 800℃, the SeO_2 adsorption rate rises with the increment of adsorption temperature, and the maximum reaches about 21μg/min at 800℃. It is necessary to point out that both SeO_2 desorption and the activity of SeO_2 adsorption sites on CaO surface increases with increasing temperature. The results clearly indicates that the influence of the latter is more predominant in this atmosphere. What the experiments shows complies with previous research to a certain degree. Ghosh-Dastida et al. found SeO_2 adsorption rate on CaO increased at 400℃—600℃, though in the atmosphere of 5% O_2 + 95% N_2 [14]. Li also found similar SeO_2 adsorption results at 300℃—400℃, though in another atmosphere – air [11].

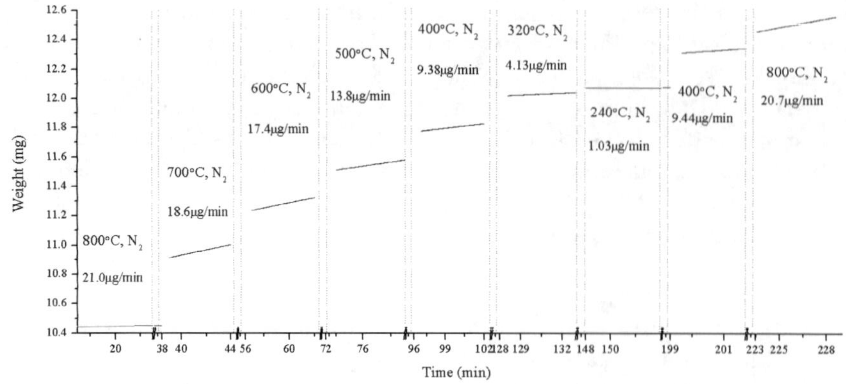

Figure 2. 240℃-800℃ without-steam SeO_2 adsorption rate.

For with-steam experiments, the adsorption rate for SeO_2 is influenced by both the steam concentration and adsorption temperature, as shown in Table 2 and Figure 3 a). In Table 2, the adsorption temperature was set to 400℃, and the actual quantity of SeO_2 adsorbed on CaO was determined by ICP-AES. As mentioned above, the SeO_2 adsorption rate on CaO would be stable for a relative long period, thus it is safe to assume that the adsorption rate of the same condition would remain stable in experiment. The SeO_2 adsorption rate could be calculated from the ratio of adsorption quantity and adsorption time as shown in (3).

$$R = W_{SeO_2} / \tau \tag{3}$$

Table 2. SeO_2 Adsorption rate influenced by steam concentration

C_{H_2O}(%)	W_0(μg)	W'(μg)	W_i(μg)	W_{SeO_2} (μg)	R (μg/min)
0	9.77×10^3	9.97×10^3	2.01×10^2	2.01×10^2	9.82
6	9.79×10^3	1.01×10^4	4.11×10^2	3.40×10^2	17.1
10	9.39×10^3	9.90×10^3	5.10×10^2	3.71×10^2	18.7
16	9.68×10^3	1.03×10^4	5.82×10^2	4.02×10^2	19.8
20	9.89×10^3	1.06×10^4	7.20×10^2	4.11×10^2	20.4

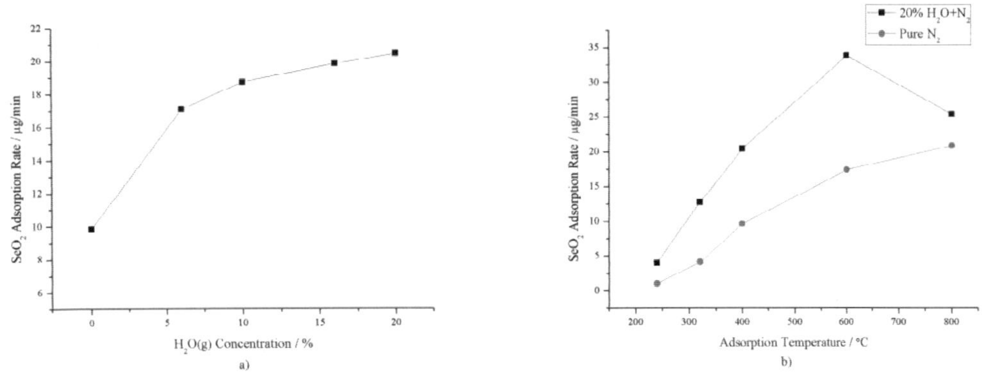

Figure 3. a) SeO_2 adsorption rate influenced by steam concentration; b) SeO_2 adsorption rate influenced by adsorption temperature.

As shown in Figure 3 a) and Table 2, steam (concentration up to 20%) can promote CaO adsorbing SeO_2 at 400℃, which is positively related with the steam concentration. The results also indicates that the promotion of the steam slows down along with the increment of steam concentration. When the steam concentration is above 10%, the adsorption rate is approaching its maximum, slightly above 20μg/min. Further increasing the steam concentration may not increase the adsorption notably.

Table 3. SeO_2 Adsorption rate influenced by adsorption temperature (with-steam experiments)

T(℃)	W_0(μg)	W'(μg)	W_i(μg)	W_{SeO2} (μg)	R (μg/min)	R_0(μg/min)
240	1.01×10^4	1.07×10^4	5.61×10^2	0.081	3.97	1.03
320	9.05×10^3	9.54×10^3	4.90×10^2	0.262	12.8	4.13
400	9.89×10^3	1.06×10^4	7.22×10^2	0.411	20.4	9.59
600	8.81×10^3	9.51×10^3	7.13×10^2	0.680	33.9	17.4
800	8.83×10^3	9.33×10^3	5.10×10^2	0.510	25.4	20.9

The influence of the temperature on SeO_2 adsorption is shown in Table 3 and Figure 3 b). The experimental conditions were similar as before. The steam concentration was set to 20% for better demonstration. The SeO_2 adsorption rates without steam are also shown in the table and figure for comparison purposes.

As shown in Figure 3 b), there is an obvious promotion for CaO adsorbing SeO_2 gas in 240℃-800℃ with 20% steam, comparing with no steam at same temperature. There is a peak at 600℃ for the fastest SeO_2 adsorption rate, about 33.85 μg/min. Before this peak in Figure 3 b), the rate rises along temperature, similar to those experiments without steam. After the peak (about 600℃), however, an inverse trend of SeO_2 adsorption rate appears (without steam). This could be explained by the diminishing of the steam adsorption ability on CaO surface along with temperature, which has been discussed by other researchers in carbonation study [19], [21].

As mentioned before, the dissociation of chemisorbed H_2O molecule on CaO solid surface could be the key to understand how steam promote CaO adsorbing CO_2. Due to similar molecular structure and reaction phenomenon, this reaction path could be also applied to explain the steam promotion of SeO_2 adsorption on CaO. In this mechanism, the oxygen vacancies on CaO solid surface are regarded as the crux of the dissociation of chemisorbed H_2O molecule. The product of H_2O molecular reaction on CaO surface is a significant factor affecting SeO_2 adsorption. Combined with the findings in previous [22], [23] and the experimental results from this study, the mechanism of steam promoting CaO adsorbing SeO_2 is illustrated in Figure 4.

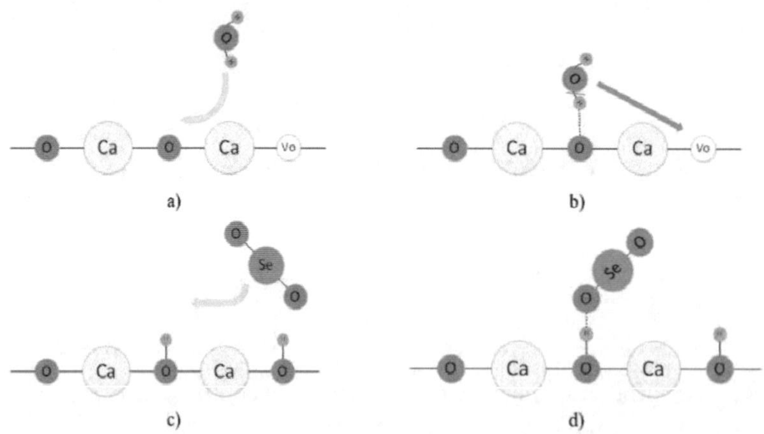

Figure 4. Schematic diagram for steam promoting CaO solid adsorbing SeO_2 gas. a) H_2O molecule adsorbed on an oxygen site on CaO surface; b) adsorbed H_2O molecule dissociated to adjacent oxygen vacancies on CaO surface, forming two hydroxyl groups; c) CaO surface with hydroxyl groups formed from H_2O molecule attracts SeO_2 molecule; d) the possible SeO_2 adsorption state on CaO surface with hydroxyl group.

As shown in Figure 4, the hydroxyl groups formed from H_2O molecule is essential for SeO_2 adsorption. However, what presented in Figure 4 was still an assumption at this stage. Due to the difficulties in direct observation of CaO surface, experiments were designed to verify step a) and b), and some characterization analysis focusing on the reaction products was employed to support step c) and d).

3. Experiments: H_2O Dissociation on Oxygen Sites/Vacancies

To confirm the role of oxygen sites/vacancies on CaO surface in H_2O dissociation [22], [23], an experimental rig was constructed as shown in Figure 5. By purging the CaO sample by a certain concentration of CO in advance, the amount of oxygen sites/vacancies on CaO surface could be qualitatively changed and the later reaction rate of ($CaO(s) + H_2O(g)$) could be adjusted accordingly. The reaction rate of kinetic-controlled stage represents the chemical change of CaO surface directly, and could be used to reflect the change of oxygen sites/vacancies. Although there is hardly any research focusing on how CO could change the CaO surface, some studies concerning the change of surface oxygen sites of metallic oxide by CO could be helpful [28]-[30].

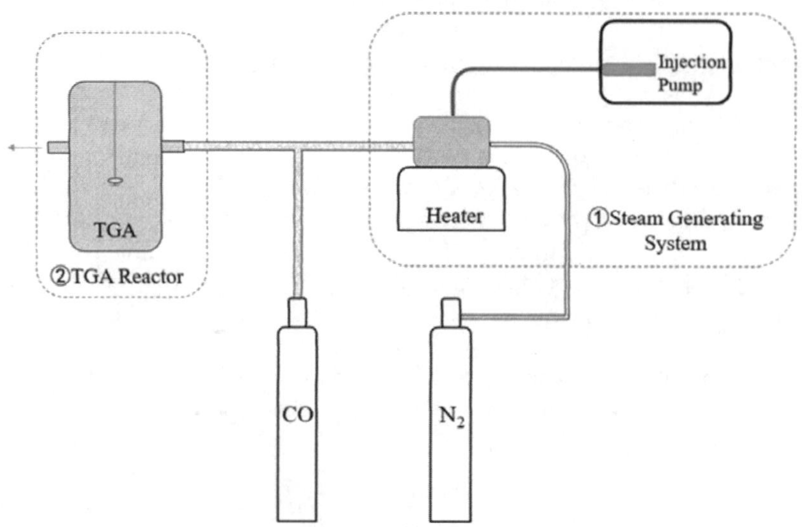

Figure 5. Schematic for CaO (s) + H_2O (g) after CO Modification.

3.1. Assemblies:

The experiment bench includes steam generating system and TGA reactor, similar as before.

3.2. Material

The same analytic pure CaO was used. CO concentration (2000ppm and 10%) was maintained by mass flow controllers. The gas mixture was balanced by N_2 and the total gas flow is 100ml/min.

3.3. Steps

For the same reasons as before, 800℃ N_2 purging as Pretreatment Ⅰ was adopted firstly. Then the CaO sample was cooled to the CO-pretreatment temperature, during which CaO surface was purged by CO for 20 min. This step was Pretreatment Ⅱ. Various temperatures and CO concentrations were chosen in Pretreatment Ⅱ. After this step, another pretreatment was used to ensure that no CO adhered on CaO surface, i.e. Pretreatment Ⅲ, in which CaO sample was heated to 800℃ again and held for 5 min. in pure N_2. After all pretreatments, CaO sample was cooled to the reaction temperature of 320℃. The different reaction rates were measured by TGA. The pressure in the reactor during the whole series of experiments was kept atmospheric.

3.4. Results and discussion

After Pretreatment Ⅱ, the reaction rates of (CaO(s) + H_2O(g)) varied obviously as shown in Figure 6 and Figure 7. In each a) part of the two figures, experimental data and both horizontal and vertical coordinates have been normalized, which means the percentage data was shown in the figure instead of actual weight variation to eliminating small weight differences between different sample. The zero of horizontal coordinates represents 5 min before the reaction beginning. For the vertical coordinates, the original weight means the sample weight after 800℃ N_2 purging (Pretreatment Ⅱ). In each b) part of the two figures, the reaction rate means the reaction rate of kinetic controlled stage in a) part for different Pretreatment Ⅱ conditions.

Figure 6 represents the influence of different CO concentrations in pretreatment Ⅱ on reaction rate. The results demonstrated that the (CaO(s) + H_2O(g)) reaction rate increased notably after low CO concentration pretreatment, but decreased after high concentration pretreatment. If put into logarithmic horizontal coordinate as shown in Figure 6 b), it is obvious that the effect of CO concentration on reaction rate has a limit.

As shown in Figure 7, comparing with the data without CO, low temperature CO pretreatment enhanced the reaction rate of (CaO(s)+H_2O(g)). However, an inhibition in high temperature was also observed. The concentration of CO in Pretreatment Ⅱ was set to 250 ppm. In general, under these experimental conditions, the reaction rates decreased along with increasing temperature.

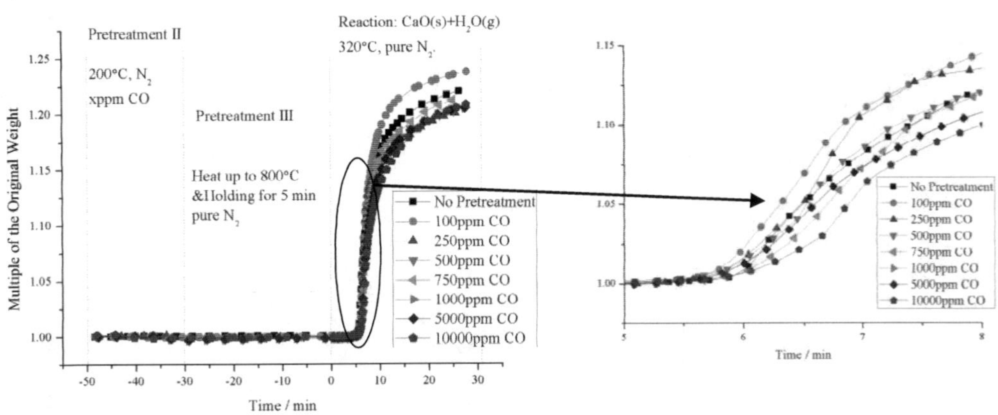

a)

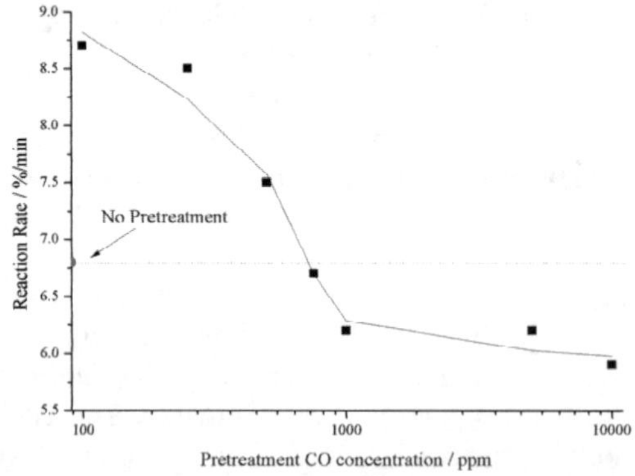

b)

Figure 6. The influence of CO concentration in Pretreatment Ⅱ on $CaO(s) + H_2O(g)$ reaction.

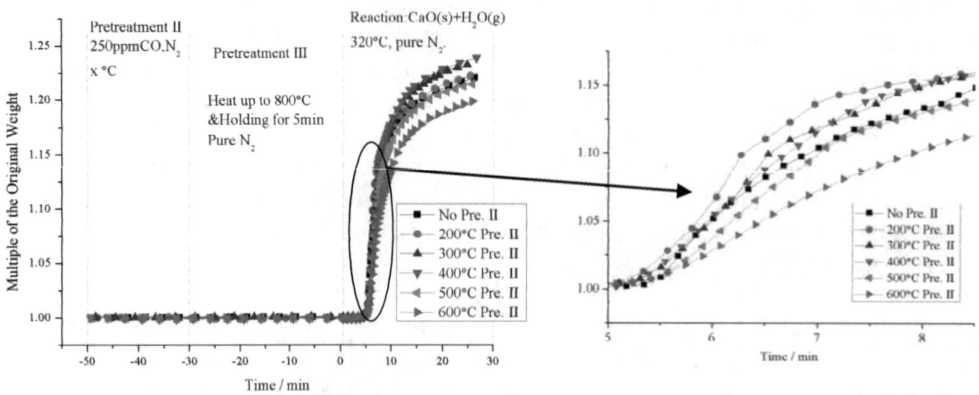

a)

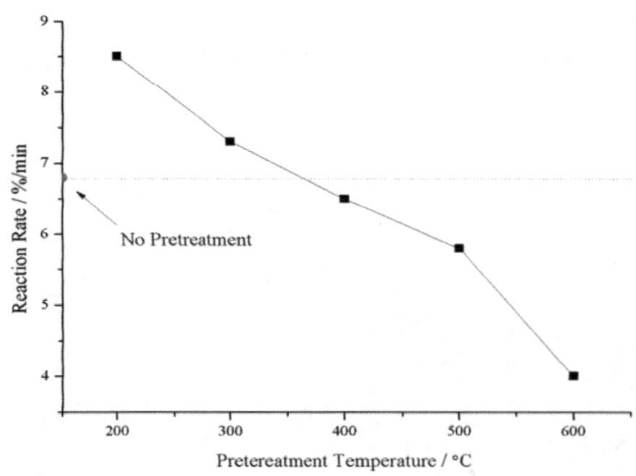

b)

Figure 7. The influence of CO concentration in Pretreatment Ⅱ on $CaO(s) + H_2O(g)$ reaction.

It seems that there is coherence in the appearance of Figure 6 and Figure 7. When temperature or CO concentration in pretreatment rises, CO should have stronger effect on CaO surface. Comparing with no Pretreatment Ⅱ, both the two figures indicates it leads an inhibition on ($CaO(s)+H_2O(g)$) reaction rates if the temperature or CO concentration in Pretreatment Ⅱ exceeds a certain limits. However, an obvious promotion on the rates could happen if the conditions in Pretreatment II are relatively moderate, which means whether the pretreatment temperature or CO concentration is within the limits. This phenomenon would hardly been explained by CO adsorption in Pretreatment II, due to no obvious weight change on TGA during in this process. Furthermore, Pretreatment III was designed to ensure that there wasn't any CO adsorption on CaO surface after Pretreatment II. As in the previous researches on steam promotion on CaO adsorbing CO_2, the most possible explanation for these results is that CO molecule changes the oxygen sites/vacancies during Pretreatment II, which could be also regarded as the key of H_2O dissociation in Figure 4. H_2O molecule would first chemisorbs on oxygen site and then decomposes by the influence of adjacent oxygen vacancies. Due to the fact that $E_{H2O, diss, ad}$ is smaller than $E_{H2O, ad}$ as mentioned before, the CaO surface with relative more oxygen vacancies reduces the energy required for H_2O molecule adsorption. Fan et al. have investigated this situation through DFT method [31] and acquired the same conclusion. Comparing with oxygen sites on CaO surface, oxygen vacancies should be in minority. In Pretreatment II, CO purging increases the amount of the oxygen vacancies and decrease the oxygen sites consequently. Whether low temperature or low CO concentration can only increase the oxygen vacancies to a finite extent, thus it makes the whole adsorption energy of H_2O molecule lower on CaO surface and promotes the reaction rate of $CaO(s)+H_2O(g)$. When temperature or CO concentration rises furthermore, oxygen vacancies increase and eventually becomes dominant on CaO surface. The amount of surface active oxygen site has been significantly reduced, which could hardly enhances H_2O adsorption and might lead to the inhibition on reaction rate. In general, these experimental results somehow illustrate the roles of the oxygen sites/vacancies played in the dissociation of chemisorbed H_2O molecule on CaO surface, and prove the reaction path and mechanism shown in Figure 4 a) and b) to certain extent.

4. Binding Energy of the Reactions Products

Figure 8 shows a distinct difference (about 0.25eV) of the Se3d binding energies from the reaction products with and without steam, which indicates that there exists different SeO_2 adsorptions on CaO surface under different reaction conditions. The peak position of binding energy without steam is consistent with previous researches [31].

The decrease of SeO_2 binding energy represents the decreases of the binding capacity of adsorption products, suggesting that SeO_2 molecule could be more easily chemisorbed in reactions. As shown in Figure 4 c), the mechanism proposed before also suggests that the adsorption capability for SeO_2 molecule on CaO surface might be promoted by CaO-Steam reaction products under appropriate temperatures. This assumption can be somewhat approved by Figure 8.

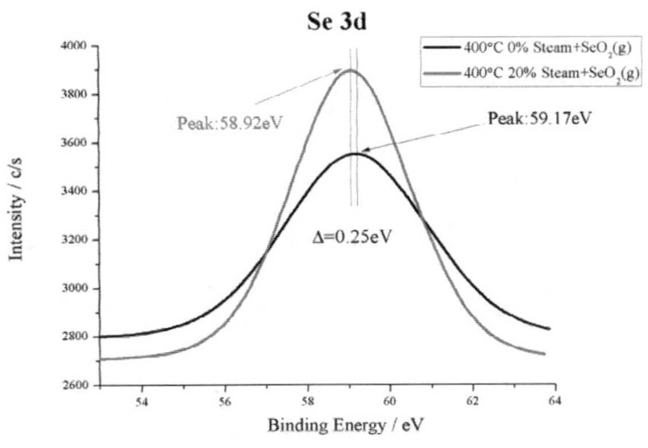

Figure 8. The Se3d binding energy of products.

The binding energies of Ca2p of three products obtained from different reaction conditions are shown in Figure 9. The results are listed in the descending order of the peak positions of Ca2p binding energy, i.e. CaO + 20% steam, CaO + 20% steam + gaseous SeO_2, and CaO + gaseous SeO_2 alone. The difference of peak position also indicates the different adsorption mechanisms, as suggested by Figure 8.

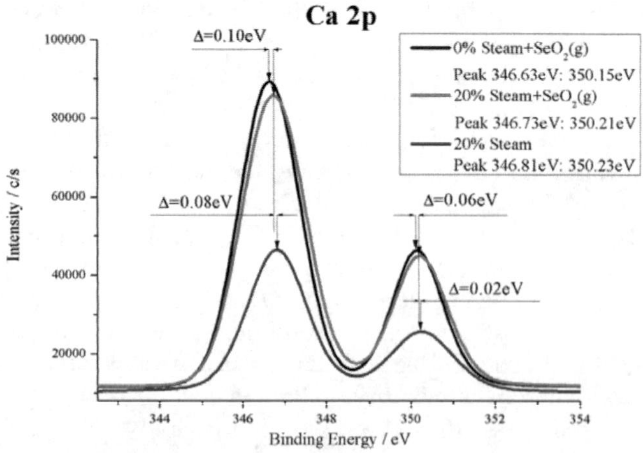

Figure 9. The Ca2p binding energy of products.

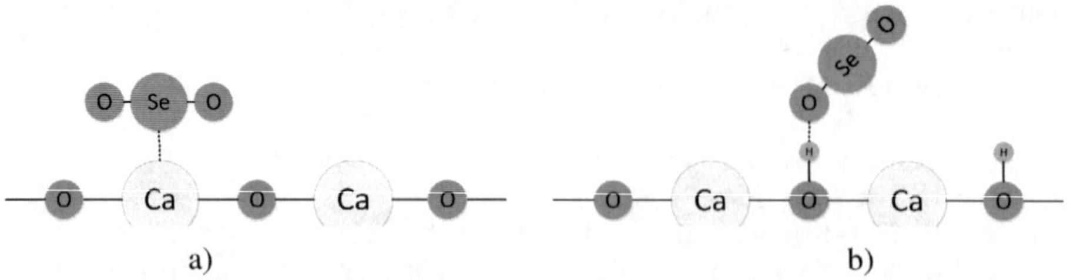

Figure 10. The structure for CaO adsorbing SeO_2 molecule: a) without steam, b) with steam.

The surface structure of CaO + 20% steam + gaseous SeO_2 may be described as Figure 10 b), which corresponds to the assumption in Figure 4 d), as indicated from the analysis of peak positions of Ca2p binding energies in Figure 9. The surface structure of CaO + 20% steam could be described as Figure 4 b), according to previous researches [22], [23]. Comparing with CaO + 20% steam, the peak positions of Ca2p binding energies of CaO + 20% steam + gaseous SeO_2 are very close to each other, indicating that Ca atom status are similar in these two cases. One reasonable explanation is demonstrated as Figure 10 b) for the condition of CaO + 20% steam + gaseous SeO_2.

The differences of the peak position of Ca2p binding energy shown in Figure 9 also illustrate that there are two totally different mechanism in CaO + 20% steam + gaseous SeO_2 and CaO + gaseous SeO_2. Due to the differences shown in Figure 9, the status of Ca atom is different, suggesting the different surface structure and different reaction mechanism. Fan et al. analysed the SeO_2 molecular adsorption processes on CaO surface without steam [32], [33]. From his work, the adsorption structure of CaO + gaseous SeO_2 could be presented as Figure 10 a), which is obviously different from Figure 10 b), i.e. the condition of CaO + 20% steam + gaseous SeO_2, and explains the difference of the peak positions of Ca2p binding energies in these two conditions.

5. Conclusion

The experimental study of this work illustrates that steam can promote CaO adsorbing SeO_2. Based on previous researches, the effect of oxygen vacancies on CaO surface is proposed as the main reason of the promotion as the dissociation of chemisorbed H_2O benefits SeO_2 adsorption. The mechanism is then verified by a set of specifically designed experiments and characterization of reaction products.

Main conclusions include: 1) For SeO_2 adsorption without steam, the adsorption ability of CaO surface positively related with temperature from 240℃ to 800℃; 2) In the experimental temperature and steam concentration range, steam could apparently promote the SeO_2 adsorption rate on CaO surface. Under relative low concentration, the promotion effect of steam is more recognizable; 3) From 240℃

to 800℃, steam could help CaO adsorb more SeO_2, the reaction rates curve has a peak at around 600℃; 4) The mechanism of the steam enhancing SeO_2 adsorption on CaO is similar to that of enhancing H_2O sorption as previous researchers indicated, similar as the role in carbonation of Calcium. Vacancy is dominant in the process of the dissociation of chemisorbing H_2O molecule on CaO solid surface and the hydroxyl groups formed from H_2O molecule is essential for SeO_2 adsorption. A series of designed experiments and reaction products characterization have confirmed the proposed mechanism.

6. Appendices

Nomenclature

Symbol	Description	Units
Vo	oxygen vacancy on CaO surface	-
Oo	oxygen atom on CaO surface	-
OHo	hydroxyl site on CaO surface which is exactly on oxygen site	-
$\mathbf{E_{H2O,\ diss,\ ad}}$	Activation energy of H_2O molecule dissolving and adsorption	kJ mol^{-1}
$\mathbf{E_{H2O,\ ad}}$	Activation energy of H_2O molecule adsorption	kJ mol^{-1}
$\mathbf{E_{OH,\ ad}}$	Activation energy of hydroxyl groups formed from H_2O molecular adsorption	kJ mol^{-1}
$\mathbf{E_{H,\ ad}}$	Activation energy of H adsorption	kJ mol^{-1}
$\mathbf{E_{H2O,\ diss}}$	Activation energy of H_2O molecule dissolving	kJ mol^{-1}
$\mathbf{SeO_2}$ Ⅰ	SeO_2 adsorption quantity during adsorption teat in with-steam experiments, determined by ICP-AES	μg
$\mathbf{SeO_2}$ Ⅱ	SeO_2 content in collected solutions, from dissolving the second-placed CaO solid and washing the exhaust gas collection system in with-steam experiments, determined by ICP-AES	μg
$\mathbf{C_{H2O}}$	steam concentration in with-steam experiments	%
$\mathbf{W_0}$	CaO sample weight after the Pretreatment Ⅰ	μg
W'	sample weight after 20min reaction in with-steam experiments	μg
$\mathbf{W_i}$	increment of sample weight in with-steam experiments	μg
$\mathbf{W_{SeO2}}$	SeO_2 weight in the sample determined by ICP-MS in with-steam experiments	μg
τ	Reaction duration in with-steam experiments	min
$\mathbf{R_0}$	The SeO_2 adsorption rate in without-steam experiments	μg/min

7. References

[1] Asokan P, Saxena M, and Asolekar S 2010 *Mater. Charact.* **61** 1342–1355.
[2] Wang L, Ju Y, Liu G, Chou C, Zheng L, and Qi C 2010 *Environ. Earth Sci.* **60** 1641-1651.
[3] Peng A, Wang Z, and Whanger P D 1995 *The environmental biology inorganic chemistry of selenium* (Beijing: Environmental Science Press of China) p156.
[4] Diehl S F, Goldhaber M B, Koenig A E, Lowers H A, and Ruppert L F 2012 *Int J. Coal Geol.* **94** 238.
[5] Zhu J, Johnson T M, Finkelman R B, Zheng B, Sykorova I, and Pesek J 2012 *Chem. Geol.* **330** 27–38.
[6] Pacyna E G, Pacyna J M, Sundseth K, Munthe J, Kindbom K, Wilson S, Steenhuisen F, and Maxson P 2010 *Atmos. Environ.* **44**(20) 2487-2499.
[7] Tian H, Wang Y, Xue Z, Qu Y, Chai F, and Hao J 2011 *Sci. Total Environ.* **409**(16) 3078-3081.
[8] Pacyna J M and Pacyna E G 2001 *Environ. Rev.* **9**(4) 269-298.
[9] Rosenberg W G 1991 *Sci. New Series* **251**(5001) 1546-1547.
[10] Rubin E S 1999 *Environ. Sci. Technol.* **33** 3062-3067.
[11] Li Y, Tong H, Zhuo Y, Chen C, and Xu X 2006 *Environ. Sci. Technol.* **40** 4306-4311.
[12] Li Y, Tong H, Zhuo Y, Wang S, and Xu X 2006 *Environ. Sci. Technol.* **40** 7919–7924.
[13] Xu S, Shuai Q, Huang Y, Bao Z, and Hu S 2013 *Energy Fuels* **27** 6880–6886.
[14] Ghosh-Dastidar A, Mahuli S, Agnihotri R, and Fan L S 1996 *Environ. Sci. Technol.* **30** 447-452.
[15] Agnihotri R, Chauk S, Mahuli S, and Fan L S 1998 *Environ. Sci. Technol.* **32** 1841-1846.

[16] Somoano D, and Martinez-Tarazona M R 2004 *Environ. Sci. Technol.* **38** 899–903.
[17] Dobner S, Sterns L, Graff R A, and Squires A M 1977 *Ind. Eng. Chem. Process Des.* **16** 479
[18] Donat F, Florin N H, Anthony E J and Fennell P S 2011 *Environ. Sci. Technol.* **46** 1262–1269.
[19] Manovic V and Anthony E J 2010 *Eng. Chem. Res.* **49** 9105–9110.
[20] Yang S J and Xiao Y H 2008 *Ind. Eng. Chem. Res.* **47** 4043–4048.
[21] Fennell P S, Pacciani R, Dennis J S, Davidson J F, and Hayhurst A N 2007 *Energy Fuels* **21**(4) 2072–2081.
[22] Li Z, Liu Y, and Cai N 2014 *Fuel* **127** 88-93
[23] R. Schaub, P. Thostrup, N. Lopez, E. Lægsgaard, I. Stensgaard, Noskov J K, and Besenbacher F 2001 *Phys. Rev. Lett.* **87** 266104.
[24] Carrasco J, Illas F, and Lopez N 2008 *Phys. Rev. Lett.* **100** 016101.
[25] Manzano H, Pellenq R J M, Ulm F J, Buehler M J, and Duin A C V 2012 *Langmuir.* **28** 4187-4197.
[26] Ochs D, Braun B, Maus-Friedrichs W, and Kempter V 1998 *Surf. Sci.* **417** 406-414
[27] Sterling R O and Helble J J 2003 *Chemosphere* **51** 1111-1119.
[28] Bhaskar D M, Gargi D, Umesh V W, and Hegde M S 2012 *Chem. Mater.* **24** 4491–4502.
[29] Wang Y, Yang X, and Li J 2016 *Chin. J. Catal.* **37** 193–198.
[30] Yang Z, Fu Z, Zhang Y, and Wu R 2011 *Ind. Eng. Catal. Lett.* **141** 7-82.
[31] Fan Y, Yao J G, Zhang Z, Sceats M, Zhuo Y, Li L, Maitland G C and Fennell P S 2018 *Fuel Process. Technol.* **169** 24-41.
[32] Fan Y, Zhuo Y, and Li L 2017 *Appl. Surf. Sci.* **420** 465-471.
[33] Fan Y, Zhuo Y, Lou Y, Zhu Z, and Li L 2017 *Appl. Surf. Sci.* **413** 366-371.

8. Acknowledgment

The authors gratefully acknowledge financial support from the National Natural Science Foundation of China (51776107).

IOP Conf. Series: Earth and Environmental Science **159** (2018) 012011 doi:10.1088/1755-1315/159/1/012011

Mapping Groundwater Vulnerability Zones in Eogenetic Karst Catchment Using Particle-tracking Method

D K S Y Klaas[1,2,3], M A Imteaz[1], A Arulrajah[1], I Sudiayem[3], E M E Klaas[3] and E C M Klaas[3]

1 Department of Civil & Construction Engineering, Swinburne University of Technology, Melbourne, VIC, Australia.
2 Department of Civil Engineering, Politeknik Negeri Kupang, Indonesia.
3 Department of Health, Science and Technology, Klinik Garuda, Kupang, Indonesia
E-mail: dklaas@swin.edu.au

Abstract. Accurate mapping of groundwater vulnerability zones based on robust groundwater model is crucial towards an appropriate development of management strategies. This study aimed at developing groundwater vulnerability zones in a tropical karst area in Rote Island, Indonesia. HGZ method was employed to accommodate the limitation of subsurface data. Using reverse particle-tracking simulation, this study combined the simulation results (travel time and pathline trajectory) with the potential groundwater contamination risk to designate three vulnerability zones. In this study preventive management measures were recommended to ensure provision of safe and good quality water from Oemau Spring and to protect the spring from contamination.

1. Introduction

Employing appropriate calibration methods that achieves the modelling objective (good fit between simulated and observed values) and properly represents the corresponding system is an important task in developing a robust groundwater model in data-limited areas [1-2]. In this study, the HGZ method was chosen to accommodate the limitation of subsurface data, e.g. the absence of spatial distribution of hydraulic conductivity and specific yield values. Therefore, the main objectives of this paper are:

1. to map groundwater vulnerability zones using advective transport code (MODPATH), and
2. to develop corresponding management strategies based on the result of particle-tracking simulation (travel time and pathline trajectory) and the potential risk of current human activities to water quality at Oemau Spring.

2. Materials and methodology

2.1. Study area

The study area is the catchment area of Oemau spring (CAOS), geographically located between latitudes 10°46'42.17"S ~ 10° 43'36.91"S and longitudes 123° 3'14.84"E ~ 123° 9'17.64"E (Figure 1). Having a catchment area of 20.11 km^2, CAOS's topography is characterised by highly rippled terrain with surface elevation ranges from around 98 m to 340 m above sea level [3-5]. Geologically, the area is typified by low karstification degree and under-developed conduits system [6] based on discharge timeseries data analysis at Oemau Spring using modified Boussinesq approach of recession curve [7]. The humidity in the area ranges from 75-92% and mean annual precipitation is 1000-2300

IOP Conf. Series: Earth and Environmental Science **159** (2018) 012011 doi:10.1088/1755-1315/159/1/012011

mm [8]. The area is governed by a monsoonal climate with two distinct seasons: dry (May-November) and wet (December-April), (Figure 1) [9-12].

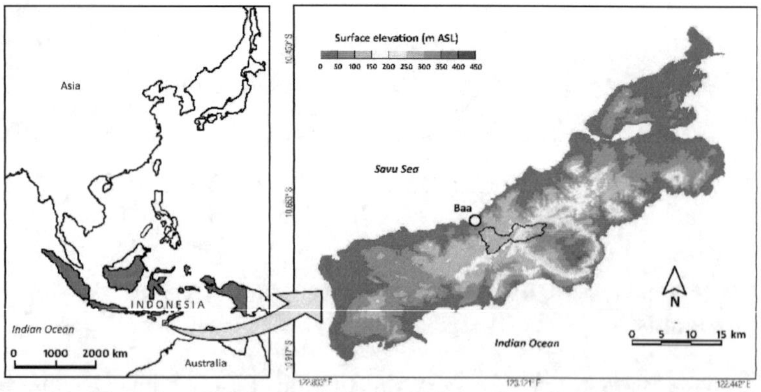

Figure 1. Study area: Indonesia, Rote Island and the recharge area of Oemau Spring.

2.2. Particle-tracking simulation

Vulnerability zones were delineated based on the time and pathline taken by hypothetical particles travelling through the simulated model to arrive at the spring. In this study, the spring vulnerability is assumed to relate to the advection function represented by travel time of particle and pathline trajectory [13]. Reverse particle-tracking method in an advective transport code, MODPATH (version 5) incorporated in GMS suite were employed to compute the pathline of the particles travelling through the simulated subsurface pathlines trajectories. Particle-tracking has been widely applied due to its competence to trace various sources of contaminants in time and space, hence it can provide a foundation for mapping zone of influence of different pollutants. In the simulation pathlines are tracked backwards through times by computing values of velocity vector based on the flow rates at every grid cell. MODPATH [14] solves three-dimensional partial differential equation as follow:

$$\frac{\partial}{\partial x}(nv_x)+\frac{\partial}{\partial y}(nv_y)+\frac{\partial}{\partial z}(nv_z)=W \tag{1}$$

where, the v_x, v_y, v_z are the principal components of the average linear groundwater velocity vector (m/day), and n is dimensionless porosity.

3. Results and discussion

3.1. Estimated flow paths and time

Using reverse particle-tracking analysis which principally calculates groundwater flow velocities, MODPATH predicted particles' flow path and tracked the particle backwards toward their origins. The distance of influence over time associated to three plausible porosity values (n = 0.15, 0.30 and 0.45) is presented in Figure 2. It shows that when the reduction of n values variably decreases the travel times to the spring. Given a distance of 4 km from the spring the travel time is reduced 30% from 24 months to around 17 months when n is reduced from 0.45 to 0.3. Reduction of n from 0.3 to 0.15 results in approximately 57% times travel reduction to about 7 months for the same distance. An intermediary value of 0.3 for n is selected in this study to draw zones of influence due to the absence of actual porosity value from sample measurement in the study area. Figure 2b illustrates simulated flow path from the source points to Oemau spring using n value of 0.3. It shows that in 6 months particles can travel for a maximum distance of around 2 km from the Oemau spring. As the travel time increases the trajectory of the pathlines extends towards the upstream area of the CAOS. Within one year the zone of influences reaches areas around Lekunik Airport; while in two years, it encompasses beyond areas in the middle of CAOS. Hypothetically, the particles are traced back to the upper areas of CAOS in three years time.

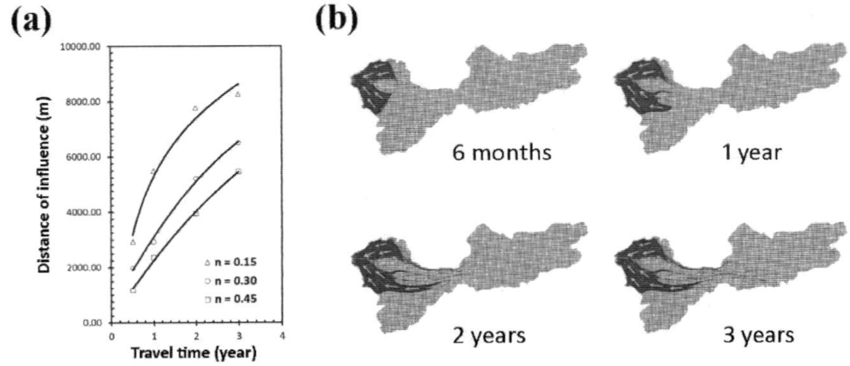

Figure 2. Results of particle-tracking simulation using MODPATH code

3.2. Potential risk and development of vulnerability zones

The division of vulnerability zones is outlined by employing the influence distance using backward particle-tracking and travel time as presented in Figure 2. Incorporating appropriate preventive management measures, the vulnerability zones design considered the potential risk of human activities to the water quality at Oemau Spring by taking into account current practice in the community and land use type in CAOS. Figure 2 shows that in a period of two years, particles can be traced back to areas in the middle of CAOS. Dominated by rice farms, this area poses potential hazards to the quality of water at Oemau Spring. Chemical pollution from fertilizers and pesticides discharges, e.g. organochlorine, nitrate, residues, and various metals, such as lead, cadmium and arsenic, applied during cultivation period in wet season could infiltrate into the aquifer and travel to the spring.

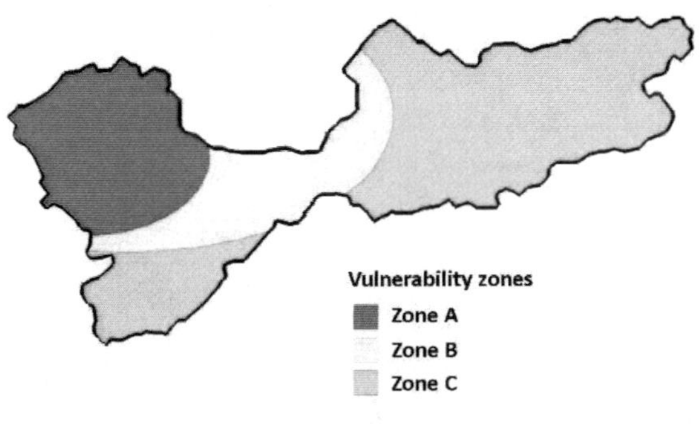

Figure 3. Recommended vulnerability zones.

Table 1. Vulnerability zones and corresponding management strategies.

Zone	Level of vulnerability	Management strategies
A	High	1. Health promotion on rural sanitation 2. Promoting lining septic tanks 3. Waste collection from households 4. Monitoring of water quality at spring and nearby dug wells
B	Medium	1. Management of pesticide and fertilizer application 2. Monitoring of water quality at dug wells
C	Low	1. Reforestation to increase recharge into aquifer

Figure 3 shows the three proposed vulnerability zones and vulnerability level description of each zone is presented in Table 1. Zone A covers an area within 3 km from the spring. This area is categorised as highly vulnerable due to potential contamination from current and anticipated dispersion of settlement and government building. Therefore, it requires immediate management strategies. Having an intermediary level of vulnerability, zone B encompasses mostly areas of rice farm situated in the middle of CAOS. Area of zone C is designed as main reforestation area mainly to boost recharge to the spring.

4. Conclusions

Three vulnerability zones were proposed based on reverse particle-tracking simulation in this study. The corresponding management strategies were presented to ensure provision of safe and good quality water from Oemau Spring. Given the main purpose of zonation, comparatively small catchment size, groundwater flow characteristics, potential pollutants and their sources, the vulnerability zones are considered effectually representative to be used as a management instrument for the local decision makers. The measures were designed by accounting the simulation results (travel time and pathline trajectory) and potential groundwater contamination risk from human activities (land use type and current practice). The recommended measures can be used by the regional decision makers to safeguard the spring from contamination.

5. References

[1] Klaas D K S Y and Imteaz M A 2017 Investigating the impact of the properties of pilot points on calibration of groundwater models: case study of a karst catchment in Rote Island, Indonesia *Hydrogeol J* **25**(6) 1703–1719 doi: 10.1007/s10040-017-1590-4

[2] Klaas D K S Y, Imteaz M A and Arulrajah A 2016 Evaluating the impact of grid cell properties in spatial discretization of groundwater model for a tropical karst catchment in Rote Island, Indonesia *Hydrol Res* doi: 10.2166/nh.2016.250

[3] Klaas D K S Y, Imteaz M A, Sudiayem I, Klaas E M E and Klaas E C M 2017 On the development of a new methodology in sub-surface parameterisation on the calibration of groundwater models *Proc. 1st Int Conf on Adv Environ Eng (AEE 2017). IOP Conf. Series: Earth and Environmental Science* **92** 1-5

[4] Klaas D K S Y and Imteaz M A 2017 Development of a groundwater model for a tropical eogenetic karst aquifer in Rote Island, Indonesia *Proc. 37th Hydrol and Water Res Symp Queenstown, New Zealand* 228-235

[5] Klaas D K S Y, Imteaz M A and Arulrajah A 2017 Development of groundwater vulnerability zones in a data-scarce eogenetic karst area using Head-Guided Zonation and particle-tracking simulation methods *Water Res* **122**, 17-26 doi: 10.1016/j.watres.2017.05.056

[6] Vacher H L and Mylroie J E 2002 Eogenetic karst from the perspective of an equivalent porous medium *Carbonates Evaporites* **17**(2), 182-196 doi: 10.1007/bf03176484

[7] Malik P 2015 *Karst aquifers: Characterization and Engineering*. Stevanović Z. (ed), Springer, Cham, Switzerland

[8] Klaas D K S Y, Imteaz M A, Sudiayem I, Klaas E M E and Klaas E C M 2017 Novel approaches in sub-surface parameterisation to calibrate groundwater models *Proc. 3rd Int Conf on Water Res and Environ (WRE 2017). IOP Conf. Series: Earth and Environmental Science* **82** 1-8 doi: 10.1088/1755-1315/82/1/012014

[9] Klaas D K S Y, Imteaz M A, Sudiayem I, Klaas E M E and Klaas E C M 2017 On the development of a new methodology in subsurface parameterisation on the calibration of groundwater models. *Proc. 1st Int Conf on Adv Environ Eng (AEE 2017). IOP Conf. Series: Earth and Environmental Science* **92** 012026 doi: 10.1088/1755-1315/92/1/012026

[10] Klaas D K S Y, Imteaz M A, Sudiayem I, Klaas E M E and Klaas E C M 2017 Assessing the impact of grid cell properties in spatial discretization of groundwater model. *Proc. 1st Int Conf Adv on Eco Eng Dev (ICEED 2017). IOP Conf. Series: Earth and Environmental Science* **109** 012037 doi: 10.1088/1755-1315/109/1/012037

[11] Klaas D K S Y, Imteaz M A and Arulrajah A 2017 Application of head-guided zonation method to delineate groundwater vulnerability zones in a data-scarce catchment. *Europ Water* **57**:395-398
[12] Klaas D K S Y 2008. *Indigenous Water Management: Sustainable Water Conservation Strategies in a Karstic Dominated Area in Rote Island, NTT Province, Indonesia* (MEngSc Thesis). Monash University, Melbourne, Australia.
[13] Maxe L and Johansson P-O 1998 Assessing groundwater vulnerability using travel time and specific surface area as indicators. *Hydrogeol J* **6**(3), 441-449 doi 10.1007/s100400050166.
[14] Pollock D W 1988 *User's Guide for MODPATH/MODPATH-PLOT, Version 3: A particle tracking post-processing package for MODFLOW*, the U. S. Geological Survey finite-difference ground-water flow model, 242 p, Reston, Virginia

Acknowledgments
The authors wish to acknowledge the Australian government for supporting the research through the Australia Awards Scholarship (AAS) and Allison Sudrajat Prize Award in the Swinburne University of Technology, Australia.

IOP Conf. Series: Earth and Environmental Science **159** (2018) 012012 doi:10.1088/1755-1315/159/1/012012

Land Unit Mapping and Evaluation of Land Suitability for Agro – forestrye in Thua Thien Hue province – VietNam as an Example

Nguyen Ngoc Dan[1,2] and Lei Guo Ping[1] and Le Phuc Chi Lang[3]

1 Land Managenment Institute, Northeastern University, Shenyang, Liaoning 110004, China
2 Mientrung Institute for Scientific Research 530000, VietNam
3 Hue University of Education, Hue University 530000, VietNam
E-mail: ngocdan859@gmail.com

Abstract. This study covers 1 city, 2 towns and 6 districts with the total area of 503,320.53 ha. Its outcome is eight decentralization criteria including soil type, slope, terrain elevation, soil layer thickness, mechanical composition, humus content, bio-climatic conditions and irrigation. 8 single maps have been built using GIS. These 8 maps are merged into a singular map of land unit. The results show that the study site has 130 land mapping units (LMUs), of which the LMU 49 has the smallest area of 100,88 ha and LMU 111 has the largest area of 35,423.23ha. Based on the land quality of LMUs, it can be said that in the study site, 80,282.02 hectares is suitable for annual crops, 2225.80 ha for Thanh Tra pomelo planting, 15528.11 ha for rubber, and 159377.1 ha for Acacia mangium planting.

1. Statements

While presented as a natural thing, land is not only an outcome of labour, but also a special irreplaceable means of agricultural production. In the course of agricultural production, human has made artificial ecosystems to replace natural ecosystems, thus degrading their sustainability [1]. Moreover, rapid urbanization and industrialization in tandem with population growth have imposed increasing pressures on agricultural land, which is progressively narrower. In addition, irrational exploitation and use of land has led to degradation and infertility of soil in many areas [2].

For effective use, protection and management of land resources, it is of great importance to evaluate the land as a basis for maximizing the land potential [3]. According to the FAO land evaluation framework, land unit mapping is among key contents. A land unit map is a thematic map which displays all land units, characteristics and properties of the soil. For our nation, it is of great significance to conduct provincial land suitability evaluation, which is highly practical as it improves the quality of agricultural land use planning and plays as a basis for synthesizing and orientating crop restructure in association with land use restructure. [3], [4].

Recent time witnesses the Thua Thien Hue province's soaring urbanization in tandem with the transformation of agricultural land into non-agricultural and urban construction. Agricultural production area has decreased and the vogue is inclined to commodity urban agriculture, clean and market-driven agriculture. Research on land unit mapping and evaluation of land suitability for agro-forestry in Thua Thien Hue Province remains a new issue of concern mentioned nowhere, whereas it is essential for the structural transformation of crops, development of provincial agro-forestry, enhancement of land use sustainability and protection of ecological environment and improvement of living standards of Thua Thien Hue people.

2. Overview of the study site

Thua Thien Hue province has coordinates of 16 – 16.8^0 N Latitude and 107.8 – 109.2^{0}E Longitude. It features various terrain types (mountains, hills, plains, and coastal areas), tropical monsoon climate along with many extreme weather phenomena, diverse flora and dense hydrological networks, all reflecting regional typical characteristics. It has a modest total area of natural land, only 503320.53ha (of which 471313.07 ha is non-submerged), but the soil cover is susceptible to many natural and socio-economic factors, hence a highly complicated differentiation. Recent times see the city's steady shrinkage of agricultural area. The province's approved land use planning toward 2020 implies that the agricultural land area would narrow to 385552 ha in the time to come. Therefore, the remaining land fund for agricultural production must be rationally and efficiently used, so as to meet the demand for food without depleting the land resources. The area of land under the survey (agricultural production land) is 760.62 ha, accounting for 45.02% of the natural area (Table 1).

3. Methodologies

This research involves the following methodologies: *Secondary data collection*: Collecting agrochemical-soil maps, 2014 land use status map of the study site, statistics, land analysis data, reports and available local projects within the study site; *Land map correction*: Sampling more 30 soil profiles and 15 agrochemical specimens in districts of Thua Thien Hue Province; in combination with the 2014 land use status map, zoning and adjusting the land map*; Construction of information layers (single maps):* with help of ArcGIS software preparing 8 information layers including soil type, slope, terrain elevation, soil layer thickness, mechanical composition, humus content, bio-climatic conditions and irrigation; *Land unit map*: using ArcGIS to merge single maps into a land unit map of the study site (Figure 1); Summary and statistics of survey and research results enabled by Excel software; *Comparison*: Comparing land use requirements corresponding to land utilization types in respect to land unit characteristics, so as to determine the suitability of LMUs;

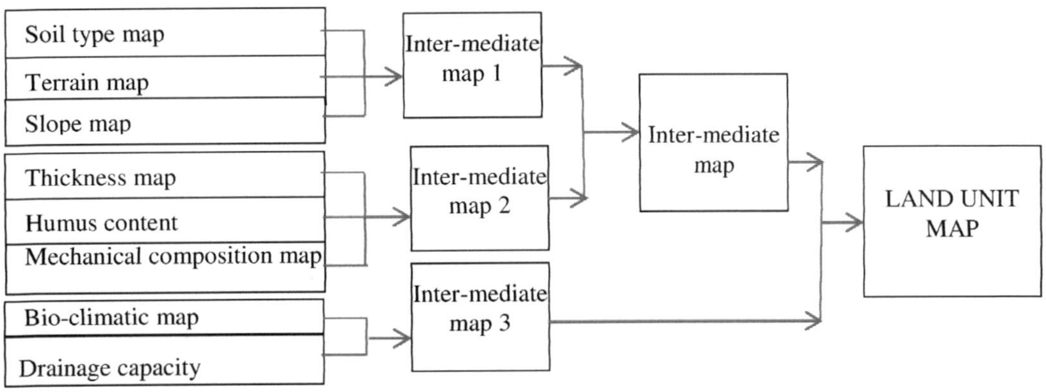

Figure 1. Overlayering single maps into a land unit map using GIS

Method for evaluation and classification of land suitability: *The land evaluation method proposed* by FAO (1976) [5] is a prominent method that for many years has been successfully applied in many developing countries as a tool for land use planning. Given the objectives and tasks set out, the researcher uses a synthetic scale and average multiplication proposed by D. L. Armand (1975) for evaluating the land suitability for development of agro-forestry in Thua Thien Hue Province. The problem looks like this

$$M_0 = \sqrt[n]{a_1 . a_2 . a_3 ... a_n}$$

Where: M0: Score of land units. a1, a2, a3 ... an: Score of indicators from 1 to n. n: number of indicators used for evaluation

4. Research findings

4.1 Land unit mapping

4.1.1. Identification of land classification criteria

Land unit is a plot of land having such soil characteristics and properties suitable for each land utilization type (LUT), bearing the same conditions of land management and same potential for soil Improvement and production. Land unit map is a collection of land units in the evaluated area [6]. Land units are identified by overlaying information layers. FAO suggests that to evaluate land in areas of modest scope and similar climatic features, it may be helpful to delve into soil factors such as: soil properties (soil type, physical and chemical properties of the soil), topographic features (slope, shape, relative topography, and elevation), water properties (irrigation, drainage and inundation), and distribution of plants and animals. Given the purposes of land unit mapping and based on an appropriate evaluation of land suitability for agro-forestry, selected criteria for classification include soil type, terrain elevation, slope, soil layer thickness, mechanical composition, humus content, climatic conditions and irrigation regime (Table 2).

Table 1. Area of the study site (ha)

Land type	Area (ha)	Percentage (%)
Total area of the administrative land	**503320.53**	**100%**
Agricultural land	390889.77	77.66
Non-agricultural land	91026.20	18.15
Unused land	21.40355	4.19

Source: People's Committee of Thua Thien Hue Province, 2016

Table 2. Indicators and categorization of indicators for land use mapping of Thua Thien Hue Province

NO	Indicators	Categories
1	Soil types	1. White sand dunes (Cc); 2. Sandy soil (C) 3. Highly saline soil (Mn) 4. Medium saline soil (M) 5. Deep active acid soil, medium saline (Sj2M) 6. Alluvial soil accreted annually (Pb) 7. Alluvial soil not accreted (P) 8. Glay alluvial soil (Pg) 9. Alluvial soil with reddish-brown spotted layers (Pf) 10. Stream alluvial soil (Py) 11. Alluvial soil covering sandy soil (P/C) 12. Wetland (J) 13. Gray soil on magma acid rocks (Fa) 14. Yellowish red soil on clay stone (Fs) 15. Yellowish red soil on metamorphic clay stone (Fj) 16. Reddish yellow soil on magma acid rocks (Fa) 17. Light yellow soil on sandy stone (Fq) 18. Yellowish brown soil on old silt (Fp) 19. Yellowish red soil mutated by wet rice cultivation (Fl) 20. Red muddy soil on metamorphic rock (Hj) 21. Red muddy soil on magma acid rock (Ha) 22. Valley soil generated from slope accretion (D) 23. Eroded soil revealing stones and gravel (E)
2	Slope	1. $<8^0$ 2. $8 - 15^0$ 3. $15 - 25^0$ 4. $>25^0$

NO	Indicators	Categories
3	Soil layer thickness	1. >100cm 2. 70 - 100cm 3. 50 - 70cm 4. <50cm
4	Mechanical composition	1. Thick loam 2. Medium loam 3. Light loam 4. Light sandy 5. Sandy
5	Humus content	1. >3% 2. 2 - 3% 3. 1 - 2% 4. <1%
6	Elevation	1. <50m 2. 50 - 100m 3. 100 - 200m 4. 200 - 700m 5. >700m
7	Bio-climate	1. IA1a, IA1b 2. IIA1a, IIA1b 3. IIIA2a 4. IVA3a
8	Drainage capacity	1. Very good 2. Relatively good 3. Poor 4. Very poor

4.1.2. Construction of single maps

a. Soil type map: Soil type is an indicator generalizing characteristics of land plots. Soil type reflects a variety of basic physical, chemical, and biological properties of soil, while giving us the perception of land use potential and land quality in respect to the growth and development of the crops [7]. Outcomes of land mapping at the scale of 1/100.000 reveal that the study site can be classified into 23 soil types under 10 groups [8].Results of land map check and revisions in accordance with the land use status map of the study site are shown in Figure 2.

b.Terrain elevation map: The diverse topography of Thua Thien Hue features. The study site is topographically is categorized into 5 ranges of altitude as follows: <50m, 50 - 100m, 100 - 200m, 200 - 700m, and >700m. Statistics show that the landform lower than 50m occupies the major part (176688ha) accounting for 35.0% of the natural area. It is followed by the 50-100m range (42040.672 ha) accounting for 8.3 % of total natural area; 100 - 200m range (53301.023ha) accounting for 10.5%, 200-700m range (168145.8192 ha) accounting for 33.4% and lastly, >700m landform accounting for 12.8%, (Figure 3).

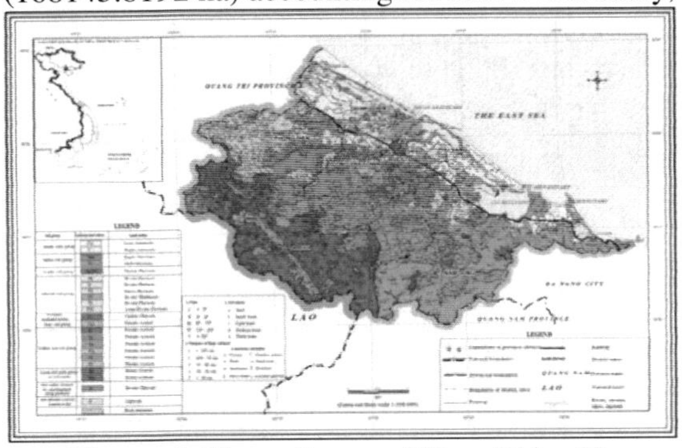

Figure 2. The soil map of Thua Thien Hue Province

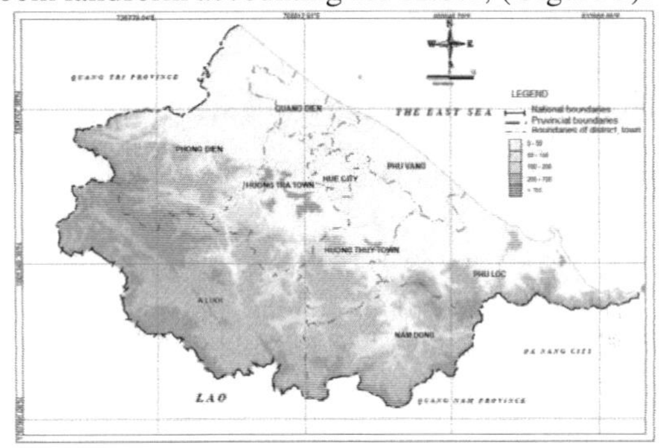

Figure 3. Terrain map of Thua Thien Hue Province

c. Slope map: Given the slope map of the study site, 4 ranges of slope can be defined as follows: <8^0, 8^0 - 15^0, 15^0 - 25^0, and >25^0. It is statistically pointed out that: land of slope less than 8^0 occupies an area of 135795.105 ha (making up 27% of the natural area of the study site), the range of 8^0 - 15^0occupies 62524.053 ha (12.4%), the range of 15^0 - 25^0occupies 122606.46 ha (24.4%), while land of slope above 25^0 has the largest area of 182,092.9 ha (36.2%), (Figure 4).

d. Soil layer thickness map: The thickness of soil layer, reflecting the soil's capacity of nutrient reserve, is related to topographic slope, vegetation cover and farming methods. Soil layer thickness presents as an important indicator for land unit mapping. The study site's soil layer thickness is defined in 4 levels: >100cm, 70-100cm, 50-70cm, <50cm. Statistics show that the soil layer of thickness less than 50cm accounts for the largest area of 269,125.918 ha (53.2% of the total natural area), thickness of 50-70cm accounts for 29865.538 ha (5.9%), thickness 70-100cm occupies 41283.505 ha (8.2%) and thickness >100cm occupies 165385.778 ha (32.7%), (Figure 5).

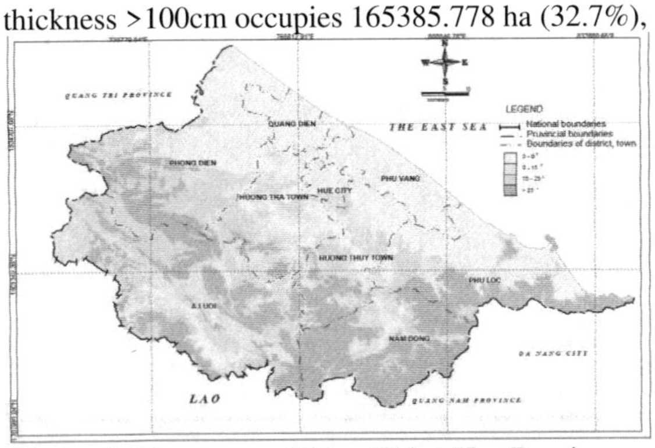

Figure 4. Slope map of Thua Thien Hue Province

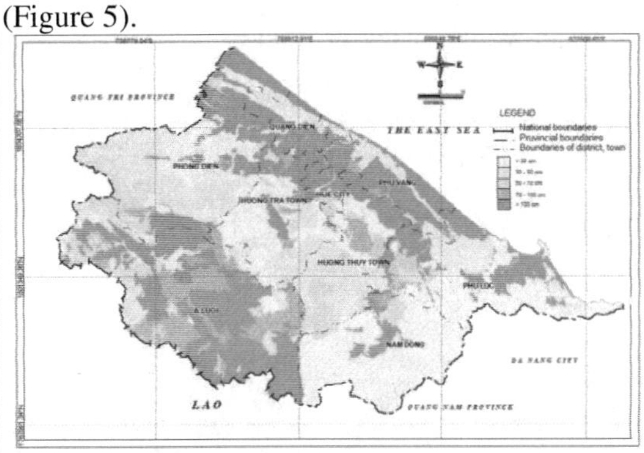

Figure 5. Soil layer thickness map Thua Thien Hue Province

e. Mechanical composition map: A given crop may adapt to soil of a certain mechanical composition. Mechanical composition determines the physical, chemical and biological properties of the soil. That is why need should arise for identification of soil mechanical composition so as to match a specific plot of land with the right crops and appropriate technical measures. In the study site, 5 levels of mechanical composition are defined: thick loam, medium loam, light loam, light sandy and sand. Their proportions are respectively 2.4%, 26.7%, 32.4%, 25.7%, and 12.8%, (Figure 6).

f. Bio-climate: Since they substantially affect the growth, development and productivity of crops, these factors help to determine the appropriate crop structure that is economically beneficial in terms of agro-forestry. It is suggested by the bio-climate mapthat Thua Thien Hue Province features 4 types of bio-climatic conditions:High precipitation, very hot, short to medium dry season (IA1a, IA1b), area of 194308.2 ha; High precipitation, hot, short-to-medium dry season (IIA1a, IIA1b), area of 165388.2 ha; High precipitation, warm, short dry season (IIIA2a), area of 82149.23 ha; High precipitation, cool, short dry season (IVA3a), area of 63505.84 ha; (Figure 7).

g. Humus content map: Humus content indicates the nutrient reserves in the soil which greatly affects crop yields. The humus content of Thua Thien Hue Province is classified into 4 levels as follows: very rich> 3%, rich 2 - 3%, medium 1 - 2%, and poor <1%, (Figure 8).

h. Drainage capacity: Thua Thien Hue territory is characterized by four levels of drainage capacity: good in mountains and hills; relatively good in mountain bases and foothills of relatively high slope; poor in flat valleys; and very poor in flat areas, ponds, lakes, lagoons, etc, (Figure 9).

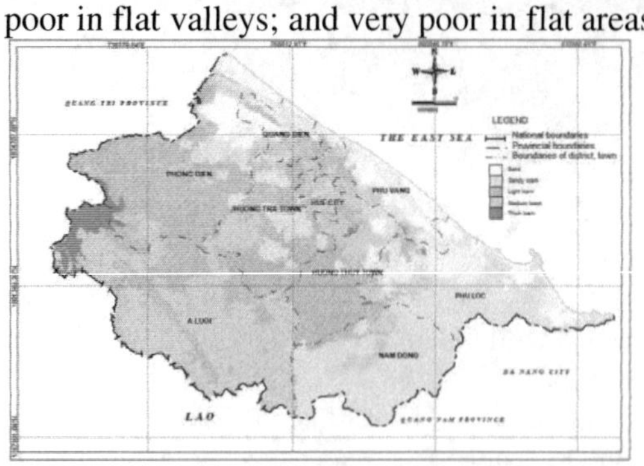

Figure 6. Soil mechanical composition map of Thua Thien Hue Province

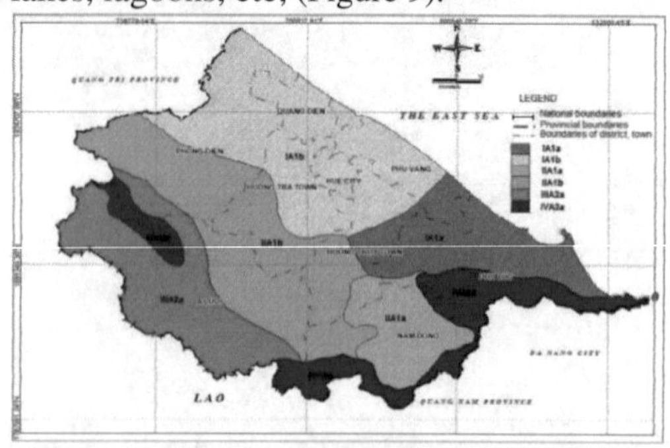

Figure 7. Bio-climate map of Thua Thien Hue Province

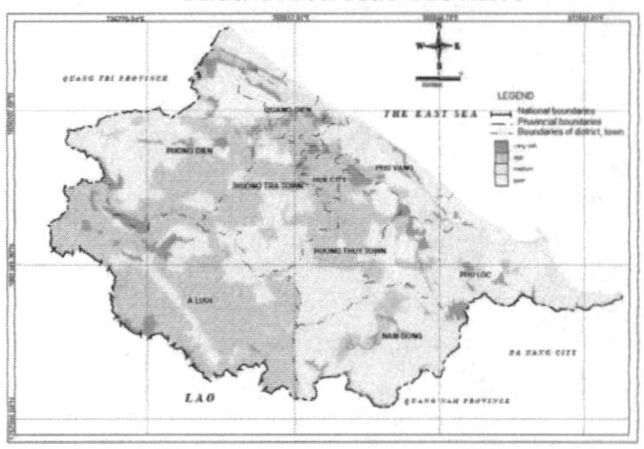

Figure 8. Humus content map of Thua Thien Hue Province

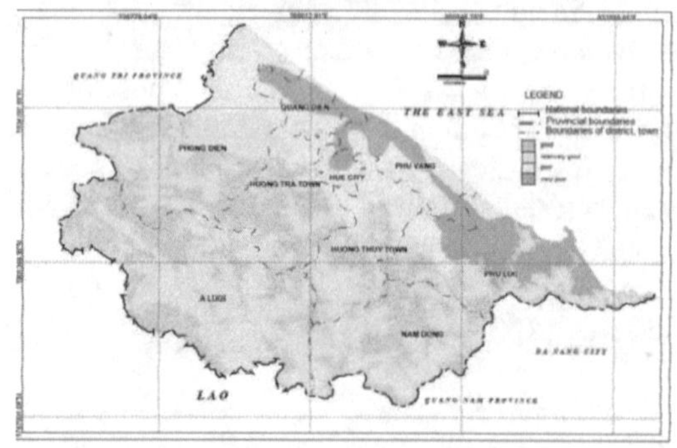

Figure 9. Drainage capacity map of Thua Thien Hue Province

4.1.3. Land unit mapping

Land unit map is the combination of information layers (single maps). Each land mapping unit contains all information expressed in single maps and is distinguished from other units by the discrepancy of categorized indicators. After single maps are constructed in respect to soil characteristics and properties, they are over-layered using ArcGIS software [4]. This comes up with an integrated land unit map (Figure 10) consisting of 130 land units. The smallest LMU is LMU 49 with an area of 100.88ha. The largest one is LMU No. 111, area of 35423.23ha.

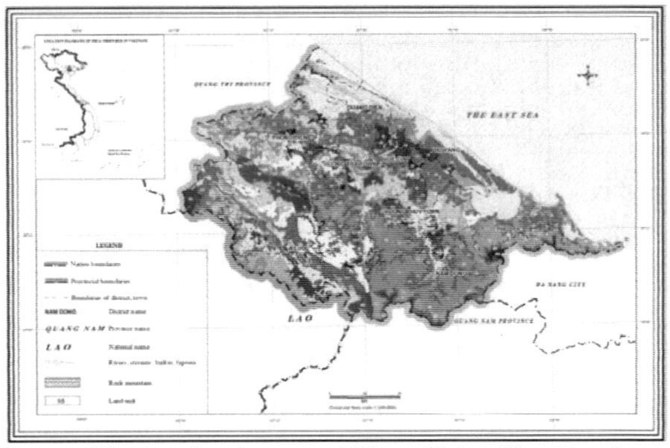

Figure 10: Land unit map of Thua Thien Hue Province

4.2. *Evaluation* of land suitability

4.2.1. Land use requirements per land quality

In order to evaluate the land suitability for agriculture development in Thua Thien Hue Province, we selected the following crops as the basis for evaluation: annualcrops (peanut, peas, sesame), fruit (Thanh Tra pomelo), and industrial crop (rubber). Grown widely in Thua Thien Hue Province, the above-mentioned plants have appeared compatible with the local production practices, giving high economic value and high potential for improvement, soil and environmental protection. They are spearheading crops that are prioritized for expansion under the Province's agricultural development strategy toward commodity production and farmer income boost.

On the basis of summing up the research results from the project: [9], [10], [11] projects and Consultation of experts. can determine the ecological needs of all kinds of plant groups (Table3).

Table 3. Summary on ecological needs of crops

Indicators	Soil types	Slope $(^0)$	Soil layer thickness (cm)	Mechanical composition	Humus content (%)	Elevation (m)	Bio-climate	Drainage capacity	classyfy
annual crops	C, Pb	<8	> 100	Light sandy	>3	<50	IA1a, IA1b	Very good	S1
	P, Pf, D, Fp, Py	-	70-100	Medium loam	2 - 3	50-100	IIA1a, IIA1b	Relatively good	S2
	Fa, Fs, Fq	8 - 15	50 - 70	Sandy, Thick loam	1 - 2	100-200	IIIA2a	poor	S3
	Remaining	>15	<50	-	<1	> 200	IVA3a	Very poor	N
ThanhTra pomelo	Pb	<8	> 100	Light sandy, Sandy phase	>3 , 2-3	-	IA1b	Very good	S1
	P, Pf, Py, D	8-15	70-100	Medium loam	1 - 2	-	IA1a,IIA1b	Relatively good	S2
	Fs, Fa, Fq, Fp	15-25	50 - 70	Thick loam	<1	-	IIA1a	poor	S3
	Remaining	>25	<50	Sandy	-	-	IIIA2a,IVA3a	Very poor	N
rubber	Fs, Fj	<8	> 100	Thick loam	>3	>100	IA1b	Very good	S1
	Fa, Fq	8-15	70-100	Medium loam	2 - 3	100-200	IA1a, IIA1b	Relatively good	S2
	Fp, D	15-25	50 - 70	Light loam	1 - 2	200-700	IIA1a, IIIA2a	poor	S3
	Remaining	>25	<50	Sandy	<1	>700	IVA3a	Very poor	N
Acacia mangium	Fs, Fj, Fq, Fa, Fp, Ha, Hj	<15	> 100	-	-	-	IIA1a, IIA1b	Very good	S1
	P, Pf, P/C, C, D,	15-25	70-100	-	-	-	IA1a, IA1b	Relatively good	S2
	E, Cc, Pb	-	50 - 70	-	-	-	IIIA2a	poor	S3
	Remaining	>25	<50	-	-	-	IVA3a	Very poor	N

Note: S1: Very suitable; S2: Suitable; S3: Hardly suitable; N: Not suitable

4.2.2. Land suitability evaluation

In consideration of land use requirements for different types of land use and land units' quality, level of LMUs' suitability will be compared and identified (Table 4).

Table 4. Summary on levels of land suitability

Crops	LMU	Land suitability	Arable area (ha)
Annual crops	2, 14, 16, 18, 20, 27, 32, 35, 38, 41	S1	41369.39
	9, 10, 12, 13, 22, 25, 31, 33, 39, 40, 53, 59, 68, 69, 70, 75	S2	38912.17
	50, 56, 57, 71, 74, 78 and 90	S3	10043.89
	47 remaining land units	N	119728.83
ThanhTra pomelo	14, 16, 20, 28, 38 and 53	S1	41369,39
	12, 18, 31, 33, 41 and 75	S2	38.912,17
	9,25,34,35,40,56,58,67,68, và 70	S3	10.043.89
	58 remaining land units	N	119728.83
Rubber	31, 56 and 62	S1	41369.39
	25,27,34,53,58,70, 107 và 109	S2	38912.17
	9,67,68,75, 100, 105, 106 và 114	S3	10043.89
	61 remaining land units	N	119728.83
Acacia mangium	9, 10, 13, 25, 27, 31, 33, 34, 38, 50, 53-59, 62, 67, 68, 70, 74, 75, 76, 78, 79, 109 and 114	S1	41369.39
	2, 6, 14, 16, 18, 20, 22, 26, 28, 32, 35, 37, 40, 41, 43, 63, 71, 95, 96, 100, 101, 103, 105-107, 119, 121 và 122	S2	38912.17
	5, 12, 19 and 39	S3	10043.89
	20 remaining land units	N	119728.83

Table 5. Categorization of land units for agro-forestry development in Thua Thien Hue Province

Land use types	Land suitability				Arable area (ha)
	S1	S2	S3	N	
1. Annual crops	41369.39	38912.17	10043.89	119728.83	90325.45
2. Thanh Tra pomelo	15305.78	6,952.30	17233.78	170562.20	39491.86
3. Rubber	4584.06	10,944.05	34344.50	160181.67	49872.61
4. Acacia mangium	68266.47	91,110.58	23651.75	27025.48	183028.80

Note: S1: Very suitable; S2: Suitable; S3: Hardly suitable; N: Not suitable

Table 5 reveals Thanh Tra pomelo, followed by rubber, features the largest "not suitable" area. This is explained by the fruit's narrow ecological amplitude and strict conditions for growth and development. Rubber hardly adapts to the stormy climate of the study site. Acacia mangium, on the other hand, enjoys the largest cultivable area. The second rank belongs to annual crops (peanut, peas, sesame), which are potential for expansion. They are not picky about the soil they are grown, but rather capable of improving the soils. Therefore, it is crucial to have an optimal layout of crops in conformity to the land potential, territorial advantages and socio-economic development strategy of the province.

Table 6. Summary on areas of different land suitability for annual crops

District	Classified by administrative units			
	S1(ha)	S2	S3	S4
Huong Thuy	1413.617	9089.616	1671.01	33535.45
Huong Tra	4977.179	4408.241	730.507	41939.01
Phong Dien	3926.623	11682.13	2279.082	77275.51
Quang Dien	315.15	-	-	16058.73
Quang Dien	12732.32	1355.105	601.229	57246.61
Nam Dong	-	12.67867	12.37523	225.0433
Phu vang	10803.24	-	-	14612.80
A Luoi	-	-	1434.724	121058.1
Hue City	3234.687	1674.428	-	2223.327
Total area	37402.82	28222.2	6728.927	364174.6

Table 7. Summary on areas of different land suitability for Thanh Tra pomelo

District	Classified by administrative units			
	S1(ha)	S2	S3	S4
Huong Thuy	1078.44	3625.474	4243.123	36762.65
Huong Tra	3879.901	2900.119	639.778	44635.14
Phong Dien	2158.942	4031.366		88973.04
Quang Dien	2.76103	-		16371.12
Quang Dien	351.039	-	5170.149	66414.08
Nam Dong	-	7.9775	3.809391	238.310
Phu vang	3378.059	-		14,612.80
A Luoi	-	2016.14		120476.7
Hue City	3234.687	-	957.181	2940.574
Total area	14083.829	12581.0765	11014.0404	391424.414

Table 8. Summary on areas of different land suitability for Rubber

District	Classified by administrative units			
	S1(ha)	S2	S3	S4
Huong Thuy	2982.22	2029.61	1659.22	39038.6
Huong Tra	-	2193.358	1881.462	47980.12
Phong Dien	-	-	3511.519	91651,83
Quang Dien	-	-	-	16373.88
Quang Dien	601.229	-	299.49	71034.55
Nam Dong	4.32186	6.795319	0.670861	238.3091
Phu vang	-	-	-	27831.17
A Luoi	935.947	4781.94	26535.67	90239.24
Hue City	-	957.181	-	6175.261
Total area	4523.71786	9968.88432	33888.0319	**390562.96**

Table 9. Summary on areas of different land suitability for Acacia mangium

District	Classified by administrative units			
	S1(ha)	S2	S3	S4
Huong Thuy	11733.25	5876.022	2630.26	25470.15
Huong Tra	20745.41	4000.988	1334.97	25973.57
Phong Dien	18866.76	4403.985	16814.03	55078.57
Quang Dien		312.389	3837.03103	12224.46
Quang Dien	3070.557	12456.92	5507.24	50900.55
Nam Dong	21.72968	-	-	228.3675
Phu vang	-	7425.18	5793.188	21732.51
A Luoi	7875.461	18572.43	-	96044.9
Hue City	1659.037	177.8683	3234.687	2060.85
Total area	63972.2047	53225.78	39151.409	**289713.9**

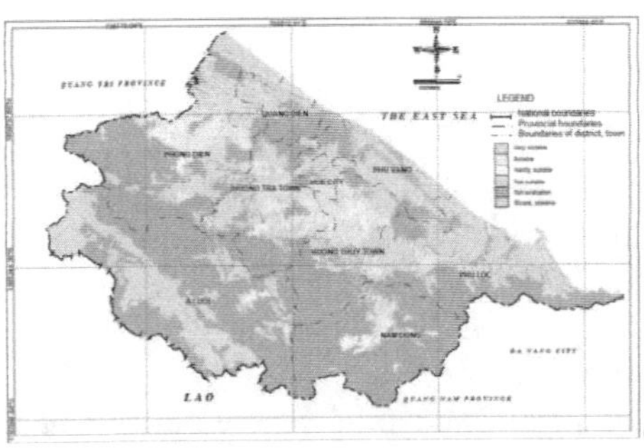

(a)

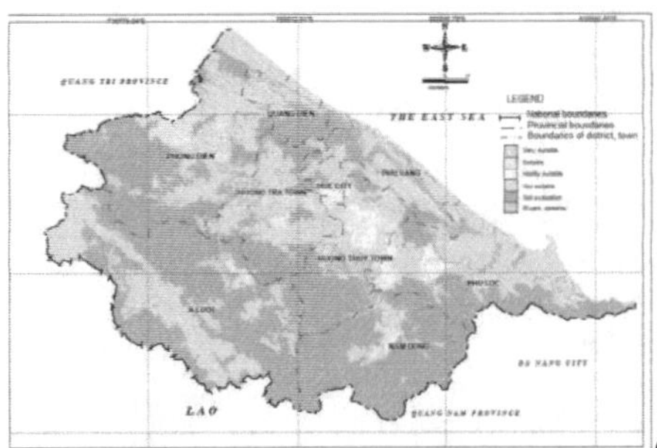

(b)

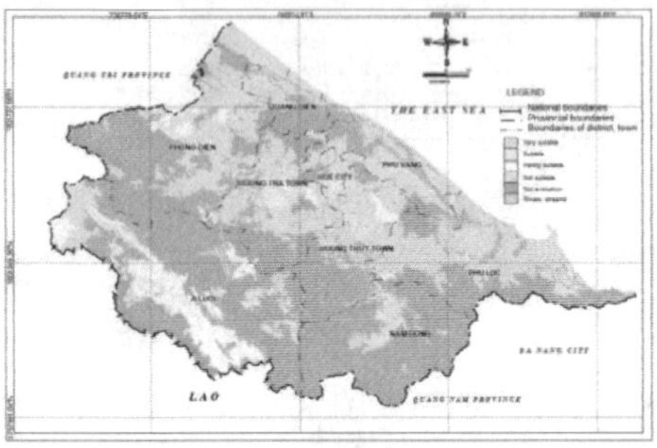
(c)

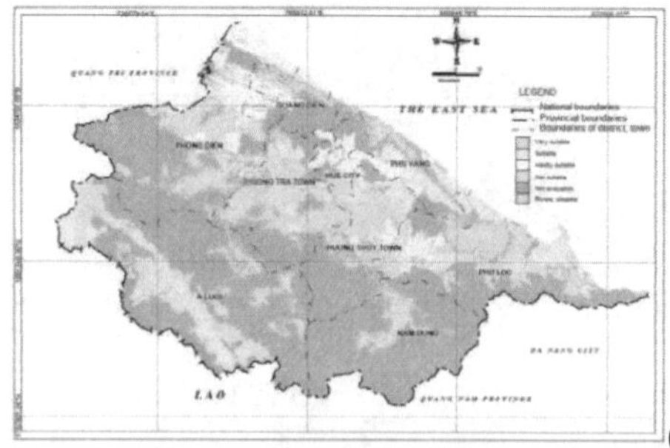
(d)

Figure 11. Map of land suitability for (a) annual crops, (b) Thanh Tra pomelo (c) rubber (d) Acacia mangium

It can be seen from tables 6, 7, 8, and 9 that the area of land suitable for annual crops is 80,281.56 ha, accounting for 15.9% of the study site. Pomelo has 22,258.08 ha of cultivable land, accounting for 4.4% of the study site. Arable area for rubber is 15,528.11 ha, representing 3.08%. Lastly, acacia mangium enjoys some 159,377.1 ha of cultivable land, occupying 31.6% of the study site. Notwithstanding, the actual numbers may be much larger, which prompts the need for soil improvement measures, irrigation systems completion and proper fertilization in order to improve crop yields.

5. Conclusion

The land unit map of Thua Thien Hue Province is constructed from 8 single maps including: soil type, terrain elevation, slope, soil layer thickness, mechanical composition, humus content, bio-climate and drainage capacity. The project works out the total number of land units in Thua Thien Hue Province, which is 130. LMU No. 49 is the smallest with an area of 100.88 ha. LMU No. 111 is the largest with an area of 35423.23 ha.

Given the land quality of LMUs, 80,282.02 ha is identified suitable for annual basis, 2,225.80 ha suitable for Thanh Tra pomelo and 15,528.11 ha suitable for rubber.

It is necessary to conduct an in-depth evaluation of the possibility to expand pomelo growing, especially rubber plantation, in typical conditions of Thua Thien Hue Province. As in agro-forestry it is the vogue to diversify crops and livestock, many more types of land uses should be added to evaluation objectives. At the same time, agro-forestry as the main production method should lay the basis for establishing a rational eco-economic model which brings about high economic efficiency and boosts the protection of land resources.

6. References

[1] Doan Cong Quy (2000). Textbook of Land Use Planning. Hanoi Agricultural Publishing House.

[2] Nguyen Tu Siem, Thai Phien (1999). Vietnam's Mountainous Land Degradation and Restoration. Agricultural Publishing House

[3] Ton That Chieu, Le Thai Bat, Nguyen Khang, Nguyen Van Tan (1999). Handbook for investigation and classification of land evaluation. Hanoi Agricultural Publishing House.

[4] Le Thi Giang, Nguyen Khac Thoi (2010). Application of GIS in evaluating land suitability for agricultural production in Son Dong District, Bac Giang Province. Science and Development journal, 8(5): 823-831

[5] FAO (1976), A framework for land evaluation, Rome

[6] Dao Chau Thu, Nguyen Khang (1998). Land Evaluation. Hanoi Agricultural Publishing House. People's Committee of Ha Tinh City (2010)

[7] Tran Thi Thu Hien, Dam Xuan Van (2012). Study on land unit mapping for agricultural land evaluation in Dong Hy District, Thai Nguyen Province. Vietnam Journal of Science and Technology, 97(9): 57-62

[8] Thua Thien Hue People's Committee (2016). 2015 Land Inventory Data of Thua Thien Hue Province

[9] Nguyen Duc Cuong (2009), *Peanut Cultivation Techniques*, Publishing House for Science and Technology, Hanoi

[10] Nguyen Duc Cuong (2010), *Cultivation Techniques for Orange, Tangerine and Pomelo*, Natural Science and Technology Publisher, Hanoi
[11] Center of Agro-Forestry Planning and Design (2011), *Statement of Rubber Cultivation Planning in A Luoi District, Thua Thien Hue Province for the period of 2011-2020*, Report archived at the Department of Agriculture and Rural Development of Thua Thien Hue Province.

IOP Conf. Series: Earth and Environmental Science **159** (2018) 012013 doi:10.1088/1755-1315/159/1/012013

Preliminary Study on the Application of Temperature Swing Adsorption in Aqueous Phase for Pesticide Removal

B Aumeier[1], H Q A Dang[1,2] and M Wessling[1,3]

1 Chair of Chemical Process Engineering, Aachener Verfahrenstechnik, RWTH Aachen University, Forckenbeckstr. 51, 52074 Aachen, Germany
2 Department of Mechanical and Process Engineering, The Sirindhorn International Thai – German Graduate School of Engineering, King Mongkut's University of Technology North Bangkok, 1518 Pracharat 1 Road, Wongsawang, Bangsue district, 10800 Bangkok, Thailand
3 DWI – Leibniz Institute for Interactive Materials, Forckenbeckstr. 50, 52074 Aachen, Germany
E-mail: anh.h-cpe2014@tggs-bangkok.org

Abstract. Temperature swing adsorption (TSA) is a well-established process for gas purification. In this work, the feasibility of TSA in aqueous phase was studied. This concept could enable *in situ* adsorbent regeneration and thus fostering sustainable decentralized adsorption processes applied to water treatment. The adsorption processes with the use of granular activated carbon (GAC) have been widely applied to remove the residual amounts of pesticides in water treatment. Amitrole was chosen as a typical pesticide in this study, GAC was selected as the main adsorbent for amitrole removal. Adsorption isotherm experiments were conducted at different temperatures of 20°C, 57°C and 94°C to identify the most appropriate sorptive – sorbent system for dynamic adsorption and TSA research. Once the isotherm experiments were accomplished, breakthrough curve experiments were subsequently investigated. Finally, TSA process was conducted with the activated carbon regeneration at the elevated temperature of 125°C. Consequently, initial obtained results proved the feasibility of the proposed TSA technique for pesticide removal in aqueous phase.

1. Introduction

Since the middle of the twentieth century, because of the rapid rise as well as the dynamic transition of different agricultural societies around the world, significant amount of pesticides were consumed. In recent years, because of the application of amitrole for industrial agriculture, negative environmental problems as well as human health issues have been detected. It has been well-known that heterocyclic organic compound of amitrole (3-amino-1,2,4-triazole) is a systemic and toxic reagent, which can also cause several environmental effects and health problems such as skin rash, vomiting, diarrhea, nosebleeds [1]. The applications of physicochemical processes, such as the adsorption by activated carbon is a positive and promising way to approach and mitigate the problem of pesticides.
Adsorption techniques have a long history and widespread applications on various aspects of human life, especially in different industrial production and treatment activities. It is also widely applied in the field of drinking and wastewater treatment. Practically, different hazardous materials such as colour forming reagents and synthetic organic chemicals (SOCs) can be completely removed by adsorption [2].

In gas phase adsorption, the state-of-the-art of *in situ* regeneration of adsorbents is primarily represented by pressure swing adsorption (PSA), where the chemical substances are removed from the adsorbent under reduced pressure or vacuum [3]. In addition, temperature swing adsorption (TSA) has also been applied for gas phase adsorption, especially for CO_2 capture and storage [4]. As shown in figure 1, typical 4-step of TSA operational procedure for gaseous phase normally includes the adsorption at low temperature, heating up, purging gas and finally the cooling process. Regarding to the thermodynamic characteristics, in gaseous phases the adsorption processes always appear to be exothermic – meaning better adsorption at lower temperature and exergonic, where reactions are spontaneous. In most cases, the sorption processes in gaseous phases are likely to occur at lower temperatures than higher temperatures [5]. As a result, it has been observed that higher adsorption capacities were obtained at lower adsorption temperatures in case of gaseous adsorption [6].

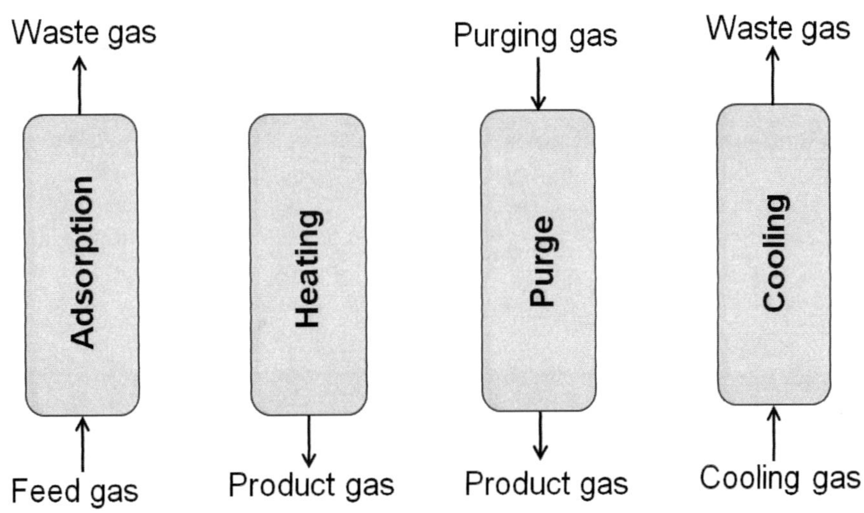

Figure 1. Sequence of a temperature swing adsorption cycle in the gas phase

However, the performance of an adsorption process in aqueous phase is different. Theoretically, it depends on the nature of sorptive – sorbent, solvent – sorbent and solvent – sorptive interactions. Generally, in aqueous phase lower sorption capacities has been recognized at lower temperatures and *vice-versa*. On the contrary, in several occasions the adsorption processes can behave differently. Specific cases of amitrole and diuron removal at different temperatures can be regarded as typical examples. Significant liquid amount of diuron was better-adsorbed and removed by granular activated carbon (GAC) at the elevated temperatures, meaning an overall endothermic process. In contrast, at similar conditions the adsorption efficiency was decreased in case of amitrole [7-8].

The aim of this study is to investigate a temperature swing adsorption (TSA) in aqueous phase for amitrole adsorbed to activated carbon and foresee the future applications of TSA, especially the benefits of *in situ* regeneration of adsorbents during water treatment processes.

2. Methodology for TSA experiments in aqueous phase

To analyse the feasibility of TSA in aqueous sorptive solutions, three important types of experiment were conducted: adsorption isotherm analysis, breakthrough curve study and subsequently the adsorption – desorption cycles for TSA research. In case of adsorption isotherm analysis, Norit® GAC 830W was utilized as the main carbon adsorbent and it was supplied by Cabot GmbH (Germany). The original GAC (particle size: 0.6 – 2.36 mm) was applied for adsorption isotherms, while the milled GAC (particle size: 0.4 – 0.8 mm) were applied for breakthrough curve investigation. Adsorption experiments were investigated in form of batch mode until the equilibrium was reached, with the concentration of GAC were varied from 400 to 4000 mg/L and the original amitrole solution was prepared up to 20 mg/L. Temperature conditions of adsorption were set at 20°C, 57°C and 94°C. The

IOP Conf. Series: Earth and Environmental Science **159** (2018) 012013 doi:10.1088/1755-1315/159/1/012013

performance of amitrole adsorption and removal were represented by the adsorption capacity (q_e) with the unit of mg/g. Furthermore, Freundlich equation was applied to describe the adsorption results.
Once the amitrole isotherm data were obtained, dynamic adsorption and breakthrough curve experiments were subsequently carried out using the stainless steel column containing approximately 4.0 g GAC. The height of carbon bed (H) was determined at 8.0 cm and the inner diameter (D_i) was 1.0 cm. Hence, total bed volume of the stainless steel column was 6.28 cm^3. Moreover, two perforated plates with mesh widths of 0.25 mm were utilized to hold the adsorbent inside the column. Amitrole solution was fed into the column in up-flow mode by a peristaltic pump. The breakthrough condition was defined as $C_t/C_F = 0.7$ in this study.
The most important experiments in this research project were the adsorption – desorption of a TSA system. In general, the adsorption was conducted similarly to the previous dynamic adsorption investigation, while the *in situ* regeneration was performed at elevated temperature of 125°C, with the application of an autoclave (CV-EL 18, CertoClav Sterilizer GmbH, Austria).
UV-VIS spectrophotometer (Varian Cary 50 at λ = 202 nm) was utilized for amitrole quantification during the research.

3. Initial results and feasibility of the research
From the adsorption isotherm results, amitrole and GAC were subsequently selected as an appropriate system for TSA research in terms of thermodynamic behaviours. Obviously, higher adsorption capacity (*qe*) of approximately 10.68 mg/g was measured at lower temperature as shown in figure 2. The adsorption of amitrole by GAC was fitted with the Freundlich equation for each temperature and it was considered as an exothermic as well as a spontaneous process.

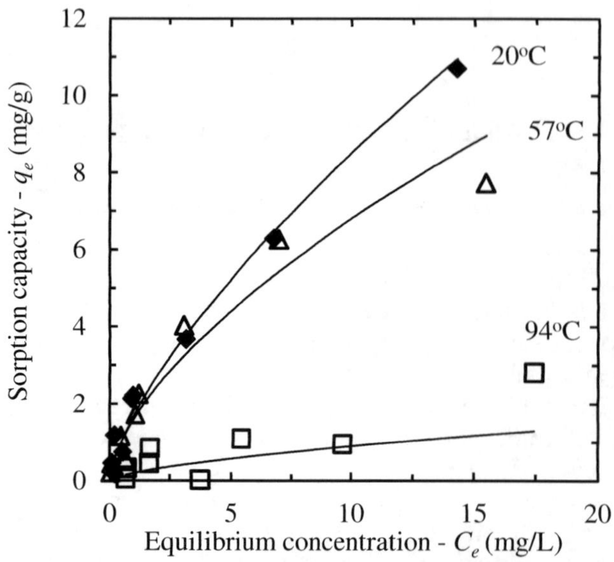

Figure 2. Adsorption isotherms of amitrole by 400 & 4000 mg/L NORIT GAC 830W

On the other hand, particle sizes of the adsorbents also contribute their important roles to the experimental shape of breakthrough curves. In this research, samples of the original NORIT GAC 830 W (particle size: 0.6 – 2.36 mm) as well as milled activated carbon of the same type (particle size: 0.4 – 0.8 mm) were utilized for 10 mg/L amitrole adsorption to clarify the effect of different particle sizes. The obtained data were subsequently demonstrated in figure 3. Specifically, milled samples of NORIT GAC 830 W have shorter mean diffusion length, hence, the equilibrium was obtained faster with smaller sorbent particles. Therefore, significant amount of amitrole solution can completely purified with residual amitrole concentration C_t being virtually zero.

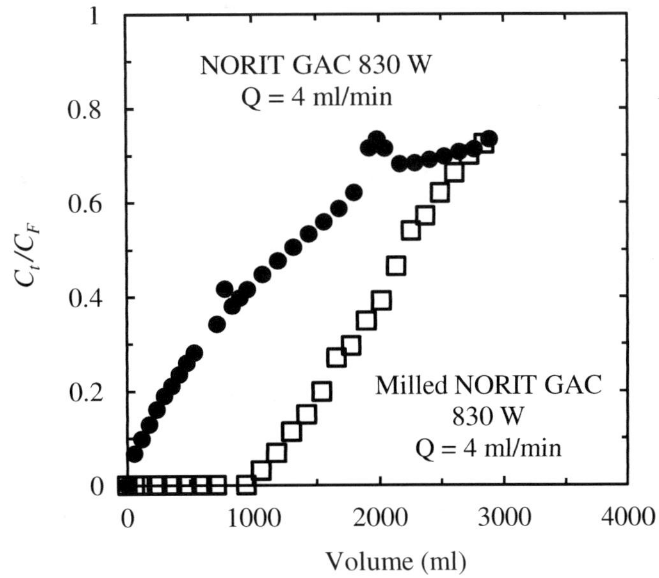

Figure 3. Effects of different particle sizes of GAC to the dynamic adsorption of 10 mg/L amitrole

The most important data of TSA experiments were demonstrated in figure 4. TSA system was successfully operated in aqueous phase with the continuous running of numerous adsorption–desorption cycles. In the displayed adsorption cycles, the removal of amitrole can be high after effective regeneration by temperature swing method. Breakthrough curves suggest that cyclic steady-state was established. However, more studies have been aimed to be conducted in the future to gain the deeper look and improve this process of TSA in aqueous phase.

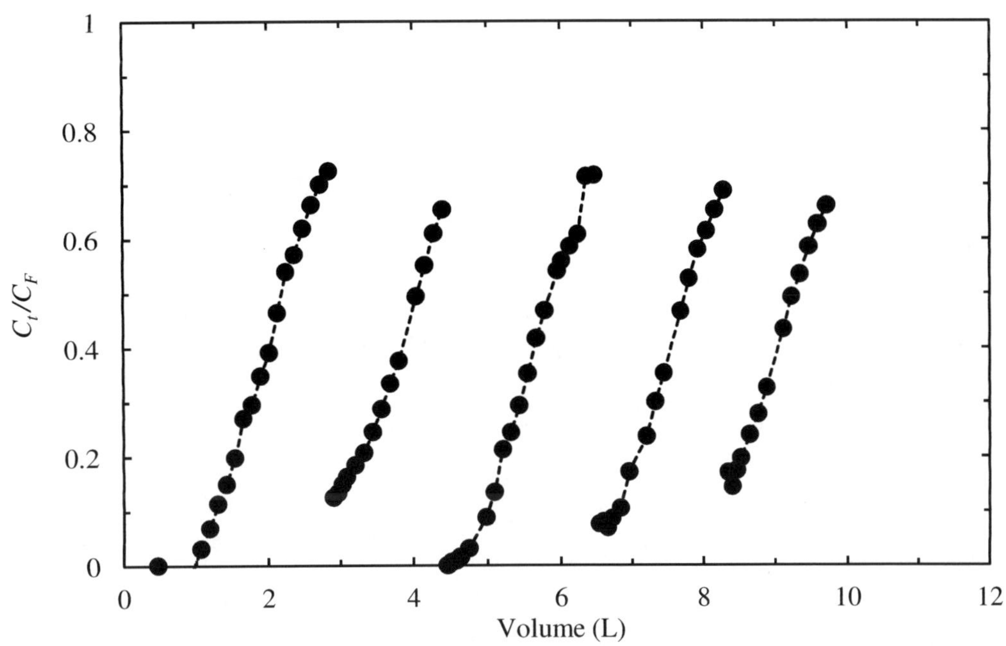

Figure 4. TSA adsorption – desorption cycles for *in situ* adsorbent regeneration at 125°C

4. Conclusions

In this research project, the removal of the amitrole as an important pesticide by adsorption on granular activated carbon (GAC) was thoroughly studied. Initial data was obtained and the feasibility of a TSA system by the use of heat in order to establish a temperature swing adsorption in the liquid phase was proven. Future works will emphasize on different chemical compounds to identify other suitable sorptive – sorbent systems for TSA application. Moreover, TSA technique for household drinking water treatment will be studied thoroughly and in line with the current trend of the combined usage of membrane technology with TSA in liquid phase [9]. On the other hand, breakthrough curve modelling should be implemented to compare with the obtained experimental results to investigate the transport processes happened within the carbon bed depth. Mathematic models to predict the adsorption behaviours which are naturally based on thermodynamic properties should also be proposed. Moreover, the obtained results from this research can be utilized as the input data for adsorption column scaling-up and pilot model design in the future.

5. References

[1] Kamrin MA 1997 *Pesticide Profiles: Toxicity, Environmental Impact, and Fate* (New York: Lewis Publishers)
[2] Crittenden J C 2012 *MWH's Water Treatment: Principles and Design* (Hoboken: John Wiley & Sons)
[3] Usman M R 2015 *Comprehensive Dictionary of Chemical Engineering* (New York: Lulu Publishing)
[4] Ntiamoah A, Ling J, Xiao P, Webley P A, Zhai Y 2016 *Ind. Eng. Chem. Res.* **55** 703–713
[5] Ruthven D M 1984 *Principles of adsorption and adsorption processes* (New York: John Wiley & Sons)
[6] Chowdhury S, Balasubramanian R 2016 *Ind. Eng. Chem. Res.* **55**(29) 7906–7916
[7] Fontecha-Cámara M A, López-Ramón M V, Álvarez-Merino M A, Moreno-Castilla C 2006 *Langmuir* **22** 9586–9590
[8] López-Ramón M V, Fontecha-Cámara M A, Álvarez-Merino M. A, Moreno-Castilla C 2007 *Water Res.* **41** 2865–2870
[9] Crawford S E *et al.* 2017 *Environ. Sci. Eur.* **29**(1) 1–10

Acknowledgments

The authors would like to express sincere thanks to the assistance from Aachener Verfahrenstechnik (AVT) and Thai – German Graduate School of Engineering (TGGS) colleagues, the research funding from the European Research Council (ERC), the financial support of the European Commission (Water4India Project, grant number 308496) and by German Research Foundation DFG Strategy Funds (Project House Water).

IOP Conf. Series: Earth and Environmental Science **159** (2018) 012014 doi:10.1088/1755-1315/159/1/012014

Pollutants Emission Characteristics of an Ultra-low Coal-fired Power Unit

Li Bing[1], Song Yongqiang[2], Zhang Qilong[1], Zhou Can[1] and Duan Haoran[1]

1Shandong Branch of Huadian Electric Power Research Institute, Jinan,China
2Huadian Power International Corporation Limited Zouxian Power Plant, Jining, China
E-mail: 13869187893@163.com

Abstract. The concentration of particulate matter (PM), SO_2, NOx, SO_3, HCl, HF, and PM2.5 were measured in a 330MW ultra low emission coal-fired power unit. The pollutants emission characteristics and the effect of the conventional pollution control devices on non-conventional pollutants were investigated. The test results show that the concentration of NOx, SO_2 and PM are 26.9 mg/m^3, 18.1 mg/m^3, 2.9 mg/m^3 respectively. The coal-fired power plant, equipped with WFGD and WESP could well control SO_3, HCl, HF, and PM2.5 emission from the flue gas. SO_3 removal efficiency of WFGD and WESP are 42.2% and 82.8% respectively. HCl and HF removal efficiency by WFGD are 90.8% and 91.5% respectively. PM2.5 removal efficiency of WESP is 86.9%.

1. Introduction

Coal-fired power plants are a major source of power generation in China so the pollution emission from coal-fired power plants has been paid more and more attention. In 2014, with much attention paid to the pollutant emitted from the coal-fired power plants, Chinese government has put forward the ultra-low emission (ULE) for coal-fired power units. The emissions of dust, sulfur dioxide (SO_2) and nitrogen oxide (NOx) are below 10 milligrams per cubic meter, 35 milligrams per cubic meter and 50 milligrams per cubic meter, respectively. Before 2020, coal-fired power plants should be retrofitted to meet ULE standards.

With the successful cases of ultra-low emission from domestic coal-fired power plants in China, the technology route for ULE retrofitting has been formed gradually [1]-[3]. Taking the dust ultra low emission as an example, fine particulate matter (PM) is a very difficult to be captured. Improvements, such as retrofitting the electrostatic precipitator (ESP) and installing a wet electrostatic precipitator (WESP), could help solve this problem. A low-low temperature ESP or a WESP is the primary dust ULE technology route. At the same time, people have paid more attention to the emission of SO_3, PM2.5 and other unconventional pollutants besides dust, SO_2 and NOx [4].

This paper has measured the emission concentration of particulate matter, SO_2, NOx, SO_3, HCl, HF, and PM2.5 in a 330MW ultra low emission coal-fired power unit, which has been retrofitted to meet ULE standards with low NOx burner (LNB) and selective catalytic reduction (SCR) technology to control NOx emission, a dry ESP and a WESP to control particulate matter emission and wet limestone-gypsum flue gas desulfurization (WFGD) to control SO_2 emission. The pollutants emission characteristics and the effect of the conventional pollution control devices on non-conventional pollutants were investigated.

2. Experiment

2.1. Unit overview

The field measurements were conducted in a 330MW coal fired power plant. This unit adopted LNB and SCR to control NOx emission. The catalyst was arranged in three layers and the reducing agent was liquid ammonia. WFGD was retrofitted in a single tower with a tray and five spraying layers to control SO_2 emission. The design inlet SO_2 concentration was 3000 mg/m^3 and the design desulphurization efficiency was no less than 98.85%. ESP of horizontal type and four electric fields and one WESP arranged in the top of WFGD were retrofitted to control dust emission.

2.2. Test method

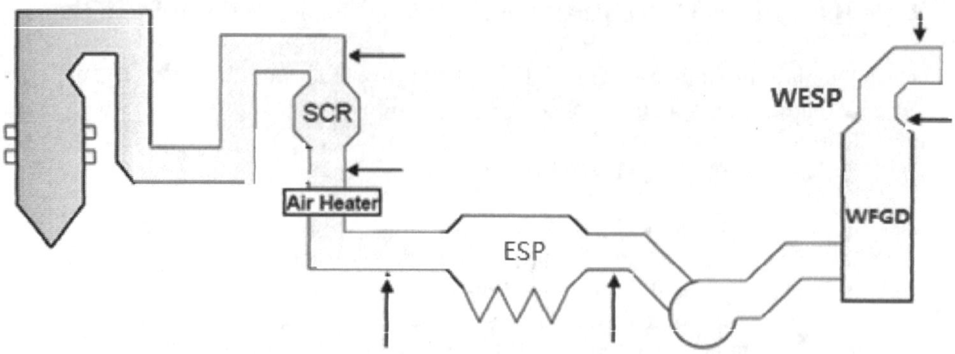

Figure 1. The diagram of measure point.

In this paper, the pollutants has been measured according to National Standards, such as GB/T 16157-1996 (Determination of particulates and sampling methods of gaseous pollutants emitted from exhaust gas of stationary source), DL/T 414-2012 (Specification for environmental monitoring of thermal power plants) and GB/T 13931-2002 (Methods of performance tests for electrostatic precipitators). The measuring points were arranged as shown in figure 1.

2.3. Test condition

During the test, the unit load was at 100% BMCR and remained stable, no more than 5% or minus 5% BMCR. The coal quality remained stable and the results of coal analysis during the test are shown in Table 1.

Table 1. Coal proximate analysis.

Moisture M_{ad} (%)	Ash A_{ad} (%)	Volatile matter V_{daf} (%)	Fixed carbon FC_{ad} (%)	Total sulfur $S_{t,ad}$ (%)	Calorific value $Q_{net,ad}$(MJ·kg^{-1})
1.7	25.84	38.92	44.26	1.05	20.67

3. Results and Discussion

3.1. Emission characteristics of NOx,SO_2 and dsut

Table 2. NOx, SO_2 and dust emission characteristics.

Unit	NOx		SO_2		Dust	
	Concentration (mg·m^{-3})	Performance (g·(kW·h)$^{-1}$)	Concentration (mg·m^{-3})	Performance (g·(kW·h)$^{-1}$)	Concentration (mg·m^{-3})	Performance (g·(kW·h)$^{-1}$)

330MW	26.9	0.079	18.1	0.053	2.9	0.009

The emission concentration of NOx, SO_2 and dust at the chimney inlet are shown in table 2. The NOx concentration is 314.9 mg/m^3 at the SCR inlet and that at the chimney inlet is 26.9 mg/m^3. The SO_2 concentration is 2500.3 mg/m^3 at the WFGD inlet and that at the chimney inlet is 18.1 mg/m^3. The dust concentration is 33.2 g/m^3 at the ESP inlet and that at the chimney inlet is 2.9 mg/m^3. The emission concentration of NOx, SO_2 and dust has met ULE standards respectively. The emission performance of NOx, SO_2, and dust is 0.079 g/ (kW·h), 0.053g/ (kW·h) and 0.009 g/ (kW·h), respectively.

3.2. Emission characteristics of non-conventional pollutant

3.2.1.SO_3. Sulfur trioxide (SO_3) is formed in the furnace and convective pass of coal-fired boilers, and the rate of formation depends strongly on coal properties, boiler design and operation conditions. The parameters that influence the amount of SO_3 formation in the boiler are fuel sulfur content, ash content and composition, excess air level, convective pass surface area, and tube metal surface temperature distribution[5, 6]. For boilers equipped with SCR reactors to control NOx emission, additional amounts of SO_3 are formed in the SCR reactors, with rate of formation depending on parameters such SCR design and reactor temperature. However, there are undesirable impacts of SO_3, including corrosion and fouling of air preheater, corrosion of flue duct and formation of acid mist in the stack plume [7].

The collaborative control of SO_3 by WFGD and WESP is shown in figure 2. The results show that SO_3 can be simultaneous removed by WFGD and WESP. The SO_3 concentration is 32.2 mg/m^3 and 18.6 mg/m^3 at the WFGD inlet and outlet, respectively. The SO_3 removal efficiency by WFGD is 42.2%. Once formed in the boiler, the SO_3 reacts with H_2O vapor to form H_2SO_4 vapor in the cold end of the air preheater. When the H_2SO_4 vapor-containing flue gas passes through a WFGD system, it is rapidly cooled below the acid dew point. Because the rate of this cooling is greater than the rate of adsorption of H_2SO_4 vapor in the scrubber liquid, that results in H_2SO_4 mist with sub-micron droplets. In a WFGD system, the mass transfer between the scrubbing liquid and H_2SO_4 mist is lower, which occurs through inertial impaction, gravitational settling, Brownian diffusion, and other effects. In general, larger H_2SO_4 mist droplets can be removed by WFGD, but a significant portion of sub-micron mist droplets are not removed and emitted from WFGD, so, ~30% of H_2SO_4 may be removed by WFGD.

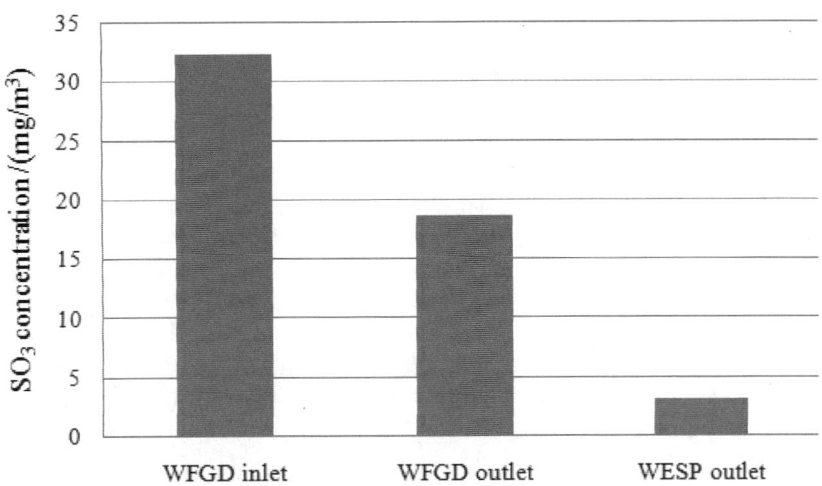

Figure 2. The collaborative control of SO_3 by WFGD and WESP.

For sub-micron H_2SO_4 mist droplets with diameters less than 0.05 μm, Brownian diffusion is known to

IOP Conf. Series: Earth and Environmental Science **159** (2018) 012014 doi:10.1088/1755-1315/159/1/012014

be the primary component of mass transfer. In a WFGD scrubber, the mass transfer depends on the relative velocity between the H_2SO_4 mist droplets and the scrubbing liquid. In the WFGD system of this 330MW unit, a tray was retrofitted. The uniformity of flue gas distribution is improved and the mass transfer is enhanced, and then the SO_3 removal efficiency is improved, reached 42.2%.

WESP has been developed to control a wider variety of particulate pollutants and exhaust gas conditions compared to dry ESP, especially for particles that are sticky, corrosive, or have high resistivity. The problem of acid plume in coal-fired power plant can be solved by installing a wet ESP after a FGD system.The SO_3 concentration is 3.2 mg/m^3 at the WESP outlet and the SO_3 removal efficiency by WESP is 82.8%. The wet environment in the WESP lowers particle resistivity and allows high power input levels to enhance the removal of sub-micron mist, so the sub-micron H_2SO_4 mist is collected by electrostatic forces in a WESP [6].

3.2.2.HCl and HF. The collaborative control of HCl and HF by WFGD are shown in figure 3.Chlorine and fluorine are among the most volatile elements according to trace elements partitioning behavior during coal combustion. The release rate of chlorine and fluorine from coal during combustion in boiler is above 96.6% [8]. Chlorine and fluorine in coal are released as hydrogen chloride and hydrogen fluoride during coal combustion, respectively [9]. The removal rate of hydrogen chloride and hydrogen fluoride by WFGD are 90.8% and 91.5% respectively, because hydrogen chloride and hydrogen fluoride are water-soluble and chemical reaction occurs with the scrubbing liquid in WFGD system, and then chloride and fluorine are most transferred into desulfurization gypsum and desulfurization wastewater.

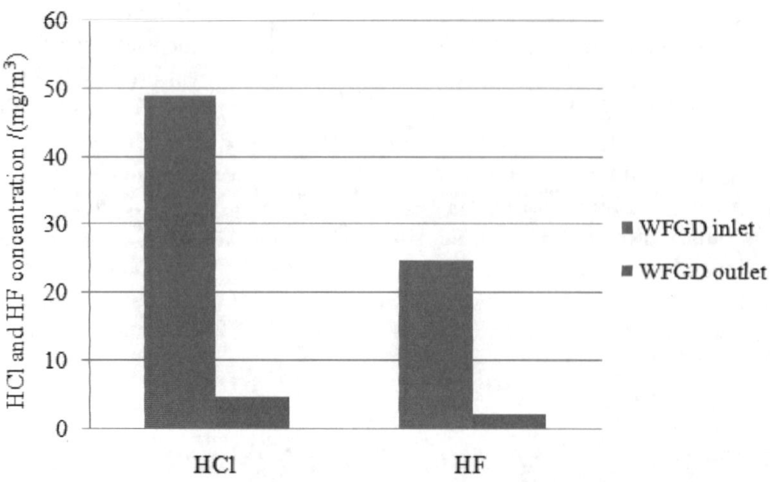

Figure 3. The collaborative control of HCl and HF by WFGD.

3.2.3.PM2.5. The ESP is the primary PM control devices in Chinese coal-fired power plants. ESP can effectively capture large particles, but they can do little for fine particles such as PM2.5 or lower. Dry ESPs use electrostatic forces, but charging of fine particles smaller than 1 μm is inefficient, and the collection efficiency sharply drops with decreasing particle size. The fine particles need to be effectively captured due to stronger corona power generated than in traditional ESP. The limits on charging mechanisms for fine particulates are exaggerated in dry ESP due to lower corona power levels caused by the resistivity of the ash layer that accumulates on the collecting surfaces. The re-entrainment of fine particles from the last field collection electrode can also be observed, which also lower the dust collection efficiency

The flue gas condition after WFGD is that the moisture content in flue gas is high and the temperature of flue gas is close to the dew point temperature. Compared with a dry ESP, WESP is particularly suitable in the flue gas condition after WFGD. WESP may provide much higher average corona power levels compared to dry ESP, because they are constantly being cleaned of resistive ash by flowing water to prevent ''back corona''. This effect enhances collection of sub-micron particles through more effective particle charging. At the same time, the periodic or continuous scrubbing water flow, used to wash the collection electrode surfaces, is found to prevent particle re-entrainment caused by rapping, which occurs in dry ESPs.

The collaborative control of PM by WESP is shown in figure 4. The proportion of PM2.5 to PM at inlet and outlet of WESP are 67.7% and 75.9%, respectively. The proportion of PM2.5 increases as flue gas passes through a WESP system. The removal efficiency of PM and PM2.5 by WESP are 88.4% and 86.9% respectively. The wet environment in the WESP lowers particle resistivity and allows high power input levels to enhance the removal of micron-diameter particulate matter by electrostatic forces in a WESP

.

4. Conclusion

The pollutants emission characteristics and the effect of the conventional pollution control devices on non-conventional pollutants in a 330MW coal-fired power unit were investigated. The test results show that the concentrations of NOx, SO_2 and dust have met ULE standards. The coal-fired power unit equipped with WFGD and WESP could well control SO_3, HCl, HF, and PM2.5 emission from the flue gas. SO_3 removal efficiency of WFGD and WESP are 42.2% and 82.8% respectively. HCl and HF removal efficiency of WFGD are 90.8% and 91.5% respectively. PM2.5 removal efficiency of WESP is 86.9%.

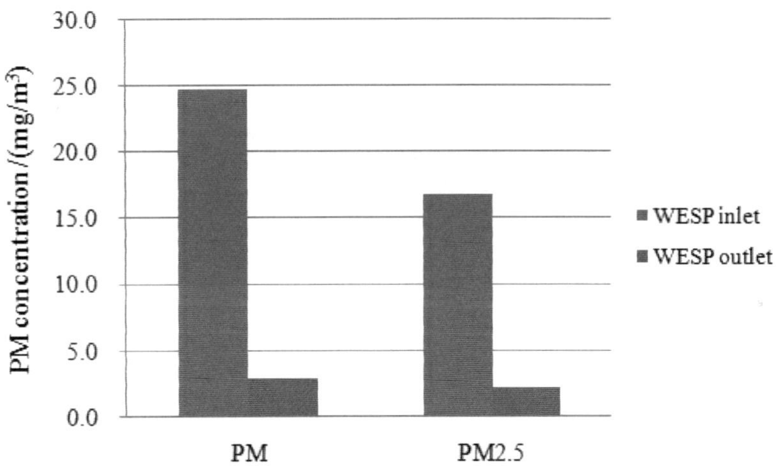

Figure 4. The collaborative control of PM and PM2.5 by WESP.

5. References

[1] Zhu F H 2017*Electric Power* **50** 11
[2] Zhao Y C, Ma S M,Yang J P,Zhang J Y and Zheng C G 2015 *J. Chin Coal. Soc.* **40** 2629
[3] Zhang J, Zheng C H, Zhang Y X, Wu G C, Zhu S Q, Meng W, Gao X and Chen K F 2016 *Proc. Chin. Soc. Electrical. Eng.* **36** 1310
[4] Xu Y Y, Xue J M, Wang H L, Li B, Guan Y M and Chen S J 2014 *Proc Chin Soc Electrical Eng.* **34** 3924
[5] Lou Q G 2008 *Energy Engineering* **6** 46
[6] Srivastava R K, Miller C A, Erickson C and Jambhekar R 2004 *Air & Waste Manage. Assoc.* **54** 750
[7] Flerg D, Alzueta M U, Normann F, Abian M, Andersson K and Johnsson F 2013 *Combustion. Flame* **160** 1142
[8] Deng S, Liu Y, Zhang C, Wang X F, Cao Q, Wang H M and Zhang F 2014 *R Environ* Sci. **27** 225
[9] Li B, Wang H L, Xu Y Y, Xue J M and Guan Y M 2015 *J. Chin Coal. Soc.* **40** 2479

Acknowledgments
This work has been supported by HUADIAN POWER INTERNATIONAL 2017 Science and Technology Project (HDPIKJ17-02-12).

Pretreated Fruit Peels as Adsorbents for Removal of Dyes from Water

Nguyen Thi Hong Nhung[1*], Bui Thi Phuong Quynh[2*], Pham Thi Thu Thao[1], Hoang Ngoc Bich[1] and Bach Long Giang [3]

1 Faculty of Food, Chemical and Environmental Sciences, Nguyen Tat Thanh University, Ho Chi Minh City, Vietnam
2 Faculty of Chemical Technology, Ho Chi Minh City University of Food Industry, Ho Chi Minh City, Vietnam
3 NTT Institute of Hi-Technology, Nguyen Tat Thanh University, Ho Chi Minh City, Vietnam
Email: nthnhung@ntt.edu.vn; quynhbtp@cntp.edu.vn

Abstract. This research focused on the development of adsorbents based on cheap, abundant and locally available agricultural wastes in Viet Nam to remove dyes from aqueous solution. In details, the effects of initial concentration (100-600 ppm), initial dose of adsorbent (0.1-1.5 g) and contacting time on the adsorption behaviors of Rhodamine B onto different kinds of pretreated fruit peel wastes, including orange peels, pomelo peels, passion-fruit peels, were evaluated. It was found that these bio-peels efficiently adsorbed in a short time, within one hour. The dye removal efficiency of orange peels and passion fruit peels ranged from 82-92% while that of pomelo peels was faster with 91-94% depending on the initial concentration (200 ppm). The efficiency of dye removal increases significantly as adsorbent dosage increases, passion-fruit peels 43-98%, pomelo peels 67-94%, orange peels 57-93%. Equilibrium adsorption results obeyed well the Freundlich isotherm; according the adsorption capacity was determined in the order: passion-fruit peels > orange peels > pomelo peels.

1. Introduction

One of the major problems faced by many countries around the world is the increase in industrial activities releases effluents containing pollutants such as heavy metal ions, organic dyes and pharmaceuticals into the aquatic environment, which cause significant health hazards to living organisms and overall deterioration of the environment [1]. Special, the treatment of dyed water is not only important but also challenging task that many countries in the world have to face. In Viet Nam, progressive activities of dye consuming and manufacturing such as textile, leader, paper, petroleum, printing, cosmetics, paint, rubber, plastic and food processing has generated increasing colored effluents containing organic dyes and other toxic compounds [2]. Because most of the organic dyes molecular pollutants are stable, small in size and not biodegradable, it is difficult to eliminate them from wastewater [3]. Many wastewater treatment methods such as chemical coagulation, photodegradation, precipitation, flocculation, activated sludge, adsorption, absorption, membrane separation and ion exchange processes have been tested for removing the pollutants [1], [4]-[6].Among various treatment methods suggested for colored effluents [2], [7], the adsorption is most favored due to its easy operation, sufficient efficiency, reasonable cost and feasibility for large-scale applications.

There are a variety of adsorbents developed for dye removals that have been studied. Activated carbons are used most widely but their applications are quite limited due to its relative high cost. In recent years, its has been found that the cheap, highly abundant agriculture wastes could be transformed to very efficient adsorbents to purify polluted water. Adsorption capacities of many different kinds of agriculture wastes, such as fruit peels, sugar cane bagasse, avocado, hami melon, dragon fruits and rice husk for a number of hazardous substances have been widely study [7]-[9]. Owing to the multifunctional characteristics of bio-peels, the adsorption of pollutants is facilitated by a combination of processes, involving ion exchange, complexion and electrostatic interactions [10], possibly low-cost and efficient adsorbents targeted for remediation of dye-polluted water. In this study, we investigated the possibility of removing organic dyes, the rhodamine B, using three kinds of tropical fruit peels including orange, pomelo, passion-fruit peels. The adsorption behaviors were described using Langmuir and Freunchlich isotherm models.

2. Materials and methods

2.1. *Chemicals and instruments*

The synthetic dyes were purchased from HiMedia Laboratories Pvt. Ltd. Sodium hydroxide (NaOH) and hydrochloric acid (HCl) were purchased from Xilong Chemmical Co, Ltd. The morphology of material was identified by scanning electron microscope (SEM) technique on the instrument S4800, Japan. The IR spectra were recorded in the range of 4000-400 cm^{-1} using a Bruker ALPHA FT-IR (Fourier transform infrared) spectrophotometer using KBr as a matrix. The concentrations of organic dyes in the solution were analyzed using an UV-VIS spectrophotometer (Shimadzu-1601 PC spectrophotometer).

2.2. *Preparation of adsorbents*

The bio-peel wastes, including orange peels, pomelo peels, passion-fruit peels were collected from the local market in Ho Chi Minh City, Viet Nam. The raw peels were cut into small pieces of about 2x2mm, washed with distilled water to remove adhering substances and dust followed by drying in the oven 8°C. The peel wastes were then treated by 0.01 M NaOH and 2-propanol solution [11]. After all, the pretreated bio-peels were washed thoroughly with distilled water and dried in the oven at 80°C for 24h.

2.3. *Batch adsorption studies*

In batch adsorption experiments, the treated peels were added into 250-mL Erlenmeyer flasks containing 100mL solution of known dye concentrations. The dye concentrations were studied from 100 to 600 ppm at a constant dried adsorbent dosage of 10g/L. All adsorption experiments were performed at ambient temperature (30 ± 2°C) using an orbital shaker operating at 180 rpm until obtaining equilibrium state in 1 hour. The mixture after adsorption was separated by centrifugation at 4000 rpm. The dye concentration (ppm) was determined via absorbance at a constant wavelength (554nm) using an UV-VIS spectrophotometer. The calibration curves were constructed by plotting the absorbance of each single dye against dye concentration at the fixed wavelength.
The percentage removal of pollutant was calculated as follows:

$$Removal\ efficiency\ \% = \left(1 - \frac{C}{C_o}\right).100 \quad (1)$$

Where C_o and C are the initial and residual dye concentration (ppm), respectively.
The dye uptake at equilibrium q_e (mg dye/g sorbent) was calculated as follow:

$$q_e = \frac{C_o - C_e}{m_a}.V \quad (2)$$

Where C_o and C_e are the intial and equilibrium dye concentrations (ppm), respectively, V is the volume of solution (L), and m_a is the adsorbent dosage (g).

2.4. *Adsorption isotherms*

The adsorption models are powerful tools to give some insights into adsorption process and to derive important parameters revealing adsorption mechanism. Among a number of isotherm models, Langmuir and Freundlich isotherms are the most commonly used models [12].

The Langmuir model is based on the assumption that monolayer adsorption occurs on a homogeneous surface with a finite number of adsorption sites and negligible mutual interactions between the adsorbed molecules [13, 14]. The linear form of Langmuir equation is given as:

$$\frac{1}{q_e} = \frac{1}{q_m K_L C_e} + \frac{1}{q_m} \tag{3}$$

Where: C_e (mg/L) and q_e (mg/g) are equilibrium concentration and adsorption capacity, respectively. q_m (mg/g) and K_L (L/mg) are the maximum adsorption capacity and rate of adsorption (Langmuir constants), respectively.

The essential characteristics of the Langmuir isotherm can be interpreted via the dimensionless equilibrium parameter (R_L) which is defined by:

$$R_L = \frac{1}{1 + K_L C_o} \tag{4}$$

Where: K_L (L/mg) is the Langmuir constant and C_o (mg/L) is the highest initial concentration. The type of the adsorption isotherm is determined by the value range of R_L: unfavorable ($R_L > 1$), linear ($R_L = 1$), favorable ($0 < R_L < 1$) and irreversible ($R_L = 0$) [9], [15].

The Freundlich model assumes that molecules are adsorbed on the heterogeneous surfaces of adsorbent based on different sites with different adsorption energies [16]. This model takes into account the mutual interaction between adsorbed molecules. The application of the Freundlich equation also suggests that sorption energy exponentially decreases upon the completion of the sorption centers of the adsorbent. The linear form of Freundlich equation is given as:

$$ln q_e = ln K_F + \frac{1}{n} ln C_e \tag{5}$$

Where: $1/n$ and K_F [(mg/g).(L/mg) $1/n$] are Freundlich constants related to the favorability of adsorption process and the adsorption capacity of the adsorbent, respectively. The heterogeneity factor, $1/n$ expresses adsorption intensity of adsorbent; the adsorption bond is stronger with decreasing $1/n$ value.

3. Results and discussion

3.1. *Characterization of biosorbents*

The functional groups on the peels were studied using FT-IR spectrophotometry in the range of 4000 to 400 cm^{-1} (Figure 1). The broad band observed at 3076-3580 cm^{-1} for all three peels matches the O-H stretching frequency, it can originate from intermolecular hydrogen bonds linking the compounds and alcohols, phenols and carboxylic acids. The strong adsorption of about 763-1776 cm^{-1} in accordance with the frequency of functional groups lasted aromatic C-H and C=O groups linked compounds aldehydes, ketones, esters, amides, carboxylic acids. The peak at 1055cm-^{1} clearly shows the presence of the C-O group in the carboxylic and alcoholic. Peaks observed at around 2920 cm^{-1} can be attributed to the C-H stretching frequency. The relatively wide central peak of about 1664 cm^{-1} stems from the extension of the C-C aromatic ring. The adsorption peaks in the range of 1000-1300 cm^{-1} can be considered for angular deformation in the bonding plane of C-H aromatic rings. In addition, consecutive peaks in the 3672-4000 cm^{-1} range may be assigned to the N-H bond. It can be seen that the surface of biosorbents contains functional groups

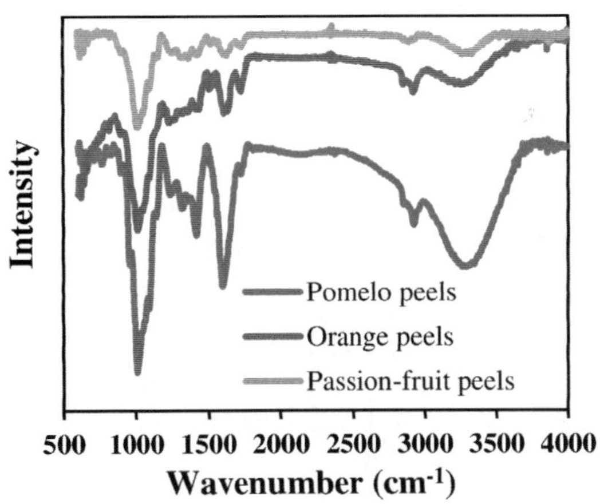

Figure 1. FT-IR spectra of pomelo, orange and passion-fruit peels.

such as hydroxyl and carboxyl groups necessary for the adsorption of dye molecules [17], [18]. In the SEM micrographs, all bio-peels possessed the typical structure of plant based cellulosic materials with highly rough, and irregular structure according to SEM analysis (Figure 2).

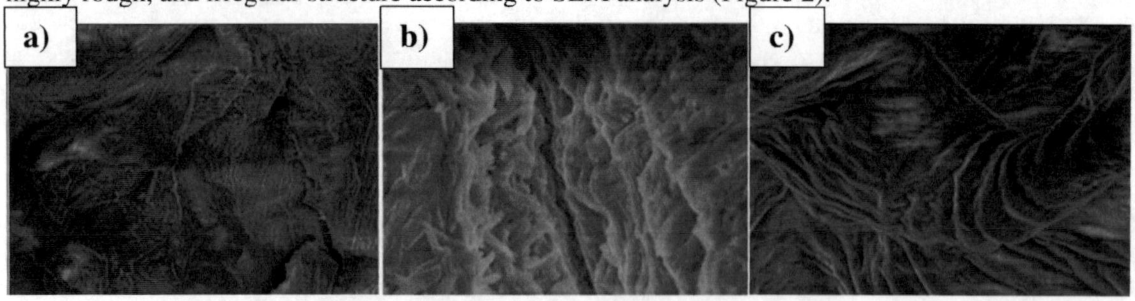

Figure 2. SEM images of the surface treated peels: (a) orange, (b) passion-fruit and (c) pomelo.

3.2. Effect of initial adsorbent dosage

Rhodamine B dye solutions (100 mL) of concentrations (200ppm) were mixed with passion-fruit, pomelo, orange peels of different dosage (0.1-1g) in separate vials. Experimental factors such as the adsorbate concentration used, agitation speed and temperature were kept the same. Figure 3 show for all three kinds of peels, the percentage removal reached the highest level at 1.5g, decreased slightly at 1g and 0.5g but rapidly dropped upon further decrease in dosage to 0.1g. The passion-fruit peels appeared as the best adsorbent among three kinds of bio-peels; the removal efficiencies were determined to be 42.4%, 94.3%, 96.8% and 97.9% at 0.1g, 0.5g, 1g and 1.5g respectively. The pomelo peels and orange peels were as efficient as passion-fruit peels in removal effect of rhodamine B dye over 90% with initial adsorbent dosage over 0.5g.

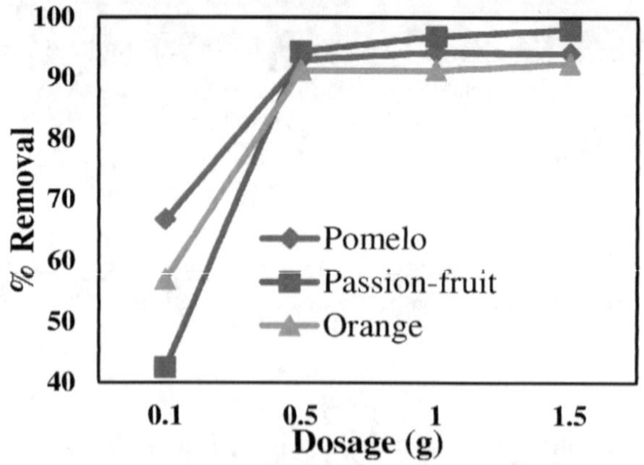

Figure 3. Effect of initial adsorbent dosage on removal efficiency using treated fruit peels

3.3. Effect of initial adsorbate concentration

In this study, rhodamine B dye solutions (100 mL) of different concentrations (100-400ppm) were mixed with three kinds bio-peels (1g) in separate vials. Effect of initial rhodamine B concentration on removal efficiency using treated fruit peels is presented in Figure 4. For all three kinds bio-peels in removal effect over 90% with initial adsorbate concentration 100-400 ppm, highest at 100 ppm (pomelo peels 98.7%, passion-fruit peels 98.3%, orange peels 92.4%) and decrease slightly at 200-400 ppm. The efficiency of the fruit peels decreased sharply when the concentration increased to 600 ppm (pomelo peels 57.4%, passion-fruit peels 56.2%, orange peels 55.9%)

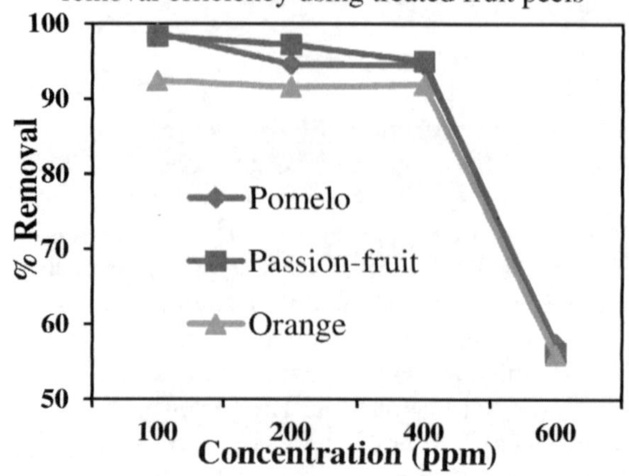

Figure 4. Effect of initial rhodamine B concentration on removal efficiency using treated fruit peels

3.4. Effect of contact time

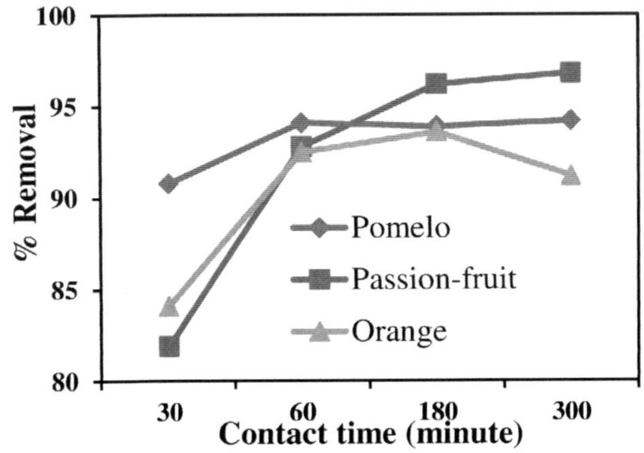

Figure 5. Effect of contact time on removal efficiency using treated fruit peels

This experiment was conducted with different contact time (30-300 minutes), rhodamine B dye solutions (100 mL) of concentrations (200ppm) were mixed with passion-fruit, pomelo, orange peels of different dosage (1g) in separate vials. Figure 5 show effect of contact time on removal efficiency using treated fruit peels. The highest efficiency of passion-fruit peels is 96.8% at 300 minutes, pomelo peels is 94.2% at 300 minutes and orange peels is 93.6% at 180 minutes. Effective treatment of passion-fruit peels increases over contact time, but effective treatment of pomelo and orange peels is unstable over contact time. When the contact time over 60 minutes the removal efficiency of three kinds fruit peels is over 90%.

3.5. Effect of pH on the adsorption of different dyes.

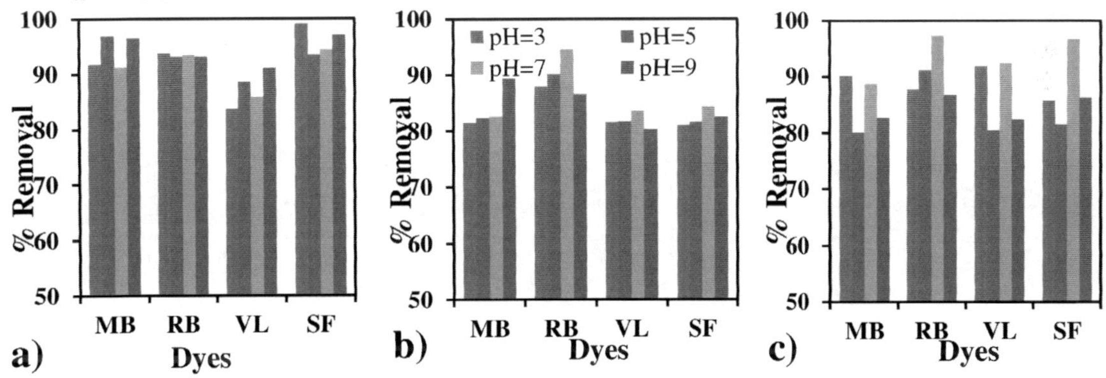

Figure 6. Effect of pH on the adsorption of different dyes:(a) orange, (b) pomelo, (c) passion-fruit peels with MB (Methylene Blue), RB (Rhodamine B), VL (Methyl Violet), SF (Safranine) dyes.

The extraction data for pollutants using different peels at various solution pH values are shown in Figure 6. The percentage removal of pollutants changed when changing pH value, owing to the efficient exchange of cations with functional groups on the surface treated peels. All three bio-peels were effective toward removing dyes from water, reaching removal rates of 80–99%.

3.6. Isotherm studies

The experimental data were fitted with Langmuir and Freundlich isotherm models; the model constants were determined as provided in Table 1. It was found that Freundlich model best describes the adsorption behavior of rhodamine B on the bio-peels with high R^2 values (0.9131-0.9967) while the fitting to Langmuir model exhibited lower compatibility. The plot of $1/C_e$ against $1/q_e$ in Langmuir model is shown in Figure 6a with the highest correlation coefficient (R^2=0.9967) obtained for the orange peels. According to the Langmuir isotherms, the maximum adsorption capacities were found in the order: orange peels (312.5 mg/g) > passion-fruit peels (51.81 mg/g) > pomelo peels (38.31 mg/g). Moreover, for all three kinds of bio-peels, the R_L values were lower than 1 revealing favorable adsorption processes.

IOP Conf. Series: Earth and Environmental Science **159** (2018) 012015 doi:10.1088/1755-1315/159/1/012015

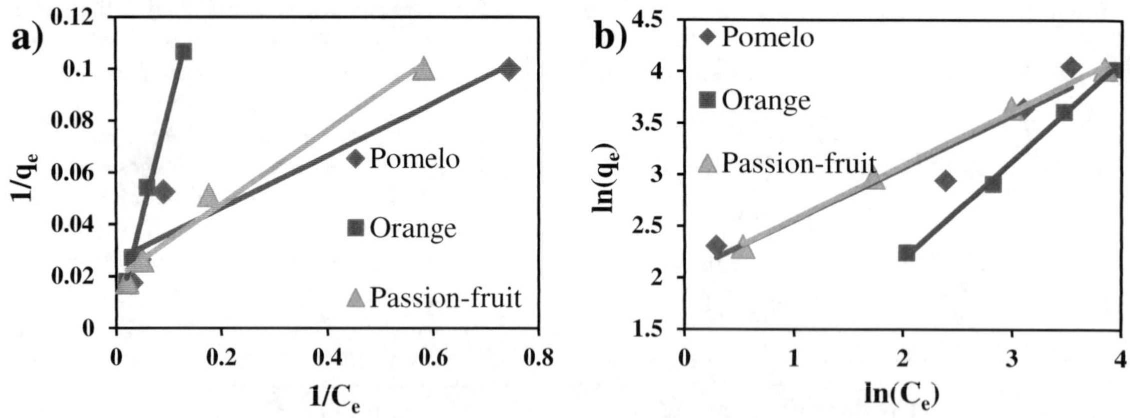

Figure 7. (a) Langmuir isotherm plots; (b) Freundlich isotherm plots

Table 1. Langmuir and Freundlich model constants

Types of peels	Langmuir isotherm				Freundlich isotherm		
	q_{max} (mg/g)	K_L(L/mg)	R^2	R_L	K_F	n	R^2
Pomelo	38.31	0.2587	0.8895	0.0064	7.7037	1.9573	0.9131
Passion-fruit	51.81	0.1359	0.9834	0.0121	7.7029	1.9091	0.998
Orange	312.5	0.0040	0.9967	0.2947	1.2896	1.0413	0.9979

3.7. Characterization of pretreated fruit peels after the removal dyes

To identify the mechanism of adsorption of dyes ions on the adsorbent, FTIR spectra were recorded before and after rhodamine B dyes adsorption as presented in Figure 8. The spectrum of three kinds bio-peels reveals the presence of characteristic peaks as follows: 3076-3580 cm^{-1} (stretching vibration of OH), 2920 cm^{-1} (stretching vibration of C-H in CH, CH_2 and CH_3), 1664 cm^{-1} (stretching vibration of C=O), 1385 cm^{-1} (blending and scissor vibration of CH_3 and CH_2), 1048 cm^{-1} (stretching vibration of C-O), 790 cm^{-1} (deformation vibration of the CH_2 bond). During adsorption process, there were some interactions between the bio-peels and dyes ions in the aqueous phase, consequently resulting in shifts of the characteristic peaks. As observed in Figure 8c, the peak at 3309 cm^{-1} before adsorption shifted towards 3460 cm^{-1} after adsorption possibly due to the bonding of OH^- with dyes ions. Moreover, in Figure 8b the shifts of 848, 1647, 2931, 3479 cm^{-1} to 1035, 1652, 2920, 3344 cm^{-1}, respectively, can be easily recognized. These shifts might be ascribed to electrostatic interaction between the π electrons of bio-peels layers and dyes ions.

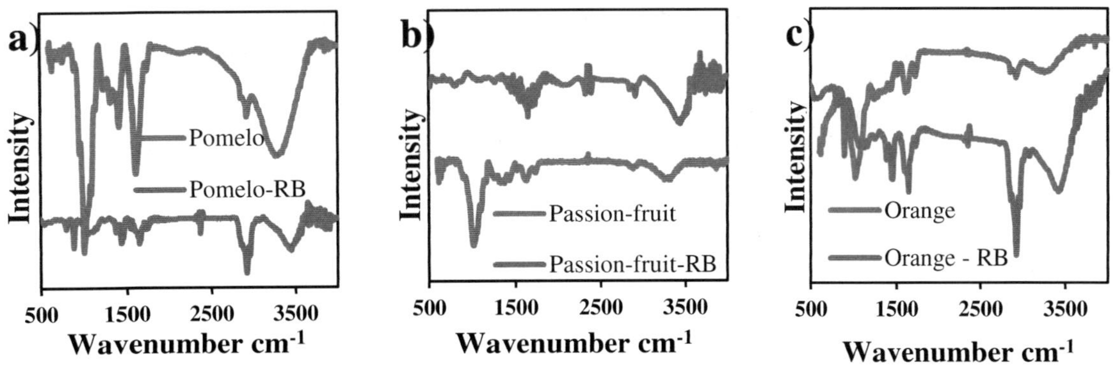

Figure 8. FT-IR spectra of pomelo, passion-fruit and orange peels after removal rhodamine B dye.

4. Conclusions

Three kinds of tropical fruit peel wastes, pomelo, orange and passion-fruit peels, were investigated as potential bioadsorbents for the extraction of dyes in water. All three natural materials were effective toward removing dyes from water, reaching removal rates of 80–99%. Orange peels showed the highest extraction efficiency, close to 99%, toward safranine at pH=3 values. The adsorption capacity of the dyes increased with increasing contact time and a plateau was reached at equilibrium at 180 minute, implying that the adsorption process was rapid in the beginning, gradually decreased as time progressed and finally attained saturation when equilibrium was reached. The Freundlich isotherm model fitted best the adsorption process, the Langmuir isotherm model shown the maximum adsorption capacities were found in the order: orange peels (312.5 mg/g) > passion-fruit peels (51.81 mg/g) > pomelo peels (38.31 mg/g). The data indicate a monolayer adsorption occurrence at binding sites on the surface of the peels. All three peels could be used as nontoxic, renewable, low cost, efficient and effective adsorbents for water purification.

5. References

[1] Mohanty, K., D. Das, and M.N. Biswas, *Adsorption of phenol from aqueous solutions using activated carbons prepared from Tectona grandis sawdust by ZnCl2 activation.* Chemical Engineering Journal 2005. **115**(1-2): p. 121-131.

[2] Roy, A., et al., *Adsorption of Anionic-Azo Dye from Aqueous Solution by Lignocellulose-Biomass Jute Fiber: Equilibrium, Kinetics, and Thermodynamics Study.* Industrial & Engineering Chemistry Research, 2012. **51**(37): p. 12095-12106.

[3] Castro, E., A. Avellaneda, and P. Marco, *Combination of Advanced Oxidation Processes and Biological Treatment for the Removal of Benzidine-Derived Dyes.* Environmental Progress & Sustainable Energy, 2014. **33**(3): p. 873-885.

[4] F.Veglio'* and F.Beolchini, *Removal of metals by biosorption: a review.* Hydrometallurg, 1997. **44**(3): p. 301 -316.

[5] Kumar*, U. and M. Bandyopadhyay, *Sorption of cadmium from aqueous solution using pretreated rice husk.* Bioresource Technology, 2006. **97**(1): p. 104-109.

[6] Nawar, S.S. and H.S. Doma, *Removal of dyes from effluents using low-cost agricultural by-products.* The Science of the Total Environment, 1989. **79**(3): p. 271-279.

[7] M.A.Barakat*, *New trends in removing heavy metals from industrial wastewater.* Arabian Journal of Chemistry, 2011. **4**(4): p. 361–377.

[8] AdibYahya, M., Z.Al-Qodah*, and C.W.ZanariahNgah, *Agricultural bio-waste materials aspotential sustainable precursors used foractivated carbon production: A review.* Renewable and Sustainable Energy Reviews, 2015. **46**: p. 218–235.

[9] Mallampati, R., et al., *Fruit Peels as Efficient Renewable Adsorbents for Removal of Dissolved Heavy Metals and Dyes from Water.* ACS Sustainable Chemistry Engineering, 2015. **3**(6): p. 1117–1124.

[10] Guangfu, X., S. Zhaoxi, and G. Ruixin*, *The kinetics studies for the adsorption of furadan from aqueous solution by orange peel.* Advanced Materials Research, 2014. **842**: p. 187-191.

[11] M.A.Martín*, et al., *Biomethanization of orange peel waste.* Bioresource Technology, 2010. **101**(23): p. 8993–8999.

[12] Ghorban, F., et al., *Application of response surface methodology for optimization of cadmium biosorption in an aqueous solution by Saccharomyces cerevisiae.* Chemical Engineering Journal, 2008. **145**(2): p. 267–275.

[13] Esfandiar, N., B.N. *, and T. Ebadi, *Removal of Mn(II) from groundwater by sugarcane bagasse and activated carbon (a comparative study): Application of response surface methodology (RSM).* Journal of Industrial and Engineering Chemistry, 2014. **20**(5): p. 3726-3736.

[14] Asfara, A., et al., *Rapid removal of Auramine-O and Methyleneblue by ZnS: Cu nanoparticles loaded on activated carbon : Aresponse surface methodology approach.* Journal of theTaiwan Instituteof Chemical Engineers, 2015. **53**: p. 1-12.

[15] C.Namasivayam, et al., *Removal of dyes from aqueous solutions by cellulosic waste orange peel.* Bioresource Technology, 1996. **57**(1): p. 37-43.

[16] Han, Y., et al., *Adsorption behavior of Rhodamine B on nanoporous polymers.* RSC Advances, 2015(127).

[17] Sugumaran, P., et al., *Production and Characterization of Activated Carbon from Banana Empty Fruit Bunch and Delonix regia Fruit Pod.* Journal of Sustainable Energy & Environment, 2012(3): p. 125-132.

[18] Chandra, T.C., et al., *Activated carbon from durian shell: Preparation and characterization.* Journal of the Taiwan Institute of Chemical Engineers, 2009. **40**(4): p. 457-462.

IOP Conf. Series: Earth and Environmental Science **159** (2018) 012016 doi:10.1088/1755-1315/159/1/012016

Synergistic Studies on the Removal of Dyes by Waste (*Metroxylon sagu*) Based Activated Carbon: an Optimization Study

JO Amode[1], Md Z Alam[2], AH Mirza[1], CC Mei[1] and JH Santos[1*]

1 Faculty of Science, Universiti Brunei Darussalam, Jalan Tungku Link, Gadong, Brunei Darussalam
2 Bioenvironmental Engineering Research Center (BERC), Faculty of Engineering, International Islamic University Malaysia (IIUM), Malaysia
Email: jose.santos@ubd.edu.bn

Abstract. Physically activated carbon prepared from untreated *(Metroxylon sagu)* waste was used as adsorbents for the adsorption studies of methylene blue (MB) and gentian violet (GV) dye in aqueous solution. Response Surface Methodology (RSM) was employed to investigate the different operating parameters on the uptake of the dye molecule. The adsorbent before and after activation process were characterized to determine the specific surface area, iodine value, textural structure, and surface functional group. The results predicted using CCD model showed high values of regression coefficients R^2 (MB= 0.9944 and GV= 0.9845) indicating good agreement with experimental data. Characterization of PAC produced also showed that the activation conditions would find good-quality adsorbent with the highest BET surface area of 1423.5 m^2/g, iodine value of 1275.2 mg/g and well-forming pores distribution.

1. Introduction

The discharge of dyes in the environment is worrying for both toxicological and esthetical reasons for the fact that the adverse effect of these dyes may be responsible for permanent injury to humans and animals because they are found to be a mitotic poisoning agent, which is recalcitrant and carcinogenic, and thus regarded as a biohazard [1], [2]. Thus, the removal of these dyes from effluents is essential not only to protect human health but also for the protection of water resources. Hence attempts are made in this study to develop an inexpensive adsorbent for removal of MB and GV from aqueous solutions using low cost activated carbon produced from agricultural waste.

There are basically two methods for preparing activated carbon, namely physical and chemical activation. Generally, physical activation which involves carbonization of a carbonaceous material followed by activation of the resulting char at elevated temperature in the presence of suitable oxidizing gasses such as carbon dioxide, steam, air or their mixtures. The chemical activation is performed by the thermal decomposition of the raw material impregnated with certain chemicals such as $ZnCl_2$, NaOH, KOH, H_2SO_4, and H_3PO_4. Then, heat treatment at moderate temperatures (400–600 °C) in a one-step process [3], [4].

Application of physical activation using steam has shown the best potential for producing activated carbons with high surface area and good pore ratio [5]. The main advantage of physical activation over chemical activation is that it avoids the incorporation of impurities coming from the activating agent, it's cheap, eco-friendly and economical for production. Although various agricultural wastes have been used for preparing activated carbons by other chemical methods, the major concerns resulting from secondary pollution and chemical consumption must be considered. However, there are no reports up to date related to the production of activated carbon from *(Metroxylon spp)* waste by one step thermal activation using steam as an oxidizing agent. This method is hoped to enhance the absorption properties of the obtained activated carbons.

The aim of this paper was first to prepare the activated carbons from sago waste and to measure their physicochemical properties such as the yield, iodine number, BET surface area, pore size distribution, total pore volume, surface morphology and the elemental analysis. The environmentally friendly steam as an oxidizing agent was used in this work for activation processes. The amounts and rates of adsorption of dyes using the activated carbon from *(Metroxylon spp)* waste were then measured to investigate various parameters viz: agitation speed, pH, temperature on the removal of MB and GV using a two-level 3 factor (2^3) Central Composite Design (CCD) with the help of Design Expert 6.0.8 (Stat-Ease, USA).

2. Materials and methods

2.1. Production of porous carbon by steam activation

Precursor, *(Metroxylon sagu)* waste also known as sago waste was used for the production of PAC by thermal pyrolysis and in the presence of steam by one step activation. A known amount of raw material was carbonized and activated in a cylindrical quartz tube (OD: 95 mm, ID: 90 mm, L: 1,100 mm) using a rectangular horizontal high-temperature CSC— Split furnace (Lenton Thermal Design, UK). The system setup is shown in Figure 1. Once the tube furnace was switched on and set at required temperatures 900 °C, nitrogen gas (N_2) was supplied to the system at 2.5 l/min [6]. This was continued for one hour at a heating rate of 23 °C/min. During this process, the impurities purged was passed through and trapped in a container of water in the fume hood.

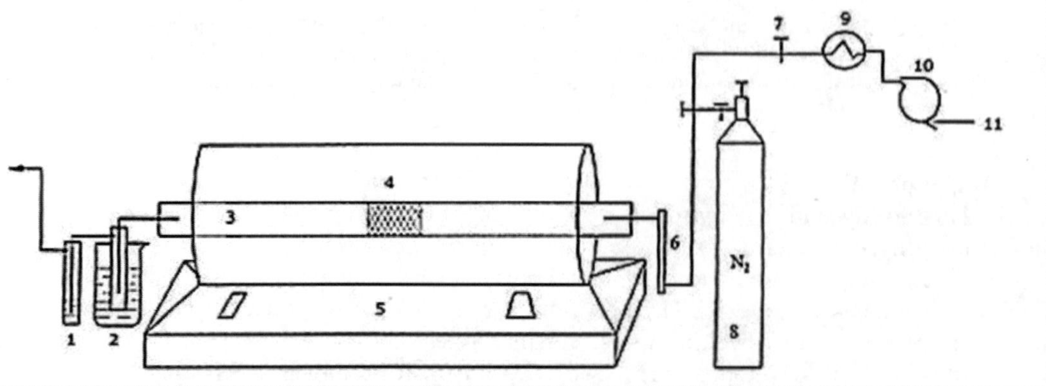

Figure. 1 Experimental set-up for the production of activated carbon. 1. Trap. 2. Cooling bath. 3. Quartz tube. 4. Carbon sample. 5. Furnace. 6. Flow meter. 7. Valve. 8. N_2 -cylinder. 9. Steam generator (Heat exchanger) 10.Dosing pump 11.Water inlet.

Upon the completion of carbonization, the N_2 gas flow was stopped and steam was introduced for activation of 15 min according to [6]. At the end of the operation, the steam generator was stopped to terminate the activation and switched back to the inert gas and the system was allowed to cool to below 50 °C before removing the AC from the quartz tube inside the furnace. The AC produced was ground and sieved to size fractions less than 250 μm [7]. The weights of samples were taken before and after production and the AC yield was calculated by the following expression:

$$Y_{Ch}\ (\%) = \frac{\textbf{Weight of Activated Carbon}}{\textbf{Weight of Precusor}} \times \mathbf{100} \qquad (1)$$

2.2. Characterization of activated carbons

The iodine number of activated carbon was obtained on the basis of the Standard Test Method, by titration with sodium thiosulphate (ASTM, D 4607-94) [8]. The concentration of iodine solution was thus calculated from the total volume of sodium thiosulphate used and volume dilution factor. The BET surface area (SBET) and pore features of PAC, and sago waste were measured under N_2 adsorption/desorption isotherms at 77 K using a surface area analyzer (Quantachrome Corporation, USA). The surface morphology of the PAC based sago waste adsorbents before and after activation was observed by scanning electron microscopy (SEM) (Hitachi S-520, Japan). Functional groups present in the raw material i.e. SW and PAC were evaluated by FTIR spectra, which were recorded between 4000 and 400 cm^{-1} (resolution of 4 cm^{-1} and acquisition rate of 32 scan min^{-1}) using an IR prestige-21, FTIR-84005, SHIMADZU Corporation (Kyoto, Japan).

2.3. Design of experiments

In order to develop an adsorption process for the design of the experiment, a number of factors influencing this process such as pH, agitation speed, and temperature are to be studied. CCD via face-centered was selected to model the effects of the methylene blue and gentian violet onto PAC based sago waste preparation variables; pH (5-9), agitation speed (100-400 rpm), temperature (35- 55 ^oC) towards their responses; MB (Y_1) and GV (Y_2) adsorption value. These three variables together with their respective ranges were chosen based on preliminary studies as given in Table 3. CCD requires three variables; 8 factorial points, 6 axial points and 6 replicates at the center points, indicating that altogether 20 experiments for this procedure, as calculated from Eq. (2) [9]:

$$\mathbf{N = 2^n + 2n + n_c = 2^3 + 2(3) + 6 = 20} \qquad (2)$$

where N is the total number of experiments required. A weighed amount (0.1 g) of activated carbon and 500 ppm of dyes concentration solution was used for all experiment

3. Results and discussion

3.1. Characteristics of the activated carbon produced

3.1.1. The effect of activation mechanism

The mechanism of preparation process for the carbon by steam activation can be explained as follows: (1) carbonization is a strong phase of chemical reaction which enhances organic matter degradation; (2) steam activation is a multiple reaction in gas/solid phase system and is divided into two processes: physical and chemical phase. These are water vapor spreading into the surface of granule, gasification reaction on the surface of carbon, decomposition of intermediate product and stripping of the reaction product. As an activating agent, steam can enter into the carbon particulate and react with it. The activation reaction mechanism of a carbon atom with steam as an activating agent has been illustrated below:

$$C + H_2O \rightarrow CO + H_2 \qquad \Delta H = +117\ kJ/mol \qquad (3)$$

$$CO + H_2O \rightarrow CO_2 + H_2 \qquad \Delta H = -41\ kJ/mol \qquad (4)$$

$$C + CO_2 \rightarrow 2CO \qquad \Delta H = +159\ kJ/mol \qquad (5)$$

In the above equations, C is the carbon atom activation sites. Steam activation was undergone under inert condition and temperature greater than 700 ^oC, and the prepared activated carbon can avoid the erosion of oxygen. The reaction of Eq. (3) will produce carbon monoxide, which will then react with H_2O to form CO_2. Then the CO_2 will react with carbon and consume more carbon and hence higher burn-off [10], [11]. Maximum yield of 12.6 % was recorded for PAC at an activation temperature of 900 ^oC, carbonization time of 1hour in the presence of steam flow for 15 mins. In general, greater the aromaticity and molecular weight of the precursor, greater will be the char yield. The low carbon yield in the case of cellulose is due to the fact that significant amount of carbon is lost from the glucose derivatives due to volatilization and as a result, the char yield is low.

3.1.2. Surface area and pore size distribution of the PAC

The N_2 adsorption-desorption isotherms and the pore size distribution by Barret, Joyner, and Halenda (BJH) method for PAC was shown in Fig. 2(a) and 2(b). As can be seen from Figure 2(a), increasing the relative pressure led to the appearance of a hysteresis loop as a result of capillary condensation, signifying a substantial development of mesoporosity. The total pore volumes were estimated to be the liquid volume of adsorbate (N_2) at a relative pressure of 0.98. All surface areas were calculated from the nitrogen adsorption isotherms by assuming the area of a nitrogen molecule to be 0.162 nm [12]. Total surface area of 1432.5 m^2/g, micropore surface area (1524 m^2g^{-1}), total pore volume (0.906 cm^3g^{-1}), micropore volume (0.542 cmg^{-1}) and average pore radius (1.72 nm) respectively was obtained after production. The BET surface area and pore volume of the PAC based SW were greater than that of the untreated sago waste as indicated in Table 1.

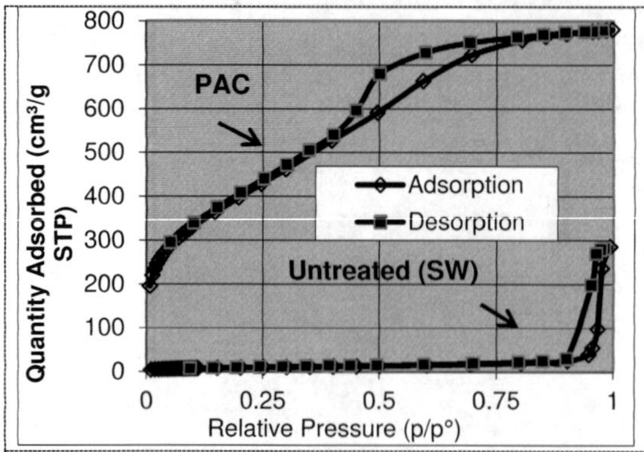

Figure 2(a). Nitrogen adsorption/desorption isotherms for PAC and untreated waste.

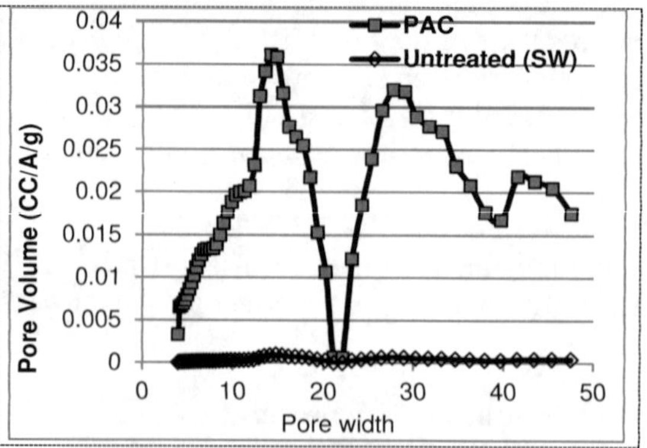

Figure 2(b). Size distributions for PAC and untreated waste.

Table 1. Physiochemical characteristics of PAC based sago waste and raw sago waste

	Iodine Value (mg/g)	Total Surface Area (BET) (m^2g^{-1})	Micropore Surface Area (m^2g^{-1})	Total Pore Volume (cm^3g^{-1})	Micropore Volume (cm^3g^{-1})	Average pore radius (nm)
PAC based SW	1275.2	1423.5	1524.1	0.906	0.542	1.724
Untreated	83.7	32.01	39.48	0.151	0.014	-

The AC development porosity is associated with gasification reaction. The pore size distribution is shown in Fig. 2(b), substantiate that the PAC comprise micro-mesopores with 2 sharp peaks pointing to pore diameters of 14.08 and 27.7 nm for the micropore and mesopores size ranges, respectively.

Each 1.0mg of iodine adsorbed is ideally considered to represent 1.0m^2 of the activated carbon internal area [13]. The iodine number of 1275.2 mg/g was obtained for PAC after production. Activated carbons with iodine numbers of about 550 mg/g can be attractive for wastewater treatment from the user's viewpoint [13]. The iodine number value is an indication of the surface area of the activated carbon [14].

Fig. 3 (a), and (b), respectively shows the SEM micrographs obtained before and after activation at an accelerating voltage of ~10 kV and 1000× magnification. At such magnification, distinct differences in the surface structure and porosity were clearly identifiable. It is evident that the novel activated carbon and its raw state affects the number of charged functional groups on the adsorbent surface. The microporosity is opened and widened with even a shift to meso- and macro-porosity as the removal of the exterior of the particle is significant at high burn-offs. This shows that steam was effective in creating well-developed pores on the surfaces of the precursor, hence leading to the activated carbon with large surface area and porous structure which had high adsorption capacity on dyes. The surface of the raw material was dense and planar without any cracks and crevices [6]. Due to these well-developed pores, the activated carbon possessed high surface area and adsorptive capacity. The pore size distribution of the activated carbon indicated that the diameter of the pores is greater than 2 μm which show that mesopores developed in the activated carbons.

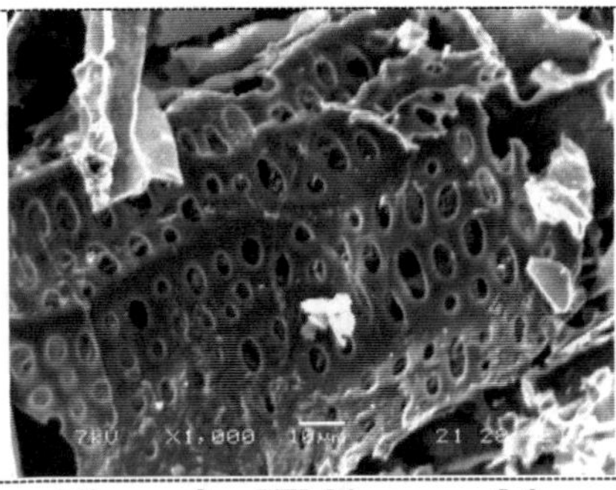

Figure 3a. SEM images of the Untreated waste.

Figure 3b. SEM images of the developed activated carbon (PAC).

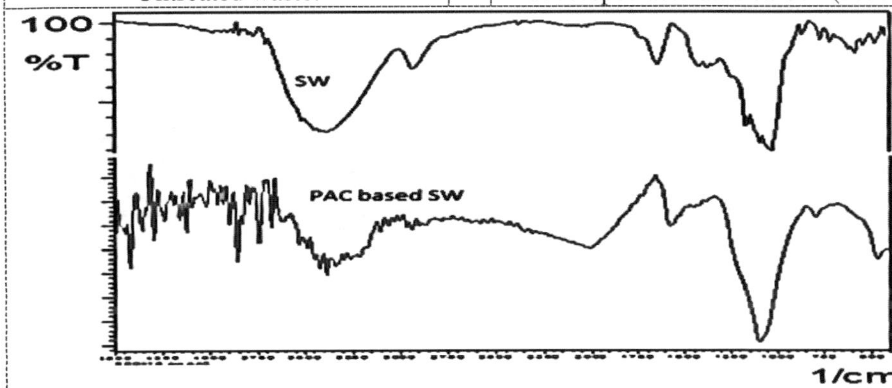

Figure 4. FTIR of the powdered activated carbon-based SW and untreated waste.

The changes in the spectrum in Figure 4 clearly indicate the possible involvement of carbonization and steam activation to amend the surface chemistry of the prepared activated carbon than the raw precursor. The new broadband at 3855 - 3423 cm^{-1} respectively found in the spectrum of the activated carbon can be assigned to O-H stretching vibration of hydroxyl functional groups including hydrogen bonding. Other peaks detected on the sago waste after activation with steam were located at 2934, 2005.5, 1574.4, 1093.0, 795.6, 651.5 cm^{-1} and 462.92 cm^{-1}, respectively assigned to C-H stretching in alkanes/alkyl groups, C-N stretching, C-O-C stretching, C-H out-of-plane bending in benzene derivatives, C-O stretching and O-H bending modes of alcoholic, phenolic, carboxylic groups, O-H stretching and C-O-H twist. C-H out-of-plane bending in benzene derivatives was commonly found on the surface of various activated carbons [15].

3.2. *Statistical analysis*

The complete design matrices together with both the response values obtained from the experimental work are shown in Table 2. Adsorption capacity value low and high of MB and GV were found to range from 216.1 to 253.1 mg/g and 104.9 to 250.7 mg/g, respectively.

The results of the surface quadratic model in the form of analysis of variance (ANOVA) for MB and GV adsorption capacity value respectively are given in Table 3. ANOVA is required to justify the significance and adequacy of the models. According to the sequential model sum of squares, the models were selected based on the highest order polynomials where the additional terms were significant and the models were not aliased. For the linear and quadratic models of MB and GV adsorption capacity were suggested by the software. However, the quadratic model was selected in this case due to the higher order polynomial.

Table 2. Experimental design matrix for dye adsorption using activated carbon

Std	A:pH	B:Agitat.Speed (rpm)	C:Temperature (°C)	Y_1: MB (mg/g) Actual	Y_1: MB (mg/g) Predicted	Y_2: GV (mg/g) Actual	Y_2: GV (mg/g) Predicted
1	5.0	100.0	35.0	220.23	220.18	222.29	227.37
2	9.0	100.0	35.0	218.45	217.75	107.53	109.15
3	5.0	400.0	35.0	247.78	248.28	154.82	161.83
4	9.0	400.0	35.0	240.56	240.59	111	112.6
5	5.0	100.0	55.0	220.85	220.38	173.86	175.11
6	9.0	100.0	55.0	216.11	215.17	104.91	100.76
7	5.0	400.0	55.0	241.26	241.52	201.34	202.57
8	9.0	400.0	55.0	231.44	231.05	199.45	197.22
9	5.0	250.0	45.0	244.23	244	228.2	213.62
10	9.0	250.0	45.0	235.56	237.56	148.67	151.83
11	7.0	100.0	45.0	228.56	230.73	215.92	212.12
12	7.0	400.0	45.0	253.12	252.72	235.18	227.57
13	7.0	250.0	35.0	245.55	245.77	250.69	235.38
14	7.0	250.0	55.0	239.56	241.1	247.66	251.56
15	7.0	250.0	45.0	248.88	248.29	238.8	242.61
16	7.0	250.0	45.0	248.87	248.29	238.8	242.61
17	7.0	250.0	45.0	248.88	248.29	238.81	242.61
18	7.0	250.0	45.0	248.88	248.29	238.79	242.61
19	7.0	250.0	45.0	248.88	248.29	238.79	242.61
20	7.0	250.0	45.0	248.87	248.29	238.82	242.61

Table 3. Analysis of variance (ANOVA) for response surface quadratic models of MB and GV

	Sum of Squares		DF	Mean Square		F Value		Prob > F	
Source	(Y_1)	(Y_2)		(Y_1)	(Y_2)	(Y_1)	(Y_2)	(Y_1)	(Y_2)
Model	2729.76	46737.05	9	303.31	5193.01	196.06	70.48	< 0.0001*	< 0.0001*
A	103.88	9545.01	1	103.88	9545.01	67.15	129.55	< 0.0001*	< 0.0001*
B	1209.12	597.22	1	1209.12	597.22	781.6	8.11	< 0.0001*	0.0173*
C	54.52	654.32	1	54.52	654.32	35.24	8.88	0.0001*	0.0138*
A^2	155.08	9859.84	1	155.08	9859.84	100.25	133.83	< 0.0001*	< 0.0001*
B^2	118.51	1424.95	1	118.51	1424.95	76.6	19.34	< 0.0001*	0.0013*
C^2	64.67	2.04	1	64.67	2.04	41.81	0.028	< 0.0001*	0.8711
AB	13.83	2380.5	1	13.83	2380.5	8.94	32.31	0.0136*	0.0002*
AC	3.86	962.29	1	3.86	962.29	2.5	13.06	0.1451	0.0047*
BC	24.22	4325.43	1	24.22	4325.43	15.66	58.71	0.0027*	< 0.0001*
Residual	15.47	736.76	10	1.55	73.68				
Lack of Fit	15.47	15.06	5	3.09	3.01	1.16E+05	1078182	< 0.0001	< 0.0001
Pure Error	1.33E-04	6.83E-04	5	2.67E-05	1.37E-04				
Cor Total	2745.23	47473.8	19						

DF; Degree of Freedom, * Significant

The mean squares were obtained by dividing the sum of the squares of each of the variation sources, the model, and the error variance, by the respective degrees of freedom. Values greater than 0.1000 indicate the model terms are not significant. If the value of Prob. > F less than 0.05, the model terms are considered as significant. From Table 3, for Y_1 the model F-value of 196.06 and Prob. > F of <0.0001 implied that this model was significant. In this case, A, B, C, A^2, B^2, C^2, AB and BC factors were significant model terms whereas AC is insignificant to the response. In addition, Y_2 response model F-value of 70.48 and Prob. > F of <0.0001indicated that the model was significant. In this case, A, B, C, A^2, B^2, AB, AC and BC factors were significant model terms whereas C^2 is insignificant to the response. The non-significant value of lack of fit showed that the quadratic model was valid for the present study.

3.3. Development of regression model equation

The final empirical formula models for the (Y_1) and (Y_2) in terms of coded factors are represented in Table 4 were adequate to predict the Y_1 and Y_2 adsorption capacity value within the range of variables studied. The coefficients with one factor represent the effect of the particular factor, while the coefficients with two factors and those with second-order terms represent the interaction between two factors and quadratic effect, respectively.

Table 4. Terms of Coded factors

	Modified equations with significant terms	R-Squared	Adj R-Squared	Pred R-Squared	Adeq Precision	Standard Deviation	Mean	C.V.[1]	PRESS[2]
(Y_1)	= +248.29 -3.22 A+11.00 B-2.33 C-7.51 A^2 -6.56 B^2 -4.85 C^2 -1.32 AB -0.70 AC -1.74 BC	0.9944	0.9893	0.9655	42.69	1.24	238.83	0.52	94.68
(Y_2)	= +242.61 -30.90 A+7.73 B+8.09 C-59.88 A^2-22.76 B^2+0.86 C^2+17.25 AB+10.97 AC +23.25 BC	0.9845	0.9705	0.9015	24.85	8.58	201.72	4.26	4677.4

A: Initial Solution, B: Agitation speed (rpm), C: Temperature (°C),
[1]: Coefficient of variation,
[2]: Predicted residual error sum of squares

The quality of the models developed was evaluated based on the correlation coefficients, R-value. In fact, the models developed seems to be the best at high R^2 statistics which is closer to unity as it will give predicted value closer to the actual value for the responses [12]. In addition, the R^2 and adjusted R^2 values for Y_1 and Y_2 were 0.994 and 0.9844, 0.989 and 0.9705 respectively. This indicated that 99.4 and 98.4% of the total variation in the MB and GV adsorption value respectively, were attributed to the experimental variables studied. Both R^2 values were considered relatively high, indicating that the predicted values 0.9655 and 0.9014 for MB and GV removal were accurate and closer to its actual value. Adequate precision measures the signal to noise ratio. A ratio greater than 4 is desirable. The ratio of 42.69 (Y_1) and 24.85 (Y_2) indicates an adequate signal. Therefore, these model can be used to navigate the design space. At the same time, a relatively lower value of the coefficient of variance (C.V. = 0.52 and 4.26 %) indicates a better precision and reliability of the experiments carried out.

4. Conclusions

The results of this study demonstrated that low-cost activated carbon can be prepared from *(Metroxylon sagu)* waste by physical activation with steam for MB and GV adsorption. PAC with the maximum BET surface area of 1423.5 m^2 /g was obtained under the following conditions: carbonization temperature 900 °C; carbonization time 60 min; activation temperature 900 °C and activation time 15 min. The analysis of results carried out by using central composite design idea was to avoid the traditional one-factor-at-a-time experiments and the model developed resulted in an improved response, reduced process variability, a time-saving approach for studying the combined influence of process parameters on the response factor and closer confirmation of response to targeted requirements. The agitation speed, initial solution pH, and temperature significantly influenced the dye removal efficiency. The corresponding adsorptive efficiency in optimum conditions was found to be (244.5 and 239.86 mg/g) and their desirability was 0.997 and 0.991 for MB and GV respectively. The PAC based SW prepared demonstrated high iodine value, surface area, and well-developed porosity. Taking into consideration all the above-obtained results, PAC based SW showed a good potential and can be used as an effective sorbent for removal of MB and GV from aqueous systems.

5. References

[1] Metivier-Pignon Hn, Faur-Brasquet C and Le Cloirec P 2003 Adsorption of dyes onto activated carbon cloths: approach of adsorption mechanisms and coupling of ACC with ultrafiltration to treat coloured wastewaters *Separation and Purification Technol* **31**(1) p 3-11

[2] Mona S, Kaushik A and Kaushik CP 2011 Waste biomass of Nostoc linckia as adsorbent of crystal violet dye: Optimization based on statistical model *Int. Biodeterioration & Biodegradation* **65**(3) p 513-521

[3] El-Hendawy A-NA, Samra S and Girgis B 2001 Adsorption characteristics of activated carbons obtained from corncobs *Colloids and Surfaces A: Physicochemical and Engineering Aspects* **180**(3) p 209-221

[4] Sahin O and Saka C 2013 Preparation and characterization of activated carbon from acorn shell by physical activation with H_2O-CO_2 in two-step pretreatment *Bioresource Technol* **136**(0) p 163-168

[5] Ahmedna M, Marshall W and Rao R 2000 Surface properties of granular activated carbons from agricultural by-products and their effects on raw sugar decolorization *Bioresource Technol* **71**(2) p 103-112

[6] Ma'an FA, Muyibi SA and Amode JO 2011 Optimization of activated carbon production from empty fruit bunch fibers in one-step steam pyrolysis for cadmium removal from aqueous solution *The Environmentalist* **31**(4) p 349-357

[7] Alam MZ, Muyibi SA, Mansor MF and Wahid R 2007 Activated carbons derived from oil palm empty-fruit bunches: Application to environmental problems *J. Environmental Sciences* **19**(1) p 103-108

[8] ASTM D4607-94 1994 Standard Test Method for Determination of Iodine Number of Activated Carbon Part 30 (New York) NY

[9] Sahu JN, Acharya J and Meikap BC 2010 Optimization of production conditions for activated carbons from Tamarind wood by zinc chloride using response surface methodology *Bioresource Technol* **101**(6) p 1974-1982

[10] Marsh H and Reinoso FR 2006 Activated carbon (Oxford: Elsevier Ltd)

[11] Chan OS, Cheung WH and McKay G 2011 Preparation and characterisation of demineralised tyre derived activated carbon *Carbon* **49**(14) p 4674-4687

[12] Hameed BH, Tan IAW and Ahmad AL 2009 Preparation of oil palm empty fruit bunch-based activated carbon for removal of 2,4,6-trichlorophenol: Optimization using response surface methodology *J. Hazardous Materials* **164**(2-3) p 1316-1324

[13] Nwabanne J and Igbokwe P 2012 Application of Response Surface Methodology for Preparation of Activated Carbon from Palmyra Palm Nut *NY Sci J* **5** p 18-25

[14] Abdullah AH, Kassim A, Zainal Z, Hussien MZ, Kuang D, Ahmad F and Wooi OS 2001 Preparation and Characterization of Activated Carbon from Gelam Wood Bark (Melaleuca cajuputi) *Malaysian Journal of Analytical Sciences* **7**(1) p 65-68

[15] Guo J and Lua AC 2003 Textural and chemical properties of adsorbent prepared from palm shell by phosphoric acid activation *Materials Chemistry and Physics* **80**(1) p 114-119

Acknowledgments

The authors thank the Government of Brunei Darussalam and the Universiti Brunei Darussalam for the award of Graduate Research Scholarship (GRS) to JOA. The authors are grateful to the Bioenvironmental Engineering Research Center (BERC), Faculty of Engineering, International

Islamic University Malaysia for providing the laboratory facilities as well as BERC students Mohammed Nurudeen Shola and Dr. Munirat Idris for their support in the laboratory work.

IOP Conf. Series: Earth and Environmental Science **159** (2018) 012017 doi:10.1088/1755-1315/159/1/012017

The Recycling of Spent Power Battery: Economic Benefits and Policy Suggestions

Xiaoming Ma[1,2], Yuan Ma[1,2], Jiping Zhou[3] and Siqin Xiong[1,2,*]

1 School of Environment and Energy, Peking University Shenzhen Graduate School, Shenzhen, China
2 College of Environmental Sciences and Engineering, Peking University, Beijing, China
3 State High-Tech Industrial Innovation Centre, Shenzhen, China.
E-mail: xiongsiqin@pku.edu.cn

Abstract. With the rapid development of the electric vehicle industry in China, the number of spent power batteries has increased. Recycling of spent power batteries is necessary not only to eliminate the environment pollution, but also to reduce the resources usage. However, whether it is necessary for the government to take economic incentives in the recycling process has not been clearly analyzed. Combined the existing literatures and the market investigation, this paper estimates the economic profits of recycling spent power batteries, in terms of LiFePO4 (LFP) battery and Li(NiCoMn)O2 (NCM) battery. The results show that the recycling process is profitable, therefore, we suggest the government should not implement additional economic incentives, like subsides or tax reduction, but focus on the establishment of compulsory recycling mechanism and the market permission system.

1. Introduction

With the increasingly broad utilization of electric vehicles in China, power battery industry experiences a spurt of growth, and meanwhile yielding an increasing volume of spent battery [1]. In 2012, the sales volume of electric vehicle was just 11,375, but it reached 409,000 in 2016. Additionally, Ministry of Industry and Information proposed that the production volume of electric vehicles and plug-in hybrid electric vehicle should be two million by 2020[2]. As the core component of the new energy automobile, the power battery market shows a trend of outbreak after the sharply rise of new energy automobile. The power battery production jumped to 28 GWh in 2016 from 3.7 GWh in 2014, an increase of more than six times [3]. In China, under the average driving condition that the annual mileage of cars is about 16,000 kilometers, the lifetime of LiFePO4 (LFP) battery is estimated about 4-6 years, while the lifetime of $Li(NiCoMn)O_2$ (NCM) battery is about 3-5 years. Combined existing literatures and our market survey analysis, this paper estimates the quality of spent power batteries, as shown in Figure 1

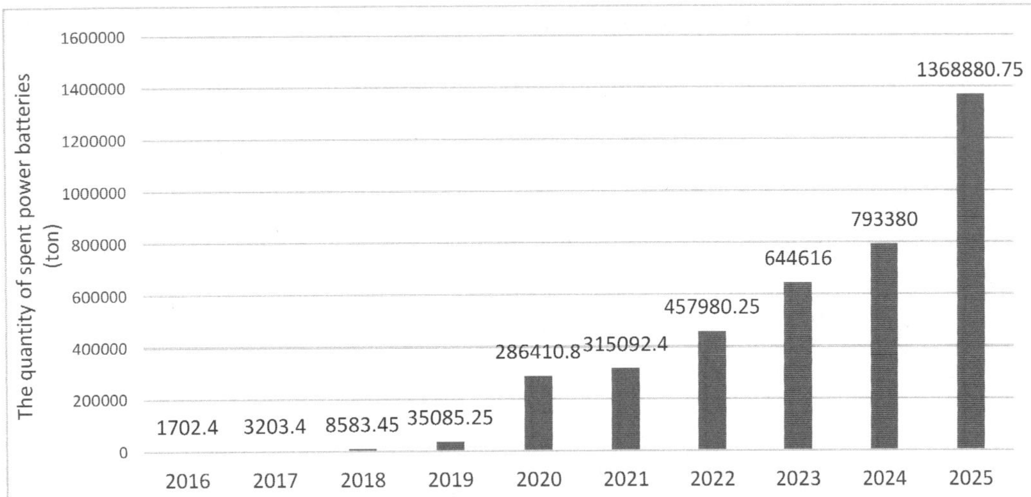

Figure 1. Estimation of spent power battery quantity.

The estimated quantity of the spent power battery is 28.6 thousand tons in 2020 and 1.36 million tons in 2025. Although the replacement from intern combustion engine to power batteries is regarded as energy saving and environmental friendly, and the LFP batteries and NCM batteries do not contain mercury, cadmium, lead or other toxic heavy metals, this does not mean that power batteries are completely non-polluting products[4]. The staggering quantity of power battery contains large amounts of nickel, cobalt, manganese, toxic electrolytes, organic chemicals and plastics, posing a risk on human health, water and soil pollution if not properly disposed of. Therefore, how to effectively recycle the vast amount of spent batteries is an urgent task to resolve.

As an emerging environmental and social issue, the recycling system of spent power battery has not been mature, therefore, government regulation is an indispensable aspect of establishing the recycling mechanism. We reviewed the international policies and found that the United States, European Union, Japan and other developed countries are actively exploring the power battery recycling mechanism, researching economic profits from recovery and technical feasibility. The U.S. government charges additional environmental cost to promote the establishment of a battery recycling network. New York and California have passed state laws that require retailers to take back batteries. Germany has established a relatively perfect legal system of recycling, stipulating that battery manufacturers and importers must assume the responsibility to organize their dealers to recycle the discarded battery and hand over to the designated agencies. European Union has regulated that member states must recycle at least 45% waste batteries in 2016, and the processing and utilization rate of lead, nickel and isolation should not less than 50%. Besides, there are also some related demonstration projects and business operations. For example, in 2011, General Motors (GM) cooperated with ABB to test how to use the scrap battery pack to collect electricity and to realize the residential and commercial power supply. Germany's environment ministry has fund for two power battery recycling demonstration projects in the research of recycling use. Japanese battery manufacturers have already set up a set of "battery production-sales-recycling-renewable processing" battery recycling system[1].

Chinese government has also implemented some policies in terms of spent battery recycling. In 2012, the State Council published "Energy Saving and New Energy Auto Industrial Plan for 2012–2020"[2], which was the first official document for power batteries and was committed to establishing a recycling mechanism. In the beginning of 2016, the National Development and Reform Commission and other four ministries jointly published "The Technical Policy for the Recovery and Utilization of Electric Vehicle Power Battery (2015Edition)"[5], proposing the extended producer responsibility system (EPR) to stipulate the battery manufacturers, importers and related businesses to assume responsibility for the recycling of spent batteries. Later in 2016, "Interim measures for recycling management of traction batteries for new energy vehicles (draft)" changed to explicate that the producers should assume main responsibility for recycling[6]. In addition, China firstly reported three national standards for power battery recycling, including “Dimension of traction battery for electric vehicles”, “Coding regulation for automotive traction battery” and “Recycling of traction battery used in electric vehicle-test of residual capacity”[7]. These standards regulate the general battery size, disassemble technical requirements, operation procedures, inspection and complementary energy test methods and the implement of them will effectively help recycling industries ensure the security, improve procedure efficiency, reduce treatment pollution, as well as maximize the recovery rate. Since 2012, some local governments have tried to issue related measures to encourage power battery recycling. For example, Shanghai Municipal government subsides 1000 yuan per electric vehicle as a special fund for the battery recycling enterprises. Shenzhen requires the local new energy vehicles manufactories to set aside special provisions for power battery recycling in the standard of 20 yuan per KWH, and then government will give subsides based on 50% of the audited provisions[1].

The related cycling policies in China is expected to stimulate the recycling of spent power batteries. However, spent power battery packs will retain approximately 80% of their energy performance, and the recycling entrepreneurs can gain some revenues by recovering the profitable mental. Whether the subsides or economic incentives are necessary concerning the recycling battery’s economic benefits and what kind of policies should be taken to effectively realize the complete recycle have not been discussed to date.

To fill this gap, this paper aims to forecast the economic profitability of recycling the spent power batteries at the current situation and then give specific policy recommendations. There are various types of power batteries, including LiFePO4 series (LFP), LiMn2O4 series, Li(NiCoMn)O2 series (NCM), Li(NiCoAl)O2 series (NCA), and LiCoO2 series[8]. Due to the safety, stability and high-cycle capacity of LiFePO4 battery and the high energy capacity of NCM battery, these two batteries account for nearly 95% in Chinese market in 2015[9].Therefore, we just take LFP battery and NCM battery as the proxy to analyse in this paper.

2. Economic Benefits

2.1. Recycling revenue

The spent power batteries contain an amount of cobalt, nickel, lithium and other valuable metal. By estimating the mental values of each battery, we predict the recovery revenue.

$$\mathrm{R} = \sum_i^m P_i * \alpha_i * \beta_i \qquad (1)$$

Where R denotes the recycling revenue of each battery; P_i is the primary commodity market price for the ith mental; α_i is the ith mental composition quantity of that kind of battery, and β_i is the recycling rate of each mental; m is the number of mental types; this paper includes the cobalt, nickel, lithium, manganese, iron, aluminum and copper as the valuable mental, thus, m equals to 7. We adopted the related parameters as Table 1:

Table 1. The recycling revenue of spent LFP batteries and NCM batteries

It is notable that the material composition in spent power batteries would vary between different

Mental	Prices (yuan/ton)	Composition (kg/ton 100KWH)		Recycling rate	Recycling revenue (yuan/100KWH)	
		LFP	NCM523		LFP	NCM523
Cobalt	443,000	0	48.48	89%	0	19,114.21
Nickel	98,275	0	80.8	62%	0	4,923.18
Lithium	860,000	8.99	9.59	80%	6,185.12	6,597.92
Manganese	10,900	0	30.13	53%	0	174.06
Iron	660	71.95	0	52%	24.69	0
Aluminum	15,380	65	87	42%	420	562
Copper	53,510	82	112	90%	3,949	5,394
Total					10,578.81	36,765.38

battery types. The main NCM type is NCM523 in China, which accounts for almost 75% in the total NCM series. Therefore, we take the NCM523 as the representative NCM type to predict the cycling revenue, cost and profits. To reflect the current situation, the value of materials is obtained from the current commodity metals prices from the website of Shanghai nonferrous metal mesh for November, 2017[10]. The composition parameter and recovery are obtained by literature reviews and our investigation of manufacturers[11-13]. The energy capacity of an electric vehicle is about 50 KWH, with its 500 kg mass. So, the recycling revenue of spent LFP and NCM batteries are about 10578.81 yuan/ton and 36765.38 yuan/ton, respectively.

2.2. Recycling costs

The recycling costs of spent power battery includes two parts: variable costs (Cv) and fixed costs (Cf), varying from power battery capacity, recycle technology level, geographic location and other factors. Variable costs are dependent on the volume of recycled batteries while fixed costs refer to the labor salaries, rents of plant, and depreciation allowances of equipment and always keep constant within a fiscal year. Because of China's power battery has not yet reached the batch scrap stage, many power battery recycling enterprises can't collect enough waste battery pack to realize large-scale operation, causing great difficulties for cost calculation. Wang etc. summarized variable cost and fixed cost of 15 international power battery recycling programs through literature reviews and reported that the average variable cost is $2800 per ton[14]. But since the sample is mainly from developed countries, such as Australia, the United States and France, it may not coincide with China's conditions. Hou etc. analyzed China's power battery recycling costs by market investigation and research and showed that the total recycling cost is about 18500 yuan/ton[15]. But when calculating the cost, he does not distinct the battery types and not take the storage cost and transport cost into consideration.

Based on the literature views and the market research of battery producers and recycling enterprises, this paper estimates variable cost and fixed cost of spent LFP batteries and NCM batteries respectively. Table 2 shows the results. text of your paper should be formatted as follows:

Table 2. The recycling costs of the spent power batteries

	Item	Content	Cost (yuan/ton) LFP	Cost (yuan/ton) NCM
Cv	Material recovery cost	Spent battery	0	1,2018
	Accessory material cost	Solution and extraction solvent	2,500	2,500
	Fuel cost	Electricity, natural gas etc.	620	620
	Pretreatment cost	crushing and separation	500	500
	Environmental handling expenses	Waste water treatment	330	330
		Residue and ash disposal cost	120	120
	Labor charges	Salary	470	470
	Transportation expenses	Fuel fees, Toll, etc.	2,000	2,000
Cf	Fixed costs	Maintenance of equipment	100	100
		Depreciation allowances	536	536
C (Total)			7,176	1,9194

The material recycle cost means the cost that recyclers pay to the battery owners. According to our survey, the recycling of LFP batteries does not need to pay fees or the fee is small enough to be ignored at current situation. The cost to pay the spent NCM batteries adopts the 50% value of the cobalt and nickel composition. The accessory material cost and fuel cost are taken from other literatures. The pre-treatment cost and environment handling expenses are obtained from other literatures and the investigation. Fixed costs include the depreciation allowances of plant construction, machinery equipment, recycling network construction, and maintenance of equipment. The residual value for fixed assets is set as 5% and the depreciable lives are set as 10 years and 30 years for equipment and plant, respectively. The labour charges and transportation expenses are obtained from market investigation in Shenzhen. At all the costs are estimated at the full load circumstances. The results indicate that the recycling cost of LFP is about 7176 yuan, of NCM is about 19194 yuan.

2.3. Recycling profits

Based on the above analysis, we can get the recycling profits by the following equation:

$$B=R-Cv-Cf \quad (2)$$

Where B presents the recycling profits of spent power battery. And the recycling profits are shown as Table 3.

Table 3. The recycling profits of the spent power batteries

Type	B (yuan/ton)	R(yuan/ton)	C (yuan/ton)
LFP	3394	10,570	7,176
NCM	17571	36,765	19,194

At current situation, the recycling of both spent LFP batteries and spent NCM batteries are profitable, with the profits of 3394 yuan/ton and 17571 yuan/ton, respectively. Note that the recycling profits are closely related to metal prices. At present, due to the high demand of new energy vehicle industry and scarcity of resources in China, especially cobalt, we believe that the metal price will not sharply decline in the short term, thus, the power battery recycling is profitable, at least, is to balance of payments. Besides, the power batteries have only been in commercial use for approximately 5 years in China and have not reached the massive scrapping stage, many power battery recycling businesses cannot realize the full capacity. So, with the large-scale spent power battery collected and the process efficiency improved, the operation costs that have been distributed to unit battery would be lower.

3. Economic benefits

3.1. Economic incentives are unnecessary

According to the above analysis, the recycling of spent power battery is profitable, therefore, there is no need for government to take no additional economic incentives, like subsides or tax reduction. Driven by interests, the entrepreneurs would automatically take on the recycling and strenuously maximize the recovery rate. However, if the added economic incentives were given, some companies were inclined to take informal measures, like subsidies deception, which is unfavorable to the market competition instead.

3.2. Establish the compulsory recycling mechanism

Without the economic incentives, it does not mean any regulation is unnecessary. The government should establish compulsory recycling mechanism by legislations or regulations, to ensure the closed loop circulation of power batteries. During a normal life of an electric vehicle, it needs at least two batteries and the battery replacement is unavoidable. This paper suggest that power battery replacement should be complied with “old change new”. Car makers and battery suppliers must collect the spent old power batteries before giving the new batteries to replace. Automotive scrappage stage is another key stage of spent battery collection. We suggest that only by the spent batteries remove from the automobiles to the recycling network, the scrap automobiles can enter the following dismantling process. Note that these regulations should be implemented on the basis of information traceback and management platform.

3.3. Stipulate the qualification for recycling enterprises

Like other electronics, power batteries also pose a potential hazard to human health and the environment if not properly treated. So, government should license the qualified recycling entrepreneurs. The permitted entrepreneurs should be equipped with secure storage or disposal facilities and all recycling process should meet the environmental requirements. Here, we suggest the electric carmakers, battery producers and the waste recycling entrepreneurs with the waste electronic processing qualification, are more favourable to be licensed.

4. Conclusion

By estimating the recycling revenue and costs under the current situation, we find that the recycling of spent power batteries is profitable, even if the material value slightly declines or the material composition changes. Therefore, we suggest the government not to take any economic incentives to stimulate the recycling. What counts more for the government is to perfect the relevant laws and regulations to establish compulsory recycling mechanism and avoid illegal recovery.

5. References

[1]. Askci, The power battery recycling development present situation and main problems. 2017.
[2]. The State Council, C., Notice of energy saving and new energy vehicles industry development planning (2012-2020). 2012.
[3]. OFweek, Analysis on power battery market pattern in 2016. 2012.
[4]. Zeng, X., J. Li and L. Liu, Solving spent lithium-ion battery problems in China: Opportunities and challenges. Renewable and Sustainable Energy Reviews, 2015. 52: p. 1759-1767.
[5]. NDRC, et al., The Technical Policy for the Recovery and Utilization of Electric Vehicle Power Battery(2015Edition). 2017.
[6]. MIIT, Interim measures for recycling management of traction batteries for new energy vehicles (draft). 2017.
[7]. OFweek, Three power battery national standards recently released. 2017.
[8]. Zeng, X., J. Li and N. Singh, Recycling of Spent Lithium-Ion Battery: A Critical Review. CRITICAL REVIEWS IN ENVIRONMENTAL SCIENCE AND TECHNOLOGY, 2014. 44: p. 1129-1165.
[9]. Mahmoudzadeh Andwari, A., et al., A review of Battery Electric Vehicle technology and readiness levels. Renewable and Sustainable Energy Reviews, 2017. 78: p. 414-430.
[10]. Market, N., The price of non-ferrous mentals. 2017.
[11]. Swain, B., Recovery and recycling of lithium: A review. SEPARATION AND PURIFICATION TECHNOLOGY, 2017. 172: p. 388-403.
[12]. Wang, X., et al., Economic and environmental characterization of an evolving Li-ion battery waste stream. Journal of Environmental Management, 2014. 135: p. 126-134.
[13]. Wang, W. and Y. Wu, An overview of recycling and treatment of spent $LiFePO_4$ batteries in China. Resources, Conservation and Recycling, 2017. 127: p. 233-243.
[14]. Wang, X., et al., Economic, Environmental and Security Implications of Plug - in Vehicles. Journal of Environmental Management, 2014. 135: p. 126-134.
[15]. Hou, B., Research on Power Battery Recycling Mode of Electric Vehicle Research on Power Battery Recycling Mode of Electric Vehicle. 2015, Chongqing University of Technology.

IOP Conf. Series: Earth and Environmental Science **159** (2018) 012018 doi:10.1088/1755-1315/159/1/012018

Information Organization of 3-D Landslides Board Game for Ecological Educations

F L Chao[1]; S L Shih[1], Y C Chien[2], and W H Kao[1]

1 Department of Industrial Design, Chaoyang University of Technology, Taichung, 436, Taiwan, R.O.C.
2 Department of Landscape and Urban Design, Chaoyang University of Technology
E-mail: slshih@cyut.edu.tw

Abstract. The 3-D board game concentrate on landslide related corresponding components include slopes board, air guns, earth and stone, trees, betel nut trees, dice, event cards, flags. The pre-learning session and simplify the information structure of cards made participant aware its relationship more natural. Simplify the terminology of cards also reduces the complexity of its association. After playing the board game, students are willing to understand more about how things are happening in our surroundings. Virtual experience through the game may help a user to explore the real world in future. The three elements present relationship more precisely and lower the barrier of using in environmental education.

1. Introduction

1.1. Natural disaster

Climate change and over-exploitation caused the natural disaster. Many rainfall-induced severe, large-scale landslides occurred during the last two decades in Taiwan [1]. Catastrophic landslides often happen without warning since the slight topographic creeping is hard to be detected. Recent studies have highlighted the ecological and geomorphological importance of wood debris to forested stream ecosystems [2]. People try to protect channels against the effect of erosion; biomaterials were installed on slopes, channel bed and along river banks [3]. Bioengineering techniques were introduced, and the vegetative check dams were found to be a practical solution for flood hazard minimization [4].

1.2. Environmental education act

The Legislative Yuan passed the Environmental Education Act in 2011 [5]. This law makes people are more environmentally conscious at school and society. People are required to attend four hours of environmental education each year, in the form of lectures, discussions, internships, and group visits. Attractive education facilities have a demand in Taiwan. This study aimed to design the eco-conscious board game including objects and rules. Landslide is a significant natural and sociological event, so we implement this board game to encourage students learning environmental problem through playing.

1.3. Environmental education

Many theoretical frameworks have been developed to explain the gap between the possession of ecological knowledge and awareness, and displaying pro-environmental behavior [6]. Factors included demographic, institutional, economic and cultural factors. The internal factors (such as environmental knowledge, awareness) affect people's attitudes, priorities, and the locus of control. Knowledge and experiences have a stronger inference of people's behavior. Indirect experiences of passive learning

IOP Conf. Series: Earth and Environmental Science **159** (2018) 012018 doi:10.1088/1755-1315/159/1/012018

lead to a weaker correlation between attitude and response [7]. The person is familiar with the environmental problem, and its causes will act to lower impact on the environment. People with pro-environmental attitudes are more likely to engage in protective behaviors.
The game enables people to experience situations in a virtual environment. Those experiences also extend knowledge domain and bring locus of control [8]. Positive environmental attitudes, ecological sensitivity could be brought into student's mind through an educational game. Board games with ecology, issues can lead to motivation to approach more related knowledge.

2. Board Game

2.1. Coverage
The board game is favorite in recent years. People stimulate thinking in the interaction between members [9]. Traditionally, board game plays in a 2-D plate. Rules of matches are present to facilitate communications between players. The sensation of fun may derive from mastering new knowledge, skills or experience. In a virtual environment, a player faces set of challenges and generate a personal strategy to win a game [10]. Story and mechanics involve the procedure that provides the structure by which user interacts with system elements. If the controls are not intuitive enough [11], the user will always be distracted by the contradictions. User invest skills to reach the goal and get the pleasure through achieving and confidence [12].

2.3. User interface
By building the environmental content into the game, we can extend student's learning span. During the playing, users have to concentrate on current situations and practice strategy requires reaching the goal. The user interface is the activities where interactions between humans and machines. The purpose of this communication is to allow efficient operation and control of the apparatus from the human end. For the board game, it included plate, game cards, tools, and the rules that provide information with which a person may interact. The high attraction allows people enjoy playing. Nielsen's mentioned principles for interaction design [13] as follow:

- Strive for consistency: users should not have to wonder whether different words, situations mean the same thing.
- Visibility of system status and offer feedback: The system should keep users informed about what is going on through appropriate feedback.
- Match the real world to yield closure: The system should speak the users' language.
- Reduce short-term memory load: recognizing something is more natural than remembering it. Minimize the user's memory load by making objects with previous experience.

3. The 3-D • Board Game Design

3.1. Concept of design
The 3-D board game concentrate on landslide related corresponding components include slopes board, air guns, earth and stone, trees, betel nut trees, dice, event cards, flags (Figure 1). Designer suggests a suitable age range of users for playing the game is 10-22. Central platform is three-dimensional landscape with a gradient. The typhoons and earth landslide disasters emulated by air-gun and rolling rock (Figure 2). Dolls design reveals their characters, babies wear a raincoat symbolize the animal guards, and attacked rabbit with an escaping posture. Wall construction and tree plant actions emulated people use bioengineering techniques to preserve the land and the animal.
Designer implemented the ecological environment theme and response to the situation in cards design. The game progress under driven of random dice (the six sides of the dice) is moving steps and the incidents such as typhoon or earth flow event. When encountered "select event card," player may get trees, Betel nut tree or building development in hillside areas. Players need to think about planting trees or moving dolls to prevent typhoon or landslide rock. Each player has three animals (such as rabbit) and running each rabbit in sequence. The first one moves all doll reaches the target area is the winner.

Figure 1. Eco-run, land conservation board game, where main table field is a curvature 3D structure to emulate the geographic landscape.

Figure 2. (a) Air guns which emulate the disturbing wind in Typhoon, (b) Rock in different size which mimics the rock impact effect in landslide.

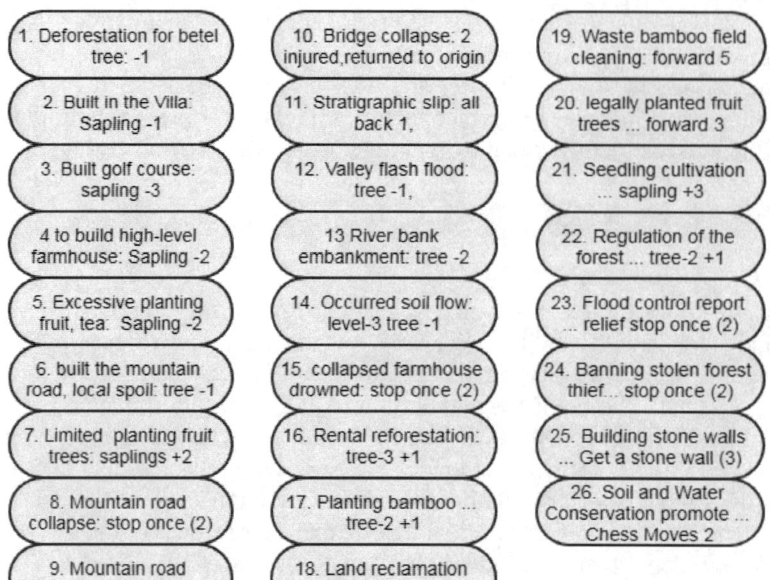

Figure 3. The 26 event cards consist of eco-related significant elements: plant, land, and water. The cards contain both positive and negative events which guiding the corresponding action within game.

All contents of the incident cards related to environmental issues (Figure 3) such as golf courses, tourist areas, farmhouses, excessive planting of fruit trees/ tea trees, mountain roads extension, destruction of forests, mountain road collapse, stratigraphic sliding and so on. Through the contents of the event card, we bring the player into a different situation. The player can acquire strategy of getting a tree to enhance the ability to sustain in disaster. The use of saplings to avoid air guns and landslide balls are other options. The events and actions prompted the importance of afforestation and soil and water conservation.

3.2. Evaluation-1

Our design aims the beginning level of the environmental education. While prototype constructed, children of different age's group are invited to play and gather user response. After participants accepted 5 minutes background introduction, the pair of users engaged together in 20 mins playing session. After that, the player filled out a questionnaire. The evaluator also recorded a log of the player's actions, failures, and coded each of these within observation record. Most of them have a positive feedback. Three of children have difficulty in understanding the content of the cards, so they proceed after evaluator's explanation. Student's reflection includes:

The attractive activities are air gun, landslide, transfer tree, and doll.
They like absorbing tutorial introduction, but some terminologies are not familiar.
One student said the game process majorly by random chance.
Children responses acquired information included: water and land conservation, caring animal and don't cut the tree from part of gameplay.

3.3. Reflections

Being lacking the beginning level of the ecological knowledge, a player did not have experience of judging the possible choices. Children have difficulty in understanding the content of cards and its relationship, so pre-learn scenarios are required. The script can be a story with the analogy which immerses the audience in imaging situation.

From evaluation-1, we found the game card needs re-organize with a suitable structure that people can follow. The knowledge structure let the student take an appropriate strategy with the environmental situation. The approach determined game process instead of chance. Information re-organization can let player a more comprehensible supporting with fewer instructor involvement.

4. Redesign of the Game Cards

Base on the interface design rules mention above; we modify the design for both logical and visual cognition by matching with user experience and reducing memory load.

Simplifies the cards and its relationship

The KJ method helps people with the right perspectives to reach consensus on priorities of subjective, qualitative data. In cards situation, different option and consequence confuse the player. Proper organizing and prioritizing cards can reduce memory load. The K-J method organizes ideas with following steps:

(1) Record each idea on cards or notes.
(2) Look for ideas that seem to be related.
(3) Sort cards into groups until all cards have been used.

Once the cards have been sorted into groups, the team may classify large clusters into subgroups for easier analysis [14]. In the first step, related or similar cards were combined so that we removed seven cardsd. The 19 remained eco-event cards were grouped as three sub-group (Figure 4).

(1) In each sub-group, we characterized as positive and negative effect
(2) Try to balance with positive and negative impact in sub-group
(3) Another seven cards were added to fulfill the missing element that reflects local problems.
(4) Reconnect the cards group with user strategy
(5) Re-drawing the cards with graphics icon (Figure 5) to improve the readability

IOP Conf. Series: Earth and Environmental Science **159** (2018) 012018 doi:10.1088/1755-1315/159/1/012018

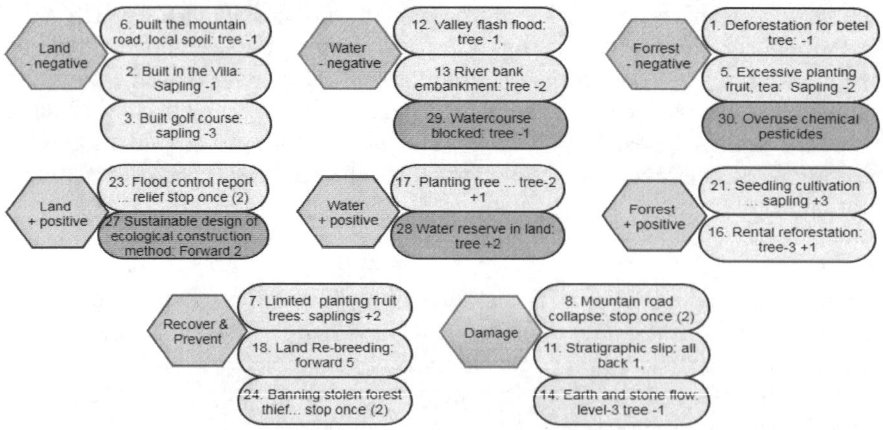

Figure 4. The 21 eco-event cards consist of eco-related elements: plant, land, and water.
In a game, a player faces set of challenges and generate a personal strategy to win a game. Story and rules involve the procedure that provides the structure by which user interacts with system elements. During the playing, users have to concentrate on current situations and practice strategy requires reaching the goal and get the pleasure through achieving and confidence.

The players build their homeland through the game activities. Two users have a chance to view other side and share their experiences through comparing the difference (Figure 6). The game enables people to experience situations in a virtual environment. Those experiences also extend knowledge domain and bring locus of control and lead to motivation to approach more related knowledge.

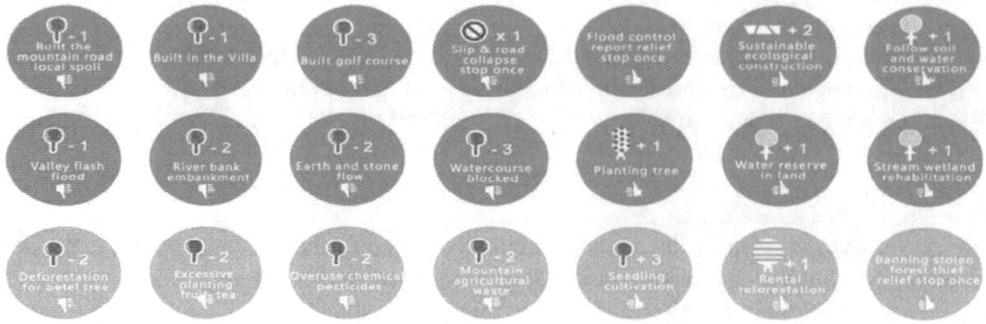

Figure 5. The design of 21 eco-event cards.

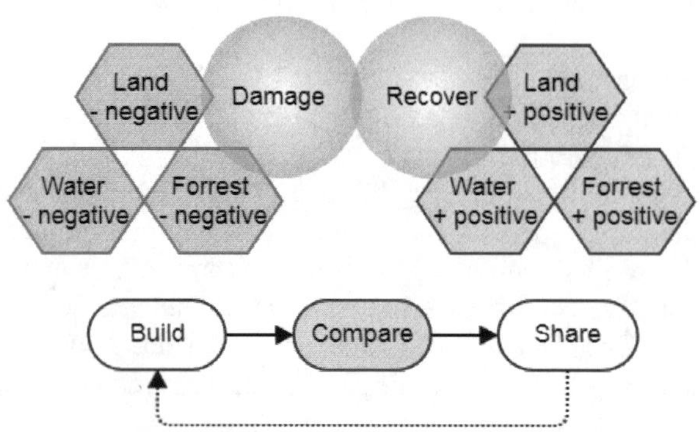

Figure 6. The event cards of positive (recover) and negative (damage) effects. Players have to concentrate on current situations, and practice strategy requires reaching the goal.

4.1. Evaluation-2

After redesigning the cards, we conducted evaluation-2 for observing: (1) The user interface (game cards, tools, and the rules) allow efficient control of the game. (2) Does the game attractive and allows people enjoy playing. (3) Do users develop personal strategy and react with appropriate feedback? Students of the year-1 group at Chaoyang University are invited to play for gathering user response (Figure 7). The significant differences between Evaluation-1 and 2 are age and knowledge level. We were uncertain about the answer of the young adults. Participants were given background story which describes the living and home country to bring the participant into the scenario.

Figure 7. The students joined evaluation-2.

After the brief introduction, the pair of users engaged together in 20 mins playing session. Then, the player filled out a questionnaire with personal comments. The observer also recorded a log of the player's actions coded in according to behaviors and emotional responses. Parts of the question are (where two numbers are average, the standard deviation of 5 points measuring scale):

(1)	Story helped me enter the situation of the board game	4.416	0.792
(2)	I found that water, land, forests are closely linked	4.750	0.452
(3)	I am willing to participate in the conservation activities	4.666	0.492
(4)	Graphic design of the game card, easy to understand	4.666	0.492
(5)	Board steps and rules are complicated	1.333	0.492
(6)	I like to use air cannon, rolling stones	4.833	0.389
(7)	I am willing to play these games again	4.750	0.452

Most of them have a positive feedback, q6 (air cannon, rolling stones) has the highest score. The results indicated the student welcomed 3-D board and physical interaction. A reverse item was placed in q5 (steps and rules are complicated), the low average score of 1.333 indicates the content of cards is understandable, and the player can practice the game without obstacle. Q7 (willing to play again) also has a high score. The results indicated the 3-D board is enjoyable for young adults.

We extracted essential phrases in the sentence collected from student's comments. A Venn diagram is drawn after the relevant group is counted (Figure 8). The size of the circle represents the strength of the phrase appears, the space between the circles represents its' relationship.

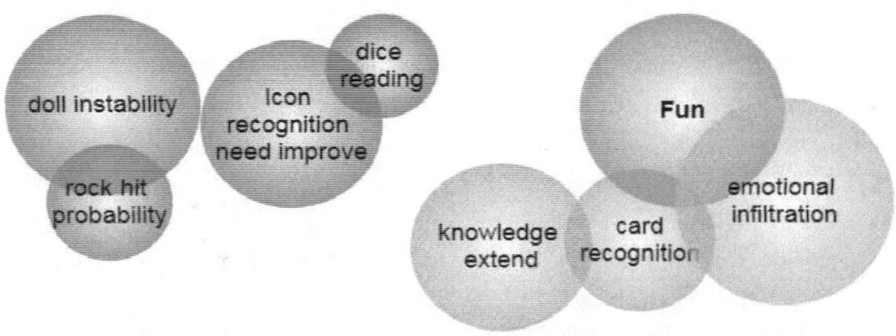

Figure 8. The user response during evaluation-2 (right: positive / left: negative).

They like absorbing the tutorial story and introduction session. The emotional infiltration and fun expression are high. After cards organization, students send several positive responses on "knowledge extension." The information structure of water, forest, and land is suitable for learning. The "Icon recognition need improve" shown the consistency between text and graphic led some confusions. Especially, the difficult of "Read the number of dice" comes from new dot pattern. It better uses traditional pattern adapt to with users' prior experience. "Rock hit probability" and "Doll instability" also need to improve to enhance the credibility. The user's investigation results correlated with observer recorded log; it provides a high correlation between user and observer.

The Pearson correlation coefficient of q2-q3 is 0.816, which is significantly correlated at a level of 0.01 (two-tailed). Question is about (2) knowing the link between water, land, and forests (b) willing to participate in land conservation. Knowledge and experiences have a stronger inference of people's behavior. Familiar with the problem and its causes support people to engage in protective behaviors.

5. Conclusions

The pre-learning session and simplify the information structure of cards made participant aware its relationship more natural. The storytelling and internal structure of the problems are used in the beginning. Simplify the terminology of cards also reduces the complexity of its association. After playing the board game, students are willing to understand more about how things are happening in our surroundings. The virtual game activities build a mutual connection with the real world. The game involves users in a specific situation and gains thinking by making own strategy, so that virtual experience through the game may help a user to explore the real world in future. More precisely knowledge structure need acquire from expert to extend the confidence of their correlation. The three elements interaction (land, water, and forest) can extend to present relationship more precisely. The adjustment of the board game can lower the barrier of using in environmental education.

6. References

[1] Chun-Hung Wu, Su-Chin Chen, Zheng-Yi Feng (2013) Formation, failure, and consequences of the Xiaolin landslide dam, triggered by extreme rainfall from Typhoon Morakot, Taiwan. Landslides, Vol. 11, issue 3, pp. 357 – 367.

[2] Chen, Su-Chin, Wu, Chun-Yi, 2014, "Debris Flow Disaster Prevention and Mitigation of Non-Structural Strategies in Taiwan," Journal of Mountain Science, 11(2): 308-322.

[3] Wu, H.L. and Feng, Z.,2006,"Ecological Engineering Methods in Soil and Water Conservation of Taiwan," Ecological Engineering, 28(4):333-344.

[4] Lo, C.M. and Feng, Z.Y. (2014) "Deformation characteristics of slate slopes associated with morphology and creep," Engineering Geology, Volume 178, pp. 132 – 154.

[5] http://law.moj.gov.tw/Eng/LawClass/LawAll.aspx?PCode=O0120001

[6] Stapp W B. The concept of environmental education, Environmental Education, 1969, 1(1): 30.

[7] Kollmuss A, Agyeman J. Mind the gap: why do people act environmentally and what are the barriers to pro-environmental behavior?, Environmental education research, 2002, 8(3): 239-260.

[8] Hsu, Shih-Jang. "The effects of an environmental education program on responsible environmental behavior and associated environmental literacy variables in Taiwanese college students." The Journal of Environmental Education35.2 (2004): 37-48.

[9] Gobet F, Retschitzki J, de Voogt A. Moves in mind: The psychology of board games. Psychology Press, 2004.

[10] Ramani G B, Siegler R S. Promoting broad and stable improvements in low- income children' s numerical knowledge through playing number board games. Child Development, 2008, 79(2): 375-394.

[11] Zagal J.P., Rick J, Hsi I. Collaborative games: Lessons learned from board games. Simulation & Gaming, 2006, 37(1): 24-40.

[12] Chawla L, Cushing D F. Education for strategic environmental behavior, Environmental Education Research, 2007, 13(4): 437-452.

[13] Nilsson, E. G. (2009). Design patterns for user interface for mobile applications. Advances in engineering software, 40(12), 1318-1328.

[14] https://en.wikipedia.org/wiki/Affinity_diagram.

IOP Conf. Series: Earth and Environmental Science **159** (2018) 012019 doi:10.1088/1755-1315/159/1/012019

Manual and Automatic Volume Measuring Methods for Industrial Timber

Z Pásztory[1], B Heinzmann[1], M-C Barbu[2, 3]

1 University of Sopron, Faculty of Wood Science, Bajcsy Zs. 4, 9400 Sopron, Hungary
2 University of Applied Sciences, Faculty of Wood Science, Markt 136a, 5431 Kuchl, Austria
3 University of Brasov, Faculty for Wood Engineering, Str. Colinei 1, 500068 Brasov, Romania
E-mail: pasztory.zoltan@uni-sopron.hu

Abstract. Given the great importance of timber measurement, various methods for industrial timber dimensional control have been established in the practice over the last decades. They allow generating the relevant results for calculation of the prices at different locations, independently of the measurement time. These methods range from simple manual measurements of the stack using a tape to photo-optical measurement systems. Also other measurement methods like water immersion are currently used in practice. The paper presents the individual manual and automatic measurement methods for industrial timber and their working principles. Furthermore, the advantages and disadvantages concerning the requirements and accuracy of the individual procedures will be explained and their practical use will be discussed.

1. Introduction

The measurement of round wood is probably the most important element of the chain connections between the forestry and the timber industry. In addition to the timber quality, timber volume is the most important parameter for the fixing of prices.

The majority of sawmilling companies have state-of-the-art round wood measurement systems. The round wood is measured either electronically in the factory, or manually and the volume determined on an individual log basis is used for the billing of the timber. Therefore, the one by one measurement of logs plays a special role when it comes to determining the prices.

In contrast to the sawmilling, pulp, paper and wood-based panel industries process round wood assortments of lower diameter and quality referred to as industrial or stacked timber. The ratio of stacked wood versus logs can be quite high, especially in case of broadleaved species. This ratio depends on more factors such as wood species, soil quality, the climate conditions and age of the stand stock. Due to the requests or qualities of the industrial wood, their low value compared to roundwood, and the absence of technical facilities, electronic measurement of these assortments on an individual log basis is not possible. Therefore, other measurement methods must be used, which are capable of representing the volume as accurately and efficiently as possible. For this reason different integrated methods were developed which try to estimate the overall volume of the piled or stacked industrial wood. The volume measurements of stacked wood last back for centuries and different solutions were developed. The methods were continuous developed according to the technical possibilities of the time. The last decade seems to be a technical revolution in the field of industrial wood volume determination. More and more innovative solutions appear on the international markets which compete with each other. Expectedly, within some years the automatic measurement methods will dominate the industrial log markets. Although the present standards and scaling manuals give instructions for manual measurements, however in some cases they mention the possibilities of use photo-optical techniques.

Present paper intends to give a short overview of the manual and automatic measurement systems for industrial logs.

1.1. Requirements and characteristics of the measurement methods

The various measurement methods for industrial timber must meet specific requirements. In general, the measurement method must be characterised by a high degree of rationalisation and at the same time should provide the highest possible level of accuracy. The measurement values must be understandable and verifiable by all persons involved in the processing and sale of timber. In addition, the sequence of the phases timber harvesting, skidding and transport must not be disturbed by the data acquisition and the cost has to be kept as low as possible.

1.2. Historical overview

Already in 1875, as the Würzburg Decisions became effective, the states of the German Empire agreed on the introduction of the units "Festmeter" (solid cubic meter) and "Raummeter" (volumetric cubic meter) as common calculation units for the sale of timber and timber acceptance [1]. An uniform regulation regarding the grading and measurement for sawmill logs and industrial timber was not established until 1936 with the "Reichsholzmessanweisung" [2] in Germany.

One of the first methods to measure the stacked wood filing factor was published by Schnur [3] who measured the tree solid volume and pilled the cut parts segmented diameter classes. He compared the filing factors of the diameter classes. To determine the converting factor of mixed oak cordwood Barrett et al. [4] used planimeter, which was the first photo optical solution. However, Barett *et al.* determined only the filing factor for manual measurements and not the solid content of every pile. The picture was taken from the end of the stack, calibrated by means of scale and put to the edge of the stack. Keepers [5] used a diagrammatic top-plan view of wood measuring device. The picture taken from the stack butt end was analyzed and calculated the ratio of wood. These studies and results prove that the photo analytical solutions offered evident and pragmatic way.

Scientific examinations however revealed that the specified measurement instructions and conversion factors resulted in variation and volume errors between - 12% and + 7% [6].

The provisions of the “Gesetzliche Handelsklassensortierung für Rohholz” [7], which entered into force in Germany in 1969 and were valid until 2009, provided a new regulatory framework for the round wood trade and measurement of industrial timber. In this context, the conversion factors were reviewed in order to calculate the timber volume as accurately as possible.

The new standard “Rahmenvereinbarung für den Rohholzhandel in Deutschland” [8], valid in Germany since 2015, suggests new conversion and reduction factors while using existing measurement methods. This suggests that a representation of the exact wood amount for industrial timber has still not been achieved and thus depends on numerous factors and sources of error.

2. Manual Measurement Methods

In the frame of this article the authors have not the possibility to use or study full overview of manual and automatic measurement methods, that is why the stream line methods are presented which are used widely in the practice or in case of automatic measurements can be a popular solution in the near future.

2.1. Sectional volumetric measuring method

The sectional volumetric method is widely used to measure industrial stacked timber also in Europe and in North and South America and also in other parts of the Earth. The method can be used for different size of the stack in width, length and height. The most common length of stacked logs is 1.0; 2.0 and 3.0 m in Europe and 2.44 m (8 ft.) in North America. The measurement method is applied to the completed stack and the determined volume is used as a measure for sale.

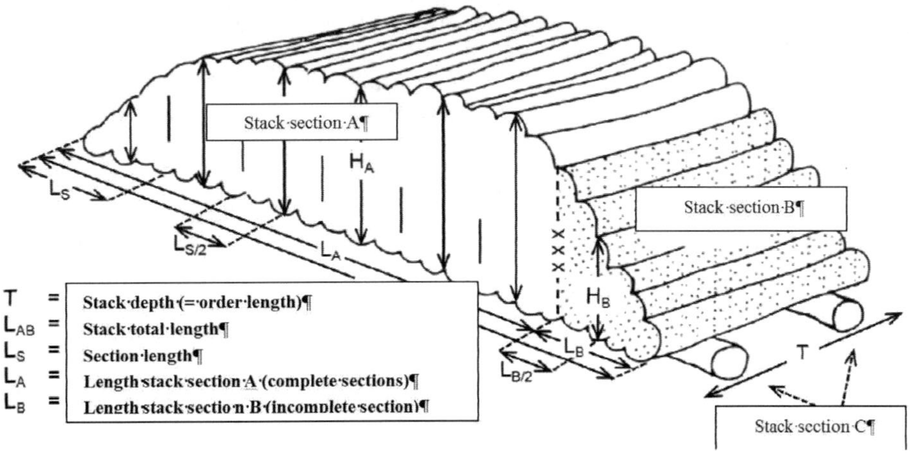

Figure 1. Measurement using the sectional volumetric method (RVR, 2015).

The length, height, and depth of the stack (corresponds to the order length of the assortment) measures are used to determine the gross volume in the unit cubic metres. Usually the measurement values are determined using a measuring tape for the length and a yard stick for the height figure 1.

Depending on the overall length of the stack (LAB), the stack is divided in defined sections of equal length (LS) (see table 1) for the length measurement. The section size can be different according the regulations in different countries. After division and marking of the respective sections, an incomplete section is obtained at the end of the stack (stack section B) with the length LB. This is measured separately and the volume is added to the determined stack volume from LA.

Table 1. Division of section lengths according RVR, 2015.

Total stack length (LAB)	Section length (LS)
up to 10 m	1 m
10 to 20 m	2 m
20 to 40 m	4 m
40 to 60 m	6 m
60 to 80 m	8 m
80 to 100 m	10 m

The respective section centres (LS/2) are relevant for the subsequent height measurement and are therefore marked using colour spray.

The section heights are determined vertically to the stack and on both sides. The measuring points at the bottom and topside of the stack must always be determined where the vertical marking leaves the end face of the last log. Height compensation within the respective sections in case of irregular placement of the logs at the top of the stack is not allowed. There are regulations that determine a recommended or maximum height [9]. There are regulations to determine height or other dimension corrections in special cases [10] [8].

The gross volume resulting from the measurement values determined (length and height of the stack) and the order length are reduced according to a fixed measure allowance of 4% and additional reduction factors depending on the wood species and the assortment length (see table 2). The resulting net volumetric measure without bark valid for the timber sale is thus calculated according to the following formula.

$$net\ volumetric\ measure\ (m^3ub) = (H \times L \times T) \times reduction\ factor\ (\%)$$

Table 2. Reduction factors depending on the timber length and type according RVR 2015.

Wood species	Assortment length	
	2m	3m
spruce/fir/Douglas fir	0.96 – 0.94	0.94 – 0.92
pine/larch	0.94 – 0.925	0.92 – 0.905

The Swedish SDC 2014 gives a similar solution a little more sophisticated way by calculating the individual section overall volume instead of average value of stack height.

2.2. Conversion factors

The calculated overall volume comply the solid wood content and the air space between the logs. The conversion factor is the relationship between staked measure and solid contents. The ratio between these two components of the pile can be different within a range of 0.50 to 0.80. The higher number means the higher solid wood content in a stacked cubic meter. Some prescription defines this ratio in percentage and do not make any diversification within an assortment [11]. Laar and Akça [12] determine the conversation factor 0.80 for pulpwood under bark and in the case of firewood, which is recovered from wood not meeting the quality standards for sawlogs, a conversion factor of 0.70 over bark. There are other official regulations make a fine diversification of converting factors for different assortments and wood species [13]. Scientific investigation was done for the more accurate determination of conversion factor function of different conditions of stacked piles [14]. The optimal cost of estimation stacked wood solid content was investigated by Smith [15].
Meyen and O'Connell [16] show a special method on how to determine the actual conversion factor using a 0.7 by 0.7 m grid and counting the numbers and diameters of the logs.

2.3. Xylometry

The difficulty of log volume measurements derives from the irregular shape of the trees and branches. The stems have bends and twist and irregular surface shapes which inhibit to develop a perfect mathematical model for calculate the real surface and volume. A measurement method which could regenerate the surface unevenness would be perfect. The log immersion to water approximates this method the closest or appropriate way. Özçelik *et al.* [17] investigated the xylometry method and compared to others. Despite of the difficulties of measurement execution the results shows very good correlation to the real volume of the stack.

2.4 Automatic Measurement Methods

The automatic measurement systems use technical background and informatics for analysis of data. By the time more and more innovative solutions come to the light and become available on the market. The electro-mechanical harvesting technology is one of the earliest ways to determine the solid contents of stacked wood. Despite of disadvantageous ~~of~~ measurement accuracy the advantage is that the measurement happens in the same time as the harvesting. No additional activity and time are requested for measurement.
Photo optical solutions strengthened in the last years and such application can be run also in smart phones. The more sophisticated technologies use other additional equipment for taking pictures and for analysing the of images.

2.5. Harvester measurement

In Germany, half of the timber harvest is already carried out fully mechanically today, which is why measurement by means of harvester is gaining increasing importance [18]. The measurement equipment contained in the harvester unit automatically captures the data of the respective processed log.

The length measurement of the sections is generally carried out by means of toothed feed rollers equipped with sensors. These rollers are pressed mechanically or hydraulically to the surface of the logs located in the processing unit and move along during processing.
With diameter measurement data acquisition is either effected by determining the distance of the feed rollers to each other or indirectly by determining the opening angle of the delimbing knife during log processing figure 2 [19].

Figure 2. Harvester unit [29].

In addition to the individual log volume determination based on the measured lengths and diameters, the quantity is required for the calculation of the total volume of a stack. This is determined fully automatically with each log cut with the chain saw and stored in the system.
Continuous calibration of the measurement systems and changes of the parameter settings according to the respective timber condition are a basic requirement for obtaining correct data [20].

2.6. Photo-optical systems
Photo-optical measurement has gained increasing importance over the past years. Currently, different systems exist on the market allowing an automatic mobile capture of individual stacks by means of image analysis [21].
A differentiation is made between two technologies with distinct principles of operation [22]. There is a largely automated measurement of stacks where the measurement is carried out by a camera system on the roof of the car with simultaneous calculation by means of special software. Other photo-optical systems are based on manual capture using digital cameras and automatic external evaluation of the photo material.
The full automated measurement technology of wood stacks allows data collecting at different viewing angles, distances, and speeds. The measurements are carried out using a stereo camera system with two cameras positioned in a certain angle to each other figure 3.
Individual images are recorded in a fixed and known scale without requiring additional reference points. The camera generates several images per second and a stereo film of the stack front. Based on this film, a 3D model with all parameters relevant for the timber volume, such as cubic measure, solid measure, quantity, diameter distribution, is calculated [23]. Once recorded, the measurement data is immediately transmitted to the server and can be imported and corrected where necessary using a monitor installed in the car [24].

Figure 3. Camera log measurement system by the company Dralle [25].

Photo-optical measurement of the timber is also possible by using a digital camera. Several frontal individual images (i.e. of the end face) of a stack are recorded. In each of these individual images, a scale (e.g. a yardstick) must be used as a reference during image analysis.

The evaluation of the image data cannot be performed on site. Instead, the obtained data sets of the individual images are sent to a central web server in an automated manner. This web server assembles the images of the individual stack sections to a total image of the stack, similar to a panoramic photo figure 4.

Figure 4. Individual steps of photo-optical stack measurement [28].

The recorded images are used to determine the log quantity, end face diameter, and the storage density of the stack. Based on these values and the uniform log lengths, the total volume of the stack in cubic metres, the average mid-diameter and the size class distribution are calculated first. By application of conversion factors it is later also possible to calculate the total volume in solid cubic metres [26].

3. Discussion

The manual and automatic measurement methods listed and explained will now be discussed with respect to their requirements, accuracy, and rationality.

3.1. Requirements and accuracy

To ensure that the determined volumetric measure is correct, certain conditions and minimum requirements must be fulfilled [8]. If these are not fulfilled, the sectional volumetric measurement method cannot be used for determination of the sales measure. Hence, it could only be used for determination of a control measure.

- Minimum stack size of 20 m³
- Stack height as uniform as possible and max. 2.50 metres
- Stacks separated according to wood species
- Stack placed in a dense and flush way
- One order length per stack of wood
- Building of stacks without inclusion of branches and foreign material (snow, dirt, etc.)
- Free access to front and rear of the stack

Even though the aggregates and the board computers are constantly further developed by the manufacturers, the harvester measurement process and the problems connected to this have not changed according to the present state of knowledge. By permanent calibration of the measurement system and changes of the parameter settings according to the respective timber condition, certain measurement accuracy can be reached, but this does not reflect the exact timber volume.

The main reason for this is the different condition during the harvesting of the timber. For example, if the logs are processed while they are producing sap, major bark portions are removed during delimbing. As a consequence, log portions with and without bark alternate during measurement making diameter determination more difficult. With length determination there is often the problem that with very knotty or bumpy logs the hydraulically pressed measurement wheels temporarily come off the log and that jamming of the length measurement wheel due to, for example, peeled of bark

pieces, leads to incorrect results. In addition, external influences such as time of the year, weather, and mechanical factors affect the measurement by harvester.
In practice, errors often occur when determining the quantity, caused by incorrect operation of the systems or software errors.
Due to these error sources the harvester measure is only used as a control measure and for logistic purposes. For calibration reasons it cannot be used as a sales measure relevant for the calculation of prices [20].
In Denmark, approximately 300.000 m^3 of roundwood has been photo-optical measured by the Federal Agency for Nature Conservation every year since 2009 [27]. It is claimed that regular calibration of the measurement equipment ensures that the specified error tolerances can be met. The accuracy ranges of these tolerances are not specified and mentioned.
The manufacturers of photo-optical equipment advertise with good and precise measurement results of their products without specifying exact tolerance ranges. According to the current state of knowledge, individual scientific studies have been conducted on the comparability of the results of photo-optical measurement and assessments on an individual log basis of sawmill timber. Tests regarding the measurement of industrial timber have not been performed yet.
In general, it can be said that the accuracy of the measurement results largely depends on the respective system, the stack quality and the operation. There are detailed regulations on the distance and the angle to the stack front in which the recording camera is to be positioned. Small changes already lead to inaccuracies and incorrect results. In addition, external influences, such as the incidence of light, affect the correctness of the measurement data as due to shadows or minor darkness individual end faces may not be detected and measured. Similar problems are caused by dark or dirty end faces. While it is possible to subsequently process theses logs with a special software, this is connected with significant extra time and efforts.

3.2. Rationality
If all conditions and requirements are met, the volumetric measurement method is characterised by a high degree of rationalisation. This is mainly beneficial in case of large stacks. Consequently, it is one of the fastest and most cost efficient manual methods [1]. In addition, the method can be reproduced by all persons involved in the wood sale at any time and the measurement result is transparent for all parties.
The harvester measurement allows an early determination of measures and therefore a quickly available measure for logistical processes. The data can be made digitally available by means of modern software and interconnected with other database systems via radio.
As explained, the photo-optical measurement methods offer the possibility of early determination of measures. In practice, the method is characterised in particular by its uncomplicated and easy way of data provision and the comprehensive documentation of the measurement. This allows for rationalisation of the complete business process of the timber sale and its transparent representation for all persons involved.

4. Conclusion
In summary, it can be concluded that although the automatic measurement methods for roundwood developed over the past years (harvester, photo-optical methods) have significantly improved with respect to the rationalisation of the timber measurement compared to the manual methods, the measurement accuracy required for the sale of timber can be achieved. It remains to be seen whether these systems can be improved further with respect to the measurement accuracy and can thus become easy and quick to calibrate.
Therefore, at present, the only alternatives available for determination of a sales measure for industrial timber are the manual measurement methods as used for several decades. The problems regarding the requirements and accuracy mentioned often give rise to disputes and arguments between the forestry and timber industry. Depending on the situation of the timber market, each try to impose the measurement method which is the best, i.e. most beneficial for the respective party.

Looking to the future, it would be necessary and desirable in the interest of the overall forestry and timber industry to generate a measurement system for industrial timber which complies with the required dimensional tolerances while ensuring a rational procedure at the same time. With today's technical possibilities and the experience from several decades in the sale of industrial timber this should be possible.

Acknowledgement
The described article was carried out as part of the „Roadmap for Structural Changes of the University of Sopron" - nr. 32388-2/2017 INTFIN. The Ministry of Human Capacities of the Hungarian Government supported the realization of this project.

5. References

[1] WILWERDING, A. (1995). Problematik der Vermessung von Profispaner-Fixlängen und Entwicklung eines Raummaßverfahrens. Freiburg im Breisgau.

[2] PLATTE, G. (1957). Untersuchungen zur Geschichte der forstlichen Rohholzsortierung und deren Weiterentwicklung. Freiburg im Breisgau.

[3] SCHNUR, G. L. (1932): Converting Factors for Some Stacked Cords, *Journal of Forestry*, Volume 30, Number 7, pp. 814-820 (7).

[4] BARETT, L. I., BUELL, J. H., RENSHAW, J. F. (1941). Some Converting Factors for Mixed Oak Cordwood in the Southern Appalachians, *Journal of Forestry* V 39:6, pp. 546-554(9).

[5] KEEPERS, C. H. (1945). A New Method of Measuring the Actual Volume of Wood in Stacks, *Journal of Forestry*. Volume 43, Number 1, pp. 16-22(7).

[6] DIETZ, P. (1966). Die Vermessung von Industrieholz nach Gewicht. Freiburg im Breisgau: Institut für Forstbenutzung und forstliche Arbeitswirtschaft.

[7] FORST-HKS. (1969). Gesetzliche Handelsklassensortierung für Rohholz, BGBl.I S.149. Ergänzungsbestimmungen für Baden-Württemberg.

[8] RVR. (2015). Rahmenvereinbarung für den Rohholzhandel in Deutschland. Deutscher Forstwirtschaftsrat e.V.

[9] Nova Scotia Scaling Manual (2007). Nova Scotia Scaling Manual 2nd Edition, Department of Natural Resources Renewable Resources Branch Forestry Division, October 2007.

[10] SDC SDC´s instructions for timber measurement (2014). MEASUREMENT OF ROUNDWOOD STACKS, sdc.se,

[11] Manitoba Conservation and Water Stewaedship (2013) Manual of Scaling Insrtuctions, fourth edition, September 2013, p 38.

[12] Laar A., Akça A. (2007): Forest Mensuration. Springer V. ISBN-13 978-1-4020-5990-2

[13] New Brunswick, Forest Management Branch Natural Resources and Energy (2003) New Brunswick Scaling Manual 3rd edition, ISBN 1-55236-207-8, Appendix A, February 2003.

[14] BARROS, M. V., FINGER, C. A. G., SCHNEIDER, P. R., SANTINI, E. J. (2008): Fator de Cubicáco Para Toretes de Eucalyptus grandis e Sua Variácio Com o Tempo de Exposicáo ao Ambiente (Cubivation Factor of Stacked Wood of Eucalyptus grandis and its Variation with the Time of Exposition to the Environment). *Ciencia Florestal*. Volume 18. pp. 109-119.

[15] SMITH, V. G. (1979): Estimating the solid wood content of stacked logs with minimum total cost. *Can. J. For. Res*. Volume 9. pp 292-294.

[16] Meyen S. and O'Connell K (2012) Stacked Timber Measurement, The 2012 ITGA Forestry & Timber Yearbook, Irish Timber Growers Association, pp. 50-56.

[17] Özçelik R, Wiant Jr H.V., John R. Brooks (2008) Accuracy using xylometry of log volume estimates for two tree species in Turkey, *Scandinavien Journal of Forest Research*, 23 2008 pp272-277

[18] WAGELAAR, R. (1997). Rundholzvermessung mit Harvestern. AFZ/ Der Wald 15 pp.

[19] SAUTER H. U. (2014). Vermessung und Qualitätsbestimmung des Rohholzes an der Schnittstelle zwischen forstlicher Produktion und Weiterverarbeitung in der Holzindustrie. FVA Baden Württemberg: Abteilung Waldnutzung - Alumnikolloquium.

[20] DIETZ, H.-U. & URBANKE, B. (2009). Harvestervermessung - Hochmechanisierte Holzernte braucht verlässliche Produktionsdaten. KWF, *Fachzeitung für Waldarbeit und Forsttechnik* 11.
[21] JORGESEN, E. R., & KRISTIANSEN, L. (2008). Digitale Fotovermessung von Industrieholzabschnitten. AFZ/ Der Wald 2/
[22] FINK, F. (2004). Foto-optische Erfassung der Dimension von Nadelrundholzabschnitten unter Einsatz digitaler, bildverarbeitender Methoden. Dissertation. Freiburg
[23] DRALLE. (2014). sScale - Measure, track and trade. Hoersholm/DK: Produktbroschüre.
[24] LWF BAYERN. (2008). Holzaufnahme im BaySF-Logistikprozess. Freising: Magazin der Bayrischen Landesanstalt für Wald- und Forstwirtschaft.
[25] Dralle A/S. (2015). Drallle A/S, Hoersholm, Denmark. http://www.dralle.dk
[26] FOVEA. (2014). Foto-optische Holzvermessung mit dem Smartphone. Uslar/DE: Produktbroschüre.
[27] HOLZ-ZENTRALBLATT NR.9. (2015). Fotovermessung in den dänischen Staatswäldern.
[28] AFoRS. (2015). Scheller Systemtechnik GmbH, Wismar, Gerrnany. http://www.afors.de
[29] Ponsse Plc. (2015). Ponsse Plc, Vieremä, Finland. http://www.ponsse.com

Acknowledgement
The described article was carried out as part of the „Roadmap for Structural Changes of the University of Sopron" - nr. 32388-2/2017 INTFIN. The Ministry of Human Capacities of the Hungarian Government supported the realization of this project.

IOP Conf. Series: Earth and Environmental Science **159** (2018) 012020 doi:10.1088/1755-1315/159/1/012020

Substation Reliability Evaluation Considering the Failure Events

Sung Hun Kim[1] and Dong Ju Lee[2]

1 Professor, Kongju National University, Republic of Korea
2 PhD candidate, Chungnam National University, Republic of Korea
E-mail: shkim3456@kongju.ac.kr; david1973.lee@gmail.com

Abstract. In this paper, bus and breaker structures of typical two substations were modelled by graph circuit to evaluate each reliability considering the various failure events and compare them each other. Passive failure event, active failure events, breaker stuck conditions and overlapping failure events in substation were conceptually explained and the reliability equations were presented to calculate the failure rate, repair rate and unavailability of each failure events. Also, deduction algorithms were implemented as a software programming to find the minimal cut sets for passive failure events and active minimal cut sets for active failure events. In the case study, reliability indices of each failure events and entire reliability indices (failure rate, annual outage time and MTTR) of typical two substations were calculated and quantitatively compared.

1. Introduction

A substation consists of buses, transformers, breakers, and diverse protective devices for safe operation of the system. When a failure has occurred in a component of a substation, relatively complicated switching actions occur. The evaluation of the reliability of a substation considering such switching actions requires clear inference of failure events including passive failures, active failures, breaker stuck conditions and overlapping failures. In addition, minimal paths are inferred from failure events in the bus system to calculate failure rates, annual repair time, and unavailability, which are reliability indices, based on the minimal cut set and the active minimal cut set.

To infer failure events of substations, a method to prepare circuit graphs for substation components and infer the minimal path from the circuit graphs to determine the minimal cut set and the active minimal cut set was proposed [1]. For reliability index calculation considering substation's switching states, a new technique was proposed that is setting a two state model and a three state model to obtain minimal cut sets for passive failures and active failures and determining active minimal cut sets among the selected minimal cut sets[2]. A technique was presented that is to prepare an array of access of components by finding and storing all minimal paths between the power supply point and all load points and determine the minimal cut set as the cut set to which power supply is stopped through combinations of the paths[3].

In this paper, bus and breaker structures of typical substations were modelled by graph circuit to infer the failure events (passive failure events, active failure events, breaker stuck conditions and overlapping failure events) in substation and these failure events are conceptually explained. Deduction algorithms were implemented as a software programming to find the minimal cut sets for passive failure events and active minimal cut sets for active failure events. As a case study, reliability indices of each failure events and entire reliability indices (failure rate, annual outage time and MTTR) in the ring bus substation and double bus double breaker substation were calculated and quantitatively compared. The results of analysis of the reliability indices of typical two substations can be utilized as basic data in the design and maintenance schedule planning of substations with diverse bus structures hereafter.

2. Failure events in substation

2.1. Graph modelling of typical substation

Figure 1 and 2 shows the system configuration and graph circuit of a substation with a ring bus structure and a substation with double bus double breaker structure, which are widely used in modern substations because of its high flexibility and reliability.

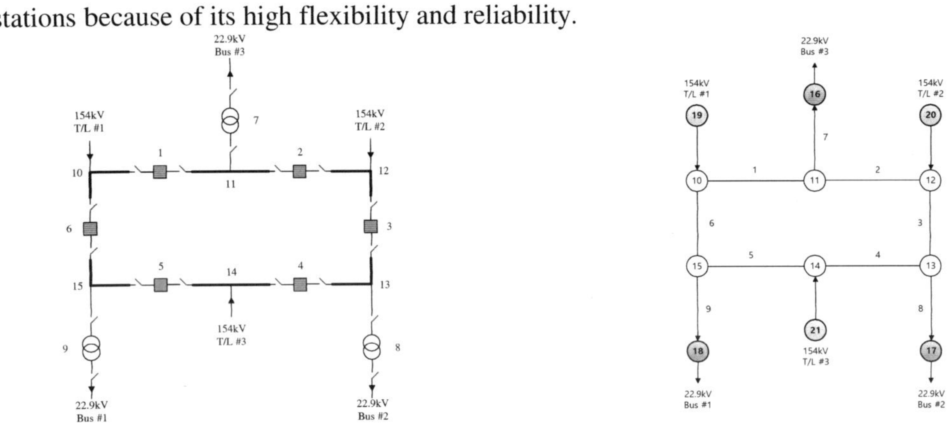

(a) System configuration (b) Graph circuit

Figure 1. Substation with ring bus

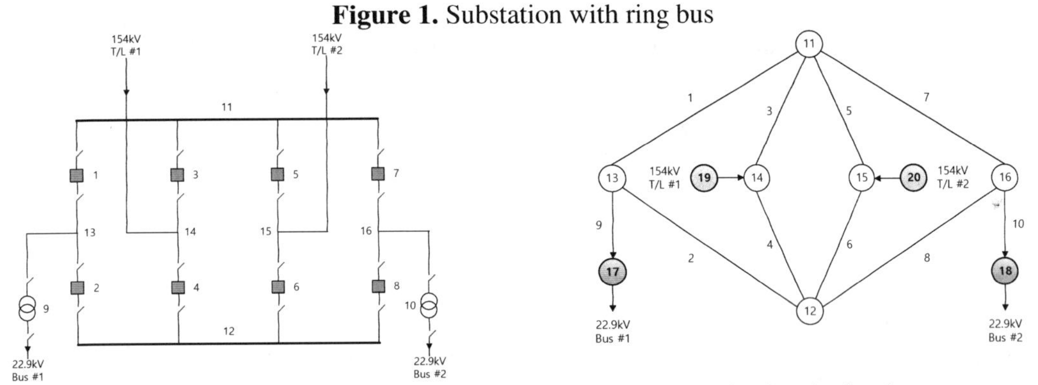

(a) System configuration (b) Graph circuit

Figure 2. Substation with double bus double breakers

The substation shown in Figure 1 consists of six breakers (branch 1~6), three transformers (branch 7~9), and ring bus (node 11~15), Three supply points (node 10, 12, 14) are connected to the transmission lines and two transformers (branch 9, 10) are connected to user load points (node 17, 18). The substation shown in Figure 2 consists of eight breakers (branch 1~8), two transformers (branch 9, 10), and two buses (node 11, 12). Two supply points (node 14, 15) are connected to the transmission line and three transformers (branch 7~9) are connected to user load points (node 16~18)

In these substations, there are various failure events of internal components for each status transitions among the normal, switching and repair status.

2.2. Passive failure events

Passive failure events mean all failures that do not induce the actions of protective breakers located around the component where failure occurred and do not affect normally operating components where no failure occurred. Since a passive failure event generally causes first order contingency, power outage occurs at one load point. The probability and frequency of occurrence are relatively low and in the case of bus and transformers, passive failures are ignored because the passive failure frequency is very close to zero.

2.3. Active failure events

Active failure events refer to all failures that induce the actions of protective breakers adjacent to the component where failure occurred and affect normally operating components where no failure occurred. For instance, when a failure has occurred due to a short-circuit accident, the component where the failure occurred may be separated or switching action may be taken and the function of the component can be continuously performed by other normally operating components. The system may be restored by repairing or replacing the component where the failure occurred.

2.4. Breaker stuck condition

A state where a breaker does not act despite that trip signals have been sent because a failure has been detected by the protective system is called stuck condition. A breaker stuck condition occurs when a breaker failed to act due to an abnormal action of the protective system or due to any trouble in the relay or the breaker per se. Stuck conditions lead to more serious effects on switching substations and provide causes of high order contingency. Therefore, the simulation and analysis of stuck conditions may be more important even if the probability of occurrence of stuck may be low.

2.5. Overlapping failure events

An overlapping failure event is a case where at failures occur sequentially in at least two components, that is, a case where a failure occurs first in a component and failures continuously occur in other components during the recovery time of the component where the failure occurred. In the case of overlapping failure events, usually up to second order overlapping failure events in two components are considered and higher order failure events may be disregarded.

3. Calculation of reliability indices

3.1. Minimal cut sets

The various efficient algorithms for finding the minimal cut sets for the output nodes of any network to find the passive failure events were presented and the algorithm in reference [3] are applied in this study.

3.2. Active minimal cut sets

The various algorithms to infer the active minimal cut sets to find the active failure events were presented and the algorithm in reference [4] are applied to find the following three failure events. The concept of these three failure events can be summarized as shown in Figure 4.

3.2.1. Active minimal cut sets

In Figure 3(a), when the first order minimal cut set is transformer 9 and the first order failure event has occurred at node 13, breaker 1 and breaker 2 correspond to the first order active minimal cut set.

3.2.2. Active failure event + total failure event

In Figure 3(b), the second order active minimal cut set for the second order minimal cut set for breaker 1 is 4A, 6A, and 8A and the second order active minimal cut set for breaker 2 is 3A, 5A, and 7A, which are in the state of active failures.

3.2.3. Active failure event + active failure event

In Figure 3(c), when the breaker 3 and the supply point (2) (node 15) are in the state of active failures, 6A and 8A are in the state of active failures and when the breaker 4 and the supply point (2) (node 15) are in the state of active failures, 5A and 7A are in the state of active failures.

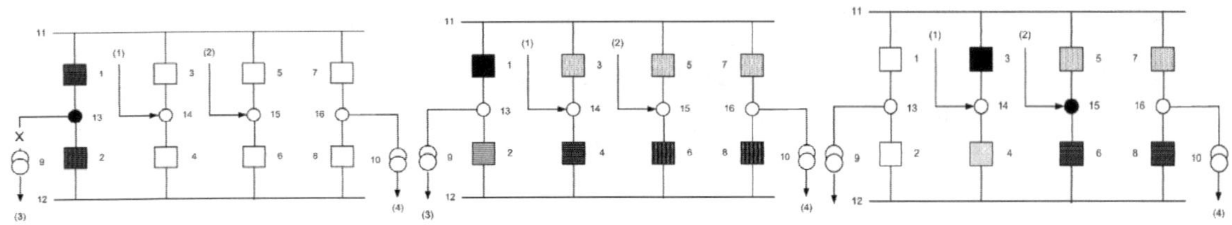

(a) (b) (c)

Figure 3. Active minimal cut sets

3.3. Calculation of reliability indices

The reliability indices for load points can be calculated using the minimal cut sets and active minimal cut sets for the load points [5].

In the case of total failure events in which a failure of one component leads to total failure, the reliability indices can be obtained from the first order and second order minimal cut sets. Reliability indices for first order passive failures of a component are given as the reliability data value of the component. On the other hand, the reliability indices for the second order passive failure events due to passive failures of two components can be calculated regarding that the two components are in the state of parallel connection.

In cases where individual maintenance works for individual components are not considered, the failure rate (λ_{pp}) for second order passive failure events can be indicated as shown by equation (1). In general, when $\lambda_i r_i \ll 1$, the repair rate(r_{pp}) and unavailability(U_{pp}) can be obtained with equations (2) ~ (3).

$$\lambda_{pp} = \frac{\lambda_1\lambda_2(r_1+r_2)}{1+\lambda_1 r_1+\lambda_2 r_2} = \lambda_1\lambda_2(r_1+r_2) \quad (1)$$

where, λ_1,λ_2: failure rates of components 1 and 2
r_1,r_2: repair time of components 1 and 2

$$r_{pp} = \frac{r_1 r_2}{r_1+r_2} \quad (2)$$

$$U_{pp} = f_{pp} r_{pp} = \lambda_{pp} r_{pp} = \lambda_1\lambda_2 r_1 r_2 \quad (3)$$

In cases where individual maintenance works for individual components are considered, the failure rate (λ_{pm}), repair rate(r_{pm}), and unavailability(U_{pm}) for second order passive failure events can be obtained with equations (4) ~ (6).

$$\lambda_{pm} = \lambda_1''(\lambda_2 r_1'') + \lambda_2''(\lambda_1 r_2'') \quad (4)$$

where, λ_1'',λ_2'' : numbers of times of maintenance of components 1 and 2
r_1'',r_2'' : numbers of times of maintenance of components 1 and 2

$$r_{pm} = U_{pm}/\lambda_{pm} \quad (5)$$

$$U_{pm} = \lambda_1''(\lambda_2 r_1'')\frac{r_1'' r_2}{r_1''+r_2} + \lambda_2''(\lambda_1 r_2'')\frac{r_1 r_2''}{r_1+r_2''} \quad (6)$$

The reliability indices for active failure events for which switching actions are considered can be calculated from the first order and second order active minimal cut sets.

The reliability indices for first order active failures of a component can be easily obtained using the switching time value for the component. That is, when one component is in the state of active failure, in cases where individual maintenance works for individual components are not considered, the failure rate (λ_{ap}) and repair rate (r_{ap}) for passive failure events are as shown by equations (7) and (8).

$$\lambda_{ap} = \lambda_1^a \quad (7)$$

$$r_{ap} = s_1 \quad (8)$$

where, λ_1^a : active failure rate of component 1
s_1 : switching time of component 1

In cases where components are in the state of second order active failures, there may be cases where an active failure of component 1 and the total failure of component 2 overlap with each other and cases where an active failure of component 1 and an active failure of component 2 overlap with each other.

First, cases where an active failure of component 1 and the total failure of component 2 overlap with each other are cases where an active failure of component 1 occurs during the repair time of component 2. The failure rate (λ_{ap}) and repair rate(r_{ap}) in such cases can be calculated as shown by equations (9) and (10).

$$\lambda_{ap} = \lambda_1^a(\lambda_2 s_1) + \lambda_2(\lambda_1^a r_2) = \lambda_1^a \lambda_2(s_1 + r_2) \tag{9}$$

$$r_{ap} = \frac{s_1 r_2}{s_1 + r_2} \tag{10}$$

Since the repair time of a component is generally sufficiently larger compared to the switching time ($r_2 \gg s_1$), the probability of failure of component 2 during the switching time of component 1 may be ignored. Therefore, equation (9) can be approximated into equation (11) and equation (10) can be approximated into equation (12).

$$\lambda_{ap} = \lambda_1^a(\lambda_2 r_2) \tag{11}$$

$$r_{ap} = s_1 \tag{12}$$

Cases where two active failures overlap with each other are cases where active failure of component 1 occurs during the maintenance time of component 2 and the failure rate (λ_{am}) and repair rate (r_{am}) in such cases can be calculated as shown by equations (13) and (14).

$$\lambda_{am} = \lambda_2''(\lambda_1^a r_2'') \tag{13}$$

$$r_{am} = \frac{s_1 r_2''}{s_1 + r_2''} \tag{14}$$

Since the maintenance time of a component is generally sufficiently larger compared to the switching time ($r_2'' \gg s_1$), equation (14) can be approximated into equation (15).

$$r_{am} = s_1 \tag{15}$$

When the active failure of a component overlaps with the breaker stuck condition (breaker stuck condition occurrence probability (P_C), the failure rate (λ_{ap}) and repair rate (r_{ap}) can be obtained with equations (16) and (17).

$$\lambda_{ap} = \lambda_1^a \times P_C \tag{16}$$

$$r_{ap} = s_1 \tag{17}$$

4. Case study

As a case study, substation with ring bus and substation with double bus double breakers were selected to analyse the reliability indices of each failure events and total reliability indices (failure rate, annual outage time and MTTR) in these substations

4.1. Minimal cut set and active minimal cut set

Based on the results of implementation of the software programming, the minimal cut sets and active minimal cut sets can be summarized as shown in Table 1 and Table 2.

4.2. Reliability indices

For the evaluation of the reliability of a ring bus substation as shown in Figure 1 and evaluation of the reliability of a double bus double breaker substation as shown in Figure 2, in the states of overlapping total outage and maintenance outage based on the load point, a passive failure model, an active failure model, an active failure and passive failure model, indices for failure rates, repair time, and unavailability considering the passive failure model and the active failure model were obtained. To calculate the reliability of substations, input data for the accident rate, repair time, and switching time of major facilities are important. However, in reality, since failure rate data recorded at the site were insufficient, the reliability data as shown in Table 3 from Hamze's paper, which is the most frequently cited, were used[6]. In addition, individual maintenance works for components were assumed to be performed for 8 hours every year for each component.

IOP Conf. Series: Earth and Environmental Science **159** (2018) 012020 doi:10.1088/1755-1315/159/1/012020

Table 1. Minimal cut set at Bus # 3 (node 16) in ring bus substation.

Order of failure events	Number of events	Components
1st order total failure	2	7, 11
2nd order total failure	4	1+2, 1+12, 2+10, 10+12
Active failure	2	1A, 2A
Active failure + Total failure	4	6A+2, 3A+1, 6A+2, 3A+10
Active failure + Active failure	1	3A+6A

Table 2. Minimal cut set at Bus # 1 (node 17) in double bus substation

Order of failure events	Number of events	Components
1st order Total failure	2	9, 13
2nd order Total failure	5	1+2, 1+12, 11+2, 11+12, 14+15
Active failure	2	1A, 2A
Active failure + Total failure	16	3A+12, 5A+2, 7A+2, 4A+1, 6A+1, 8A+1 3A+12, 5A+12, 7A+12, 4A+11, 6A+11 8A+11, 3A+15, 4A+15, 5A+14, 6A+14
Active failure + Active failure	6	3A+6A, 3A+8A, 5A+4A, 5A+8A, 7A+4A, 7A+6A

Table 3. Substation components reliability data.

Component Name	Total Failure Rate λ [failure/yr]	Repair Time r [hour]	Active Failure Rate λ_a [failure/yr]	Switching Time s [hour]	Stuck Breaker Probability P_c
Breaker	0.010	12	0.010	1	0.06
Transformer	0.040	40	0.040	1	-
Bus	0.010	4	0.010	-	-

Note 1. Uncoordinated Maintenance outage rate λ'' = 1 [time/year], Uncoordinated Maintenance time r'' = 8 [hours]

As shown in Table 4 and Table 5, the failure rate in cases where the individual maintenance works for individual components were not considered is 0.05[failure/yr], and the failure rate in cases where individual maintenance works were performed is 0.000457 [failure/yr] in case of ring bus substation and is 0.000283 [failure/yr] in case of double bus double breaker substation indicating that the failure rate drastically decreased when individual maintenance works were performed so that the improvement of overall operation reliability can be quantitatively identified.

Table 10 and Table 11 show a summary of the reliability indices for events where two active failures overlapped with each other and the total failure rate of those events was calculated as 0.0000000806 [failure/yr] in case of ring bus substation and 0.000000137 [failure/yr] in case of double bus double breaker substation. From these results, it can be quantitatively identified that the events shown in Table 6 and Table 7 are first order active failure events and have larger effects on the reliability of the entire system than the other overlapping failure events shown in Tables 8 ~ 11.

Table 4. Overlapping total outages and a maintenance outage in ring bus substation.

Failure Event	λ_{pp} [failure/yr]	r_{pp} [hour]	U_{pp} [hour/yr]	λ_{pm} [failure/yr]	r_{pm} [hour]	U_{pm} [hour/yr]
7	4.00E-02	40.00	1.60E+00	-	-	-

11	1.00E-02	4.00	4.00E-02	-	-	-
1 + 2	2.74E-07	6.00	1.64E-06	1.83E-05	4.80	8.77E-05
1 + 12	3.15E-07	6.00	1.89E-05	1.14E-04	4.99	5.70E-04
2 + 10	3.15E-07	6.00	1.89E-05	1.14E-04	4.80	5.48E-04
10 + 12	3.62E-05	6.00	2.17E-04	2.10E-04	4.80	1.01E-03
Total	5.00E-02	32.78	1.64E+00	4.57E-04	4.85	2.21E-03

Table 5. Overlapping total outages and a maintenance outage in double bus substation

Failure Event	λ_{pp} [failure/yr]	r_{pp} [hour]	U_{pp} [hour/yr]	λ_{pm} [failure/yr]	r_{pm} [hour]	U_{pm} [hour/yr]
9	4.00E-02	40.00	1.60E+00	-	-	-
13	1.00E-02	4.00	4.00E-02	-	-	-
1 + 2	2.74E-07	6.00	1.64E-06	1.83E-05	4.80	8.77E-05
1 + 12	1.83E-07	3.00	5.48E-07	1.83E-05	3.33	6.09E-05
11 + 2	1.83E-07	3.00	5.48E-07	1.83E-05	3.33	6.09E-05
11 + 12	9.13E-08	6.00	5.48E-07	1.83E-05	2.00	3.65E-05
14 + 15	3.62E-05	6.00	2.17E-04	2.10E-04	6.00	1.26E-03
Total	5.00E-02	32.78	1.64E+00	2.83E-04	5.32	1.51E-03

Table 6. Active failure in ring bus substation

Failure Event	λ_{ap} [failure/yr]	r_{ap} [hour]	U_{ap} [hour/yr]
1A	1.00E-02	1.00	1.00E-02
2A	1.00E-02	1.00	1.00E-02
Total	2.00E-02	1.00	2.00E-02

Table 7. Active failure in double bus substation

Failure Event	λ_{ap} [failure/yr]	r_{ap} [hour]	U_{ap} [hour/yr]
1A	1.00E-02	1.00	1.00E-02
2A	1.00E-02	1.00	1.00E-02
Total	2.00E-02	1.00	2.00E-02

Table 8. Active failure + total failure in ring bus substation

Failure Event	λ_{pp} [failure/yr]	r_{pp} [hour]	U_{pp} [hour/yr]
6A + 2	1.48E-07	0.92	1.37E-07
3A + 1	1.48E-07	0.92	1.37E-07
6A + 12	1.71E-06	0.92	1.58E-06
3A + 10	1.71E-06	0.92	1.58E-06
Total	3.71E-06	0.92	3.42E-06

Table 9. Active failure + Active failure in double bus substation

Failure Event	λ_{ap} [failure/yr]	r_{ap} [hour]	U_{ap} [hour/yr]
3A + 6A	2.28E-08	0.50	1.14E-08
Total	2.28E-08	0.50	1.14E-08

Table 10. Active failure + total failure in ring bus substation

Failure Event	λ_{pp} [failure/yr]	r_{pp} [hour]	U_{pp} [hour/yr]
3A + 2	1.48E-07	0.92	1.37E-07
5A + 2	1.48E-07	0.92	1.37E-07
7A + 2	1.48E-07	0.92	1.37E-07

Table 11. Active failure + Active failure in double bus substation

Failure Event	λ_{ap} [failure/yr]	r_{ap} [hour]	U_{ap} [hour/yr]
3A + 6A	2.28E-08	0.50	1.14E-08
3A + 8A	2.28E-08	0.50	1.14E-08
5A + 4A	2.28E-08	0.50	1.14E-08

4A + 1	1.48E-07	0.92	1.37E-07
6A + 1	1.48E-07	0.92	1.37E-07
8A + 1	1.48E-07	0.92	1.37E-07
3A + 12	5.71E-08	0.80	4.57E-08
5A + 12	5.71E-08	0.80	4.57E-08
7A + 12	5.71E-08	0.80	4.57E-08
4A + 11	5.71E-08	0.80	4.57E-08
6A + 11	5.71E-08	0.80	4.57E-08
8A + 11	5.71E-08	0.80	4.57E-08
3A + 15	1.71E-06	0.92	1.58E-06
4A + 15	1.71E-06	0.92	1.58E-06
5A + 14	1.71E-06	0.92	1.58E-06
6A + 14	1.71E-06	0.92	1.58E-06
Total	8.06E-06	0.92	7.40E-06

5A + 8A	2.28E-08	0.50	1.14E-08
7A + 4A	2.28E-08	0.50	1.14E-08
7A + 6A	2.28E-08	0.50	1.14E-08
Total	1.37E-07	0.50	6.85E-08

Table 12. Reliability indices

Reliability indices	Figure 1	Figure 2
Failure rate [failure/year]	0.070503	0.07005
Annual outage time [hour/year]	23.58013	23.7023
MTTR [hours/failure]	1.662474	1.66023

Table 12 shows the results of calculation of reliability indices for the entire system by synthesizing all the failure events shown in Tables 4 ~ 11 and the failure rate, annual down time, and MTTR of the entire system were calculated as 0.070503 [failure/yr], 23.58013 [hour/year], and 1.662474 [hrs/failure] in case of double bus double breaker substation and calculated as 0.07005 [failure/yr], 23.7023 [hour/year], and 1.66023 [hrs/failure] in case of double bus double breaker substation
The reliability indices calculated as described above are the results of qualitative calculations to see how much the reliability levels of internal components affect the reliability indices of entire system, which can be used as basic data for the establishment of maintenance plans for internal components to most economically maximize the reliability of the entire system.

5. Conclusion

Reliability indices of each failure events and entire reliability indices (failure rate, annual outage time and MTTR) in the ring bus substation and double bus double breaker substation were calculated and quantitatively compared. From these results, it can be quantitatively identified that the single active failure events have larger effects on the reliability of the entire system than the other overlapping failure. The reliability evaluation results of typical two substations can be utilized as basic data in the design and maintenance schedule planning of substations with diverse bus structures hereafter.

6. References

[1] Dong-Li Duan, Xiao-Yue wu and Hong-Zhong Deng, “Reliability Evaluation in Substations Considering Operating Conditions and Failure Modes” , IEEE Trans. on Power Delivery, Vol. 27, No. 1, Jan. 2012
[2] J. J. Meeuwsen, W. L. Kling, “Substation Reliability Evaluation including Switching Actions with Redundant Components” , IEEE Trans. on Power Delivery, Vol. 12, No. 4, October, 1997.
[3] D. Mgee, a. Refsum, RESIN, A Desktop -Computer Program for Finding Cut-Sets “, IEEE Trans. on Reliability, Vol. R-30, No. 5, Dec. 1981
[4] R. Billinton, G. Lian, “A New Technique for Active Minimal Cut Set Selection used in Substation Reliability Evaluation” , Microelectron, Reliab., Vol. 35, No. 5, pp. 797-806, 1995
[5] R. Billinton and R. N. Allan, Reliability Evaluation of Power Systems, 2nd ed. New York: Plenum, 1996.
[6] Hamze Hajian-Hoseinabadi, "Impacts of Auto- mated Control Systems on Substation Reliability", IEEE Trans. on Power Delivery, Vol. 26, No. 3, July 2011.

Acknowledgments

This work was carried out in 2017 with the support of selected research and development projects based foundation funded by the Electric Power Research Institute, Korea Electric Power Corporation. [Project No. : R15XA03-44]

Study of the PWM Control Strategies for Three Switches Isolated Boost DC-DC Converter

Truong-Duy Duong[1], Minh-Khai Nguyen[2], Young-Cheol Lim[1] and Joon-Ho Choi[1]

1 Department of Electrical Engineering, Chonnam National University, Gwangju 500-757, Korea.
2 Department of Electrical Engineering, Chosun University, Gwangju 61452, Korea.
E-mail: nmkhai00@gmail.com

Abstract. This paper presents the PWM control strategies for three-switch isolated boost DC-DC converter. By inserting the extra states into the conduction time of switches, the current input is still continuous; unchanged primary and secondary voltage waveforms of the high-frequency (HF) transformer. The parameter design guideline with the different PWM control strategies is discussed. A 400 W prototype has been tested in the laboratory to verify the performance of the three-switch DC-DC converter with two PWM control strategies and a PI controller was used to clarify the DC output voltage at 400 V. The simulation and experimental results are shown to validate the theoretical analysis.

1. Introduction

In recent year, the development of further industrial technology and the enormous consumption of electrical energy lead to exhaust the natural resources, many environmental problems occur. So, the development clean energy are the most-effective solution, such as wind, solar, and fuel cells... The distributed generation technologies is used to generate the energy sources which have a low-voltage DC and instability. In order to connect them to the residential loads, the power-conversion system is widely used, as shown in Figure 1. In the DC-DC converter stage, a high step-up DC-DC converter is utilized to convert the low voltage of the renewable sources into 200V or 400V DC voltage [1], [2]. In the second-stage, the DC-AC inverter is used to provide the AC voltage 110 Vrms or 220 Vrms to connect grid for the industrial application.

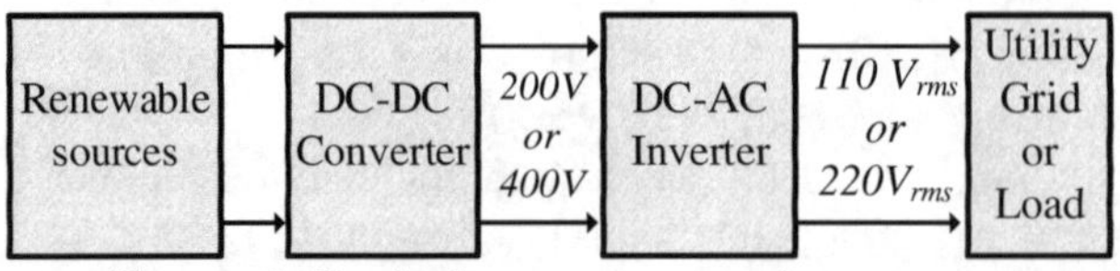

Figure 1. Typical power-conversion system.

High step-up DC-DC converters have been proposed and consider to obtain a high-voltage gain, such as isolated and non-isolated topologies. For non-isolated topologies, many researchers have proposed a lot of converter, it consists of a cascade topology [3], couple inductor [4], a switched-capacitor, a switched-inductor [5], [6] and the mixture of them [7], [8]. The non-isolated topologies are without HF transformer, so the system volume and cost is cheap. However, to have the isolation between input and output source, achieve the high voltage step-up gain, many isolated topologies have been presented in [9]-[11]. For isolated topologies, the voltage-fed and current-fed topologies are widely used. The

IOP Conf. Series: Earth and Environmental Science **159** (2018) 012021 doi:10.1088/1755-1315/159/1/012021

voltage gain of the voltage-fed DC-DC converters [11], [12] is provided mainly by a HF transformer. In order to achieve a high gain, the turn ratio of the transformer must be large. Nevertheless, the current-fed converter have a boost ability by using the boost inductor connect directly to the input voltage, and it is suitable for applications with high step-up ratio. Hence, the input current ripple and turn ratio of the HF transformer in current-fed topologies is lower than those in voltage-fed topologies in [13]-[15].

Compared to the conventional current-fed full-bridge (CFFB) DC-DC converter in [16], [17], the three-switch isolated boost DC-DC converter has some advantages, which are as follows: 1) the input current is continuous with low ripple; 2) it uses one fewer active switch and 3) the snubber circuit is not used. In this paper, the PWM control strategies for three-switch isolated boost DC-DC converter are proposed in this paper. To verify the analysis, simulation and experimental results are provided.

2. Proposed PWM control strategies for three-switch DC-DC converter

The three-switch isolated DC-DC converter, as shown in Figure 2, it's consists of a boost inductor (L_1) , three switches, one boost capacitor (C_1) , one diode (D_1) in the primary winding of the transformer, voltage doubler rectifier (VDR) circuit include two diodes (D_2 and D_3) and two output capacitors (C_2 and C_3), and resistor load (R).

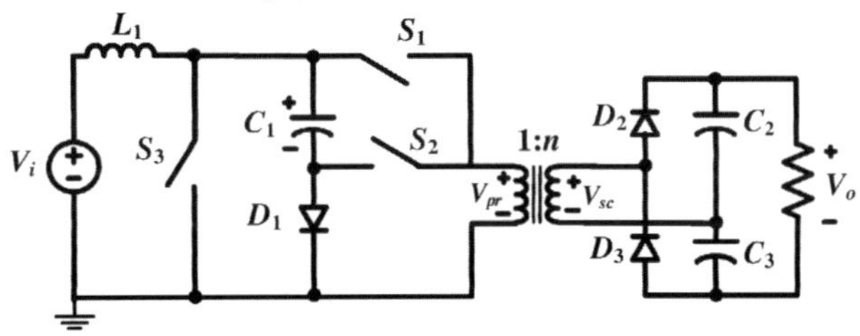

Figure 2. The three-switch isolated boost DC-DC converter.

2.1. PWM Control Strategies

The pulse-width-modulation (PWM) control methods for the three-switch DC-DC converter are shown in Figure 3. In Figure 3(a), the PWM control method 1, the extra states are inserted into gate signal of switches S_1 and S_3 to extend the ON time interval of two switches S_1 and S_3. The extra states are inserted into the zero state of the period, thus the positive and negative of the primary voltage is only kept at $0.3T$. We can change the gain voltage by changing the ON time interval of the switches, the primary and secondary voltage waveforms is unchanged.

Similarly, the PWM control method 2 is shown in Figure 3(b), the extra states is also inserted into the zero state of the period. So, the primary and voltage waveforms is similar the PWM control method 1.

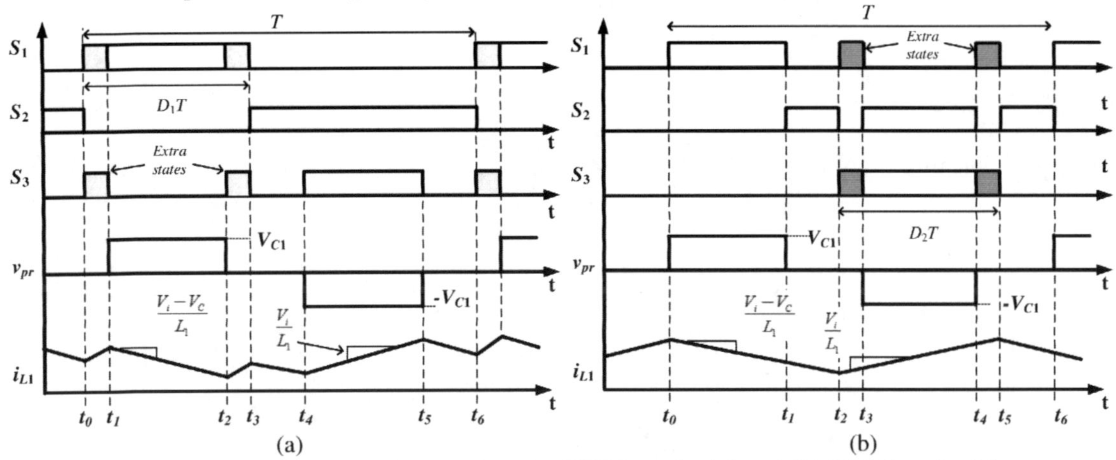

Figure 3. PWM control strategies. (a) PWM method 1 and (b) PWM method 2.

IOP Conf. Series: Earth and Environmental Science **159** (2018) 012021 doi:10.1088/1755-1315/159/1/012021

2.2. Operating principle of the three-switch DC-DC converter with PWM control method 1

For the operation analysis of circuit, the performed conditions are following as: 1) operate in continuous conduction mode (CCM); 2) all devices are ideal; 3) the high-frequency transformer is modeling by an ideal transformer; 5) the capacitance of the capacitors is sufficient to keep a constant capacitor voltage; the AC pulse primary and secondary voltage waveforms of the HF transformer is 30% positive, 20% zero, 30% negative, and 20% zero sequentially, respectively.

Stage 1–[$t_0 - t_1$, $t_2 - t_3$, Figure 4(a)]: At t_0, the extra state is inserted into the on time interval of S_1 and S_3. Furthermore, S_1, S_3 are turned on and S_2 is turned off. The inductor L_1 is charged by the input voltage.

$$L_1 \frac{di_{L1}}{dt} = V_i, \tag{1}$$

Stage 2–[$t_1 - t_2$, Figure 4(b)]: S_1 is turned on, S_2 and S_3 are turned off. The inductor L_1 is discharge. The primary voltage of the transformer is V_{C1}. We have:

$$L_1 \frac{di_{L1}}{dt} = V_i - V_{C1}, \tag{2}$$

Stage 3–[$t_3 - t_4$, $t_5 - t_6$, Figure 4(c)]: In this time, only S_2 is turned on. The diode D_1 is forward-bias and the inductor L_1 current is through diode D_1. The primary voltage of the transformer is shorted by diode D_1. Thus, the secondary voltage of the transformer is zero; the D_2 and D_3 diodes are reverse-biased. We get:

$$L_1 \frac{di_{L1}}{dt} = V_i - V_{C1}, \tag{3}$$

Stage 4–[$t_4 - t_5$, Figure 4(d)]: At t_4, S_3 and S_2 is turned on, S_1 remains off. The inductor L_1 is charged, when the capacitor C_1 is discharged. The primary voltage of the transformer is $-V_{C1}$. Hence, the secondary voltage is negative in which passed through the step-up transformer. The voltage of the inductor L_1 is calculated as

$$L_1 \frac{di_{L1}}{dt} = V_i, \tag{4}$$

By using the volt-second balance law to the inductor L_1, from (1)-(4), the capacitor C_1 voltage and output voltage gain can be determined as

$$\begin{cases} V_{C1} = \dfrac{1}{1-D_1} V_i \\ G = \dfrac{V_o}{V_i} = \dfrac{2n}{1-D_1}, \end{cases} \tag{5}$$

Where D_1 is the time on of switch S_1 in the PWM method 1.

The operating principle for PWM method 2 is analyzed similarly with the PWM method 1. And we get the capacitor C_1 voltage (V_{C11}) and output voltage gain (G_1), respectively, can be calculated as

$$\begin{cases} V_{C11} = \dfrac{1}{1-D_2} V_i \\ G_1 = \dfrac{V_o}{V_i} = \dfrac{2n}{1-D_2}, \end{cases} \tag{6}$$

Where D_2 is the time on of switch S_1 in the PWM method 2.

IOP Conf. Series: Earth and Environmental Science **159** (2018) 012021 doi:10.1088/1755-1315/159/1/012021

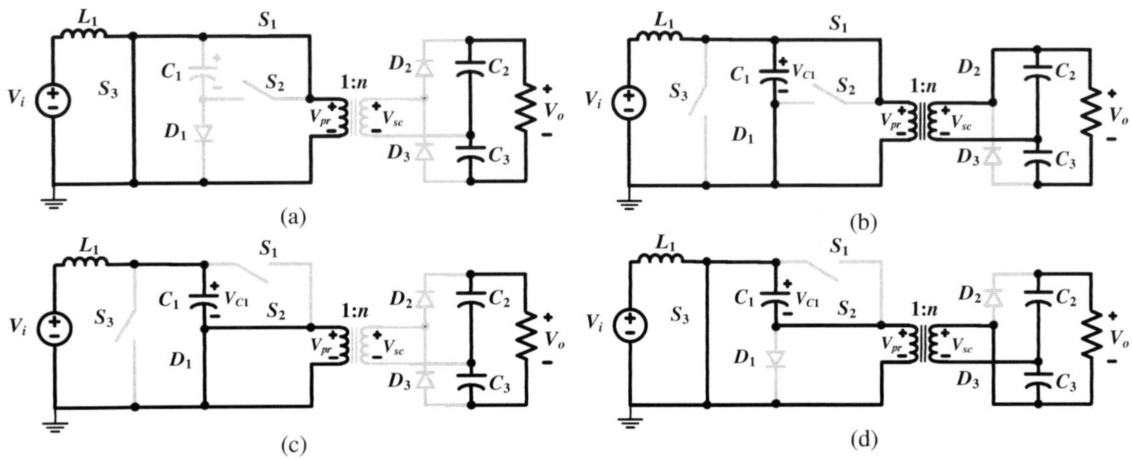

Figure 4. Key waveforms of the three-switch converter with PWM control method 1.

2.3. *The PI control algorithms for the three-switch converter*

The PI controller is used to keep the output voltage is constant at 400 V, and the diagram is shown in Figure 5. The error signal between the output voltage and the reference voltage of 400 V is through the PI controller. The PI control output is determined by the limiter module. The PWM generator is generated by the PWM control method 1 or method 2, and the gate signals of the S_1, S_2 and S_3 switches is created from the PWM generator.

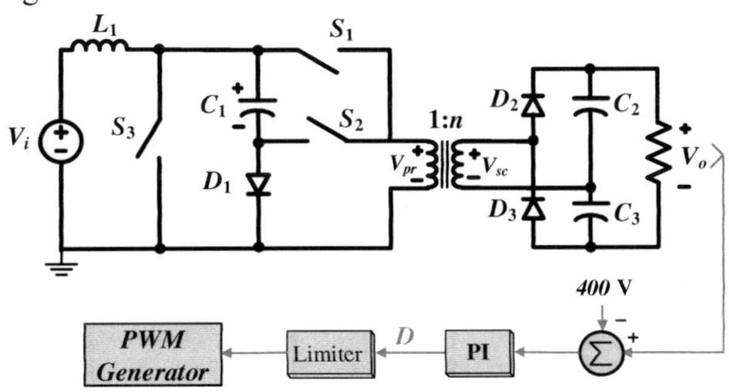

Figure 5. PI control diagram for the three-switch DC-DC converter.

3. Parameters Selection

3.1. *Inductance Selection*

Based on the PWM control method 1 is shown as in Figure 3(a). When the secondary voltage is negative, the peak-to-peak inductor L_1 current with $D_1 < 0.5$ is determined as

$$\Delta I_{L1} = \frac{0.3V_i}{f \cdot L_1}, \tag{7}$$

Where f is the switching frequency.

When $D_1 \geq 0.5$, the peak-to-peak inductor L_1 current is calculated

$$\Delta I_{L1} = \frac{0.3D_1V_g}{(1-D_1)fL_1}. \tag{8}$$

The inductor current ripple of the three-switch converter for PWM method 1 is calculated as

$$\Delta I_{L1} = r_{L1}\% \cdot I_{L1}, \tag{9}$$

The L_1 inductance is defined in

$$L_1 = \begin{cases} \dfrac{0.3V_i^2}{r_{L1}\% fP_o}, & D_1 < 0.5 \\ \dfrac{0.3D_1V_i^2}{(1-D_1)r_{L1}\% fP_o}, & D_1 \geq 0.5. \end{cases} \quad (10)$$

For the PWM control method 2 is shown as in **Figure 3(b)**. The peak-to-peak inductor L_1 current is calculated as

$$\Delta I_{L1} = \frac{D_2V_i}{f \cdot L_1}, \quad (11)$$

The L_1 inductance in this case is determined as

$$L_1 = \frac{D_2V_i^2}{r_{L1}\% fP_o}, \quad (12)$$

3.2. Capacitance Selection

Both the PWM control method 1 and the PWM control method 2, the negative state of the primary voltage waveform is $0.3T$. Thus, the capacitor C_1 is designed base on the current flow to it. We have the current C_1 is

$$C_1 \frac{\Delta V_{C1}}{0.3T} = \left|n\bar{i}_{sc}\right|. \quad (13)$$

The capacitor C_1 is defined as

$$C_1 = \frac{0.3(1-D)^2 P_o}{r_{C1}\% fV_{dc}^2}. \quad (14)$$

Where $r_{C1}\%$ is the capacitor voltage ripple.

The capacitor C_2 and C_3 of the VDR module can be selected in the zero mode time. If the ripple voltage of the C_2 and C_3 capacitors is $r_C\%$.

$$\begin{cases} C_2 = C_3 = \dfrac{0.4(1-D_1)^2 P_o}{4f \cdot r_C\% n^2V_i^2}, & for\ PWM\ method\ 1 \\ C_2 = C_3 = \dfrac{0.4(1-D_2)^2 P_o}{4f \cdot r_C\% n^2V_i^2}, & for\ PWM\ method\ 2. \end{cases} \quad (15)$$

4. Simulation and Experimental Results

4.1. Simulation Results

The three-switch DC-DC converter with two method PWM control is verified by using PSIM 9.1 simulation. The frequency of switching is 10 kHz. The simulation parameters is chosen from Table 1. These simulation results are agreeable with the theoretical analysis. In Figures 6-7, the output voltage of 400 V is boosted from the input voltage of 40 V with two methods PWM control. The input current is continuous with two PWM methods. The primary and secondary voltage waveforms of the transformer are not changed, when the PWM methods are applied.

Table 1. Parameters for simulation and experiment.

Parameter		Value
Output voltage		400 V
Output power		400 W
Input voltage		40 V
Inductors (L_1)		1 mH
HF transformer		1:2.5
Capacitors	C_1	330 uF
	$C_2 = C_3$	82 uF
Switching frequency		10KHz

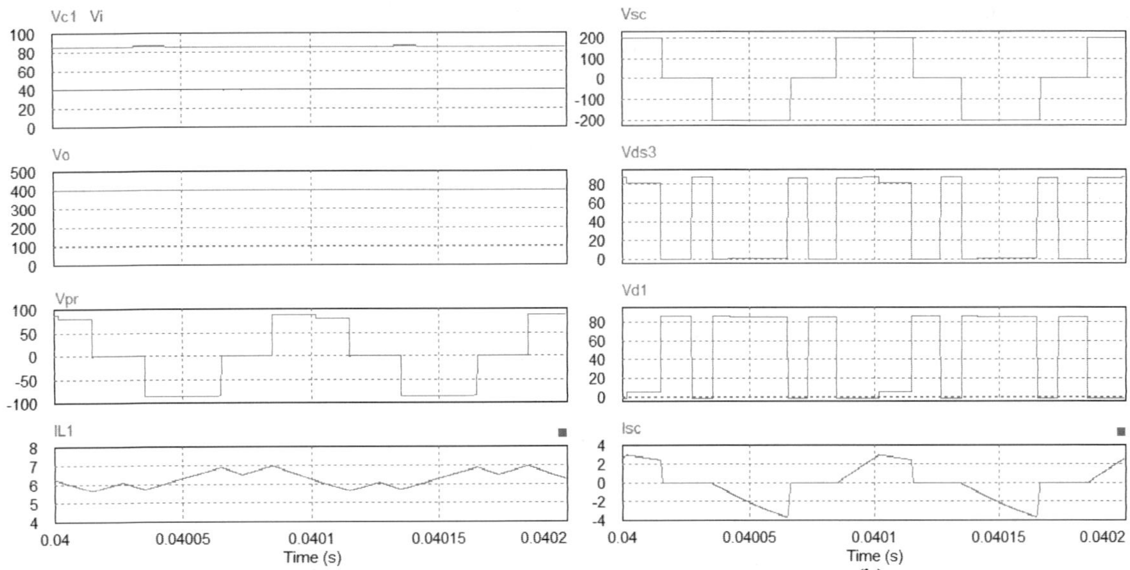

Figure 6. Simulation results with PWM method 1 when $V_i = 40$ V.

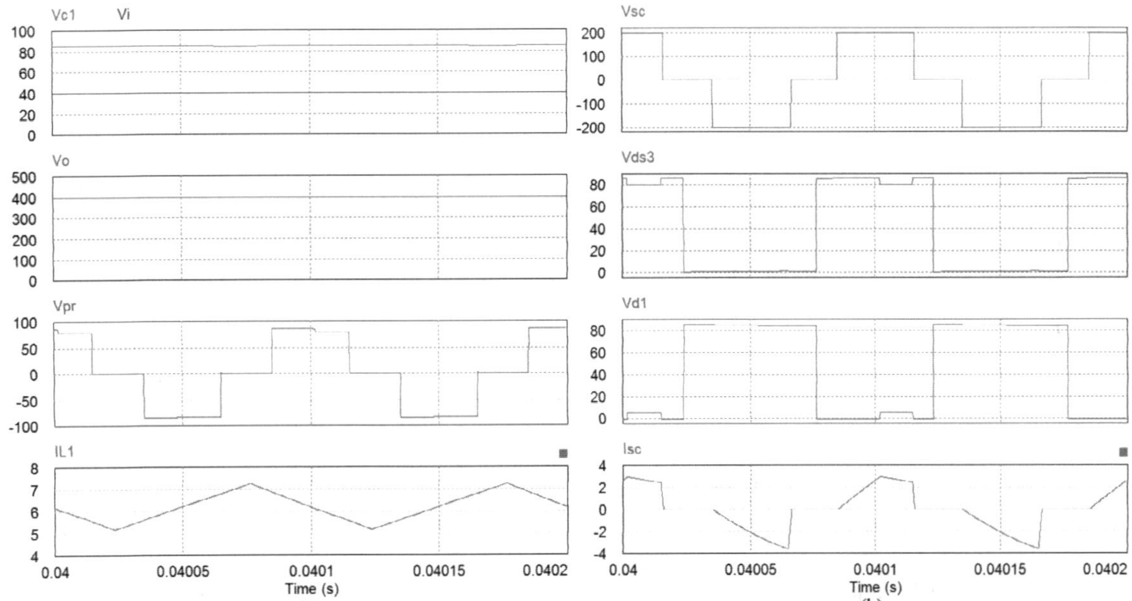

Figure 7. Simulation results with PWM method 2 when $V_i = 40$ V.

4.2. Experimental Results

In order to demonstrate the PWM control strategies are proposed in Section II, the experiment results with 400 W are implemented by using DSP kit TMS320F28335. The parameters experiments are used, as shown in Table 1.

One boost inductor is 1 mH/ 20 A is used to connect to the input voltage by direction, one boost capacitor is 330 μF/ 200 V. Two 82 μF/ 450 V capacitors were used for the secondary side of the HF transformer. The switches and diodes are IRFP460 MOSFETs and DSEP30-12A diodes. The PWM signals are generated by TMS320F28335 through the isolated amplifiers (TLP250) and the gating signal to drive all MOSFETs.

Figure 8 shows the experimental results for the PWM control method 2 when the input voltage and the power load are 40 V and 260 W, respectively. An extra states are inserted into the PWM signals generation like to the PWM method 1 and method 2. The voltage C_1 voltage was 85 V for the PWM control method 1 and 86 V for the PWM control method 2. The input current of two PWM control method are continuous. But the peak-to-peak ripple inductor L_1 current for PWM method 1 is lower than that the peak-to-peak ripple inductor L_1 current for PWM method 2. These simulation and experimental results are suitable for the theoretical analysis.

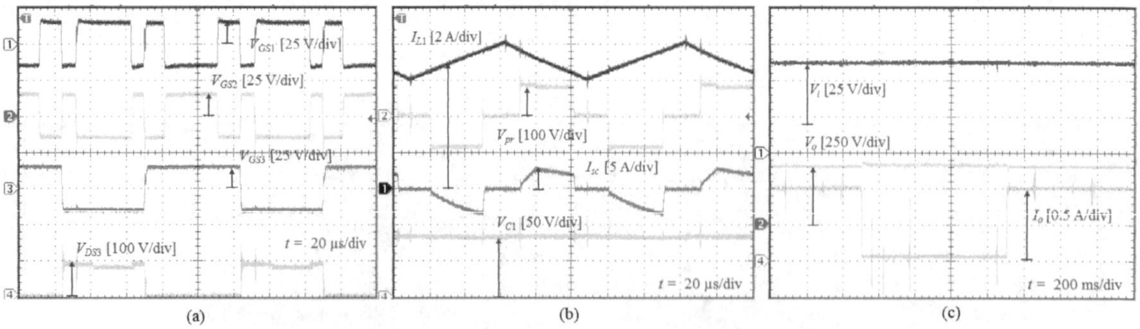

Figure 8. Experimental results with PWM method 2 when $V_i = 40$ V.

In Figure 8, the key waveforms are the gating signals of MOSFETs S_1-S_3 voltage, drain-source MOSFETs S_3 voltage; the inductor L_1 current, primary voltage, the secondary current, and load current. The transient response of PI controller of the three-switch DC-DC converter is shown in Fig. 8(c), the waveforms from top to bottom are the input voltage, the load voltage, and the load current, respectively. In this case, the input voltage is generated at 40 V when the step change of power load occur from full load (400 W) to light load (21 W). The stable operation in case that the graph has a very small obvious overshoot.

5. Conclusion

PWM control strategies for three-switch isolated boost DC-DC converter has been presented in this paper and had the following characteristics: the current input is continuous with two PWM strategies; unchanged primary and secondary voltage waveforms of the HF transformer. The inductance and capacitance selection of the circuit are shown. The operating principles and analysis of converter are depicted with PI controller. The proposed converter is applicable for fuel-cell applications with low-dc input voltage.

6. References

[1] S. M. Chen, T. J. Liang, L. S. Yang, and J. F. Chen, "A boost converter with capacitor multiplier and coupled inductor for ac module applications," *IEEE Trans. Ind. Electron.*, vol. 60, no. 4, pp. 1503–1511, April 2013.

[2] G. Wu, X. Ruan, and Z. Ye, "Nonisolated high step-up DC-DC converters adopting switched-capacitor cell," *IEEE Trans. Ind. Electron.*, vol. 62, no. 1, pp. 383–393, January 2015.

[3] Y. J. A. Alcazar, D. S. Oliveira, Jr., F.L. Tofoli, and R.P. Torrico-Bascopé, "DC–DC Nonisolated boost converter based on the three-state switching cell and voltage multiplier cells," *IEEE Trans. Ind. Electron.*, vol. 60, no. 10, pp. 4438–4449, Oct. 2013. Y. P. Hsieh, J. F. Chen, T. J. Liang, and L. S. Yang, "Novel high step-up DC-DC converter with coupled-inductor and switched-capacitor techniques," *IEEE Trans. Ind. Electron.*, vol. 59, no. 2, pp. 998–1007, Feb. 2012.

[4] A. I. Bratcu, I. Munteanu, S. Bacha, D. Picault, and B. Raison, "Cascaded DC–DC converter photovoltaic systems: power optimization issues," *IEEE Trans. Ind. Electron.*, vol. 58, no. 2, pp. 403-411, February 2011.

[5] L. S. Yang and T. J. Liang, "Analysis and implementation of a novel bidirectional DC-DC converter," *IEEE Trans. Ind. Electron.*, vol. 59, no. 1, pp. 422–434, Jan. 2012.

[6] B. Axelrod, Y. Beck, and Y. Berkovich, "High step-up dc–dc converter based on the switched-coupled-inductor boost converter and diode-capacitor multiplier: steady state and dynamics," *IET Power Electron.*, vol. 8, no. 8, pp. 1420–1428, 2015.

[7] R. Y. Chen, T. J. Liang, J. F. Chen, R. L. Lin, and K. C. Tseng, "Study and implementation of a current-fed full-bridge boost dc–dc converter with zero-current switching for high-voltage applications," *IEEE Trans. Ind. Appl.*, vol. 44, no. 4, pp. 1218-1226, July/Aug. 2008.

[8] M. Nymand, and M. A. E. Andersen, "High-efficiency isolated boost dc–dc converter for high-power low-voltage fuel-cell applications," *IEEE Trans. Ind. Electron.*, vol. 57, no. 2, pp. 505-514, Feb. 2010.

[9] Z. Zhang, O. C. Thomsen and M. A. E. Andersen, "Soft-switched dual-input DC-DC converter conbining a boost-half-birdge cell and a voltage-fed full-bridge cell" *IEEE Trans. Ind. Electron.*, vol. 28, no. 11, pp. 4897–4092, Feb. 2013.

[10] C. Yao, X. Ruan, and X. Wang, "Automatic mode-shifting control strategy with input voltage feed-forward for full-bridge-boost dc–dc converter suitable for wide input voltage range," *IEEE Trans. Power Electron.*, vol. 30, no. 3, pp. 1668–1682, March 2015.

[11] A. Mohammadpour, L. Parsa, M. H. Todorovic, R. Lai, R. Datta, and L. Garces, "Series-input parallel-output modular-phase dc–dc converter with soft-switching and high-frequency isolation," *IEEE Trans. Power Electron.*, vol. 31, no. 1, pp. 111–119, Jan. 2016.

[12] Z. Zhang, O. C. Thomsen, M. A. E. Andersen, and H. R. Nielsen, "Dual-input isolated full-bridge boost dc-dc converter based on the distributed transformers," *IET Power Electron.*, vol. 5, no. 7, pp. 1074–1083, Aug. 2012.

[13] U. R. Prasanna, and A. K. Rathore, "Extended Range ZVS Active-Clamped Current-Fed Full-Bridge Isolated DC/DC Converter for Fuel Cell Applications: Analysis, Design, and Experimental Results," *IEEE Trans. Ind. Electron.*, vol. 60, no. 7, pp. 2661-2672, July 2013.

[14] M. Nymand, and M. A. E. Andersen, "High-efficiency isolated boost dc–dc converter for high-power low voltage fuel-cell applications," *IEEE Trans. Ind. Electron.*, vol. 57, no. 2, pp. 505-514, Feb. 2010.

[15] M. K. Nguyen, L. C. Lim, J. H. Choi, and G. B. Cho "Isolated High Step-up DC-DC Converter Based on Quasi-Switched-Boost Network". *IEEE Trans. Ind. Electron*, vol. 63, no. 12, pp. 7553-7562, Dec. 2016.

[16] A. Mohammadpour, L. Parsa, M. H. Todorovic, R. Lai, R. Datta, and L. Garces, "Series-input parallel-output modular-phase dc–dc converter with soft-switching and high-frequency isolation," *IEEE Trans. Power Electron.*, vol. 31, no. 1, pp. 111–119, Jan. 2016.

[17] Z. Zhang, O. C. Thomsen, M. A. E. Andersen, and H. R. Nielsen, "Dual-input isolated full-bridge boost dc-dc converter based on the distributed transformers," *IET Power Electron.*, vol. 5, no. 7, pp. 1074–1083, Aug. 2012.

IOP Conf. Series: Earth and Environmental Science **159** (2018) 012022 doi:10.1088/1755-1315/159/1/012022

Power Quality Feature Modelling of Waveform Disturbances for Distribution System Diagnosis

Sang-Keun Moon[1,2], Jong-Hyuk Lee[2], Hong-Rok Lim[2], Sang-Won Oh[2], Dae-Won Seo[2], Jin-O Kim[2,3]

1 Korea Power Electric Cooperation Research Institute (KEPCO), Daejeon, Korea
2 Department of Electrical Engineering, Hanyang University, Seoul, Korea
E-mail: sk.moon@kepco.co.kr1, jokim@hanyang.ac.kr3

Abstract. This paper presents a Power Quality (PQ) modelling and filtering processes for the distribution system disturbance recognition learning. typical PQ waveforms with mathematical applications and gathered field data are applied to our models. The process has modules to obtain the waveforms of incipient failures or abnormal operations, trouble condition waveforms in the distribution system through feature extraction and trigger condition techniques. In this regard, we found that proposed extraction criteria of determination rules can interact suitably with our parameter update method to detect and containing the waveforms. Therefore, the recurring analysis can classify the event conditions and ensures disturbance and fault recognitions which containing the waveform information.

1. Introduction

Power Quality (PQ) monitoring to control are emerging issues in perspective of the real-time intelligent power distribution networks. Distributed energy resources (DERs) has become widespread and other bi-directional power sources can lead us to a new environment for the future distribution power system operation as well. Including renewable resources that have grid connection capabilities claim that the necessity of multi directional power flow operations especially in the network management aspect. One of the reasons that the PQ has become an emerging issue is to avoid equipment damage and to determine the cause of the disturbances from newly using power network installations. Essentially, PQ can be used to measure and evaluate the state of distribution system corresponding to effects, stability, abnormal equipment operation, and even fault conditions [1].

In recent years, engineers and researchers in the field of power quality, power system protection, and equipment testing have realized that useful information can be extracted from the waveforms. For the purpose of the equipment condition monitoring including DERs, the methodology to analyse the waveform-type power disturbance data and extract information has attracted a good interest from industry and academia recently [2]. The waveform signature and the feature study have concurrently been conducted with the aim of improving the distribution networks with respect of the system state prediction [3].

With respect to PQ data analytics, the primary goal is to share the signatures of various equipment failures so that researchers can develop appropriate algorithms to identify equipment abnormality. Equipment failures and system faults are infrequently occurred and that possibly be observed from the voltage and current waveforms associated with the devices and locations [4]. The acquired data, however, is not always sufficient to determine the exact nature of problems [5].

In order to make sense of waveform data available, many approaches have been made to model disturbances, classify, and estimate the waveforms [6]. In this regard, we modelled typical PQ

IOP Conf. Series: Earth and Environmental Science **159** (2018) 012022 doi:10.1088/1755-1315/159/1/012022

waveforms with mathematical applications and used gathered field data to our models. For three phase meters which measure both voltage and current waveforms, the data is minimum 8 times much including neutral and binary data.

More practical solutions developed in recent years are to store waveform data as long as an PQ event occurs at which particular cycles and samples are allocated or to use more signal processing methods such as frequency transformed data compression and sparse signal decomposition methods. These applied techniques provide signal and feature insights for the PQ recognitions as well [7]-[9]

Thus, our objective is analysing PQ data with respect to monitoring, discriminating, and evaluating the waveform of power disturbances to ensure the system preventative system failure protections and complex system problem estimations. Suggested signal filtering techniques are used for the field waveform noises and feature extractions. Using extraction and learning classification techniques in this paper, we focus on interactive modelling methods to suitably verify the PQ disturbances in this paper. The waveform features and further learning analytic basis are proposed as well [10], [11].

2. Modelling waveforms of disturbances

As short-term and long-term duration of waveforms contain certain information of the system state. Accordingly, voltage and current measurements in field devices at monitoring locations are treated by control, and operation systems such as Distribution Management System (DMS) and Power Quality Management System (PQMS). The systems are basically operating on monitoring acceptable voltage, current, frequency levels and conditionally obtaining intentional waveforms. In practical applications, the system attempts to improve the reliability and quality of distribution network operating conditions as to providing empirical and practical criteria to acquire power quality waveforms.

The steady state components of voltage and current waveforms $w(t)$ with the magnitude of A (p.u.) are modelled as sampling duration having certain cycles of the device performances. Therefore, the ideal state of single phase waves which has angular velocity $\omega = 2\pi f_n$ with the nominal operating frequency $f_n = 60Hz$ is:

$$w(t) = A \cdot \sin\{\omega t + \delta(t)\} + \eta(t) \tag{1}$$

where, t is the time instant as sampled signals; $\delta(t)$ is the disturbance phase angle; $\eta(t)$ is the modelled noise denoted by the estimated uniform distribution function where $p(\gamma) = 1/(b-a)$ with the parameter (detection set) $\gamma, a \le \gamma \le b$; and the ranges a and b are upper and lower outliers respectively of DMS and PQMS systems (sampled) because of measurement errors and piled up (superposition) noises from the field device. The magnitude of the signal is selectively decided as $A = 1 \pm [u(t-t_1) - u(t-t_2)] \cdot \gamma$ with respect to disturbance signal types that are modelled and generated. t_1 and t_2 represent a lapse of the disturbance duration determined with estimated PQ signal characteristics. Our mathematical model uses general standard criteria based on the PQ signals and the representative disturbance waveforms referenced by IEEE-1159-2009 PQ standard [12, 13]. Through consistently modifying parameters, the actual distribution line signals which are common PQ disturbances are determined as sag, swell, interruption, flicker, oscillation, notch, spike, and harmonics. In addition to the waveform generation, the proposed model has feedbacks and update processes so that the model varies in accordance with newly obtained data and improve the signal identification ability on its own. Accordingly, the modelled wave generation parameters are incorporated into randomize processes within the PQ criteria after obtaining learning weight from the actual event signals. Real distribution line event waveforms obtained from the field devices are combined with the dataset and the proposed model. The data contains not only PQ waveforms and disturbances but also system fault and operation-related waveforms such as temporary / permanent faults, inrush / switching operations, and incipient fault waveforms as illustrated in Fig. 1(a)-(c).

IOP Conf. Series: Earth and Environmental Science **159** (2018) 012022 doi:10.1088/1755-1315/159/1/012022

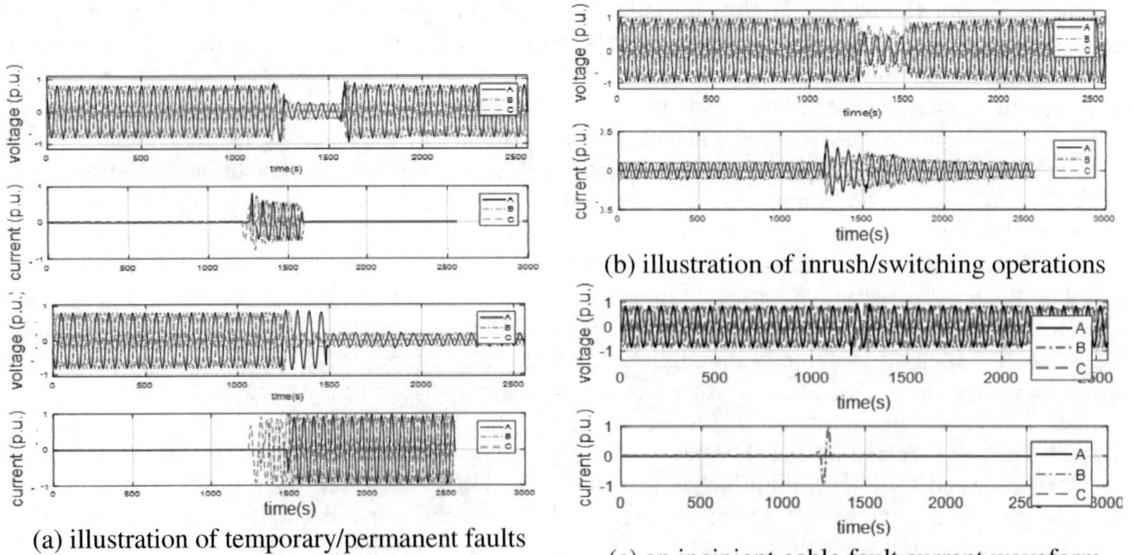

(a) illustration of temporary/permanent faults

(b) illustration of inrush/switching operations

(c) an incipient cable fault current waveform

Figure 1. Types of field event waveforms for signal feature variations.

In accordance of the waveform acquiring scheme, various types of measurements (e.g., voltage, current, switch status, temperature, and oil level) are recorded in the field at the distribution transformer and feeder devices. Further studies will have combined to the conventional data to make state inference of predictive fault detections.

3. Waveform pre-processing

PQ event signals are detected and identified by means of real time measurement and calculation processes as sampled waveforms. The processes have modules to obtain the waveforms of incipient failures or abnormal operations, trouble condition waveforms in the distribution system through feature extraction and trigger condition techniques. We found that proposed trigger and extract criteria of determination rules interact suitably with our parameter update method to detect and containing the waveforms so that the recurring analysis can classify the event conditions.

3.1. Symmetrical component processing

Since the distribution line operate in the balanced state of three phases, the PQ disturbances disrupt the synchronization therefore the distortion occurs concurrently. Symmetrical components are primarily used in the power system analysis in terms of evaluating the unbalanced state phasors of the balanced three phase system. The approach is to extract unbalances of the phase angles in the vaguely scattered event signals compared with other phases, currents, and voltages. Moreover, due to the real-time measurement burdens on the line devices, we minimize data by extracting a single phase symmetrical value instead. Simultaneity of the disturbance phase and other steady states is derived as an index of disturbance trigger conditions. We use $v_d^-(t)$, $i_d^-(t)$ of the negative sequence components where the disturbance phases (or certain any phase) $d \in a,b,c$ and $v_{abc}^{-,+}(t)$, $i_{abc}^{-,+}(t)$ of three phase of sequence states to estimate disturbances because these components represent differences of each phase and only returns near zero values in the steady state otherwise. The negative sequence component matrix is obtained as follows:

$$\mathbf{v}_{a,b,c}^- = \begin{bmatrix} v_a^-(t) \\ v_b^-(t) \\ v_c^-(t) \end{bmatrix} = \frac{\sigma^c}{3} \begin{bmatrix} 1 & \alpha_2 & \alpha_1 \\ \alpha_1 & 1 & \alpha_2 \\ \alpha_2 & \alpha_1 & 1 \end{bmatrix} \cdot \begin{bmatrix} v_a(t) \\ v_b(t) \\ v_c(t) \end{bmatrix} \qquad (2)$$

where, $\alpha = e(j \cdot 2\pi / 3)$ is the Fortescue operator. Here, as it is a frequency domain, the angle can shift into the time domain concept based on the time-dependent variables [14].
The α is converted as a time shifting operator by calculating the of samples in a third of cycle samples as $1/3f$ at a determined sampling frequency f . This is a domain change process from frequency domain of $e^{j2\pi}/3$ to $1/3$ sampling rate for the digital measurements and the measurements are subsequently scaled and normalized when the waveforms are acquired by the field devices. The normalization parameter σ^{c} is determined as:

$$\sigma^{c} = \begin{cases} 1 & \text{modelled} \\ \dfrac{v_{r}}{\lambda_{2}^{c}} \cdot \lambda_{m}^{c} & \text{field obtained} \end{cases}, \tag{3}$$

where v_{a}^{r} represents the rating ratio value; λ_{2}^{c} and λ_{m}^{c} are the secondary value ratio and the channel multiplier of COMTRADE (Common format for Transient Data Exchange for power systems) format waveform [15] respectively when it comes to the field obtained waveforms. In addition to that, the sum of negative sequence components $v_{t}^{abc-} = (v_{t}^{a-} + v_{t}^{b-} + v_{t}^{c-})/3$ is applied for the magnitude of distortions for the saved events.
The Phase calculation is based on the phase locked loop (PLL) modulation with real-time phase shifts. Despite of the fact that sequence components are decomposed successfully obtained, actual samplings are usually not separable by exact 1/3rd of samples that makes sequence component noises. We handle the noise and gaps following feature extraction steps such as additional filtering.

3.2. *Signal filtering (smoothing / de-noising)*

The noise especially on field measured waveforms has a particular measurement error which occurred by following reasons. One is due to measurement errors from the installed field devices (like switches) using potential and current transformers which have physical (hardware) accuracy limitation that related with sensor performances.
The other one is calculation deviations by PLL, even though the problem can deal with increasing the samples for the measurement cycle, the three phases signal basically make separating deviation by phase shifting on the symmetrical calculations. The PLL phase shifting is not exactly be separated by 1/3rd cycle of the nominal frequencies due to sampling can be archived typically 64 ~ 128 samples in field devices and that are not the times of thirds. The noises are normally recognizable when tracking the waveform patterns since majority of overlapped data is not generally in steady state. When it comes to symmetrical negative sequences, the frequencies are distinctively different near the zero region so that if we shrink the filter based on statistical signal process methods, the noises can be eliminated except the waveform features are mostly remained smoothly.
We suggest the filter that updated intermittently for cancelling the sampling noises using the histogram based analysis. The noise smoothing was achieved by generating a cumulate filter proposed in this paper which is similar to opposite Gaussian mask of negative symmetrical values. Thus, if we treat signal $h(t)$ as a negative sequence value of $\mathbf{v}_{a,b,c}^{-}$ with disturbance phase $v_{d}^{-}(t)$, then $h(t)$ apparently has a detecting characteristic in which the unbalance exists, otherwise the values are zero. Therefore, the $h(t)$ has majority values of zero except the disturbance, and the noises for the histogram analysis, the zero values are not necessary so that we can filter the sampling noises. The histogram that has bins with commonly used Square root rule and Sturges (Sequence value histogram) method as:

$$\mathrm{b} = \begin{cases} \left\lceil \sqrt{h(t)} \right\rceil, & \text{square root} \\ \left\lceil 1 + \log_{2} h(t) \right\rceil, & \text{sturges, where, } h(t) \neq 0, \end{cases} \tag{4}$$

where, the braces indicate the ceil function and Sturges method implicitly assumes an approximately normal distribution. From the histogram, we can model a statistical distribution of quantified histogram $H(h)^{\mathrm{b}}$ with b bin levels at a set of the region. The relation is given between the measured value and distribution and the noise can be perceived with applying threshold at which the number of $H(h)^{\mathrm{b}}$ considerably exceeds the expected distribution. The gaussian fitting is derived as the histogram stochastic signal $p(b) \sim N(\mu, \sigma^2)$ for normal distribution and the upper, under thresholds are

$$\begin{aligned} &\delta^{\mathrm{under}} = H(h)^{axis}_{\mathrm{first}}, \\ &\delta^{\mathrm{upper}} = H(h)^{axis}_{\mathrm{last}}, \\ &\text{where, } H(h)^{\mathrm{b}} > C \cdot p(\mathrm{b}). \end{aligned} \tag{5}$$

If histogram $H(h)^{\mathrm{b}}$ frequencies exceed the probability fitting value, the points where criteria with the weight parameter C are applied to the filtering range starting from the first value to the last exceed point. Consequently, the filtering conditions are reflected on time series waveform symmetrical values as shown in Fig. 2, and the filtered values $h(t)_F$ are obtained as follows:

$$h(t)_F \begin{cases} 0, & \delta^{\mathrm{under}} < h(t) < \delta^{\mathrm{upper}} \\ h(t), & \text{otherwise} \end{cases} \tag{6}$$

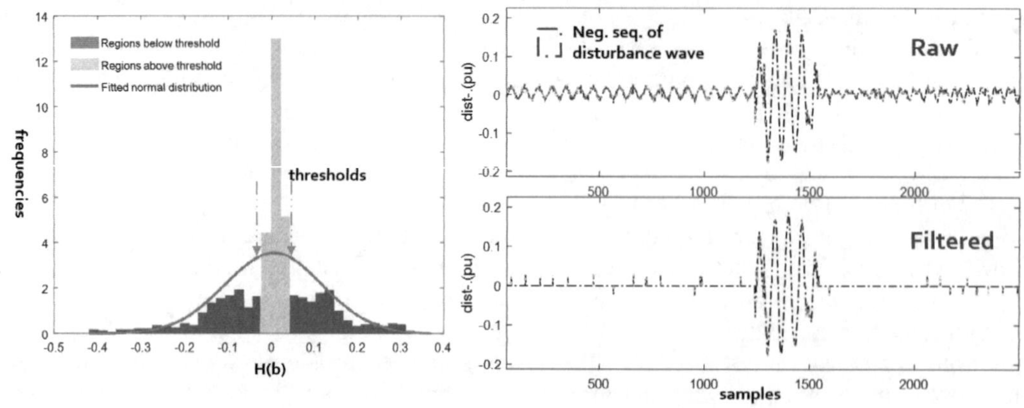

Figure 2. Illustration of the sequence value histogram and the applied noise filtering example

4. Feature extraction models and trigger conditions for waveforms

PQ waveforms illustrate diverse pattern on dense samplings which closely sustaining 60Hz speed in real time. Accordingly, the waveform need to be transformed to classify the noticeable disturbances so that the extracted features are used for PQ detection and for triggering the signal measurement previously at the field devices. In addition, we suggest criteria for the feature extractions to discover desired signal from the patterns. When we detect changing points of the unbalanced signal as disturbance occurring, proposed triggering condition assists to acquire the waveforms for the further signal learning processes. Some of basic features are derived to measure the grouped features as following Table 1 for the further pattern recognition processes.

Table 1. Features of converted waveforms

Features	Equation
1-norm (magnitude)	$f_1 = \omega_1 \cdot \lVert h(t)_F \rVert_1, \ \forall t$
Signal deviation	$f_2 = \omega_2 \cdot \sum_{t=1}^{T} \left(h(t)_F - \bar{h}_F \right)^2, \ \forall t$
Disturbance duration	$f_3 = \omega_3 \cdot \sum_{t=1}^{T} \left| \mathrm{sign}\left(h(t)_F \right) \right|, \ \forall t$
Zero crossing counts	$f_4 = \sum_{t=1}^{T} \left| \mathrm{sign}\left(h(t-1)_F \right) - \mathrm{sign}\left(h(t)_F \right) \right| \cdot \omega_4, \ \forall t$

Furthermore, we combine features which readily represents the distinctive feature regions as per PQ waveform of extracted signals. Therefore, the features with respect to the type of PQ pattern are determined as:

$$f_E^{\mathrm{m}} = (f_1 + f_3)/\sigma_{\mathrm{s}}^{\mathrm{m}}$$
$$f_E^{\mathrm{f}} = \sigma_{\mathrm{s}}^{\mathrm{f}} \cdot (f_2 + f_4) \tag{7}$$

Consequently, the triggering conditions are determined in accordance with feature criteria of desired power quality signals. measuring devices consistently solves following values whether the signal would exceed the trigger condition in which we consistently have set for sparingly occurred disturbances.

5. Results

Our model suggests the PQ feature extraction processes with regard to the waveform modelling and field obtain PQ data by means of waveform signal processes. The waveform of selected 8 disturbances are modelled with randomized parameters of IEEE-1159 PQ ranges. The range, parameters and weights are updated regarding field waveform obtained. Along with voltages, currents have same process to obtain the waveform features, the voltage is normalized apart from some of ratings and filters. For the reason that changing loads are causing the distortion in the voltage waveform, the different patterns can be obtained from the current variation as shown in Fig. 3 and Fig 4.

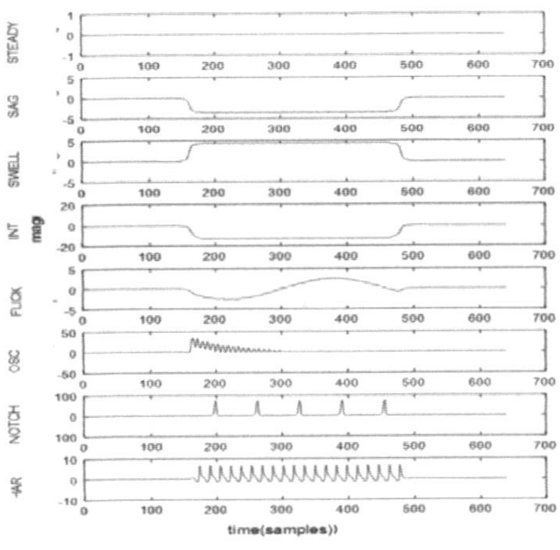

Figure 3. Filtered symmetrical component values (Neg.) of modelled PQ waveforms.

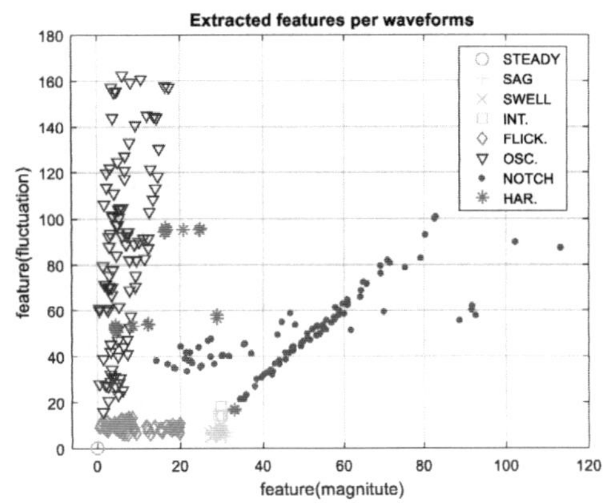

Figure 4. Extracted signal features of modelled PQ waveforms.

6. Conclusion

PQ disturbances in the voltage and current waveforms indicate different types of patterns and variations. The modified technique based on the symmetrical components in time domain was proposed in this paper for the PQ disturbance detections and classification. Our method is based on the fact that obtained waveforms from the suggested trigger conditions contain potential information for abnormality detections. The extracted features will sequentially be applied to estimation and recognition learning modules of further studies.

7. References

[1] Santoso, Surya, et al. "Power quality disturbance waveform recognition using wavelet-based neural classifier. I. Theoretical foundation." IEEE Transactions on Power Delivery 15.1 (2000): 222-228.

[2] "Electric Signatures of Power Equipment Failures" IEEE PES T&D, Power Quality Subcommittee WG, PES-TR(Draft), (2015)

[3] Valtierra-Rodriguez, Martin, et al. "Detection and classification of single and combined power quality disturbances using neural networks." IEEE Transactions on Industrial Electronics 61.5 (2014): 2473-2482.
[4] Wischkaemper, Jeffrey A., et al. "Application of Waveform Analytics for Improved Situational Awareness of Electric Distribution Feeders." IEEE Transactions on Smart Grid 6.4 (2015): 2041-2049.
[5] Madhusudan, Shertukde, Hemchandra (2014). Distributed photovoltaic grid transformers. p. 91. ISBN 1482247194. OCLC 897338163
[6] Dutta, Papiya, Ahad Esmaeilian, and Mladen Kezunovic. "Transmission-line fault analysis using synchronized sampling." IEEE Transactions on Power Delivery 29.2 (2014): 942-950
[7] Mishra, Sukumar, C. N. Bhende, and B. K. Panigrahi. "Detection and classification of power quality disturbances using S-transform and probabilistic neural network." IEEE Transactions on power delivery 23.1 (2008): 280-287.
[8] Biswal, Milan, and Pradipta K. Dash. "Detection and characterization of multiple power quality disturbances with a fast S-transform and decision tree based classifier." Digital Signal Processing 23.4 (2013): 1071-1083.
[9] Manikandan, M. Sabarimalai, S. R. Samantaray, and Innocent Kamwa. "Detection and classification of power quality disturbances using sparse signal decomposition on hybrid dictionaries." IEEE Transactions on Instrumentation and Measurement 64.1 (2015): 27-38.
[10] Borges, Fábbio AS, et al. "Feature extraction and power quality disturbances classification using smart meters signals." IEEE Transactions on Industrial Informatics 12.2 (2016): 824-833.
[11] Sriram, S., et al. "Signal denoising techniques for partial discharge measurements." IEEE Transactions on Dielectrics and Electrical Insulation 12.6 (2005): 1182-1191.
[12] "IEEE recommended practice for monitoring electric power quality." IEEE Std (2009): 1159-2009.
[13] Kumar, R., Singh, B., & Shahani, D. T. (2016). Symmetrical Components-Based Modified Technique for Power-Quality Disturbances Detection and Classification. IEEE Transactions on Industry Applications, 52(4), 3443-3450.
[14] Karimi-Ghartemani, Masoud, and Houshang Karimi. "Processing of symmetrical components in time-domain." IEEE Transactions on Power Systems 22.2 (2007): 572-579.
[15] "IEEE standard on transitions, pulses, and related waveforms", IEEE Std (2011).181-2003.

A Single Phase Quasi H-Bridge Boost Inverter with Common Ground for Grid connected Applications

TT Tran[1], MK Nguyen[2], YC Lim[3] and JH Choi[4]

1 PhD. Student, Chonnam national University, KOR
2 Professor, Chosun University, KOR
3 Professor, Chonnam national University, KOR
4 Professor, Chonnam national University, KOR
E-mail: nmkhai00@gmail.com

Abstract. This paper presents a single-phase quasi H-bridge boost inverter with common ground for grid-connected applications. In this topology, the ground of ac grid and input dc voltage source is connected together directly. So, the ground leakage current is significantly limited. Furthermore, by combining a switched-capacitor boost converter with quasi H-bridge inverter, the RMS of output voltage can be greater than input dc voltage source. With the introduced simple PWM strategy, single phase quasi H-bridge boost inverter with common ground is applicable to photovoltaic applications where a low dc voltage source have to be inverted to a desired ac output voltage without transformer. The operating principles and circuit analysis for the single phase quasi H-bridge boost inverter with common ground are described. Simulation results by PSIM software are implemented to validate the performance of the introduced topology.

1. Introduction

In the recent, because of energy crisis, using the renewable energy sources is effective remedy. Renewable energy has some merits, i.e. its clean and cheap. But, most renewable energy sources generates a low dc power output. To feed the dc electric energy from the renewable energy sources into ac utility grid, an inverter is used. The inverters can operate in two mode: stand-alone and grid-connected. The voltage source inverters (VSIs) [1]-[5] are popularly applied in high power applications such as distributed power systems, AC motor drives, etc. But, the traditional VSIs are a step down power conversion, where the amplitude of output voltage is not able to be higher than the value of input dc voltage source.

In addition, a grid connected system without transformer produces a ground leakage current. For safe reasons, both the grid and the input dc voltage source should be grounded together. To solve the problems, inverters with common ground were proposed in [6]-[8].

Fig. 1 shows the single phase inverter topology with common ground in [8]. It is comprised of a single capacitor, a single diode , four switches and a single-phase filter. Compared to traditional H-bridge inverter, this topology use one more capacitor and one more diode for sharing a common ground between the utility grid and input dc voltage source.

With the aforementioned merits of the single phase inverter with common ground [8], this paper introduces a single phase quasi H-bridge boost inverter with common ground (qHB2I-CG). It is a combination of a switched-capacitor boost converter [9] with quasi H-bridge inverter [8]. The proposed qHB2I-CG is comprised of an inductor, three capacitors, four diodes and six switches. The operating principles and analysis of single phase quasi H-bridge boost inverter with common ground

IOP Conf. Series: Earth and Environmental Science **159** (2018) 012023 doi:10.1088/1755-1315/159/1/012023

for grid-connected applications are introduced. Also, a simple PWM strategy for the introduced topology is demonstrated. Simulation results are illustrated to verify the operating principle of the introduced inverter. The paper is organized as follows: Operating principles and circuit analysis for proposed topology are discussed in section II. The control system algorithm for grid connection is discussed in section III. Simulation results and the conclusion come in section IV and in V section, respectively.

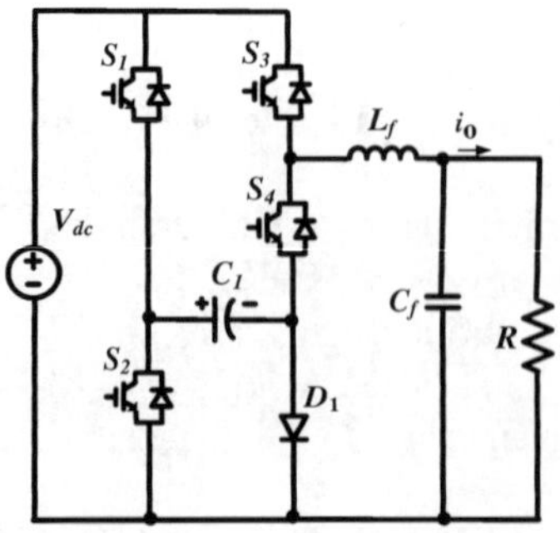

Figure 1. Single phase inverter with common-ground [8].

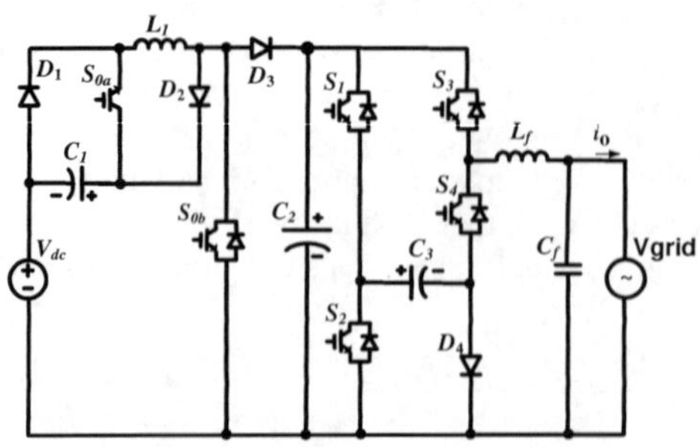

Figure 2. Single phase quasi H-bridge boost inverter with common ground for grid connected applications.

2. System description

Fig. 2 shows the single phase quasi H-bridge boost inverter with common ground for grid connected applications. It used a inductor (L1), three capacitors (C1, C2 and C3), four diodes (D1, D2, D3 and D4), six switches and a single-phase filter.

In the boost state, switches S0ab and S0b are turned on simultaneously as shown in Fig. 4(a). Both diodes D1, D2 and D3 are blocking. The inductor L1 is charged. The capacitor C1 and C2 are discharged. The time interval in the boost state is DT, where T and D are the switching period and the duty ratio, respectively. We have:

$$\begin{cases} L_1 \dfrac{di_L}{dt} = V_{dc} + V_{C1} \\ C_1 \dfrac{dv_{C1}}{dt} = -I_L. \\ C_2 \dfrac{dv_{C2}}{dt} = -i_{PN}. \end{cases} \quad (1)$$

For the non-boost state, switches S0ab and S0b is turned off simultaneously as shown in Fig. 4(b). Both diodes D1, D2 and D3 are conducting. The inductor L1 is discharged. The capacitor C1 and C2 are charged. The time interval in the boost state is (1-D)T. We get:

$$\begin{cases} L_1 \dfrac{di_L}{dt} = -V_{C1} \\ C_1 \dfrac{dv_{C1}}{dt} = I_L - I_{\text{in}} \\ C_2 \dfrac{dv_{C2}}{dt} = I_{in} - i_{PN}. \end{cases} \tag{2}$$

Applying the volt-second balance principle to the inductor L1 and the amp-second balance principle to the capacitor C1, C2 in steady state, from (1) to (2), we obtain

$$\begin{cases} V_{C1} = \dfrac{D}{1-2D} V_{dc} \\ I_L = \dfrac{1-D}{1-2D} I_{\text{in}} \\ I_{in} = \dfrac{1}{1-D} i_{PN}. \end{cases} \tag{3}$$

From (3), we have:

$$I_L = \frac{1}{1-2D} i_{PN} \tag{4}$$

The Bus voltage (VPN) that crosses the inverter bridge is expressed as

$$V_{PN} = V_{C2} = V_{dc} + V_{C1} = \frac{1-D}{1-2D} V_{dc}. \tag{5}$$

The boost factor of single phase quasi H-bridge boost inverter with common can be expressed as

$$B_s = \frac{V_{PN}}{V_{dc}} = \frac{1-D}{1-2D} \tag{6}$$

A simple PWM Strategy for single phase quasi H-bridge boost inverter with common ground for grid connected applications is introduced in Fig 3. The signal, V_{SH}, are used to compare with a high frequency carrier waveform, v_{tri}, to produce the boost control signal for controlling for the switches S_{0a} and S_{0b}. Also, a control waveform, $v_{control}$ is also used to compare to V_{tri}, to generate control signals for switches of quasi H-bridge as shown in Fig 3.

From (6), it should be noted that the maximum duty ratio of the boost state is less than 0.5. To guarantee a good quality of the output voltage waveform, the amplitude of $v_{control}$ is not set higher than *1*. Note that the constant voltages V_{SH} is determined as (1- D).

The peak ac output voltage is determined as

$$v_{ac} = M \cdot V_{PN} = \frac{M \cdot (1-D) V_{dc}}{1-2D}. \tag{7}$$

The dc-ac inversion voltage gain is expressed as

$$G = M \cdot \frac{1-D}{1-2D} = \frac{v_{ac}}{V_{dc}}. \tag{8}$$

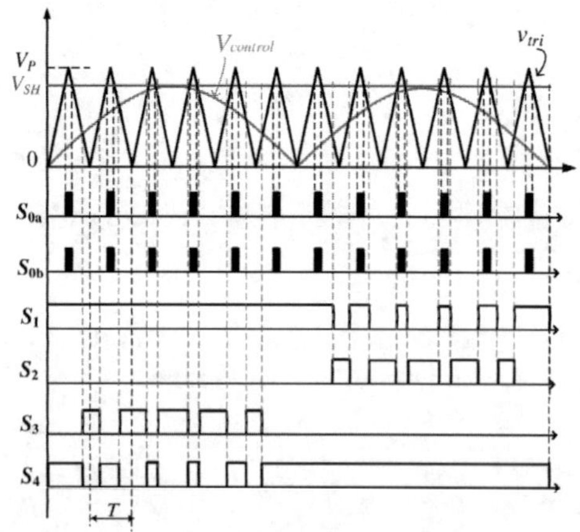

Figure 3. PWM Strategy for single phase quasi H-bridge boost inverter with common ground for grid connected applications.

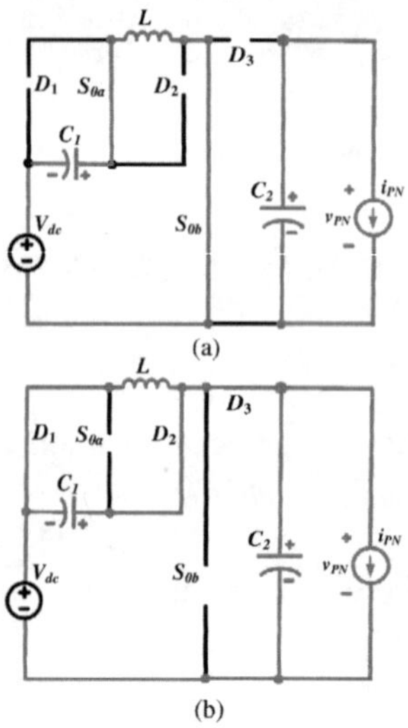

Figure 4. Operating states of single phase quasi H-bridge boost inverter with common ground. (a) boost state and (b) non-boost state.

3. Control system algorithm for grid connection

To deliver energy from dc voltage source to the ac grid, a inverter is used. Moreover, the output voltage of inverter and ac grid bear a close resemblance in terms of the phase and frequency. So, a grid synchronization technique is demanded to get the phase angle of the utility grid voltage. The synchronization strategies must behave significantly to varies of the utility grid. A lot of synchronization techniques have been introduced to get the phase angle of the grid such as zero detection [10], phase locked loop (PLL) [11]-[14]. A PLL is a closed loop control where an internal oscillator is controlled by using a feedback loop to maintain the time and phase of an external periodical signal. Fig. 5 shows structure of PLL controller. A simple current control for single phase quasi H-bridge boost inverter with common ground for grid connection is introduced in Fig. 6.

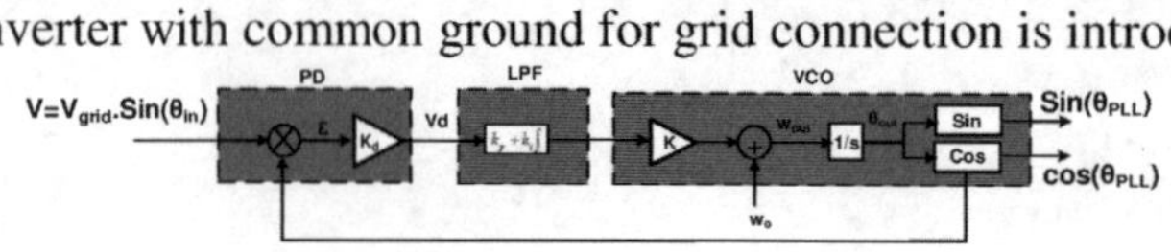

Figure 5. Block diagram of the PLL.

IOP Conf. Series: Earth and Environmental Science **159** (2018) 012023 doi:10.1088/1755-1315/159/1/012023

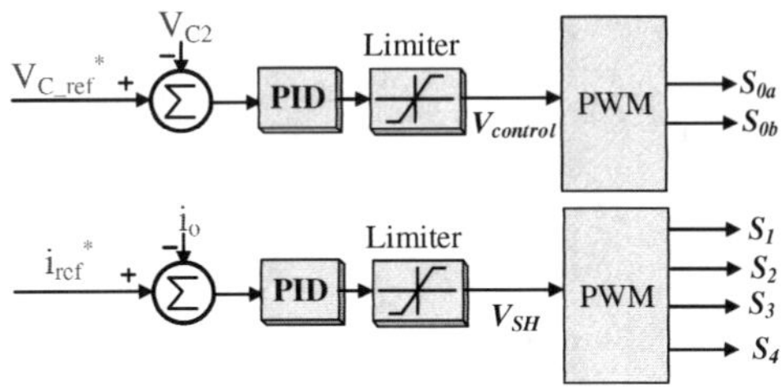

Figure 6. Overall control configuration of the grid connected system.

Table 1. Parameters of the proposed inverter

Parameter	Value
DC input voltage	60V
ac output voltage	110 Vrms /50Hz
Grid voltage	110 Vrms /50Hz
Inductors (L_1)	2 mH
Capacitors C_1, C_2	2200 μF
Capacitors C_3	1000 μF
Inductor (L_f)	3 mH
Capacitor (C_f)	10μF
Switching frequency	20KHz

4. Simulation results

Please follow these instructions as carefully as possible so all articles within a conference have the same style to the title page. This paragraph follows a section title so it should not be indented.

To confirm the operating principle of single phase quasi H-bridge boost inverter with common ground, the simulation results with PSIM simulation software is used. Table I lists the simulation parameters for single phase quasi H-bridge boost inverter with common ground.

The input dc voltage source is set to 60 V in the proposed system in order to test properties of single Fig. 7, Fig. 8 and Fig.9 show the simulation waveforms of the single phase quasi H-bridge boost inverter with common ground when Vdc = 60 V.

As shown in Figure 5, the simulation result of PLL illustrates that the PLL can successfully extract, without errors, the phase angle of the grid voltages, which allows for synchronization with the grid.

We can see that the output voltage of the proposed inverter is 110Vrms with 3-level as shown in Fig. 8. The average inductor current is 10.3

Fig. 9 illustrates the simulation waveform of current and voltage of the grid. From Fig. 9, the current inject to the grid (io) is sinusoidal. The single phase quasi H-bridge boost inverter with common ground is tied with the grid at 110 Vrms. The inverter injects a real power of 390 W to the grid as shown in Fig 9.

IOP Conf. Series: Earth and Environmental Science **159** (2018) 012023 doi:10.1088/1755-1315/159/1/012023

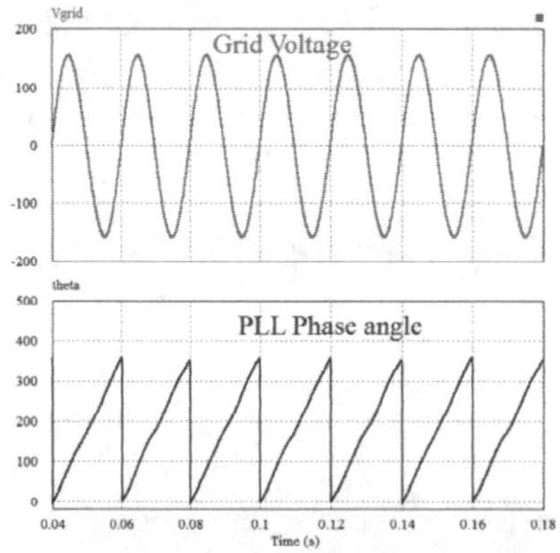

Figure 7. Simulation results of Grid Synchronization.

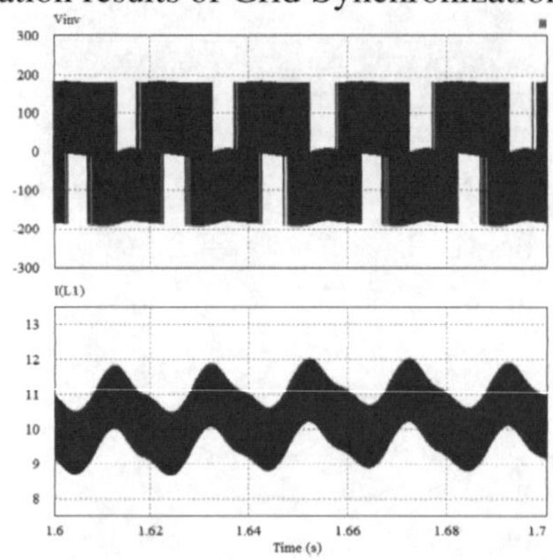

Figure 8. Simulation results for single phase quasi H-bridge boost inverter with common ground. From top to bottom: 3-level output voltage, inductor current.

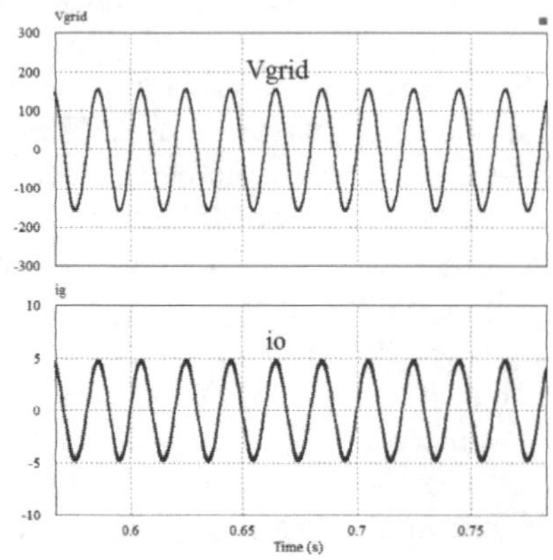

Figure 9. Simulation results for single phase quasi H-bridge boost inverter with common ground. From top to bottom: voltage and current of the grid.

5. Conclusion

This paper presents the operating principles and analysis of the single-phase quasi H-bridge boost inverter with common ground. With sharing the same ground with grid, the ground leakage current is significantly limited. Furthermore, by combining a switched-capacitor boost converter with quasi H-bridge inverter, the RMS of output voltage can be greater than input dc voltage source. Therefore, the single-phase quasi H-bridge boost inverter with common ground is applicable to photovoltaic applications where a low dc voltage source must be inverted to a desired ac output voltage without transformer. Also, the single-phase quasi H-bridge boost inverter with common ground is in line with connecting directly to grid without using transformer. The system is tested for grid connection with the introduced PWM technique and PLL technique. Simulation results for the single phase quasi H-bridge boost inverter with common ground are presented.

6. References

[1] Y. Xue, L. Chang, S. B. Kjær, J. Bordonau, and T. Shimizu, "Topologies of single-phase inverters for small distributed power generators: an overview," IEEE Trans. Power Electron., vol. 19, no. 5, pp. 1305–1314, Sep. 2004.

[2] J. Jiao, J. Y. Hung and R. M. Nelms," Gain scheduling control strategy for a single-phase grid-connected inverter," 2017 IEEE 26th International Symposium on Industrial Electronics (ISIE), UK, 2017, pp.723-728.

[3] T. K. S. Freddy, N. A. Rahim, W. P. Hew, and H. S. Che, "Comparison and analysis of single-phase transformerless grid-connected PV inverters," IEEE Trans. Power Electron., vol. 29, no. 10, pp. 5358–5369, Oct. 2014.

[4] Kerekes T., Teodorescu R., Liserre M., Klumpner C., and Sumner M., "Evaluation of three-phase transformerless photovoltaic inverter topologies", IEEE Trans. Power Electron., vol. 24, no. 9, pp. 2202–2211, 2009.

[5] F. Blaabjerg, Y. Yang, K. Ma, and X. Wang, "Power electronics – the key technology for renewable energy system integration," in 2015 International Conference on Renewable Energy Research and Applications (ICRERA), 2015, pp. 1618–1626.

[6] V.Abinaya, B.Prabakaran and R.B.Selvakumar, "An Improved Transformerless Inverter For Grid-Connected Photovoltaic Power Systems," International Journal of Engineering and Innovative Technology, vol. 2, no. 12, pp.5-9, 2013.

[7] N. Vazquez,J. Vazquez, j. Vaquero, C. Hernandez,E. Vazquez and R. Osorio"Integrating Two Stages as a Common-Mode Transformerless Photovoltaic Converter," IEEE Trans. Ind. Electron., vol. 64, no. 9, pp. 7498-7507, 2017.

[8] Y. P. Siwakoti and F. Blaabjerg, "Common-Ground-Type Transformerless Inverters for Single-Phase Solar Photovoltaic Systems," IEEE Trans. Ind. Electron., DOI: 10.1109/TIE.2017.2740821,2017.

[9] M. K. Nguyen, T. V. Le, S. J. Park, and Y. C. Lim, "A class of quasi-switched Boost inverters," IEEE Trans. Ind. Electron., vol. 62, no. 3, pp. 1526-1536, Mar. 2015.

[10] Nasrudin A. Rahim, Krismadinata Chaniago, and Jeyraj Selvaraj," Single-Phase Seven-Level Grid-Connected Inverter for Photovoltaic System", IEEE Trans. Ind. Electron., vol. 58, no. 6, june 2011

[11] P. Rodriguez, J. Pou, J. Bergas, J. I. Candela, R. P. Burgos, and D. Boroyevich, "Decoupled double synchronous reference frame PLL for power converters control," IEEE Trans. Power Electron., vol. 22, no. 2, pp. 584–592, Mar. 2007.

[12] M. A. Perez, J. R. Espinoza, L. A. Moran, M. A. Torres, and E. A. Araya, "A robust phase-locked loop algorithm to synchronize staticpower converters with polluted ac systems,"IEEE Trans. Ind. Electron.,vol. 55, no. 5, pp. 2185–2192, May 2008

[13] Qi Zhang; Xiang-Dong Sun; Yan-Ru Zhong; Matsui, M.; Bi-Ying Ren, "Analysis and Design of a Digital PhaseLocked Loop for Single-Phase Grid-Connected Power Conversion Systems," Industrial Electronics, IEEE Transactions on , vol.58, no.8, pp.3581-3592, Aug. 2011

[14] L. G. Barbosa Rolim, D. R. Rodrigues da Costa, Jr., and M. Aredes, "Analysis and software implementation of a robust synchronizing PLL circuit based on the pq theory,"IEEE Trans. Ind. Electron., vol. 53, no. 6, pp. 1919–1926, Dec. 2006.

IOP Conf. Series: Earth and Environmental Science **159** (2018) 012024 doi:10.1088/1755-1315/159/1/012024

A Simulink Model of Photovoltaic Modules under Varying Environmental Conditions

Innocent Nkurikiyimfura[12], Bonfils Safari[1] and Emmanuel Nshingabigwi[1]

1 Physics Department, School of Science, College of Science and Technology, University of Rwanda, Po. Box: 3900 Kigali, Rwanda
2 African Center of Excellence in Energy for Sustainable Development (ACEESD), College of Science and technology, University of Rwanda, Po. Box: 3900 Kigali, Rwanda
E-mail: inkurikiyimfura@ur.ac.rw

Abstract. In this paper, a model for predicting the effects of environmental conditions on the performance of photovoltaic panels is developed and analysed. A one-diode model taking into account effects of environmental conditions such as solar irradiance, ambient temperature and wind was developed and simulated in Simulink/Matlab environment. The accuracy of the developed model was confirmed *via* a comparison of the simulated results with the output characteristics of two polycrystalline photovoltaic (PV) modules from different manufacturers. Effects of various environmental parameters on the cell temperature and output characteristics of PV modules were analysed using the developed model. The results revealed that environmental condition variations have significant effects on the electrical performances of PV modules and should be taken into consideration in PV system design. The developed model could also be useful in the design of optimum cooling systems for photovoltaic/thermal (PV/T) systems since it takes into consideration the cooling effects of the wind.

1. Introduction

Photovoltaic (PV) solar energy conversion has received continuous attentions for decades as a clean, inexhaustible, reliable and cheap renewable energy technology [1–3]. In a PV system, solar energy is directly converted into direct current (DC) electricity on the smallest units known as PV cells. For higher output voltages, PV cells are generally grouped in series and/or parallel. On the surface of a PV solar cell, incident solar energy is converted into electricity via the so-called photovoltaic effect. The operating temperature and thus the overall performances of any PV solar module are generally affected by environmental conditions such as solar irradiance intensity, wind speed and direction, ambient temperature, etc. [4–6]. While higher ambient temperature and the solar irradiance have detrimental effects on the overall performance of solar modules, high speed wind could lower the cell temperature leading to an enhancement in the overall performances of the PV module. In previous studies, effects of ambient temperature and solar irradiance on the performances of solar cells/modules were extensively investigated using experimental and mathematical modelling [2,7].

Various models accounting for the ambient temperature and solar radiation have been thoroughly investigated based on the single diode [3,8,9], two-diode [10–12] and recently, three-diode [13,14] models. The developed models are generally simulated using various software programs including C-language programs, excel, Matlab, Simulink, SPICE, etc. [5,15]. Although two-diode and three-diode based models have shown more accurate results, the use of one-diode model has received more attention due to its simplicity and ease of simulation using various software packages such as

IOP Conf. Series: Earth and Environmental Science **159** (2018) 012024 doi:10.1088/1755-1315/159/1/012024

Matlab/Simulink [11]. Using one-diode models, various authors have accurately simulated effects of ambient temperature and solar radiation using Simulink [2,16]. They confirmed that the P-V and I-V characteristics of PV modules could be strongly affected by the ambient temperature and solar radiation. While increased solar radiation increases the output characteristics of the solar modules, higher temperatures have detrimental effects on them. Ayaz et al. [17] developed an improved matlab/simulink model considering the ambient conditions. The developed model was tested on various PV technologies including monocrystalline silicon, polycrystalline silicon, and thin films and the results were validated experimentally. The comparison of the simulated and experimental results under real atmospheric conditions showed good agreement for monocrystalline silicon and thin films. However, the prosed model showed lower performances for polycrystalline silicon.
Models that take into consideration the effects of environmental condition variations on the output characteristics on PV module are rare in the literature. Therefore, this work is devoted to developing a MatLab/Simulink model that could be used to analyze the combined effects of environmental conditions on the performances of PV modules. A model that accounts for wind speed, ambient temperature, and solar irradiance intensity was developed and used to analyze the output characteristics of polycrystalline PV modules.

2. Photovoltaic models

2.1. Photovoltaic cell model

A photovoltaic (PV) cell is the smallest unit of a photovoltaic module, with power output of less than 2 W at 0.5 V [4]. A PV cell is physically modelled using the equivalent circuit of Figure 1 (a) [4,8]. It consists of a photocurrent and a diode with an internal resistance connected to a parallel resistor that simulates a leakage current [4].

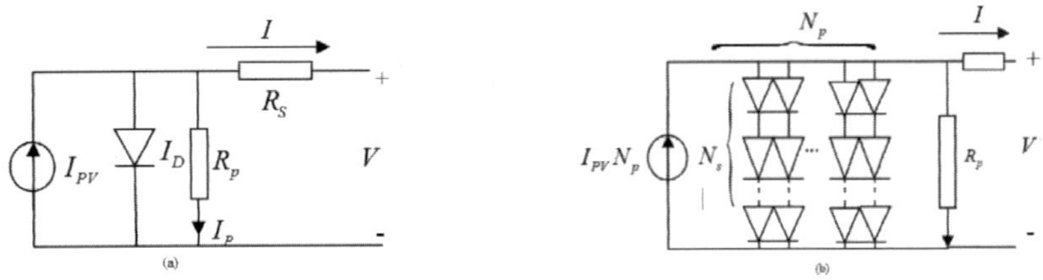

Figure 1. Physical model of (a) photovoltaic solar cell and (b) photovoltaic solar module

According to the well-known Kirchhoff's current law, the current flowing in the circuit (*I*) can be modelled by the following equation [14]:

$$I = I_{PV} - \underbrace{I_0\left[\exp\left(\frac{qV}{Ak_BT_{cell}}\right)-1\right]}_{I_D} - \underbrace{\frac{V+IR_s}{R_p}}_{I_p} \qquad (1)$$

where I_{PV} is the light-generated current also referred to as photocurrent, I_0 is the dark saturation current, T_{cell} is the cell temperature, $k_B = 1.3806503\times10^{-23} JK^{-1}$ represents the Boltzmann constant and $q = 1.6\times10^{-19} C$ is the electron charge. *V*, *A*, R_s and R_p are the voltage, ideality factor, series resistance and shunt resistance, respectively.

2.2. Photovoltaic module model

A photovoltaic module consist of an array of several PV cells electrically connected in a series-parallel configuration [4]. An equivalent circuit of a photovoltaic module is depicted in Figure 1 (b). In the circuit, PV cells are arranged in N_p parallel and N_S series branches. Similarly to the PV cell model, the current flowing generated by a PV module could be modelled as [2]

$$I = N_p I_{PH} - N_p I_0 \left[\exp\left(\frac{V + IR_s}{AV_t} \right) - 1 \right] - \frac{V + IR_s}{R_p} \tag{2}$$

where the parameter V_t, also known as thermal voltage, is calculated using the following formula

$$V_t = \frac{N_s k_B T_{cell}}{q} \tag{3}$$

In practice the shunt resistance can be set equal to infinity. Thus, last term of equation (2) can be omitted and the current equation can be written in its simplified form as follows [2,18]

$$I = N_p I_{PH} - N_p I_0 \left[\exp\left(\frac{V + IR_s}{AV_t} \right) - 1 \right] \tag{4}$$

The photocurrent is calculated as [18]

$$I_{PH} = I_{PH,STC} + K_I \left(T_{cell} - T_{STC} \right) \frac{G}{G_{STC}} \tag{5}$$

where $I_{PH,STC}$ is the photocurrent at standard testing conditions, which is approximately equal to the short circuit current (I_{sc}) and K_I is the temperature coefficient of the current which is generally provided by the PV cell manufacturers. G and G_{STC} stand for the solar insolation intensity and solar insolation intensity under standard testing conditions (i.e., $T_{STC}=298K$, $G=1000Wm^{-2}$, and air mass, $AM=1.5$), respectively.

The reverse saturation current is calculated as

$$I_0 = I_s \left(\frac{T_{STC}}{T_{cell}} \right)^3 \exp\left[\frac{qE_g}{Ak_B} \left(\frac{1}{T_{STC}} - \frac{1}{T_{cell}} \right) \right] \tag{6}$$

where I_s is the saturation current, $T_{STC}=298K$ is the standard test temperature and E_g is the band gap of the semiconductor, which is estimated at $1.12\ eV$ for silicon semiconductors. The saturation current is calculated as

$$I_s = \frac{I_{sc}}{\exp\left(\frac{qV_{oc}}{N_s AkT_{STC}} \right) - 1} \tag{7}$$

where I_{sc} and V_{oc} are the short circuit and open voltage, respectively.

2.3. *Cell temperature under varying environmental conditions*

The cell temperature is a key parameter for PV module performance evaluation. In previous studies, various models were proposed. A comprehensive review of cell temperature models and their comparison is provided in [19]. Most of the proposed models account for solar irradiance and temperature as environmental parameters. In this work, a model combining effects of solar irradiance, wind, and ambient temperature is used. On-cell temperature is conveniently modeled using the following equation for polycrystalline silicon PV cells [19,20].

$$T_{cell} = \frac{U_{PV}(v) T_a + G\left[\tau\alpha - \eta_{STC} \left(1 - \beta_{STC} T_{STC} \right) \right]}{U_{PV}(v) + \eta_{STC} \beta_{STC} G} \tag{8}$$

where G and T_a are the irradiance and ambient temperature, respectively. η_{STC} and β_{STC} are the efficiency and the temperature coefficient of maximal power under standard test conditions. The parameter $U_{PV}(v)$ is the heat exchange coefficient for the total surface of the cell. For polycrystalline silicon PV cells, $U_{PV}(v)$ is estimated using the following equation [19,20].

$$U_{PV} = 26.6 + 2.3v \tag{9}$$

where v is the wind speed.

3. Environmental conditions effects modeling and simulation

3.1. Photovoltaic module parameters

In this study, two polycrystalline silicon PV modules, i.e., KT260-6GC-A (Kyocera ltd., Japan) and ND-220(ELF) (Sharp, Sharp Energy Solution Europe, German) were used to assess the accuracy of the developed Matlab/Simulink model. The key specifications of these modules are summarized in Table 1.

Table 1. Solar modules specifications (1000Wm-2, 298 K, AM=1.5)

Characteristics	Specifications	
Type	KT260-6GC-A	ND-220(ELF)
Maximum power (P_{max})	260 Wp	220 Wp
Maximum power current (I_{mp})	8.39 A	7.54 A
Maximum power voltage (V_{mp})	31 V	29.2 V
Short circuit current (I_{SC})	9.09 A	8.2 A
Open circuit voltage (V_{OC})	38.3 V	36.5 V
Module efficiency	15.8%	13.4%
Temperature coefficient of (I_{sc})	+0.06%/K	+0.053%/^{0}C
Temperature coefficient of (V_{oc})	-0.36%/K	-0.13 %/^{0}C
Temperature coefficient of (P_{max})	-0.45%/K	-0.485% ^{0}C
TOCT	45 ^{0}C	47.5^{0}C

3.2. Matlab/simulink model building and performances analysis

A MatLab/Simulink model taking into account effects of solar irradiance intensity, ambient temperature, and wind as environmental parameters was built and used to analyze I-V and P-V characteristics of polycrystalline PV modules.

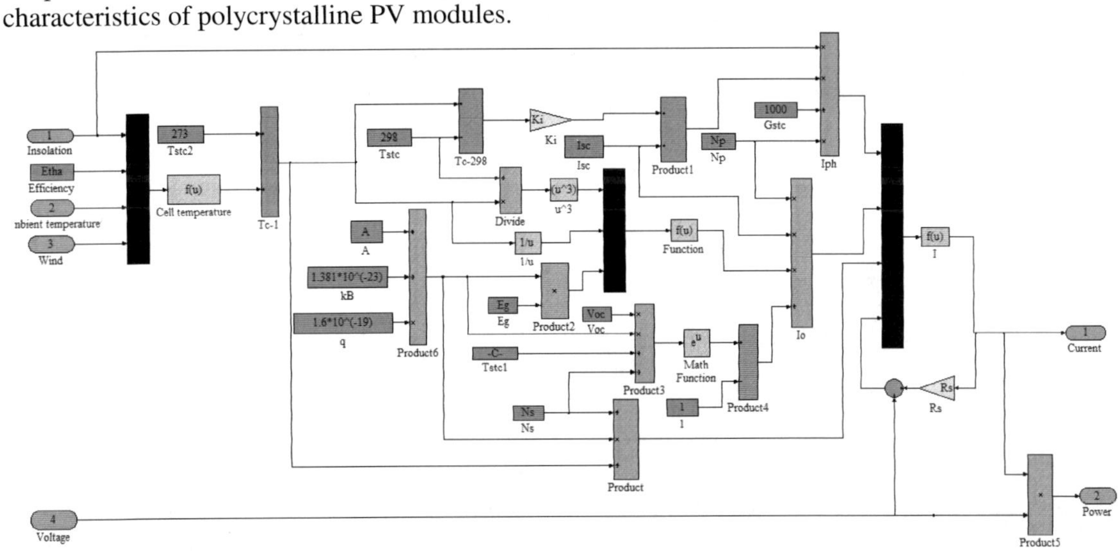

Figure 2. Simulink subsystem implementation of a PV module under varying environmental conditions

The block diagram of the developed model is shown on Figure 2. In this model, various environmental parameters including solar irradiance intensity, ambient temperature, and wind were used as input parameters (Figure 3 a). The current and power were estimated at various operating voltages which was modelled using a ramp block with 0.95 as input parameter. The cell temperature and PV models (equations 3 through 8) were modelled using appropriate simulink built-in functions and various parameters were fed into the model *via* the developed dialog box (Figure 3 b).

Performances of the developed model were assessed via a comparison of the simulated results and the I-V and P-V characteristics provided in the datasheets of two polycrystalline PV modules taken as

IOP Conf. Series: Earth and Environmental Science **159** (2018) 012024 doi:10.1088/1755-1315/159/1/012024

example (**Table 1**). Figure 4 and 5 show the simulated I-V and P-V output characteristics of the two modules.

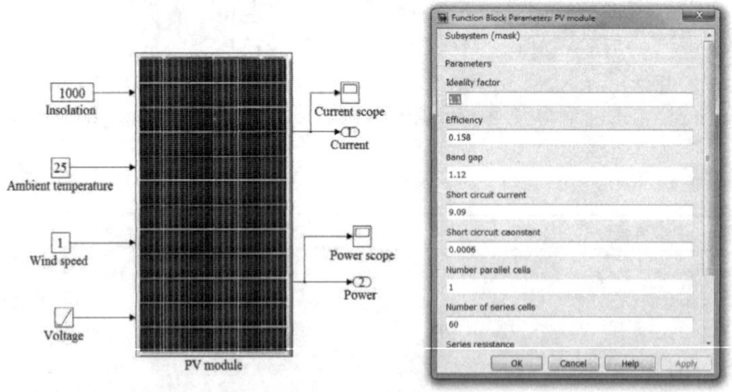

Figure 3. Masked PV module and (b) dialog box of the developed MatLab/simulink model

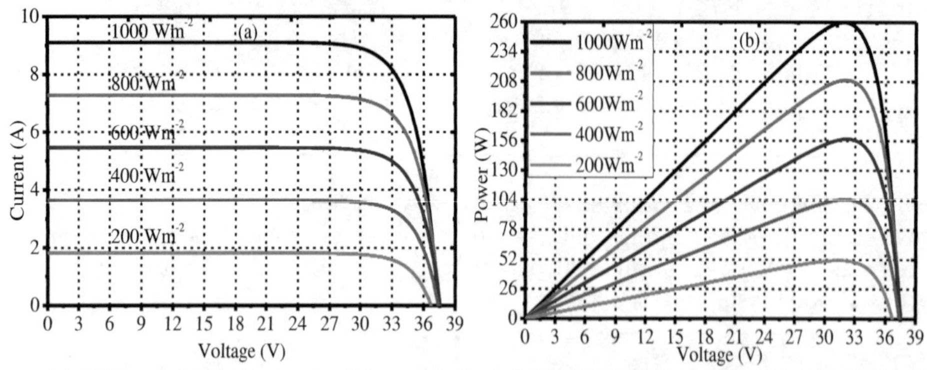

Figure 4. (a) I-V and (b) P-V output characteristics of KT260-6GC-A at various solar irradiance intensities

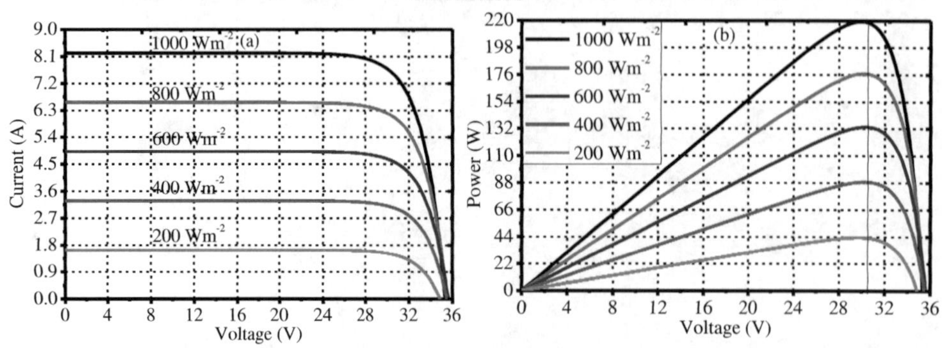

Figure 5. I-V and P-V output characteristics of ND-220(ELF) at various solar irradiance intensities

Interestingly, the simulated I-V and P-V output characteristics were comparable to those provided in the datasheets within acceptable error, confirming the accuracy of the proposed model. For instance, the simulated output powers for the two modules under the standard testing conditions were 260.31 W at 31.98 V and 219.52 W at 29.64 V for KT260-6GC-A and ND-220(ELF) PV modules, respectively. Comparing the latter values with those provided in the datasheets (**Table 1**), the relative errors are estimated at +0.12% and -0.22% for KT260-6GC-A and ND-220(ELF) PV modules, respectively. It is also interesting to note the effects of solar irradiance intensity on the performances of PV modules as presented on Figure 5Figure 6. The I-V and P-V characteristics were simulated at standard conditions of the temperature and wind speed (i.e., $T_{STC} = 25°C$ and $v_{STC} = 1ms^{-1}$). Obviously, the output

IOP Conf. Series: Earth and Environmental Science **159** (2018) 012024 doi:10.1088/1755-1315/159/1/012024

characteristics are related to solar irradiance intensity. Higher output powers were recorded at higher irradiance intensity.

4. Simulation results and discussion

4.1. Effects of ambient temperature variation on output characteristics of a PV module

Effects of ambient temperature variation on output characteristics of a PV module were analysed using the developed model and the mode KT260-6GC-A was taken as example. The model was simulated by keeping the solar irradiance intensity and the wind at their standard testing conditions values and varying the temperature with a step of 25 0C. Figure 6 shows the I-V and P-V characteristics of a PV module simulated at various ambient temperatures.

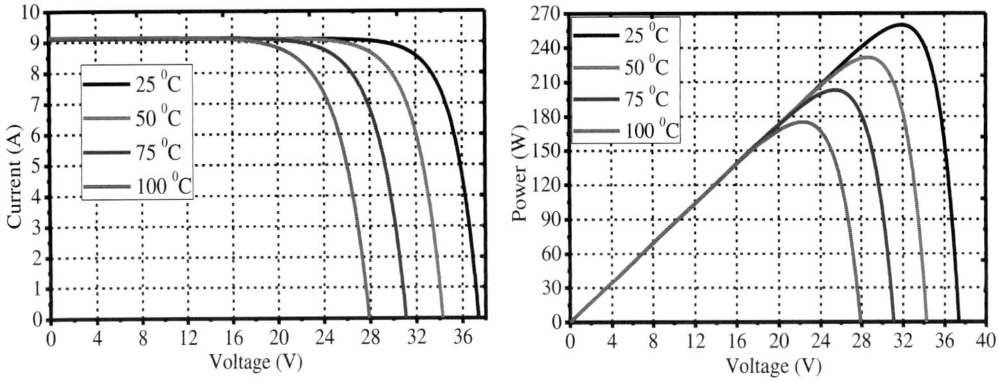

Figure 6. (a) I-V and (b) P-V output characteristics of a PV module at various temperatures

It can be seen that the maximum point voltage and power decreased with an increase of ambient temperature. The increase in ambient temperature leads to a substantive increase in the cell temperature and thus to a reduction in both I-V and P-V outputs of the module. For instance, an increase of 25 ^{0}C from the standard testing conditions led to a reduction in the maximum power output from 260.31 W at 31.98 V to 231.76 W at 28.86 V. It is clear from these results that a cooling system could be needed to keep the PV module at its optimum point.

4.2. Effects of wind speed variation on output characteristics of PV modules

Using the developed matlab/simulink model, effects of wind speed variation on the output characteristics of PV modules were analysed using the model KT260-6GC-A PV. The solar irradiance intensity and the temperature were kept at the standard testing conditions and the output characteristics were simulated at various wind speeds.

Figure 7 shows the I-V and P-V output characteristics of the KT260-6GC-A PV module under various wind speeds. Clearly, higher maximum power point was obtained at higher wind speed. In addition, an increase in wind speed leads to an increase in the heat exchange coefficient for the total surface of the cell (equation 8) and thus to a reduction in the cell temperature. It is evident from the previous analysis that the change in the cell temperature affects the performances of the PV module. The lower the cell temperature, the higher the electrical performance. As can be seen from Figure 7, the highest output power of 274.06 W at 33.54 V was obtained at 15 ms^{-1}. Compared to the standard testing conditions, the enhancement in the output power could be estimated at 5.13%. Since it takes into account effects of wind speed on the performances of PV modules, the developed model could be useful in the design of optimum cooling systems for PV/T systems.

To analyse the output characteristics of a PV module under varying atmospheric conditions, the wind was assumed to follow a Gaussian distribution defined as

$$v(t) = v_{\max} \exp\left(-\frac{(t-t_c)^2}{2\sigma^2} \right) \quad (10)$$

where v_{max} is the maximum wind speed at a given time of the day, t_c is the centre time and σ is the standard deviation of Gaussian function.

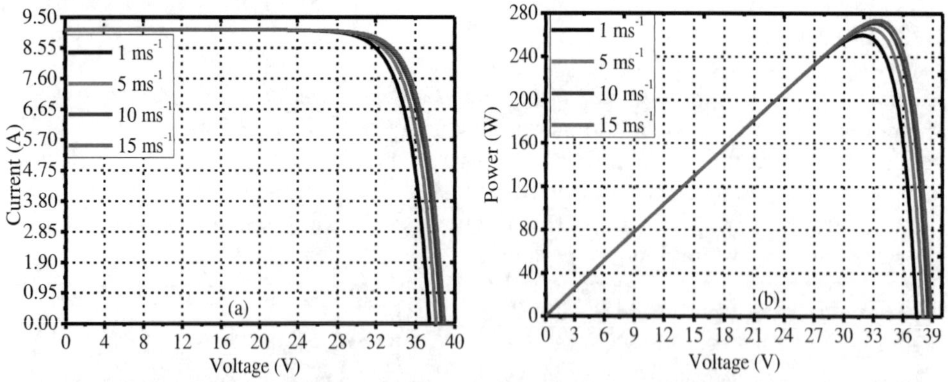

Figure 7. (a) I-V and (b) P-V output characteristics of a PV module at various wind speeds

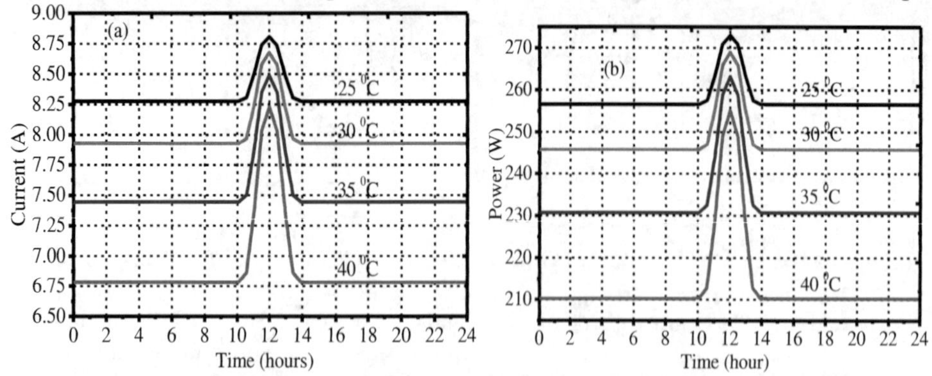

Figure 8. (a) I-V and (b) P-V output characteristics of a PV module during a sample day at various temperatures

The output characteristics of the PV module taken as example were simulated for a sample day with maximum wind speed $v_{max} = 15ms^{-1}$, centre time $t_c = 12$ and $\sigma = 0.5$. The output characteristics of a PV module for the sample day were simulated at various temperatures ranging from 25 ^{0}C to 40^{0}C for a constant solar irradiance intensity of 1000 Wm^{-2} at the maximum operational voltage (31 V). Figure 8 shows the output current and power as function of voltage at various temperatures. As can be seen from the figure, both the current and the output power are significantly affected by the wind and temperature variations. The optimum output current and output power are 8.81 A and 273.03 W, respectively. It is also interesting to note that, despite the cooling effects of the wind, a slight increase in temperature is followed by a significant reduction in the output characteristics and significant reduction was observed at relatively higher temperature.

5. Conclusion

In this work, a MatLab/Simulink model of a PV module was developed and analyzed under varying environmental conditions. The cell temperature was modeled taking into account the solar irradiance intensity, the ambient temperature and the wind speed as environmental parameters. The accuracy of the developed model was confirmed *via* a comparison of the simulated results and the data provided by the PV module manufacturers. The relative error for the modules taken as examples were estimated at + 0.12% and -0.22% for for KT260-6GC-A and ND-220(ELF) PV modules, respectively. The obtained results confirm that environmental condition variations have a significant influence on the performances of PV modules and should be taken into consideration during the systems design. In addition, the developed model could be a useful tool in designing optimum cooling systems for PV/T systems.

Acknowledgement
This work was financed by the World Bank through the East Africa Higher Education Centers of Excellence (Project ID: P151847), Center of Excellence in Energy for Sustainable Development (ACE-ESD).

6. References

[1] Armstrong S and Hurley W G 2010 A thermal model for photovoltaic panels under varying atmospheric conditions *Appl. Therm. Eng.* **30** 1488–95
[2] Bellia H, Youcef R and Fatima M 2014 A detailed modeling of photovoltaic module using MATLAB *NRIAG J. Astron. Geophys.* **3** 53–61
[3] Chander S, Purohit A, Sharma A, Arvind, Nehra S P and Dhaka M S 2015 A study on photovoltaic parameters of mono-crystalline silicon solar cell with cell temperature *Energy Reports* **1** 104–9
[4] Tsai H L 2010 Insolation-oriented model of photovoltaic module using Matlab/Simulink *Sol. Energy* **84** 1318–26
[5] Nguyen X H and Nguyen M P 2015 Mathematical modeling of photovoltaic cell/module/arrays with tags in Matlab/Simulink *Environ. Syst. Res.* **4** 24
[6] Olukan T A and Emziane M 2014 A comparative analysis of PV module temperature models *Energy Procedia* **62** 694–703
[7] Paul P S 2014 Modeling Comb bined Effect of Tempe erature and Irradiance on o Solar Cell Parame ters by MA ATLAB / Simulink *8th Int. Conf. Electr. Comput. Eng.* 512–5
[8] Villalva M, Gazoli J and Filho E 2009 Comprehensive Approach to Modeling and Simulation of Photovoltaic Arrays *IEEE Trans. Power Electron.* **24** 1198–208
[9] Bechouat M, Younsi A, Sedraoui M, Soufi Y, Yousfi L, Tabet I and Touafek K 2017 Parameters identification of a photovoltaic module in a thermal system using meta-heuristic optimization methods *Int. J. Energy Environ. Eng.* **8** 331–41
[10] Abdulkadir M, Samosir A S and Yatim A H M 2012 Modelling and simulation of maximum power point tracking of photovoltaic system in Simulink model *PECon 2012 - 2012 IEEE Int. Conf. Power Energy* 325–30
[11] Krismadinata, Rahim N A, Ping H W and Selvaraj J 2013 Photovoltaic Module Modeling using Simulink/Matlab *Procedia Environ. Sci.* **17** 537–46
[12] Ishaque K, Salam Z and Syafaruddin 2011 A comprehensive MATLAB Simulink PV system simulator with partial shading capability based on two-diode model *Sol. Energy* **85** 2217–27
[13] Vimalarani C and Kamaraj N 2015 Modeling and performance analysis of the solar photovoltaic cell model using Embedded MATLAB *Simulation* **91** 217–32
[14] Khanna V, Das B K, Bisht D, Vandana and Singh P K 2015 A three diode model for industrial solar cells and estimation of solar cell parameters using PSO algorithm *Renew. Energy* **78** 105–13
[15] Díaz-Bernabé J L and Morales-Acevedo A 2015 Photovoltaic module simulator implemented in SPICE and Simulink *2015 12th Int. Conf. Electr. Eng. Comput. Sci. Autom. Control. CCE 2015*
[16] Liu X 2015 Development of Fast Simulation Models for Photovoltaic Generation System Based on Simulink 3265–70
[17] Ayaz R, Nakir I and Tanrioven M 2014 An improved matlab-simulink model of pv module considering ambient conditions *Int. J. Photoenergy* **2014**
[18] Zainal N A, Ajisman and Yusoff A R 2016 Modelling of Photovoltaic Module Using Matlab Simulink *IOP Conf. Ser. Mater. Sci. Eng.* **114**
[19] Schwingshackl C, Petitta M, Wagner J E, Belluardo G, Moser D, Castelli M, Zebisch M and Tetzlaff A 2013 Wind effect on PV module temperature: Analysis of different techniques for an accurate estimation *Energy Procedia* **40** 77–86
[20] Mattei M, Notton G, Cristofari C, Muselli M and Poggi P 2006 Calculation of the polycrystalline PV module temperature using a simple method of energy balance *Renew. Energy* **31** 553–67

IOP Conf. Series: Earth and Environmental Science **159** (2018) 012025 doi:10.1088/1755-1315/159/1/012025

Analysis of the Constraints in the Renewable Energy Sector within a Multi-level Energy Transition Perspective (MLP): The Case of the Philippines

Phoebe Grace Saculsan[1]

1 Graduate Student, Kyoto University, Kyoto, Japan
Recipient of the Japanese Government Scholarship, 2015-2018
E-mail: pjsaculsan@gmail.com

Abstract. As an emerging economy, there are stumbling blocks for the Philippines that it needs to overcome to actualize its growth potentials. One major challenge is providing energy to its growing economy and population without sacrificing energy security and sustainability. This paper explores the country`s energy system within a multi-level decision-making energy approach (MLP) to better understand the context and nature of barriers in the renewable energy (RE) sector. Analysis show that: first, past crisis continue to have lingering effects, brought positive and negative consequences, in the electricity sector; second the energy security goal is only focused on increasing capacity generation but in reality has only led to increased share of conventional energy such as coal; and lastly, the unbundling of the electricity sector under the EPIRA Law has paved the way for oligopoly/monopoly to thrive, it has created a fragmented energy structure, and has allowed greater costs for the consumers. All of these have pose greater barriers for RE to thrive and, ultimately, is hindering the country from achieving a sustainable energy transition.

1. Introduction

The Philippines is an archipelago with more than 7,000 islands and is situated in the West Pacific coast. It is categorized as a lower middle income country by the World Bank [1] although economic projection is bright for this country with its young, dynamic population; teeming natural resources; and strong macroeconomic fundamentals. Yet these potentials can be reversed if persistent internal problems won`t be addressed. One of the major challenges identified by scholars and policymakers that might push back the country`s growth and development is in the energy sector. This is the topic the author wishes to explore in this paper.

As the economy and population expands along with it rises the demand for energy. Electricity, as the primary energy source for the households, is projected to expand at an annual average rate of 5.1 percent, from 5.9 MTOE in 2011 to 15.3 MTOE in 2030 [2]. A major portion of this supply comes mainly from conventional energy sources (e.g., coal, oil and natural gas), which is already more than 70 percent in 2014 alone. What is problematic is that mostly of these sources are imported, and historically this did not do good for the country. In turn, the majority of the poor populace is paying an expensive electricity rate, which is among the highest in Asia.

Like the Philippines, the threats brought about by overdependence on conventional energies have paved for countries to develop their own indigenous energy resources and to adopt renewable energy (RE) technologies. This is furthered by the increased commitment of countries towards climate change,

IOP Conf. Series: Earth and Environmental Science **159** (2018) 012025 doi:10.1088/1755-1315/159/1/012025

having its effects becoming more felt today. As a result, we see global energy policy shifting more and more towards cleaner and sustainable energy options.

This changing energy landscape presents opportunities for the Philippines that have vast potentials for RE resources. This is a good opportunity for the country to take advantage from given that RE technologies continues to be cost-effective with its decreasing costs; better technological capacity; and a wider recognition of the multiple benefits it can bring.

But even with such favourable prospect of renewables, developing the country`s RE sector has remained a challenge until today. This happens when conditions do not allow for RE to be at a level playing field with conventional energies. Since RE is still a relatively novel technology for emerging economies such as the Philippines, there are reservations that RE cannot provide a sufficient and stable energy supply for its huge demand for energy alongside its growing economy and population. That is why it remains a balancing act between energy security and sustainability the energy policy of the Philippines.

2. Data and analytical framework

To provide additional discussions and aid the policy-making for the successful transition of the Philippines to a cleaner and sustainable energy system, the central theme of the paper is about the constraints the RE sector is facing in the Philippines. It is because the energy transition plans towards cleaner and sustainable energy system has always been challenged and impeded by these factors.

For this paper, semi-structured interviews and surveys were conducted among renewable energy investors/developers to identify the "major" constraints as derived through their own personal experience in investing and financing RE projects in the Philippine. About 20 companies/organization participated in the interviews/survey. All the company/organization representatives hold managerial positions, either the owner and/or the RE project manager. The fieldwork was done in September to November 2016 in Metro Manila, Philippines, where most of the companies` offices are located. The findings were then further verified through extensive literature reviews.

The top 3 constraints as identified by the majority of these respondents are: 1) the difficulty in getting permits due to local oppositions; 2) the tedious administrative procedure due to in situ government policies; and 3) the difficulty in accessing finance because of the scope of small companies that cannot compete with big energy companies and/or because these companies don't have sufficient assets/guarantee to entitle them for a loan.

3. Conceptual framework: what is multi-level decision-making energy transition perspective (MLP) and its significance in analysing the constraints in the renewable energy (RE) sector

Using the conceptual tools of energy transition (ET) theory, the author develops the analytical framework of this paper in explaining the abovementioned constraints to RE.

The transitions theory per se is not an entirely new concept as this has already been used by scholars before when explaining the changes and dynamism within a particular system. Transition, as defined by Oxfords dictionary, is "the process or a period of changing from one state or condition to another" [3]. The concept of transition was later adopted to explain changes in societies, and is now also used to explain the new phenomenon in energy-modernity within a sustainability perspective. The shift in the Dutch energy system away from coal to gas and renewables, for example, is widely acknowledged and studied in this area while taking note of the "complexity, nonlinearity and the multiplicity of the actors involved" in the transition [4]. With this in mind and how this is relevant in the Philippine context, the paper focuses on the multi-level decision-making perspective (MLP) of energy transition, where the socio-political conditions are emphasized. MLP is defined as

> *"...a middle-range theory that conceptualizes overall dynamic patterns in socio-technical transitions. The analytical framework combines concepts from evolutionary economics (trajectories, regimes, niches, speciation, path dependence, routines), science and technology studies (sense making, social networks, innovation as a social process shaped by broader societal contexts), structuration theory and neo-institutional theory (rules and institutions as 'deep structures' on which knowledgeable actors draw in their actions, duality of structure, i.e. structures are both context and outcome of actions, 'rules of the game' that structure actions).*

IOP Conf. Series: Earth and Environmental Science **159** (2018) 012025 doi:10.1088/1755-1315/159/1/012025

> *[...] The MLP views transitions as non-linear processes that results from the interplay of developments at three analytical levels: niches (the locus for radical innovations), socio-technical regimes (the locus of established practices and associated rules that stabilize existing systems), and an exogenous sociotechnical landscape [5]."*

MLP emphasizes that the existence of a greater tension between the "need for energy security in the light of growing demand and the need to reduce environmental degradation and check the rise of air pollutants and greenhouse gas emissions tied to climate change at the same time" [6]—a situation that is more pronounced in emerging economies. This tension is primarily triggered by the "socio-political conditions, including the political system, civil society actors or decision-making process, and the fragmentation that exists between and among these agents" [7]. Taking Philippines a country case, this paper highlights the dominant agents/players in the energy sector that affect the energy system and its so-called transition pathways. The analysis is, thus, framed as follows: first, by providing a historical background of the current energy context/regime; and second, by discussing how the current energy context and regime has contributed to constraints in the RE sector.

4. Results and discussions

4.1 The transition in energy policy and governance structure: A historical analysis of the Philippine electricity sector

Electricity first came in the Philippines in the 1890s during the Spanish era. Then in the 1900s, the Manila Electric Light and Railroad Company (more commonly known as MERALCO), which is owned by the Lopezes, one of the richest families in the Philippines, bought the franchise and took charge in delivering electricity to Manila and its adjacent cities [8]. Now MERALCO is considered to be the largest electricity distribution company in the Philippines.

In the 1930s, when the country gained independence from the Americans, it became a priority objective of the government to electrify the whole country for nation-building. With this, the government has aimed to develop its own indigenous energy resources. That is why by virtue of the Commonwealth Act 120 and now revised RA 6395, the National Power Corporation (NPC) was created to "undertake the development of hydroelectric generation of power and the production of electricity from nuclear, geothermal and other sources" [9]. NPC was a monopoly that dominates both the generation and transmission operations. By 1956, NPC had generated about one-third of the country`s total generation capacity while the remaining two-thirds were handled by private and municipally-owned electric utilities [8]. In the 1960s, the government saw through that the national electrification program reached the countryside thereby creating the National Electrification Act and accordingly establishing the National Electrification Administration (NEA). NEA is mandated to fasten the electrification efforts by providing loans, subsidies, and affordable electric rates, among others. As of December 2011, household electrification level stands at 70.2 percent and is targeted to increase to 90.0 percent by 2017 [2].

Until the 2000s, the electricity sector is nationalized and still primarily government-owned through NPC. Crises that led one to another has resulted into multitude of problems that is still felt today. This began in the 1970s when oil crisis hit the country, thus, had led to increased prices of imported energy, decreased foreign concessional loans, and hiked the cost of borrowing used for the government`s electrification program [10]. Domestic problems had likewise amplified the crisis during the Marcos era having fraught with corruption and government agencies` mismanagement. One mothballed project that could have saved the country from dependency on energy imports was the Bataan Nuclear Power Plant. Due to allegations of corruption and fears after the Chernobyl incident, BNPP was stalled in 1985. The government spent USD 2.3 billion for its completion and up to this date is incurring yearly maintenance cost of around PhP 50 million (USD 1.06 million) for a project that was never utilized [11]. After the downfall of the Marcos regime and the installation of Corazon Aquino as the new president, the Department of Energy (DOE) was likewise abolished due to its alleged inefficiencies and corruption. This is also one of the means of the Aquino administration to reverse all the policies of the Marcos dictatorship. But the dissolution of DOE had brought more damages to the already debilitating condition of the energy sector at that time. Without an agency to plan, manage and coordinate activities within the energy sector, power crisis became worse.

"Nationwide blackouts" continued until the early administration of the newly-elected President Fidel V. Ramos (or "FVR"). This has of course negatively affected the economy and resulted into massive unemployment. In hope to end the economic and power crisis at that time, then president Ramos was granted an emergency power by virtue of the Republic Act (RA) No. 7648, otherwise known as the "Electric Power Crisis Act of 1993". Through RA 7648, the procedure to bid out government projects was fast-tracked thereby allowing the Independent Power Producers (IPPs) to immediately "construct, repair, rehabilitate, improve or maintain power plants, projects and facilities" [12]. A new policy called "take or pay" was also formulated which forced the distribution utilities to pay for a specified amount of power produced by the generation companies whether they like it or not [13]. The excess electricity was of course passed on to the consumers who now have to bear a higher electricity price.

In order to avoid a power crisis from happening again and to reduce the expensive electricity rate, the next president Gloria Macapagal Arroyo (or "GMA") signed the Republic Act 9136 or the "Electric Power Industry Reform Act" (or the EPIRA Law) in 2001. The EPIRA Law goals to (i) disaggregate the industry into generation, transmission, and distribution; (ii) introduce competition in generation and supply; (iii) introduce a wholesale electricity spot market (WESM); (iv) privatizing generation and operation of transmission by a concessionaire; (v) introduce open access to distribution networks; and (vi) set up an independent agency to regulate the industry [14]. These goals are ultimately aimed at reducing the power cost and securing the country`s energy supply.

But until this moment, the EPIRA Law has failed to delivers its promises. The Philippines is still considered to have one of the highest electricity rates in Asia, even higher than a developed country like Japan. The country has also still been suffering from bouts of blackouts from time to time. That is why some scholars have argued that rules and regulations governing the EPIRA Law has actually created a condition for "oligopoly, monopoly" to thrive, more harm than benefits to the energy sector.

4.2 The dynamics and the dominant dgents in the electricity sector under the EPIRA law

Analysing the dominant agents/players and decision-makers in each of the electricity sub-sector can shed light on the loopholes in the current electricity structure (Figure 1).

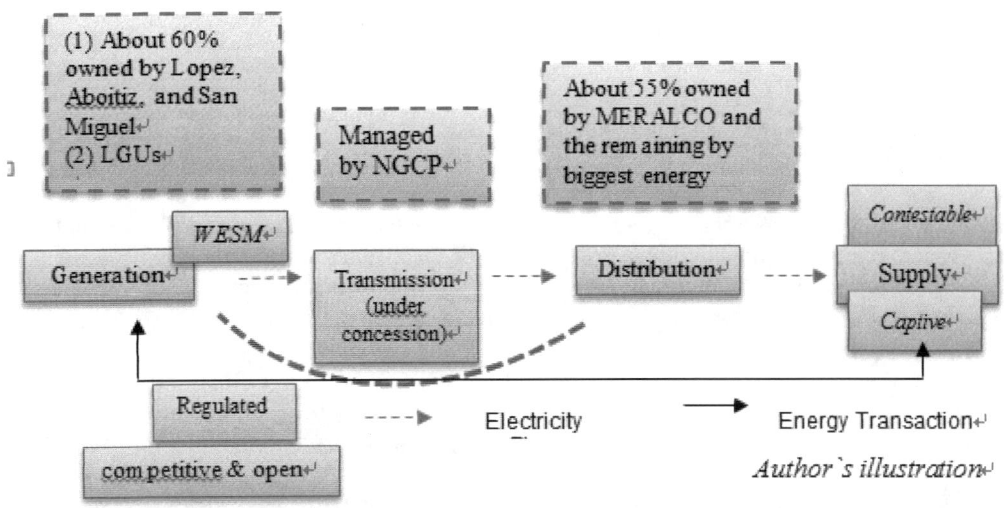

Figure 1. Dynamics and the Dominant Agents in the Electricity Sector

One, the current structure is monopolistic/oligopolistic in nature. Under Sec.45 of the EPIRA Law, a company/person or related group is allowed to supply energy demand to a maximum of 30% of the installed generation capacity of a grid (e.g., Luzon grid) and 25% of the national installed generating capacity. Generation are owned by the largest companies in the Philippines like San Miguel (22%); Aboitiz (20%); and the Lopez (18%), thus, a combine share of around 60% of the total produced energy in the country [15]. The government-owned NPC respectively only have 8% and the IPPs only 18% shares nationwide [15]. The dominance of coal as the preferred energy source has continued to pushback RE on the sideline. The generation subs-sector is regulated primarily by the Energy Regulatory Commission (ERC) to ensure that generation companies "comply with standards set forth by relevant agencies and are not engage in market abuse or anti-competitive behaviour" [16]. Contradictory, ERC is accused of corruption and collusion.

Even energy developers will have difficulty building RE facilities if they every choose to do so. The Local Government Units (LGUs) can have powers that can rule out national directives, particularly pertaining to the issuance of permits and licenses, for example [17]. This is the result to the lack of knowledge on the benefits of RE plus the loose coordination between the national and local level of government. And because the approval of these permits and licenses can ultimately make or break an RE project, this has sometimes perpetuated bribery and corruption at the local level.

Second, the gradual privatization even of the government`s transmission assets have also provided additional cost especially to small time RE developers. TRANSCO (or the National Transmission Commission) is mandated to "assume authority and responsibility for the planning, construction, and centralized operation and maintenance of the country`s high voltage transmission facilities, including grid interconnections and ancillary services" [16]. In 2007, the National Grid Corporation of the Philippines (NGCP) won the concession for 25 years, and has assumed the responsibility of TRANSCO – part of the government`s plan to privatize its electricity assets. NGCP is a consortium of three (3) companies: Monte Oro Grid Resources Corp., Calaca High Power Corporation, and the State Grid Corporation of China (ngcp.gov.ph). NGCP is 60% Filipino and 40% Chinese owned. For small time RE developers that already have to shoulder additional costs due to bureaucratic procedures, delays, etc. transmission charges and the cost of transmission infrastructure can make RE financially inviable. One of interviewee said, " one of the greatest crimes committed to the Filipino people is the privatization of the grid" [17].

Third, the EPIRA Law allows cross-ownership. One example is the largest electricity distribution company in the Philippines, MERALCO, which has about 70% market share in Luzon, and about 55% market share in the whole country. In the pretence of cross-ownership clause under Sec. 45, a distribution company is "allowed source to from bilateral power supply contracts more than fifty percent (50%) of its total demand from an associated firm engaged in generation"[16] Following this, MERALCO has its own generation subsidiary, MERALCO Power Generation (MGen), which has an energy portfolio that is primarily coal and natural gas [18].

By 2020, MGEN aims to have a total of 3,000 MW installed capacity. Two big coal power plants are already in the pipeline and is targeted to be finished by 2018: the 1,200-MW in Atimonan, Quezon; and the 500-MW San in Buenaventura, Quezon [19]. MERALCO has also 40% stake at the the Calaca coal-fired plants, among others. Another example of this cross-ownership is Abotiz Power which has three (3) distribution company-subsidiaries namely, the Visayan Electric Co., the Subic Enerzone Lima Enerzone, and the Davao Light [19]. MERALCO also has bilateral contracts with Aboitiz Power to supply the company with its energy demand, most of which are coal.

Fourth, the EPIRA Law is geared towards the privatization of all electricity assets owned by government. While this was a move to create a competitive market and to pay for the debts of NPC, this has done little, if nothing, to serve these supposed purposes. What is more infuriating the financial obligations incurred by NPC thought so-called stranded debt and stranded contract cost will be passed on to the consumers. Stranded debt (SD) refers to NPC`s "unpaid financial obligations which have not been liquidated by the proceeds from the sales and privatization of NPC`s assets"; while stranded contract cost (SCC) refers to the "excess of the contracted cost of electricity under eligible IPP contracts of NPC over the actual selling price of the contracted energy output of such contracts in the market" as a result of the "take-or pay" arrangement of the government during the power crisis in the 2000s [16]. NPC`s financial obligations reached PhP 943 billion in 2001 and climbed to an astonishing PhP 1.24 trillion in 2009 [13]. To pay for these ballooning debts, most of the assets of NPC were either sold or concession or passed on to consumers. In 2017, ERC has allowed the NPC to recover a total of PhP 37 billion (PhP 24.2 billion for SD; P12.88 billion for SCC) from electricity consumers through a P0.0265 per KwH "universal charge" for a period of nine (9) years [20]. "Universal charge" also includes costs for missionary electrification; the equalization of the taxes and royalties applied to indigenous or renewable sources of energy vis-a-vis imported energy fuels; and environmental fees meant for watershed rehabilitation and management [16].

5. Conclusions

Exploring the energy policy, context and regime within a multi-level decision-making perspective (MLP) of energy transition (ET) theory provided us insights on the challenges the RE sector in the Philippines faces. Beginning with a historical analysis has shown that past mistakes continue to have lingering effects until today. First, as the government tries to avoid the power crises and economic turmoil it suffered in the past, it centres its policy on energy security. This is more beneficial if only energy security is framed within a sustainability perspective. But in reality, this has brought about positive and negative consequences to RE development. It is positive because it has welcomed more diversity in the energy portfolio including renewables. But it is negative because energy security was framed towards the goal of only increasing generation capacity. Since conventional energies are already at an advantage to begin with, this makes it difficult for RE to compete. Thus, the current energy policy only led to more expansions on conventional energies (e.g., coal) whereas it made growth in RE sector stagnant over the years. In that sense, perhaps we can say a "false" energy security because the country remained very much dependent on imported fuels. Second, the creation of the EPIRA Law, which was supposed to make the electricity market competitive and to lower down the electricity rates, had failed. In upholding privatization and commercialization, the government had in the process withdrawn its power to "steer and manage" the energy policy that is most beneficial for the people. In the end, the EPIRA Law has paved the way for oligopoly/monopoly to thrive; has created a fragmented energy structure; and has allowed greater costs for the consumers. These abovementioned reasons we have shown in the paper poses constraints to the RE sector, and thus, is hindering the country from achieving a sustainable energy transition.

6.References

[1] World Bank (WB). (2013). *Press statement of Motoo Konishi, Co-Chair, Philippines Development Forum.* Manila: World Bank((WB).

[2] Philippine Department of Energy (DOE). (n.d.). *Philippine Energy Plan 2012-2030.* Taguig: Philippine Department of Energy.

[3] Oxfords . (n.d.). *transitions definition.* Retrieved November 2, 2017, from Oxfords Living Dictionaries: https://en.oxforddictionaries.com/definition/transition

[4] Loorbach, D. (2009). Transition management for sustainable development: a prescriptive, complexity-based governance framework. *GOVERNANCE: An International Journal of Policy, Administration, and Institution*, 161-183.

[5] Rip and Kemp, 1998; Geels, 2002, *2005 in* Ewijk, S. v. (n.d.). *Re-inventing the multi-level perspective for technological transitions: a more rigorous tool for analysis and two case studies.* Retrieved November 2, 2017, from http://www.sv.uio.no/tik/english/about/news-and-events/2013/stijn_van_ewijk_innovation_essay.pdf

[6] e.g. Brandshow 2014; Hallding et al.2009; Sovacool 2008 *in* Marquardt, J. (2017). *How power shapes energy transitions in Southeast Asia.* Routledge Studies in Energy Transition

[7] Krause et al., (2003) *in* Marquardt, J. (2017). *How power shapes energy transitions in Southeast Asia.* Routledge Studies in Energy Transition.

[8] Cham, M. R. (2007, June). The Philippine power sector: issues and solutions. *The Philippine of Review of Economics, XLIV*(1), 33-63.

[9] National Power Corporation (NPC). (n.d.). *Mandates and Functions.* Retrieved August 14, 2017, from National Power Corporation (NPC): http://www.napocor.gov.ph/index.php/menu-transparency/submenu-about-us/subsubmenu-mandates-and-functions

[10] Patalinghug, E. (2003). An Analysis of the Philippine Electric Power Industry. *International Conference on the Challenges to Development: Innovation and Change in Regulation and in Competition* . Pasig City: Ombudsman of the Philippines.
[11] Philippine Daily Inquirer. (2016, September 17). *Mothballed plant cost $2.3 billion*. Retrieved from Philippine Daily Inquirer : http://technology.inquirer.net/52355/mothballed-plant-cost-2-3-billion
[12] MAGDALO Party List Representatives introduced by Gary Alejano and Francisco Acedillo. (2014, January 27). *Philippine House of Representatives.* Retrieved from Philippine House of Representatives: http://www.congress.gov.ph/legisdocs/basic_16/HB03751.pdf
[13] Bello, W. (2012, April 23). The power mess: How we got into it and how to get out. *Philippine Daily Inquirer*.
[14] Asian Development Bank (ADB). (2016). *Philippines: Electricity Market and Transmission Development Project.* Manila: Asian Development Bank (ADB).
[15] Rappler. (July, 31 2014). INFOGRAPHIC: Top power players in the Philippines. *Rappler*
[16] R.A 9136 or the EPIRA Law document. Available online at http://www.neda.gov.ph/wp-content/uploads/2013/12/R.A.-9136.pdf
[17] Anonymous (September 2016) *Interview,* Metro Manila: Philippines
[18] MERALCO PowerGen Corporation. [*Website]* http://www.meralcopowergen.com.ph/
[19] Oplas, B JR. (2014, August 14). The Philippine electric market: monopoly and competition. *BusinessWorld*.
[20] Rivera, D. (2017, July 25). PSALM cleared to recover PhP 37 Billion to cut NPC debt. *PhilStar Global*

IOP Conf. Series: Earth and Environmental Science **159** (2018) 012026 doi:10.1088/1755-1315/159/1/012026

Energetical Examination of Pyrolysis Carbon Black and Pine Shavings Blend Pellets

V Papp[1] and D Szalay[2] and A Vágvölgyi[3]

1 PhD candidate, MSc. Environmental researcher, University of Sopron, Sopron HU
2 PhD candidate, MSc. Energy management engineer, University of Sopron, Sopron HU
3 Assistant professor, University of Sopron, Sopron HU
E-mail: papp.viktoria@uni-sopron.hu

Abstract. Pellets for firing are not only made of wood or by-products of herbaceous plants, it can also be used for energy purposes by using different waste materials or mixtures thereof. These materials do not appear in conventional pellet firing or residential sectors. Waste pellets could be used in larger industrial facilities and in waste combustion plants. During the work pellets are produced of various proportions using furniture industrial pine-based wood shavings and carbon black from the pyrolysis of waste tires. The carbon black is purely purified by the usual wood-pellet making methods, in most places the wet materials are reused in rubber industry. The energetic utilization of wood-based and carbon black mixture pellets was only slightly explored. During the tests, pellets are prepared for pine-based wood which were analyzed by energetic and mechanical properties. In the wood industry, large quantities of contaminated byproducts also appeared: surface-treated materials, glue, paint residues, which cannot be wood pellets in residential use according to EN-14961-1 standard. The study dealt with pyrolysis carbon black (up to 20% mixing) could be used with this contaminated wood.

1. Introduction

The production and trade of wood pellets have increased significantly since 2008, the production of agripellets and industrial pellets have been also increasing [1], [2]. The United States and Canada are the main producers and exporters, primarily to the European Union. The EU wood pellets "dependence" is caused by the political directives to reduce greenhouse gas emissions and to increase the share of renewable energies. Many European electricity producers have converted or are in the process of replacing coal and gas with biomass, which is often done with wood pellets [3], [4].The production of wood pellets is also becoming more well-known in Vietnam, although the amounts produced are still small and mainly for exports. 85-90% of the pellets produced in the Hungarian wood pellet plants are sold in Italy in the domestic sector [5]. According to the Hungarian Pellet Association's 2015 data, 80% of wood pellets are exported to Italy and Austria, mainly to meet the demand for basic heating and small pellet boilers [6]. The increase in multilateral demand for wood pellets has resulted in wood of different grades, sizes and compositions being incorporated into pelletization, while clear pulp and sawdust from the wood industry do not cover the need. Since the timber industry stocks are limited, priority is therefore to expand the raw material base of the pelleting process with different materials, utilization of industrial pellets to get an increasing role in the future.
During the pyrolysis breakdown of waste-tires, I investigated the energetic characteristics of the pyrolysis carbon black and produced different proportions of soot and pine shaving pellets. The utilization of waste tires has environmentally a key role. 700 million wheel tire manufactured per year

in the world and (apart from a small increase in consumption) essentially the same amount is thrown away as garbage. These are one of the most serious waste disposal problems in the world as the proportion of recycling is very low and tire is not degraded in nature. Part of it is burned, for example, in cement factories, but there are still plenty of things to bury somehow or somewhere to accumulate. [7] The EU Decree on the disposal of waste prohibits [8] tires should now be recycled 100%. The most common way to use the material, the grinding of tires, are separated from the other components, and using this grinder with additives, they try to produce various rubber elements from it [9]. Waste is so much that at present only a fraction of the waste can be utilized by the industry. Therefore, the energetic utilization of 50,000 tons of rubber waste produced annually in Hungary is of paramount importance. In addition to the use of tire in cement factory and waste incinerators, pyrolysis technologies have been gaining an increasing role. Since rubber from tires have a higher calorific value than coal as well as considerable amount of carbon black, it seems reasonable to find a way to take advantage of its high energetic and raw material potential in order to progress in the search of alternative fuels, CO_2 emission mitigation and the reduction/recycling of raw materials [10].

2. Characterization of pyrolysis and carbon black

Pyrolysis, also termed thermal distillation or thermolysis, is a thermochemical treatment that allows breaking apart chemical bonds [11] by means of a thermal decomposition under non-oxidative conditions (inert atmosphere or vacuum) and it is also the first step in any gasification or combustion process. Three products or byproducts are generated: pyrolysis gas, which is generally used for heating the equipment, as well as pyrolysis oil and carbon black (soot)[9]. From the recycling point of view, the main advantage of the process is that it can deal with wastes which are otherwise difficult to recycle, and consequently can create reusable products [12]. Pyrlolysis oil is also often burned and used to generate electricity or is processed by new technologies and diesel engines are used as fuel. Carbon black can be used as a fuel (coke) after the process because it has high calorific value. The LHV of the carbon black was 29,991.2 ± 33 kJ/kg, which is comparable with that of the good quality coal[13]. Similar results were reported by other authors using a pilot rotary kiln reactor (around 30,000 kJ/kg)[14][15] Further treatment, by increasing the active surface, it can also be used in rubber production, wholly or partly induced by the higher price industrial soot produced for this purpose. Most analyzes show that the energy balance of the pyrolysis is unprofitable without the use of by-products (carbon black and pyrolysies gas), the efficiency is 62-65%[7]. Most of the gas is used to heat the reactor. In its original form, carbon black is difficult to transport, so it would be important to make the transport more economical, before energetic utiliztaion. One option can be the wood-briquette addition or carbon black briquette production, studies are conducted since 2013 [16]. In most plants, carbon black pellets are produced in wet conditions, unlike wood pellets, generally with a smaller die hole.

In the Environmental Impact Study of the Hungarian Academy of Sciences, Homatech W has been examining Hungary's planned technology utilizing 30,000 tons of tires annually. The plant's products are pyrolysis gas (4-6.5 m / m%), sulfur disulfide gas (~ 0.2 m / m%), pyrolysis oil (20-30 m / m%), pyrolysis coke, substantially carbon black (35-45 m / m%) and steel wire (10-12 m / m%). Keeping coke wet, the dust is largely avoided. The most harmful elements in technology to health and environmental damage are mutagenicity in polyaromatic hydrocarbons (PAHs), aromatic amines in carbon black and wastewater contaminated with PAHs and amines. Among the rubber-forming polymers, chlorine and oxygen constituents, such as chlorinated or chlorosulfonated polyethylene, polyacrylate, urethane rubber, etc., may also be present. Therefore, it can not be excluded, that dioxins and furans, even trillions of parts (ppb), are generated. Potentially hazardous to health can be the inhalation of nano- and micron-sized particles of polluting carbon black substances. Successful power plant experiments have been carried out with coke. Its sulfur content is similar to the sulfur content of mineral carbon. No additional load on the coal-fired power plant flue gas treatment of tire-derived pyrolysis coke by burning. PAH compounds can be burned together with carbon [17].

In energy recovery, environmental contaminants in PM10 dust particles may be particularly dangerous. This is primarily due to the fact that the air fed to the firing can capture the smaller particles without they are sufficiently burned (especially when there is excessive air supply due to a worse setting or fire problem). It is not inconceivable that the burners used for pellet burning - due to their design - are not suited for the perfect burning of smaller crumbs [18]. Since carbon black may also be present in mutagenic compounds and during the mixing the pellet length also be significantly reduced. Therefore, these materials may be released into the environment without proper filtering equipment.

Regarding the compositon of tires (except for steel), almost all materials are flammable or pyrolysisable, with a high calorific value. The pyrolysis process also has a lot of variants and different technologies, so the output values of the different materials, main and by-products can vary considerably. The main determinant of the yield values is, of course, the composition of the input material besides the method used. At Szent István University, researchers deal with different types of pyrolysis technologies utilizing various waste materials [19]. Based on their research, the following table 1 summarizes the amount of byproduct produced by different technologies.

Table 1. The amount of solid and gas-purifying residues per technology (tons / 100000 tons) [19]

	Lurgi	Compact Power [a]	Compact Power [b]	Brightstar	Waste Gen	Foster Wheeler	Thermoselect
Technology	firing	pyrolysis	pyrolysis	pyrolysis	pyrolysis	gasification	pyrolysis-gasification
Carbon black	21353	19598	13065	23400	27090	30351	23390
Gas clearing byproduct	2265	945	630	500	2257	0	3583
Sum	23618	20543	13695	23900	29347	30351	26973

a without pretraetment
b with pretreatment

It can be seen, therefore, that the amount of coke / carbon black produced is significant, usually 20-30%, but higher in the gasification, even if the amount of carbon black (35-40%) can be higher than pyrolysis oil (30-35 %). The utilization of carbon black is therefore a crucial issue, it can determine the profitability of a technology and significantly affect the energy balance.

3. Presentation of the tests, methods

The pyrolysis carbon black and wood blend pelletization is less explored. One of the reasons for this is that the carbon black is difficult to handle and pelletize. On the other hand, due to dusting of carbon black powder, mixing is difficult to solve under operating conditions. In 2013 Liu et al. developed a special pelletizer in China for the utilization of pyrolysis carbon black and biomass blends. However, the pellet diameter of these blends is 14-20 mm, which is significantly higher than the usual 6-10 mm size in Europe [20]. There is no data on the quality of the pellets and the proportion of mixtures. An American company [21] has also developed a pelletizing technique for carbon black, with a diameter of 0.3 to 2mm of pellets of pure carbon black, which is significantly smaller than the usual sizes. In most cases, the aim was not the energy recovery, but also to facilitate delivery and reduce dusting of carbon black powder.

The process of manufacturing fuel pellets involves placing ground biomass under high pressure and forcing it through a round opening called a “die.” When exposed to the appropriate conditions, the biomass “fuses” together, forming a solid mass. This process is known as “extrusion.” Some biomass (primarily wood) naturally forms high-quality fuel pellets, while other types of biomass may need additives to serve as a “binder” that holds the pellet together. However, the creation of the pellets is only a small step in the overall process of manufacturing fuel pellets. These steps involve feed stock

IOP Conf. Series: Earth and Environmental Science **159** (2018) 012026 doi:10.1088/1755-1315/159/1/012026

grinding, moisture control, extrusion, cooling, and packaging. Each step must be carried out with care if the final product is to be of acceptable quality [22].
In the investigations, different proportions of mixtures were prepared by using a Kahl type press, experimental pelletising equipment (max. Power 22 kW, matrix thickness 28 mm, matrix bore diameter 6 mm, capacity 200-250 kg / h) for which energetic characteristics, and pellets length change are analyzed. Carbon black is not purely pelletable, at 30% shoot-mix cumbersome, it is "burned into" the press holes according to my own experience.

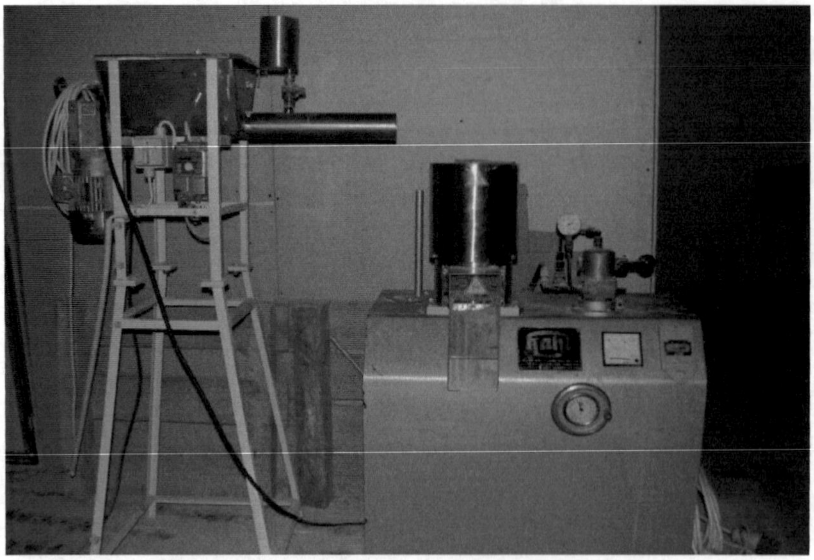

Figure 1. Kahl type pellet press with feeder.

Among the most important energetic properties, calorimetric measurements were performed to calculate the calorific value, the ash content by the muffle-furnace measurements, and the moisture content by automatic moisture measurement. The changes in pellet lengths were determined by caliper measurements. The energy measurements were made separately for the raw materials and the different mixture pellets.

3.1.Calorimetric mesurments
The essence of a calorimeter is that combustion occurs under certain conditions. To do this, by analytical balance measured fuel sample must be taken into a sample holder, while the sample burning machine measures the temperature change of the calorimeter water tank. The test was carried out in accordance with MSZ EN 14918 standard, using an IKA 2000C bomb scale analyzer. During the test, the instrument receives the direct heat of the combustion heat. Calculation of the net calorific value can then be determined by using the following formula.

$$\text{Net Calorific Value(NCF)} = \frac{\text{GCF}-\left(\frac{2447\cdot(\text{U}+9\cdot\text{H})}{100}\right)}{1+\frac{\text{U}}{100}}\left[\frac{\text{kJ}}{\text{kg}}\right]$$ where:

GCF: Gross calorific value (kJ/kg)
U : the gross moisture content of the test substance [m / m%]
H : hydrogen content of the test substance [~ 6 m / m%]
Calculation of calorific value is the most important issue from an energetic point of view. The energy content of the various materials and their compressions can be determined by energy efficiency, so the energy balance is always tested for the calorific value of the raw material and the finished product.

3.2 Ash content
Ash content is an energetically important feature due to the design of combustion plants. Wood pellets are low, generally less than 0.5%, while pellets made of herbaceous plants or industrial materials have

IOP Conf. Series: Earth and Environmental Science **159** (2018) 012026 doi:10.1088/1755-1315/159/1/012026

a higher ash content, usually between 3-10%. The unburned slag generated during the energetic utilization of biomass, as well as the higher-capacity combustion plants, can cause special operating problems. On the one hand, it is related to the destruction of a combustion plant and, secondly, to the disposal of large amounts of ash.

Test Procedure:

- Measurement of moisture content, drying of material
- Measurement of dry matter on analytical scales
- Burning in a muffle furnace
- Weighing of ash on analytical scales

Ash content calculated with the formula: AS(%)=mo/m1*100

AS(%)- ash content, m1 (g)- weight of the dry sample, mo (g)– weight of ash.

3.3.Moisture content

From a firing point of view, moisture content is one of the most important properties, since it is closely related to the calorific value [23]. Determining the moisture content is important because of the pelletisation, if it is too high or too low the pellet is crushed and disintegrated. The optimum moisture content is about 10-12% at wood pelleting process. After pelleting the moisture content is reduced, and it must be below 10% according to regulations.

Test procedure:

The BOECO SMO 01 automatic humidity tester was used during the tests. A sample is placed on the sample holder of the moisture gauge, then dried at 105 ° C to a mass flow using a heating coil in the instrument. The moisture content of samples can be determined directly by the weight before and after drying.

Theoretical definition:

$n = (m_n-m_{sz})/m_n \cdot 100$ [% v.g/g]

n-moisture content(%), mn: mass of the wet sample(g), msz: mass of the dry matter of the sample(g)

3.4.Pellet length

For pellets, it is important to have a medium length between 3.15 and 40 mm. The length of pellets may be influenced by a number of factors. Particle size, moisture, and most of all material composition. In the production of wood pellets, corn starch is often used, thereby reducing the fine fraction and adjusting the appropriate length. All pellets in 80-100 grams of samples should be tested by using a simple digital caliper.

Figure 2. Carbon black 5% and pine shaving 95% blend pellets.

The carbon black, during the work is composed of nano - and microparticles of fraction, so it was very important during the laboratory work to observe the safety regulations due to the possible dusting of

IOP Conf. Series: Earth and Environmental Science **159** (2018) 012026 doi:10.1088/1755-1315/159/1/012026

powder. Before pelleting, the optimum moisture content of 12% for wood pellets was adjusted for both raw materials. The by-product of furniture industry were pine-shavings that have a calorific value of 18.6-19.2 MJ / kg due to its high resin content. Pure wood had an ash content of 0.4%, the moisture content of the finished pellet changed from 8.9 to 9.3%. The calorific value of carbon black was high, the average of calorimetric measurements was 27,9 MJ / kg. Ash content was between 3-4%.

4. Results

By increasing the rate of soot, the pellet length apparently decreased, as measured by the caliper measurements. I assumed a linear correlation between the mean average length and the rate of blend of carbonblack. The average of the length changes and the degree of mixing ratio can be written using the least squares method, the straight line closest to the points is regression line.
Thus, by increasing the mixing rate as expected, the mean average length of the pellets decreases linearly, the slope of the regression line is -0.43.

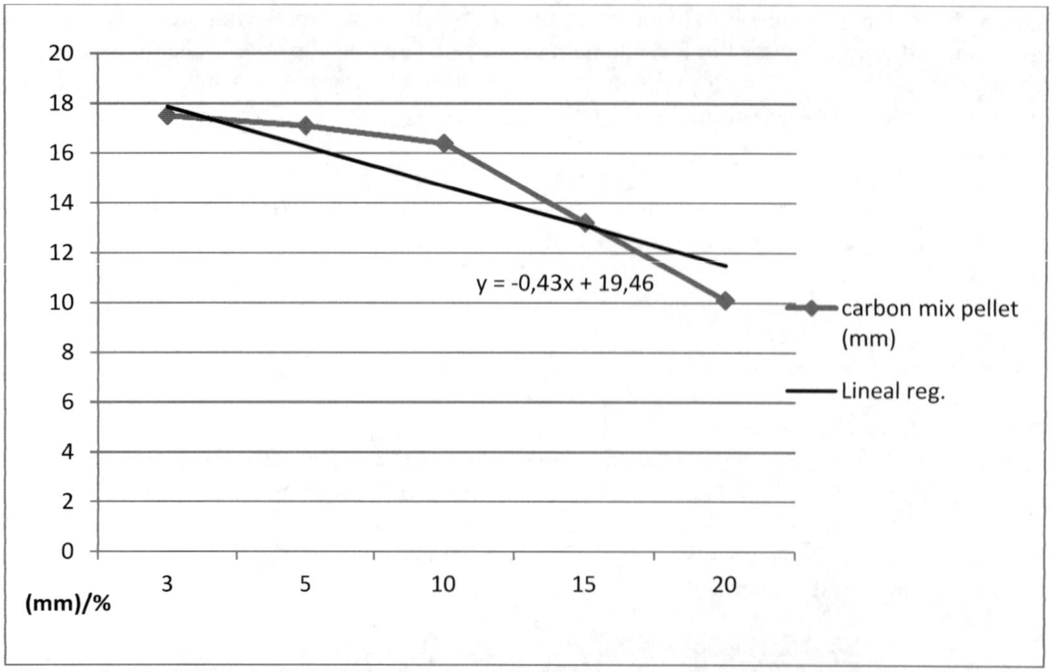

Figure 3. Medium length change of carbon black blend pellets depending of the mixing ratio.

Although the carbon black has a negative effect on the length of the pellets, due to the high 27-28 MJ / kg calorific value, the increase in the mixing rate is evident, and the calorific value of the mixture pellets also increased, which is energetically positive, as shown in table 2.

Table 2. Net calorific value changes of carbon black pine-shavings blend pellets

Carbon black ratio (%)	Calorific value (MJ/kg)
3	19,2
5	19,4
10	19,9
15	20,7
20	21,4

The ash content did not change much, the 0.4% value of wood pellets increased by 0.7% in the case of 20% mixing of carbon black. The moisture content after pelletizing process decreased from 12% to 9.2-9.5%.

5. Conclusions

During utilization, the high sulfur content of the carbon black and any mutagenic compounds (PAHs, chlorosulfonated polyethylenes, esters ppb, dioxins and furans) may be present. Therefore, it is of the utmost importance that recovery can only take place in combustion equipments with appropriate cleaning and filtering system. The reduction in the length of pellet-wood blend pellets also greatly influences combustion in pellet furnaces, as if smaller pellets are placed in the furnace - and the fine fraction is increased - it can also modify the flue gas composition and the amount of leaving solids. Based on the results, it can be said that the soot itself can not be pelletized using the usual wood pellet methods. By mixing with wood, I would recommend a maximum mixing ratio of 20% of carbon black at 6 mm diameter. Further studies could be carried out to increase the diameter, or increased the moisture content, which would probably be able to mix the soot into the wood-shavings even more. While the carbon black has a negative effect on the length of the pellets, due to the high calorific value, the heating value of the blend pellets increases as the mixing ratio increases. In the future, with the spread of the waste tire or other pyrolysis plants, in addition to industrial use and material recycling, the energetic utilization of the carbon black will be a key issue.

6. References

[1] Sikkema R. 2016 - The European wood pellet markets: current status and prospects for 2020 - *Biofuels Bioproducts and Biorefining vol.* 5. 250 – 278p.
[2] PellCert Project 2014 – *European pellet markets* – European pellet report
[3] Calderón C, Gauthier G, Jossart J.M. 2015, *AEBIOM- Statistical report-* European Bioenergy Outlook
[4] European Comission 2015 - Bioenergy Report, *Study on impacts on resource efficiency of future EU demand for bioenergy*
[5] Papp V, Szalay D, Gaál L, 2016 – Agripellet előállítás alapanyagbázis vizsgálata Magyarországon – *Journal of Central European Green Innovation* – 4. vol.
[6] MaPe 2014 – Hungarian Pellet Assosiation, *Wood pellet Trends Report.*
[7] Nagy B, 2011 – *Újrahasznosítási ismeretek –Gumiabroncs* –Digitális tankönyvtár
[8] Council Directive - 1999/31/EC- European Union directive - waste management
[9] Farkas H, 2015 Gumiabroncs hulladék újrahasznosítása, *Waste tire utilization* http://www.hirado.hu/2016/03/03/evente-otvenezer-tonna-gumihulladek-keletkezik-magyarorszagon/
[10] Martínez J. D., Puy N, Murilloa P,Garcíaa T, Navarroa A, Mastrala M, 2013 Waste tyre pyrolysis – A review. *Renewable and Sustainable Energy* Reviews, vol. 23, 179-213p.
[11] Wampler T.P, 2007 - *Applied pyrolysis handbook* (2nd ed.), CRC, 23. p.
[12] Scheirs J., Kaminsky W, 2006 - *Feedstock recycling and pyrolysis of waste plastics: converting waste plastics into diesel and other fuels*, John Wiley & Sons Ltd.
[13] Sienkiewicz M, Kucinska-Lipka J, Janik H, Balas A. 2012 *Waste Manag.* 2012 32 vol. 1742-51p. Progress in used tyres management in the European Union: a review.
[14] Li S, Yao Q, Chi Y, Yan J, Cen K, 2004 *Pilot-scale pyrolysis of scrap tires in a continuous rotary kiln reactor,* Industrial & Engineering Chemistry Research, 43 p..
[15] Galvagno S, Casu S, Casabianca T, Calabresem A, G. 2002 Cornacchia Pyrolysis process for the treatment of scrap tyres: preliminary experimental results,*Waste Manage.*, 22 v. 917-23pp.
[16] Kovács O. 2013 Pirolízis korom brikettálása, Carbon black briquette https://www.slideshare.net/eco-gumm/pirolzis-12652833
[17] MTA Environmental Impact Assessment 2016 - Scientific Opinion of the Institute of Materials and Environmental Chemistry of the Hungarian Academy of Sciences, Institute of Natural Sciences on the environmental adequacy of HOMATECH-W (TM) technology

[18] Németh G. (2014) – Kis teljesítményű faalapú pellet tüzelő berendezés környezeti hatásainak vizsgálata, Faipar, 62.vol.
[19] Lakatos K. 2013 – *A pirolízis elve és folyamata, pirolízis technológiák – Különböző alapanyagok és hulladékok pirolízise – Pyrolysys of different waste materials-* SZIE Konzorcium.
[20] Mars Mineral Ltd. 2015 *Carbon black pellets*, Company description
[21] Liu Y, 2014 *Carbon black and biomass pellets*, Product solutions, Company description
[22] USAID 2014 – Vietnam Clean Energy Program, *Woody biomass energy generation*, Final report
[23] Marosvölgyi B, 2006 *Szilárd, biomassza alapú tüzelőanyagok energetikai tulajdonságainak vizsgálata-* Egyetemi jegyzet.

Acknowledgment
The described work was carried out as part of the „Sustainable Raw Material Management Thematic Network – RING 2017", EFOP-3.6.2-16-2017-00010 project in the framework of the Széchenyi 2020 Program. The realization of this project is supported by the European Union, co-financed by the European Social Fund.

IOP Conf. Series: Earth and Environmental Science **159** (2018) 012027 doi:10.1088/1755-1315/159/1/012027

Sustainability of Refuse Derived Fuel Potential from Municipal Solid Waste for Cement's Alternative Fuel in Indonesia (A Case at Jeruklegi Landfill, in Cilacap)

W Paramita[1], D M Hartono[2], T E B Soesilo[3]

1 Graduate School of Environmental Science, University of Indonesia. Kampus UI Salemba - Gedung C (FKG) Lt. 5 dan 6. Jl. Salemba Raya No. 4 - Jakarta Pusat 10430, Indonesia
2 Department of Civil Engineering, Faculty of Technic, University of Indonesia. Kampus UI Depok 16424, Indonesia
3 School of Environmental Science, University of Indonesia. Kampus UI Salemba - Gedung C (FKG) Lt. 5 dan 6. Jl. Salemba Raya No. 4 - Jakarta Pusat 10430, Indonesia
Email: luvlie.mita@gmail.com, djokomh@eng.ui.ac.id, soesilo@indo.net.id

Abstract This paper examines Refused Derived Fuel (RDF) potential from Municipal Solid Waste (MSW) as cement industry's alternative fuel in developing country. Municipal solid waste remains an issues for city while cement industry are needed a sustainable alternative fuel resources. Jeruklegi Landfill, in Cilacap City, Central Java Province, Indonesia was selected as a research case, because its location is close to cement industry. Research's approach is using triple bottom line sustainability; economic, social and environment. It investigates RDF potential from quality aspects (calorific value), economic aspect (RDF price), environment aspect (GHG emission) and social inclusion of waste picker. It is found that MSW on Jeruklegi Landfill has a potential for processing in to RDF, since its average calorific value is 3563 Kcal/kg, price is 293,000 IDR/ton, and avoided $2.9x10^6$ Nm^3 methane emission. An inclusive approach should be taken into account in early stage to engage waste picker rather that marginalized them. This findings contributes to government as well as cement industry on RDF as a solution for waste management and fossil fuel conservation.

1. Introduction

Economic growth resulted a massive development in cities. UN predicts in 2050 three quarter of world population is living in city. In Indonesia, projection of population growth in 2025 is 271,227,000 people with 178,731,000 of them live in urban area and would have generated 151,921 ton/day Municipal Solid Waste (MSW)[1]. Forty five regency/city in Indonesia are using open dumping system on its landfill some were found using open burning[2]. Methane is generated in landfills by the anaerobic biodegradation of organic matter [3], and also categorized as Green House Gas (GHG)[4]. Thus research on MSW is remained necessary and up to date to current issues. Cilacap city is also facing an issue with MSW. They project that their final disposal land would reach its maximum capacity in 2018 if there is no different approach rather than open dumping. Referring to waste management hierarchy, they decided to develop waste into energy, by processing MSW into RDF. Under current circumstances RDF is the most probable solution since their final disposal location is about 5 km from cement plant. Cement plant is the most ready and suitable user due to no need major modification on its equipment [5]. Existing practice, cement industry has been utilized rice husk as their alternative fuel, but its intermittent supply creating another issues.

Debate centers on the issue of RDF is its benefit to industry. Would RDF viable to be used as an alternative fuel if it has small differences on price than coal, with lesser heat value and requires specific handling? The preferable production of RDF is a result of bio-dried waste fractions with particle size higher than 45 mm that mainly composed of plastics and papers[6], and its utilization in cement industries will be financially viable if its price is lower than coal price [7]. Bio-drying is also could lower investment cost[6] [7]. In developing countries are common to come across waste pickers on landfills. In Jeruklegi landfill there are approximately 100 waste pickers working there[8]. This paper differs from that of many other researchers in that it focuses on sustainability of RDF potential. Sustainability here is assessing four aspect, i.e; i). economical aspect by comparing RDF price with rice husk, ii). technological aspect by comparing its calorific value using bio-drying with minimum criteria, iii). environmental aspect by projecting Green House Gas (GHG) emission that can be avoided and iv). social aspect by analysing waste pickers perception and seek a proper engagement approach. Sustainability can only be achieved if all aspects are fulfilled.

2. Objective

This study aims to analyse a potential of RDF from Jeruklegi landfill's municipal solid waste as cement industry's alternative fuel.

The paper result will be helpful for developing renewable energy as well as integrated solid waste management strategy which is important as an input city/regency strategic plan in Indonesian local governments and can encourage cement industry to utilize RDF. Cilacap regency can extend their landfill lifetime while cement industry can reduce their fossil fuel consumption which will lead to reducing their CO_2 emission at the end.

3. Method

3.1. Location and time

The study was conducted at Cilacap City, Central Java Province for about 5 months, from June until October, 2017. The data was collected from the relevant institutions in Cilacap City, previous relevant researches cement industry and waste pickers.

3.2. Data collection

The primary data was collected by in-depth interview with Cilacap Environmental Agency officials, waste pickers at Jeruklegi landfill as well as cement industry by filling the questionnaires. Selection criteria for waste pickers are those whose permanently worked there and member of The Jeruklegi waste picker association. Interview with waste picker conducted by focus group discussion, and it consists of two different group which is women and men. Selection criteria for cement industry is an integrated cement plant located in Java and has been utilizing alternative fuel specifically rice husk for at least 3 years consecutively.

The secondary data were Study of Waste Specification in Cilacap by Ganesha Environmental and Energy Services, ITB 2015, Study of Social Mapping of Waste Pickers at Jeruklegi Landfill Cilacap by Gadjah Mada University 2016, and data from BPS (Statistical Center Agency) of Cilacap Regency.

3.3. Data analysis

This study investigates four aspect, namely; i). RDF pricing, ii) calorific value RDF, iii). avoided CO_2 emission by using RDF and iv). waste picker inclusion. The 1st and 2nd aspect were analysed by comparing results with criteria from cement industry, obtained from questionnaires and other country standards. Criteria standards is shown in Table 1. Avoided GHG emission was calculated based on projection of emission that could have been released if MSW is landfilled. Perception and inclusivity of waste picker is analyses descriptively.

Sustainability of RDF is analysed using Integrated Sustainability Waste Management (ISWM) approach[9]. Below (figure 1) is ISWM framework.

IOP Conf. Series: Earth and Environmental Science **159** (2018) 012027 doi:10.1088/1755-1315/159/1/012027

Table 1. RDF criteria

Parameter	Finland[a]	Italia	Great Britain	Indonesia[c]	Thailand
Calorific value (mJ/kg)	13-16	15	18,7	12,5	N/A
Moisture (%wet)	25-35	max 25	7-28 [b]	< 20	< 30
Ash content (%wet)	5-10	20	12	N/A	N/A
size (mm)	N/A	N/A	N/A	≤ 50	< 30

Source : Gendebien *et al.*, 2003, Nithikul, 2007

a: Restricted for household waste

b: 7-28 for densified-RDF and 28 for coarse-RDF

c: draft utilization of MSW into alternative fuel, Indonesian Environment & Forestry Ministry

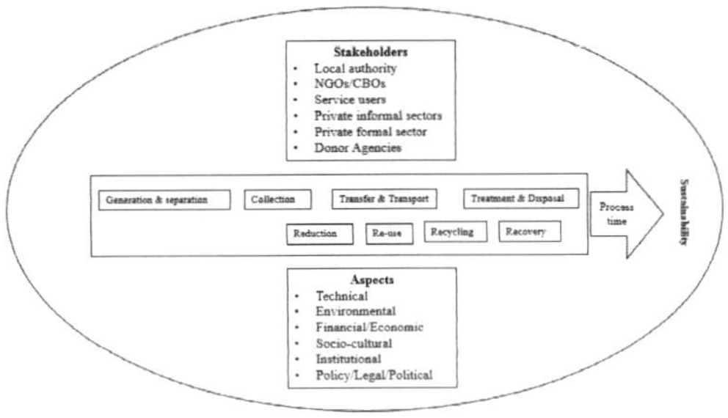

Source: Scheinberg, 2001

Figure 1. Integrated sustainability waste management model.

4. Results

Information and data from Environmental Agency were; i). capacity of RDF plant is 120 ton/day of waste, ii). bio-drying, is selected since waste has approximately 50% organic waste [10], iii) amount and scheme of funding [11], iv). amount of tipping fee [11], and v) amount of production cost of RDF plant [11]. Based on selection criteria, there 3 cement industry that eligible, however only 2 has responded. Firstly is Semen Gresik at Tuban, secondly is Semen Holcim Indonesia which covers Narogong plant, Cilacap plant and Tuban plant. Later on Semen Gresik will stated as respondent A and Holcim Indonesia as respondent B. Summary of questionnaires result is presented below(Table 2).

Table 2. Questionnaires results

Questions	Respondent A	Respondent B
Target on fuel substitution	5-10%	10-15%
Price rice husk (IDR/ton)	300.000-350.000	250.000-260.000
Willingness to pay RDF per ton (IDR/ton)	300.000-350.000	Below 300.000
Min calorific value (Kcal/kg)	3000	3000

Source: Questionnaires

4.1. RDF Pricing

Refuse Derived Fuel pricing is sum of total cost and profit[7] [9]. Total cost is not only consist of investment and operational cost but also from tipping fee. Tipping fee/gate fee is sum of money that waste processor received from government. There was a foreign funding from Danish Government for machinery of IDR45 Bio, while land cost, IDR1.8 Bio, is also grant from Cilacap Regency [11]. Thus, both cost are not included on calculation. Investment cost consist of land preparation and construction, which is divided between Public Works & Settlements Ministry and Central Java Province budget, for IDR35Bio [11]. Tipping fee for Cilacap Regency has been decided for IDR100.000 per ton incoming waste [11]. Production cost is IDR272,000/ton RDF [11]. Assumptions profit for this study is 10% of total cost and 10 years project period. Investment cost per year is IDR3.5Bio. Therefore calculation of RDF price, can be seen below:

RDF production per year	:	40% * 120 * 310 = 14,880ton
RDF price (IDR)	:	[Total investment – tipping free]+ profit
	:	293,000 per ton

Comparing RDF price with maximum willingness to pay of industry, shows that RDF has a competitive price. Refuse Derived Fuel only financially viable if less price than coal[7], but this study shows that compared to rice husk, RDF still competitive.

4.2. RDF Calorific value

Experiment has been conducted for approximately 5 months from June – October. There were 3 batches of waste. This experiment investigates effectiveness of bio-drying to increase calorific value of waste. Result of experiments is shown in Table 3.

Table 3. Result of Bio-drying experiments

Parameter	Unit	Batch					Avg
		1A	**1B**	**2A**	**3A**	**3B**	
Incoming moisture	%	58	58.2	60	58.2	59.6	58.8
Duration	days	46	41	27	27	40	36.2
Final moisture	%	26	20.3	25.1	21.5	28	24.2
Calorific value	Kcal/kg	3686	3602	3026	4064	3435	3563

Source: Holcim Indonesia, 2017

Minimum calorific value required by cement industry in Indonesia is 3000 Kcal/kg, thus bio-drying method is quite effective to increase calorific value of MSW. Compared to EU and Thailand standard from Table 1, experiment result also showing good outcome. Average calorific value is 3563 Kcal/kg[12]

4.3. Avoided GHG emission

Landfilling of MSW in certain period of time will triggered a methane gas released to air, theoretically 50 Nm^3 methane released from one ton of landfilled MSW[13]. Refers on that number, this study concluded that avoided GHG emission by RDF initiative is 6000 Nm^3/ton landfilled waste per day or 2,190,000 Nm^3/ton waste per year.

4.4. Waste picker perception

The most probable affected community with RDF initiative is waste pickers. This study conduct FGD to two different groups, both groups are consist of 10 participants. Groups are divided as per gender category, since women waste pickers tend to afraid to speak up when among men waste picker[8]. Women group stated that they have no objection for RDF planning as long as Cilacap government gives them enough time to collect valuable waste. Differs from women group, almost all participants of men group wished to keep as current practice. Both of group concerns are losing income if time slot to collect waste is reduced. They also has no interest to work as employee at RDF facility, due to non-flexible working hours and stricter regulation. Although, most of participants said that they will obey how government rules. This statement is an opportunity to engage waste picker on government program.

5. Discussions

Integrated sustainability of solid waste management must take into account aspects, stake holders and solid waste process flow[9]. This study limited on four aspects (financial, technical, environment. social), three stakeholder (Cilacap government, cement industry, waste picker) and one process of solid waste, final disposal process. This study resulted that financially RDF has a competitive price compared to rice husk, it has met minimum requirements of calorific values, which is technically accepted, and converting waste to RDF also able to avoid about 2.9 $x10^6$ Nm^3 methane released to ambient air but socially, currently waste pickers concerns about their livelihood being threatened with RDF. Waste picker is also one of key stakeholder at Jeruklegi landfill that needs to be engaged otherwise negative risk will rise. Historically there is no experienced of strike on Jeruklegi but the risk is remained. Based on Indonesian Public Works Minister Decree No. 13 year 2003, waste picker should be addressed on waste management strategy on final disposal area. However social inclusion of waste picker into structured working condition is also challenging. Partnerships are easier to set up if waste pickers are organized, and a task supported by advocacy group[14]. In Jeruklegi, waste picker has a simple organization, but has no legal entity and formal structure. Affirmative action is needed to strongly encourage waste picker to organize, in Indonesia the most common organization is cooperative. There should be a regulation to only allowed cooperative member of waste picker works at Jeruklegi, thus waste picker will agree to be organized. As Marello&Helwege, 2014 paper said that an advocacy team is needed to support, Jeruklegi also need a supporting team consist of cooperative agency, social agency and probably also labour agency to empower them. This paper suggests Cilacap government to develop a waste picking bay where waste picker cooperative could collect waste. A belt conveyor with appropriate working condition should be provided. Waste picker works on certain working shift where they are involved on deciding the group member and schedule. It will not be easy but with task supportive group, it will possible. Inclusivity program should be on priority list. Only by this, a sustainability of solid waste management is achieved.

6. Conclusions

This study concludes that municipal solid waste in Jeruklegi landfill has potential to be converted into RDF from financially aspect, RDF price is 283,000 IDR/ton, which is in a rice husk price range and under willingness to pay amount of cement industry; technically aspect, RDF calorific value is between 3026-4064 Kcal/kg and above minimum standard; environmentally sound, where can avoid approximately $2.9x10^6$ Nm^3 methane. Care should be taken for social aspect, where social inclusion of waste picker could be a solution for it.

This study would like to propose; i). Cooperative for waste picker is necessary if Cilacap government would like to include waste picker on RDF initiative, ii). Task supportive team has to be assist during cooperative developing, iii). Further research may be needed to seek most appropriate mechanism of support.

7. References

[1] Hoornweg D and Bhada-Tata P, (2012). *What a waste: a global review of solid waste management*, Urban Development Knowledge Series Paper

[2] --------, (2017). National policy, strategy, and programs of solid waste in Indonesia. Direktorat Pengelolaan Sampah, Kementerian Lingkungan Hidup dan Kehutanan

[3] Themellis N J, *Waste-to-Energy: Renewable Energy Instead of Greenhouse Gas Emissions*, Earth Engineering Center (EEC), Columbia University, New York, also available at http://www.seas.columbia.edu/earth/wtert/sofos/themelis_AD_paper_Nov19.pdf, retrieved December 14, 2017

[4] --------, United Nations Framework Convention on Climate Change, *Global Warming Potential*, available at http://unfccc.int/ghg_data/items/3825.php, retrieved December 14, 2017

[5] Reza B, Soltani A, Ruparathna R, Saidq R and Hewage K, (2013). Environmental and economic aspects of production and utilization of RDF as alternative fuel in cement plants: A case study of Metro Vancouver waste management, *Resources Conservation and Recycling*, **81**, 105-114.

[6] Li M S, Zhong H M, Hua Z, Dong Q Z and Pin J H, (2010). Bio-drying and size sorting of municipal solid waste with high water content for improving energy recovery, *Waste management Journal*, **30**, 1165–1170.

[7] Nithikul J, (2007). *Potential of refuse derived fuel production from Bangkok municipal solid waste*, Asian Institute of Technology.

[8] --------. (2015), *Laporan Akhir Studi Pemetaan Sosial Pemulung TPA Jeruklegi Kabupaten Cilacap*, PT Holcim Indonesia, Tbk & Departemen Pembangunan Sosial Dan Kesejahteraan Universitas Gadjah Mada.

IOP Conf. Series: Earth and Environmental Science **159** (2018) 012027 doi:10.1088/1755-1315/159/1/012027

[9] Şcheinberg A, (2001). *Financial and economic issues in integrated sustainable waste management*. Experiences from the Urban Waste Expertise Programme (1995-2001). website: www.waste.nl.

[10] Ganesha Environmental & Energy Services (GEES)-ITB, (2015), *Studi spesifikasi sampah di kabupaten cilacap*.

[11] Dinas Cipta Karya dan Tata Ruang Kab. Cilacap (2015). *Rencana pengelolaan sampah di TPA jeruk legi, Presentasi Dinas Cipta Karya dan Tata Ruang Kabupaten Cilacap kepada Bupati Cilacap*, 23 Mei 2015.

[12] --------. (2017). Holcim Indonesia's experiments on Biodrying.

[13] Themellis J N and Ullo A P, (2007). Methane generation in landfills, *Renewable energy*, **32**, 7, 1243-1257.

[14] Marello M and Helwege A, (2014). Solid waste management and social inclusion of waste pickers: opportunities and challenges, *Global economic governance initiative research*, Boston University.

IOP Conf. Series: Earth and Environmental Science **159** (2018) 012028 doi:10.1088/1755-1315/159/1/012028

Analysis of the Base Material Footprint of Conventional and Lignocellulosic Biofuel Production

D Szalay[1], V Papp[2], and I Czupy[3]

1PhD candidate, MSc Energy management engineer, University of Sopron, Sopron, HU
2PhD candidate, MSc Environment researcher, University of Sopron, Sopron, HU
3Associate Professor, Head of Institute of Forest and Environm. Techniq. University of Sopron, HU
E-mail: szalay.dora@uni-sopron.hu

Abstract. The base material footprint (BMF) is a measuring system that provides the theoretical collection area for a unit of biofuel base material. Eight different types of base materials of three countries were examined; four types of biofuel had been taken into account. According to the results, the collection area varied between 0.08-0.61 ha to produce 10 GJ biofuels. It required the smallest collection area the forest residues, while the largest had the corn stalk. The change of the collection area due to yield fluctuations in the last decade, was the largest in the case of corn and corn stalk for bioethanol, over 70% in a very droughty year. Further comparison was performed for the theoretical and real collection areas for corn. Under real conditions the collection of base materials needs 10 times bigger area. This can increase up to 20 due to utilization for animal feed and green manure.

1. Introduction

The global energy consumption has stagnated in the last two years, achieving the goals of COP 21 (Paris 2015) is far behind. Today, the most important energy source is petroleum. Its use in the transport sector is continuously increasing, covering 32% of global energy use in 2016 [1]. Long-term use of fossil fuels raises serious questions such as global warming and the depletion of fossil fuels. A more significant problem is that the resource of conventional hydrocarbons is locally concentrated; the market is mostly dominated by the Middle Eastern countries. The energy vulnerability of other countries can lead to geopolitical disputes.

Primary and secondary lignocellulosic energy sources will have significant importance in reducing energy dependence and in restraining of climate change. In 2016, 11% of the world's energy consumption was covered by biomass [1]. Within this, the use of biofuels in the road transport sector was 3.5% [2]. The use of cereals and oilseeds is rapidly expanding. While a hundred years ago corn and sunflower were only considered as a food plant, they are important sources of biofuels today. The most important task is to involve non-food by-products of agricultural, of forestry, of industry and of household into biofuel production. Due to the technological development in recent years, a wider range of base materials is available for the production of biofuels, see Figure 1.

The conversion of applied new lignocellulose base materials into biofuels requires a much more complex technology than the first-generation biofuel production. At present, the production of advanced fuels is not costeffective because there are a number of technical barriers that need to be overcome before their potential can be realized [3]. However, the second generation biofuels may be more advantageous for land use demand.

IOP Conf. Series: Earth and Environmental Science **159** (2018) 012028 doi:10.1088/1755-1315/159/1/012028

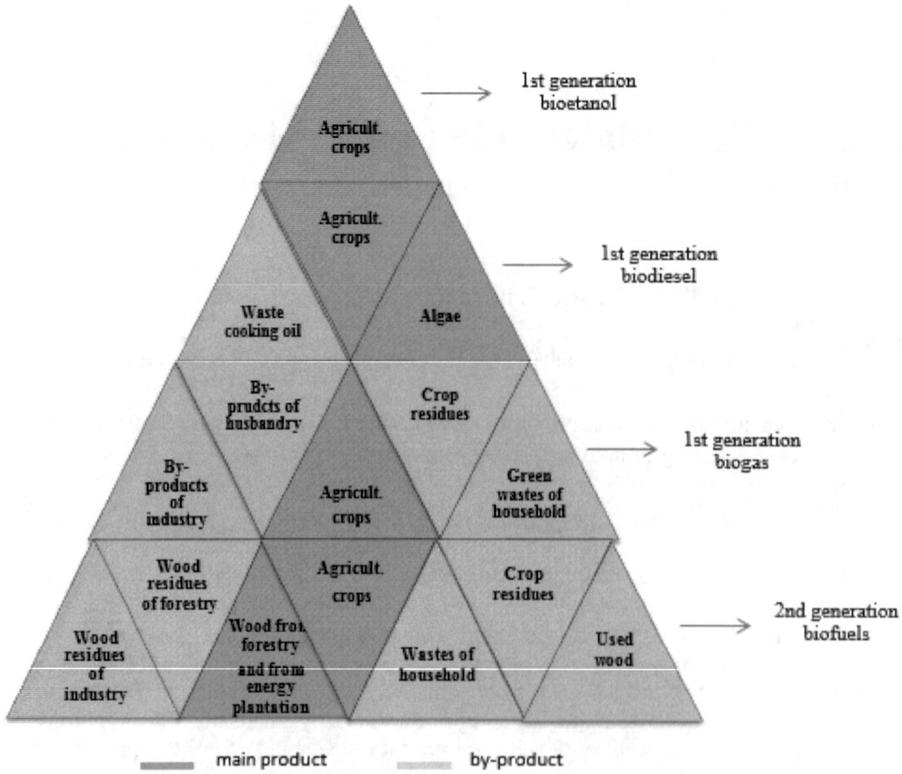

Figure 1. Expanding the choice of base materials to advanced biofuels.

According to Sanderson's (2006) research, the bioethanol yield of Miscanthus is the highest, with a maximum of 7,300 l/ha. The lowest in the case of corn 3,100-4,000 l/ha. Utilization of by-products will result in further improvement [4]. Currently, 6,000-7,000 l of ethanol is produced from one hectare of sugarcane. When bagasse, thatched, straw can also be utilized for ethanol production, the output is likely to double to 12,000-15,000 l per hectare [5,6].

In case of biodiesel, the intake of waste cooking oil (WCO) to production may improve the amount of base material collection area. WCO has great potential worldwide, about 70% of the WCO would be recoverable in the cities [7]. Another advantage is that they can be added into existing plants without their conversion. In Vietnam, fish fat has significant potential, in addition to WCO. The waste of catfish is 120,000 t/year. However, the cost of biodiesel from fatty fish is still higher than diesel due to the higher price of fish oil [8].

2. Legal Background

In the future, the European Union (EU) wants to limit the food crops for energy production. On 6 April 2009, the European Council (EC) adopted the EU Energy and Climate Package (2009/28/EC), which provides 20% renewable share in total energy consumption for the EU as a whole. A further objective is 10% share of renewable energy by 2020 in the total energy use of transport, binding for all EU Member States. The 2015/1513 Directive (iLUC) represents a major tightening of the previous standards, limits the share of biofuels from crops grown on agricultural land that can be counted towards the 2020 renewable energy targets to 7%.

In Vietnam there are two important requirements for the use of biomass. The main objectives of the national energy development strategy up to 2020, with 2050 vision (Decision No 1855/QĐ-TTg; December 27, 2007) to increase the proportion of new and renewable energies to about 5% of the total amount of commercial primary energy by 2020 and 11% by 2050. The goals of the National program for development biofuels up to 2015, with 2025 vision (Decision No 177/2007/QD-TTg on November 20, 2007) to develop biofuel, a new and renewable energy, for use as an alternative to partially replace conventional fossil fuels, contributing to assuring energy security and environmental protection. The biofuel production technology applied in Vietnam will attain the world's advanced leve1. The ethanol and vegetable oil output will reach 1.8 million tons, satisfying some 5% of the whole country's gasoline and oil demand. In Vietnam, the Ministry of Industry and Trade is not extending the January 1, 2018 deadline for petrol wholesalers to switch to E5 biofuel made from cassava.

3. Materials and Methods

During the calculation of BMF, we conducted a complex analysis. Our focus was on Europe and Vietnam, where the used materials are extremely diverse due to the climatic and ecological conditions, but there are serious parallels in the problem of supply.

3.1. Selection of examined base materials and yield analysis

Three countries were analyzed and compared. The typical characteristics for the production area of the selected countries are shown in Table 1.

Table 1. The characteristics of the selected countries [9; 10].

	Forest area (%)	Agric.area (%)	Arable land (%)	Arable land (10^6 ha)
Germany	33	48	34	11.9
Vietnam	48	35	21	6.4
Hungary	21	57	46	4.3

Our investigation was performed on four types of biofuels, four main agricultural products and three high-lignocellulose containing by-product, see Table 2. Depending on the availability of the data [11,12], the median value of the yield of the base materials was calculated considering the last 7 or 10 years in order to dampen the outlier values and determine a more realistic yield.

The amount of corn stalks and rice straw was calculated by using the crop:stalk/straw ratio. According to the literature, crop:stalk ratio is 0.8-1.5 in the case of corn, and 0.7-1.4 in the case of rice [13, 14, 15, 16]. Currently, the utilization of corn stalk approx. 93-94% of the sown area is the plowing into the soil. [17]. Approximately 80-90% of the rice straw is disposed by open-field burning; however, in some periods of the year, rice straw has a market for cattle feed and mushroom cultivation [18].

Table 2: The yield of the examined base materials.

Type of biofuel	Applied technology	Type of base material	Yield (t/ha/year)
1st gen bioethanol[a]	fermentation of plants containing sugar and starch	corn (Germany)	9.70
		corn (Hungary)	6.43
		corn (Vietnam)	4.31
		cassava (Vietnam)	17.68
1st gen biodiesel[b]	transesterification of vedetable oils and animal fats	rapeseed (Germany)	3.90
		rapeseed (Hungary)	2.67
		soya-bean (Vietnam)	14.50
2nd gen bioethanol[a] **and biodiesel**[bc]	enzymatic hydrolysis and pyrolysis of cellulose	corn stalk (Germany)	3.88
		corn stalk (Hungary	2.57
		rice straw (Vietnam)	4.84
		forest residues (Hungary)	9.88

[a]median 2007-2016

[b]median 2010-2016
[c]dry matter

3.2. Base material footprint for theoretical and real area

The basic idea of a BMF system is the well-known ecological footprint. The ecological footprint is the only metric that measures how much nature we have and how much nature we use. The lower the ecological footprint, the more consistent it is with the use of available natural resources. The BMF gives the theoretical collection area of unit of biofuel. It was determined for 1 ton of biofuel weight. Due to the different energy content of biofuels, 10 GJ of energy was determined, as well, for better comparability. In case of corn, the data on theoretical areas were extended to a real environment. Our investigations were carried out in the area of 50 km radius of two intensively utilized agricultural district in Hungary. The coordinates of the center of the districts are 47°31'64.4 "N, 18°90'84.7" E and 47°54 '57.5 "N, 21°2' 33.8" E

4. Result and Discussion

The yield of the main bioethanol and biodiesel base materials in EU shows a high fluctuation compared to Vietnam, due to the different climatic conditions. Among the analyzed biodiesel base materials, the highest yield loss (-25%) was shown by rapeseed production in 2011 in Germany comparing to the median value. In Hungary, the yield of rapeseed remained below 20% of the average in 2010 and significantly exceeded the average yield in 2014 and 2016. The yield of soybean in Vietnam was quite balanced, the maximal yield loss did not reach 2% (2014), while the highest yield was experienced in 2016 (+ 8%), see Figure 2.

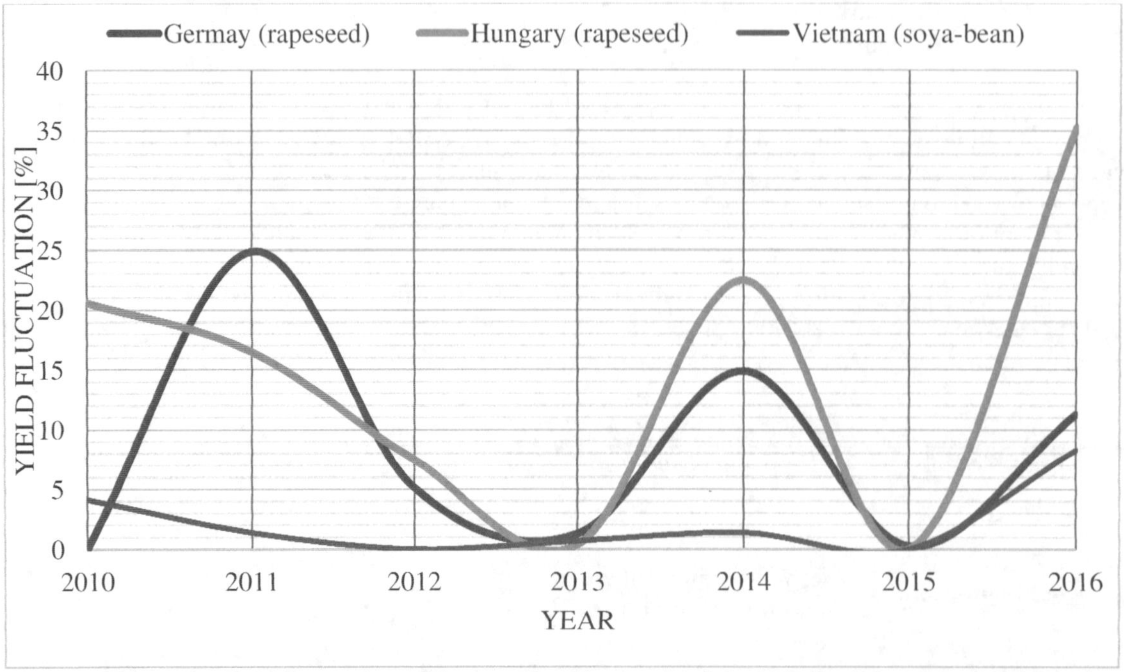

Figure 2. Yield fluctuation of the examined base material of biodiesel between 2010-2016.

The yield fluctuations of corn were particularly problematic in Hungary. It has almost disappeared by 40% of the mean value (2007, 2012, 2016), see Figure 3. This had serious consequences in the production of biofuels. In Hungary, about 20 bioethanol producer plants were under construction or authorization procedure from 2006, but none of them was successful due to the extremely low corn harvest in 2007. The reason in most cases was the strong increase in the price of base materials and the lack of continuous feedstock. In Vietnam, none of the two examined base materials exceeded 10% yield fluctuations. In Vietnam, the main problem was the lack of a domestic biofuel market and the high base material price, which made the exports of biofuel even more difficult. Because of the multiple use, there were not enough raw materials for biofuel production, even though it was located near a cassava cultivation area. Because of the limited area and manual cultivation, cassava prices have remained very high [19, 20].

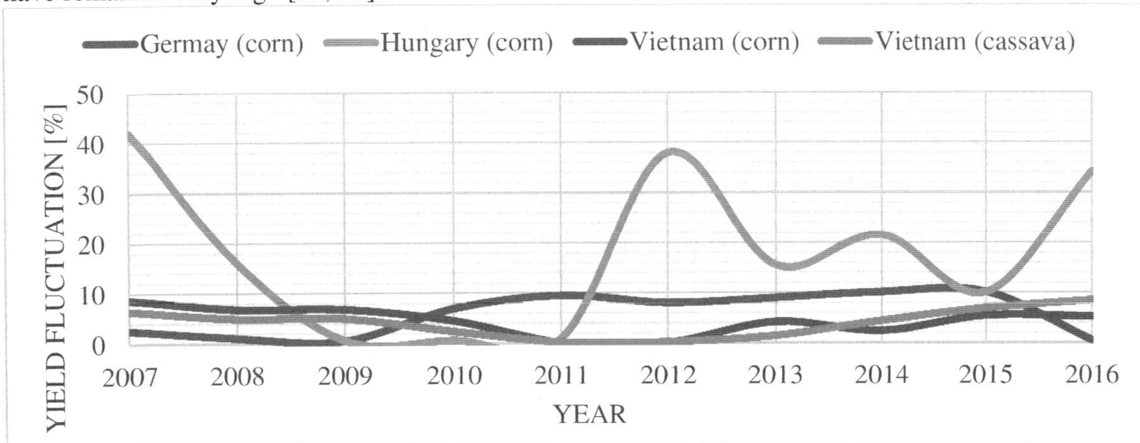

Figure 3. Yield fluctuation of the examined base material of bioethanol in the last 10 years.

Based on our research, the lowest BMF is available in the production of biooil from forest residues, see Table 3. Although the technology is in the research phase, some plants already operate in commercial scale. In Hengelo in the biooil plant a number of raw material was tested and wood residues gave the most favourable biooil yield [21]. With regard to the main agricultural products, soybean-based biodiesel has given the most optimal area. Although its oil content is much lower (~ 18%) [22, 23] than the rapeseed (~ 40%), [24] it requires a much smaller collection area, due to more than four times yield per hectare. Among the agricultural by-products, rice straw can be collected from the smallest area to produce a unit of biofuels due to high lignocellulose content and high yield per hectare. In Vietnam, rice straw has great potential, currently used only to a small extent for animal feed and other purposes. The corn stalk has the highest base material footprint. However, they also represent a large untapped potential.

Table 3: Base material footprint for 1t and 10 GJ biofuel

Type of biofuel	Type of base material	BMF for 1t biofuel [ha]	BMF for 10 GJ biofuel [ha]
1st gen bioethanol[a]	corn (Germany)	0.34	0.13
	corn (Hungary)	0.51	0.19
	corn (Vietnam)	0.76	0.28
	cassava (Vietnam)	0.61	0.23
1st gen biodiesel[b]	rapeseed (Germany)	0.61	0.16
	rapeseed (Hungary)	0.89	0.24
	Soya-bean (Vietnam)	0.37	0.10
2nd gen bioethanol	corn stalk (Germany)	1.08	0.40
	corn stalk (Hungary)	1.63	0.61
	rice straw (Vietnam)	0.83	0.31
2nd gen biooil	corn stalk (Germany)	0.59	0.34
	corn stalk (Hungary)	0.89	0.51
	rice straw (Vietnam)	0.50	0.29
	forest residues (Hungary)	0.14	0.08

From the security point of view of the base material supply, the corn and corn stalk are the most unfavourable source of biofuels, see Figure 4. The increase of BMF can exceed 72% due to yield loss in drought years in Hungary. In the case of forest residues, yield fluctuations are not relevant; however, availability of base material depends on the actual forest logging plan. Collecting them can be cumbersome and ecological problems caused by contagion and organic matter removal. However, the intensified use of first generation biofuel could endanger the worlds agricultural areas because of the increasing use of pesticides and artificial fertilizers on industrial produced monocultures [25].

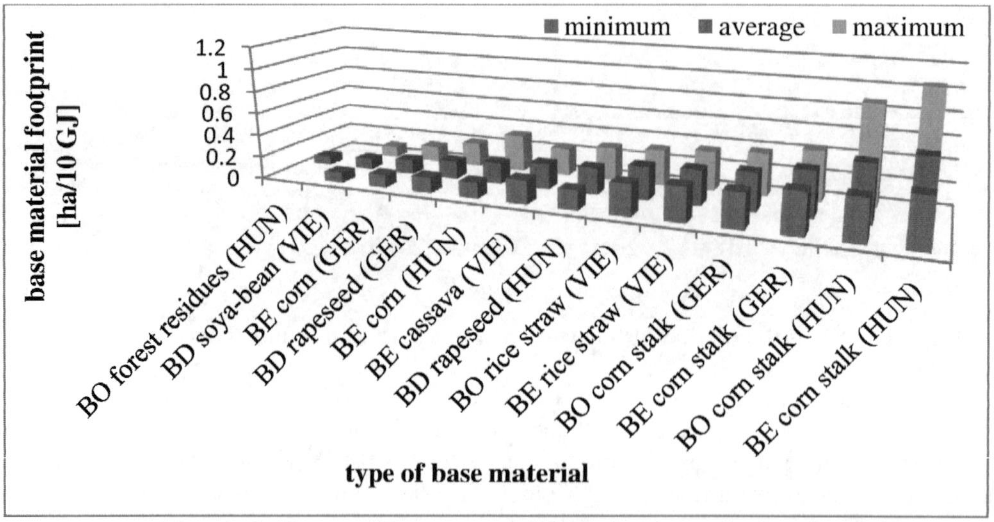

Figure 4. Change of base material footprint depending on the yield.

In the case of corn, we were investigating that how much larger area the real base material footprint covers, compared to theoretical base material footprint. We selected two areas in Hungary where intensive farming is taking place. As biomass can be transported within a max. 50 km radius [26], therefore, we have taken such a large area. The result is that the collection of the same amount of corn or corn stalk in real conditions requires 9 to 11 times bigger area. If we are taking into account the impact of drought, the required area will rise to 20 times bigger. In case of corn stalk, the BMF in real condition is more than 7 times bigger in the case of second generation biooil, and 15 times bigger if we count with the technical loss, and the need of the green fertilizer demand, and animal husbandry.

5. Conclusion

The importance of the develop of methods of base material footprint system is to facilitate the correct selection of the biofuel type and the production technology that is compatible with the available base material in the area. Improperly funded investments have serious consequences. Many bioethanol plants have not been realized or closed over the past 10 years regardless of geographical location, thanks to high raw material prices due to various reasons.

The choice of biofuel base materials has undergone a strong transformation over the last decade. The range of base materials that are less endangered for the food supply security is expanded, but still sustainable. To develop a base material footprint system, all applicable base materials would be under investigation in different geographic environments. It should also be taken into account that more than one raw material can be included in the biofuel production system at the same time.

The lower is the base material footprint of the biofuel, the smaller the collection area and the required transport distance are. According to our research, forest residues have the lowest area requirement. However, collecting them is cumbersome and a considerable by-product only occurs in case of clear cutting. Most of the by-products are concentrated in the wood processing industry, but they are still largely utilized today. Agricultural by-products, i.e. corn stalks, rice straws have a high potential, however, because of their low bulk density, transportation is not economical for long distances, and their conversion rate is low. At the same time the risk of iLUC can be reduced. During the planning of the biofuels plant, the effect of extreme weather must be taken into account, which can increase the collection area by up to 70%. Meeting the ecological needs doubles the base material footprint. The agricultural by-product is also used in animal husbandry. Harvesting and storing corn stalks with high moisture content is still cumbersome.

6. References

[1] Enerdata 2017 Global *Energy Statistical Yearbook*
[2] IEA International Energy Agency (2013): *Medium-Term Renewable Energy Market Report 2013*. Market Trends and Projections to 2018. Executive Summery.
[3] Naik S N, Goud V V, Rout P K and Dalai A K 2010 Production of first and second generation biofuels: A comprehensive review *Renewable and Sustainable Energy Reviews* **14** (2010) pp 578–97
[4] Sanderson K 2006 US biofuels: A field in ferment. To move US biofuels beyound subsidized corn will be a challenge *Nature* **444** pp 673-76.
[5] Halling P and Simms-Borre P 2008 Overview of lignocellulosic feedstock conversion into ethanol—focus on sugarcane bagasse *International Sugar Journal* **110**
[6] Santos F A, Queiróz J H, Colodett E J L, Fernandes S A, Guimarães V M and Rezende S T 2012 Potencial da palha de cana-de-açúcar para produção de etanol. *Quim Nova* **35**(5) pp 1004–10
[7] Diya'uddeen B H, Aziz A R, Daud W M A W and Chakrabarti M H 2012 Performance evaluation of biodiesel from used domestic waste oils: a review *Trans IChemE* Part B 90 pp 164–79
[8] Truong K, Huong C T, Xuan T D, Trung D M and Khanh T. D. (2016): Current Status and Future Plan of Development of Bioenergy Crops as Renewable Energy Sources in Vietnam *J of Bioenergy and Nature* **5**(1) pp 1-8.
[9] KSH Központi Statisztikai Hivatal 2017 *Statistic of landuse*
[10] WB World Bank Open Data 2015 *Statistic of land use*
[11] EUROSTAT European statistics 2017 *Corn yield, rapeseed yield*
[12] GSOVN GENERAL STATISTICS OFFICE of VIET NAM 2017
[13] Sokhansanj S, Turhollow A, Cushman J and Cundiff J 2002 Engineering aspects of collecting corn stover for bioenergy *Biomass and Bioenergy* **23** pp 347-55
[14] Boris C, Zoran S and Novic D 2011 Geographic distribution of economic potential of agricultural and forest biomass residual for energy use: Case study Croatia *Energy* **36** pp 2017-28
[15] Pappné V J 2010 *A biomassza mint megújuló energiaforrás hasznosítási lehetőségei, különös tekintettel Magyarországra*. PhD . ELTE. Budapest
[16] RKB Rice Knowledge Bank 2014 **Rice straw**
[17] Haszon Agrár 2010 Mit kezdjünk a kukoricaszárral? *Haszon Agrár Magazin* **10**
[18] Vietnam News 2011 *Rice straw starts having value* Vietnamese; April, 2011
[19] VNN Viet Nam News 2015 *VN ethanol industry faces crisis*
[20] BTG-BTL 2016 *Feedstock* https://www.btg-btl.com/en
[21] VET Vietnam Economic Times 2016 PetroVietnam bio-fuel plant closes
[22] Lee H, Cho B, Kim M S, Lee W, Tewari J, Bae H, Sohn S and Chi H 2013 Prediction of crude protein and oil content of soybeans using Raman spectroscopy *Sensors and Actuators B: Chemical* **185** pp 694-00
[23] Farooq A, Ghulam M K, Farooq N and Ghulam S 2016 Variations of quality characteristics among oils of different soybean varieties *J of King Saud University – Science* **24**(4) pp 332-38
[24] AGMRC Agricultural Marketing Resource Center 2017 *Rapeseed*
[25] Palocz-Andresen M 2013 *Decreasing Fuel Consumption and Exhaust Gas Emissions in Transportation* Green Energy and Technology Springer-Verlag Berlin Heidelberg p 318
[26] Torjai L 2015 *Optimális szállításütemezés a lágy szárú biomassza ellátási láncban* PhD Dissertation Pécs p 153

Acknowledgement
The described work was carried out as part of the „Sustainable Raw Material Management Thematic Network – RING 2017", EFOP-3.6.2-16-2017-00010 project in the framework of the Széchenyi 2020 Program. The realization of this project is supported by the European Union, co-financed by the European Social Fund.

Utilization of Fast Growing Plantation Timber as Bioenergy in Hungary

R Németh[1], S Fehér[2] and Sz Komán[3]

1 Professor, University of Sopron, Institute of Wood Science, Sopron, HU
2 Assoc. Professor, University of Sopron, Institute of Wood Science, Sopron, HU
3 Assoc. Professor, University of Sopron, Institute of Wood Science, Sopron, HU
E-mail: nemeth.robert@uni-sopron.hu

Abstract. The enhanced utilization of plant biomass for bioenergy is a key factor to contribute to a sustainable development. In this research work 3 Hungarian hardwood species *Populus x euramericana.* cv. 'I-214', *Populus x euramericana.* cv. 'Pannonia', *Robinia pseudoacacia* (2 Poplars and Black locust) were investigated on heating value and ash content. 3 age classes were defined (from 7 years up to 29 years) in order to analyse the effect of the tree's age on the investigated properties. Significant differences were found between the bark, the sapwood and the heartwood. Age influenced the heating value to, but differently for bark, heartwood and sapwood. The 10 years old Black locust's bark showed the highest heating value (21,53 MJ/kg), while the 19 years old Pannónia poplar heartwood gave the lowest value (17,68 MJ/kg). In terms of ash content dramatic differences have been proven between bark and xylem (sapwood and heartwood). The 26 years old Black locust heartwood had 0.2 % ash content only, while the 10 years old Pannonia bark 5.2%.

1. Introduction

There are more scientific evidences proving that the intensive (over) use of Earth's energy resources will lead to exhaustion of natural and fossil resources (e.g. forests biomass, coal, crude oil, natural gas). The pollution of the environment has also dramatic consequences for the society and different industries. Sustainability has not been reached yet, as the human society is facing to global problems, like overpopulation, climate change through greenhouse gas emissions. Realizing these challenges researchers are globally working on solutions developing technologies which are using renewable energies. One of the renewables is the plant biomass. The utilization of biomass as energy resource has been always a method in the history of mankind, as it was the energy solution for households and smaller farms [1]. Recent research works have definitely proved the plant biomass as a potential energy resource for the future.

Energy forests and energy plantations are the main resources for organic products from biological origin in Europe. Energy forests are forests which were planted and are managed with specific purpose, while the energy tree plantations were established for energy production [2]. The main species are: willow (*Salix*), alder (*Alnus*), improved Poplars (*Populus*) and Black locust (*Robinia*). Mostly fast growing species can be considered on the plantations with high sprouting potential.

Important characteristics of the woody biomass are: bark percentage, percent moisture, specific gravity, heating value, amount of extractives, concentration of inorganic elements including alkali metals, ash content and cellulose/lignin ratio [3]. The energetic properties of a given wood species with a genetic heritage are influenced by the micro- and the macrostructure and the age of the tree. According to

earlier investigations [4] [5] the calorific value of the bark is lower compared to the wood material, but the first shows considerable higher ash content [6][7].
The main goal of the presented research work was to determine the effect of the age of plantations of different tree species on their calorific value.

2. Experimental

Tree species suitable for energy plantations and forest plantations in Hungarian were investigated, namely two improved Poplars (*Populus* spp.) and Black locust (*Robinia pseudoacacia*). The species and age classes are summarized in Table 1.

Table 1. Forming of age classes

Wood species	Age classes (years)		
	I.	II.	III.
Populus x euramericana. cv. 'I-214'	7	10	19
Populus x euramericana. cv. 'Pannonia'	6	10	19
Robinia pseudoacacia	10	17	26

For each species and age class combinations 6 trunks were selected randomly from the different plantations. The trees were harvested of the plantations managed by the Forest Directorate "Kisalföldi Erdőgazdaság Dél-hansági Erdészete" in the north-west part of Hungary. The sites (soil, hydrology, et.) were determined as average for Hungarian conditions. The sample discs were cut from each of the trunks at breast height, from which the test pieces were produced. As the heartwood of the Black locust is rather narrow (2-5 annual rings, ca. 1 cm), thus the practical relevance is low, the investigations here were limited to the heating value.
This research work focuses on the calorific value and the ash content of the Poplars and Black locust in different age classes, with special emphasises on the different anatomic tissues. Thus from each sample disc the heartwood, the sapwood and the bark were investigated separately.

3. Results and discussion

The results of the calorific value is summarized in Table 2.

Table 2. Heating values of Poplars and Black locust

Species / Age class		Heating value (MJ/kg)		
		Heartwood	Sapwood	Bark
'Pannonia' (Poplar)	I.	18,96	19,24	17,92
	II.	18,15	18,55	17,68
	III.	17,68	18,95	18,09
'I-214' (Poplar)	I.	19,04	19,26	18,03
	II.	18,34	18,71	18,44
	III.	18,82	18,99	18,23
Black locust	I.	18,89	18,49	21,53
	II.	19,00	18,41	20,50
	III.	19,39	18,96	20,36

In case of 'Pannónia' Poplar the highest heating values were measured in the sapwood, regardless the age class, while the lowest values were determined for the bark, see figure 1. An exemption here is the oldest age class, where the heating value of heartwood was even lower. Comparing the age classes the youngest plantations showed the highest values, excluding the bark. In the age classes II and III the ranking of the heartwood and the bark is different, thus in higher ages the bark contains more energy (related to mass) compared to the younger barks. The heating value of the heartwood is declining with the age.

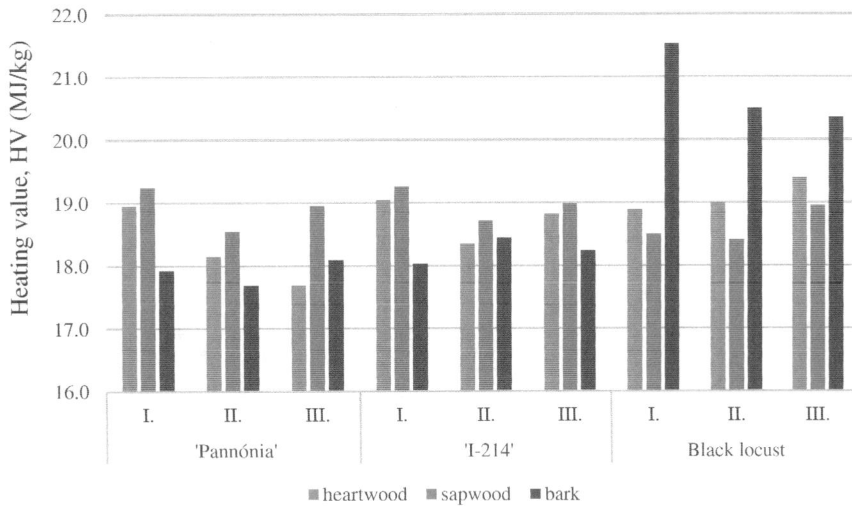

Figure 1. Heating values of Poplars and Black locust for different age classes and tissues.

The measured values and tendencies for the 'I-214' Poplar are similar to the 'Pannónia'. The highest heating values were measured for the sapwood and the lowest for the bark. However in the age class II the heartwood's value falls off the bark. The highest heating values were determined for the youngest plantation materials (Class I). Similar to the other Poplar, the oldest materials' values exceed that of the middle age (Class II). Analysing the two Poplars, it can be stated, that there are only minor differences between them. Thus, they give the same bioenergy at the same site under the same management.
In figure 1. the data for Robinia are shown. Surpassing values can be observed for the bark, followed by the heartwood and the sapwood. Studying the age classes reverse tendencies can be observed for the bark and the wood tissues (sap and heart). While the sapwood's and the heartwood's heating values are enhancing with the age, the bark's values are declining considerably.
Comparing the Poplars and the Black locust the bark of the last species shows extraordinary high values, exceeding any tissue by the investigated tree species. In case of Poplars the sapwood show higher values compared to the heartwood, while for Robinia the heartwood's values are higher.

Table 3. Ash content

Species / Age classes		**Ash content (%)**		
		Heartwood	**Sapwood**	**Bark**
'Pannonia'	I.	-	0,36	4
	II.	1,2	0,32	5,2
	III.	0,78	0,48	4,6
'I-214'	I.	-	0,4	4,6
	II.	1,2	0,38	4,3
	III.	1,2	0,43	4,7
Black locust	I.	0,43	-	3,9
	II.	0,24	-	4,3
	III.	0,2	-	4,2

Table 3. shows the ash content for the investigated species, for all three age classes and for the different tissues.
The ash content of different tissues show similar for the two poplars (figure 2.). The lowest values could be detected by the sapwood, while the heartwood has 3x and the bark 10x more ash content. Regarding the influence of age, the sapwood's ash content is increasing, while the heartwood's value is decreasing in case of poplars. No clear tendency could be observed for the barks. The two poplars do not show significant differences to each other.

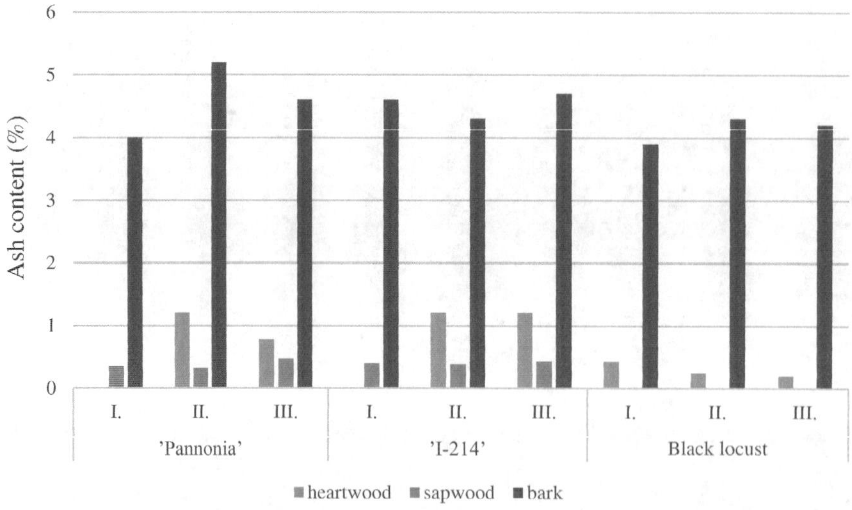

Figure 2. Ash content of Poplars and Black locust for different age classes and tissues.

In figure 2. it is well demonstrated that in case of Robinia the bark's ash content is much higher compared to the heartwood. The bark's heating value is slightly increasing from the younger Class I to the oldest Class III, while the heartwood shows opposite tendency.
Comparing the two species it can be stated, that the bark Black locust has slightly lower ash content compared the two Poplars, while in case of heartwood Poplars show much higher values.

4. Conclusions

In terms of heating value no significant differences could be found between the investigated tree species' xylem (sap and heart). Considering the age classes usually the youngest samples showed the highest heating values in case of Poplars. Excluding the bark, the oldest materials delivered higher values in case of Black locust.
While the Poplars' and Black locust's sapwood heartwood values do not differ considerably, remarkable differences could be found for the barks, as the Black locust's bark is significantly superior compared to the other tissues' values.
The ash content of bark is several times higher compared to the sap and to the heart. The ash content of the heartwood is similar by the two investigated Poplars despite to the age class, and in all cases higher than by Black locust. Poplars do not differ significantly in their sap woods' ash content. The Poplars' bark usually contains more ash compared to Black locust.

5. References

[1] Bai A, Lakner Z, Marosvolgyi B, and Nabradi A, 2002. A biomassza felhasználása. Szaktudás Kiadó Ház, Budapest
[2] Marosvölgyi B. 2002 Az energetikai ültetvények létesítése és hasznosítása. In: Körmendi P, Pecznik P (szerk.): Megújuló energiaforrások hasznosítása. FM Műszaki Intézet, Gödöllő, GATE, 49–

[3] Kenney WA, Sennerby-Forsse L, Layton P, 1990 A review of biomass quality research relevant to the use of Poplar and willow for energy conversion. Biomass 21:163–88.
[4] Klašnja B,, Orlovi, S. and Galić Z, 2013 Comparison of Different Wood Species as Raw Materials for Bioenergy. South-East Eur For. 4(2) 81-88.
[5] Jamnická G, Petrášová V, Petráš R, Mecko J, Oszlányi J, 2014 Energy production of Poplar clones and their energy use efficiency. iForest 7: 150-155
[6] Passialis C, Voulgaridis E, Adamopoulus S, Matsouka M, 2008 Extractives, acidity, buffering capacity, ash and inorganic elements of Black locust wood and bark of different clones and origin. European Journal of Wood and Wood Products 66, 395-400.
[7] Nosek R, Holubcik M, Jandacka J, 2016 The impact of bark content of wood biomass on biofuel properties Bioresources 11(1):44–53.

Acknowledgments
The described work/article was carried out as part of the „Production with nature - Agroforestry as a new perspective", EFOP-3.6.2-16-2017-00018 project in the framework of the Széchenyi2020 Program. The realization of this project is supported by the European Union, co-financed by the European Social Fund.

Renewable Energy and Smart Hybrid Strategies for High Performance Architecture and Planning in Case of Tehran, Iran

F Heidari [1] and M Mahdavinejad [2] and SH Sotodeh [3]

1 Senior Production Editor, M. Sc. Student, Department of Architecture
Tarbiat Modares University, Tehran, Iran
2 Senior Production Editor, Associate Professor, Department of Architecture
Tarbiat Modares University, Tehran, Iran
3 Senior Production Editor, M. Sc. Student, Department of Architecture
Tarbiat Modares University, Tehran, Iran
E-mail: F-Heidari@modares.ac.ir

Abstract..It is significant to adopt renewable energy systems in building industry as one of the most important sectors in energy consumption. Level of energy consumption in building industry in contemporary architecture of developing countries such as Iran enjoys much more interest than other parts especially developed countries. Literature review of the research show that adoption of renewable energy systems has a lot to do with politics, policies, law, finances and market. While renewable energy finance and energy economics is in need of building sector to focus on adoption of renewable energy systems more than other sectors. It is very important to explain that high-performance architecture and planning requires some kinds of combined strategies to enhance the level of energy efficiency. The main scope of the research is to show concept of smart hybrid strategies for high-performance architecture and planning in case of Tehran, as one of the most important and efficient factors toward efficient adoption of renewable energy systems. The most important questions of the research are: 1- How socio-cultural and economic aspects of developing countries affect the idea of energy efficient architecture and planning? 2- What are the most important characteristics of high-performance architecture and planning for developing countries? Regarding to the hypothesis of the paper: 1- The idea of energy efficient architecture and planning in developing countries requires combined and smart hybrid strategies to meet energy efficient architecture and planning. 2- Smart hybrid strategies for high-performance architecture and planning consists of smart ventilation, photovoltaic systems, heat pumps, thermal storage, isolated building systems, low-e windows, efficient envelope and hyper - controlled openings.

1. Introduction

Climate change is a serious global environmental Concern; it is primarily caused by the building up of green gases in the atmosphere. The global increases in carbon dioxide concentration and primarily due to fossil fuel use and land use change, while those of methane and nitrous oxide are primarily due to agriculture. [1] global warming is specific example of the broader term ''Climate Change'' and refers to observed increase in the average temperature in the air near Earth's surface and ocean in the recent decades. According to the united nations, cities consume two third of global energy use. It was emphasized that 76% of the world's energy-related Carbon dioxide are also by cities by transport, industry, and building and construction related activities. Studies have shown that buildings and construction activities use 40% energy, 30% mineral resources and 20% water of world's resources. It also accounts for 40% CO2 emissions, 30% solid wastes and 20% water pollution in the world. Energy and environment are critical constituents of physical development of the built environment. Energy conversation and environmental quality have been widely acknowledged as important parameters of building design. It is necessary to understand the strong interaction between architecture and physical development and establishes its importance with environment. [2] Therefore the energy conscious strategy for buildings emphasizes integration of several environmental factors, to achieve reductions in the use of conventional energy by desirable ventilation, utilizing day lighting and using appropriate thermal design of building. Today the subject of decrease in natural resources is one of the main issues in the policy of different countries. This subject is of one the main branches of each field of science and very important and inseparable part of architecture. Architecture and designers in the past considering their facilities and technology at that time had a very important role in this subject like form of the building, increasing or decreasing of the height of the building, and the stuff that they have used. The people in the house considering the seasons they used to live in different rooms and places of the building and this action led to using less fuel and more comfort. Today the issue of overpopulation of everyday life, lack of non-renewable energy resources and using renewable resources has an important role [2]. Part of this usage considering the chart-1 reaches to the highest level in residential buildings. Most of the energy usage includes lighting and heating, this subject has a straight connection with the environment, it puts the sensitive arts of the buildings like surface and sidewall in danger. So that the light transfer, air conditioning and the most heating exchanging happens from these sidewalls. If we consider these usages and overpopulation these resources will be gone in a short time. As a result, production and consumption need more energy and resources and it increases pollution.(Fig.1)

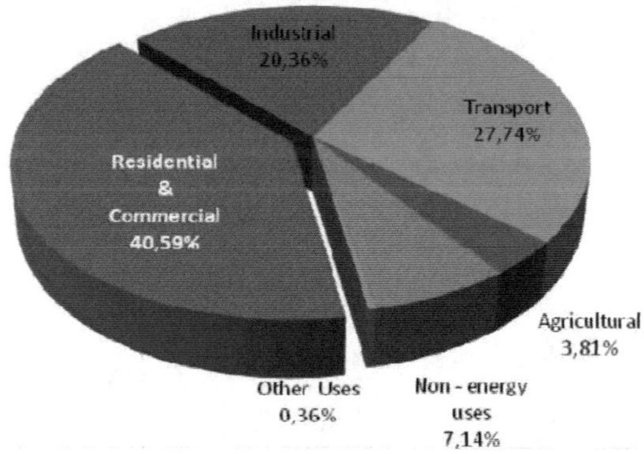

Figure 1. Energy consumption by sectors [3]

During the past decade's environmental issues were given a great attention from different aspects. The 1960s was the beginning of this attention and the main focus was on the pollution because of industrial development of that era was considerable [4]. In the late 1970s the issues of business and environment raised and the environmentalists held a lot of discussions all over the world against the development of business and its harmful effects on environment. In their opinion because of business freedom, increase in exports, pollutant activities and economic activities the usage of resources and energy will raise in an appropriate way [5]. In addition to energy crisis as the consumption increases so does pollution. Today pollution is one of the main managing struggles in a way that countries follow these issues internationally in addition to their inland policies. One of the examples of pollution is air pollution and because of its nature of development is noticeable all around the world. Without any doubt the air pollution is dependent

IOP Conf. Series: Earth and Environmental Science **159** (2018) 012030 doi:10.1088/1755-1315/159/1/012030

on the economic progress of countries. This subject in economic literature (environment literature) follows as EKC graph which shows the process of nature destruction considering the nature and different levels of economic progress.(Fig.2)

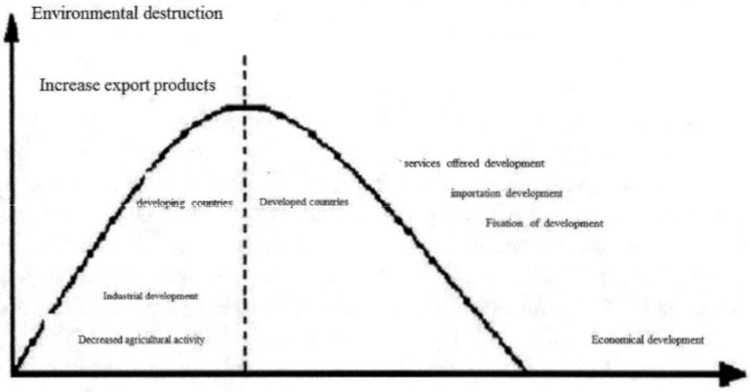

Figure 2. Environmental destruction According to Economic Development [6]

2. Energy Consumption

2.1. Energy consumption in Iran

The building and housing sector with more than 40 percent of energy consumption is the largest energy consumer in Iran. The energy consumption of buildings in Iran is more than 5.2 times average global consumption. Along with major cities such as: Tehran has a high air pollution. That is often caused by the consumption of fossil energy. While more than 98 percent of the energy consumption of buildings in Iran will be provided of oil and gas products. The housing sector is one of the main sources of pollution. This sector allocated about 26.4 percent of the carbon dioxide emissions itself in Iran. Buildings in Tehran produces over 40 percent of carbon dioxide. Reduce energy consumption in the building and housing sector will have a significant impact on the total energy consumption of the whole country. [7](Fig.3-4)

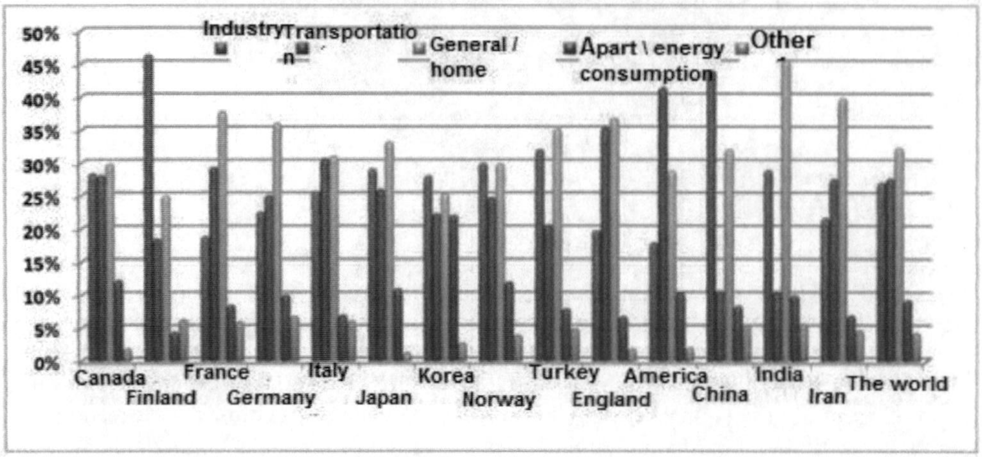

Figure 3. Graph of energy consumption in different areas of the world-IEA [8]

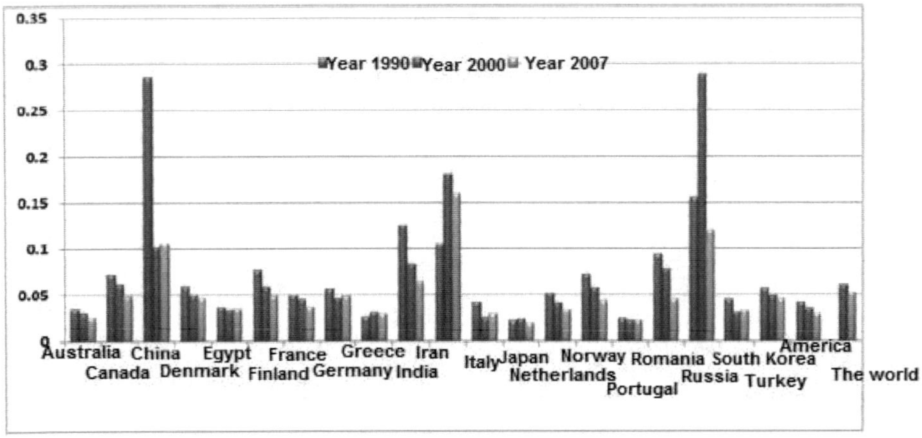

Figure 4. the energy intensity in the household sector than in sector private consumption Koe / $ ppp [8]

Reduce energy consumption of buildings is imperative in terms of economic and environmental. Remove energy subsidies and increase in energy prices added to the importance of saving energy in buildings. In an architecture with high efficiency in energy consumption set rules and policies of government and organizations plays an important role, relevant organizations and government agencies shall provide scientific context and legal work and practical methods for achieving energy savings in buildings. The potential of energy saving in buildings housing more than other sectors and reduce energy consumption in this sector easier and with less investment available than other sectors. [9] believes the high demand for housing in Iran and the high potential of energy efficiency in new buildings and relatively low useful life of buildings in Iran all and all can be disabled using hybrid intelligent systems to save energy in their buildings. As the leading countries in this field, such as: Germany emphasis on reducing emissions in the building sector including new buildings and existing buildings. Establish the standard for energy efficient buildings in Iran need to technical and economic conditions specific to the use of standards and indirectly expertise organized in the relevant field for application of the principles. Iran Need to its own strategy in this field. [10]

The energy efficiency strategies in building and housing in Iran according to economic conditions, political, social, could be as follows: 1) Reduce energy consumption of buildings with architectural design, 2) Standards and energy labels for buildings, 3) Promotional support for the construction of energy efficient buildings, 4) Making culture to reduce energy consumption in buildings, 5) Construct pattern buildings, 6) The standards for energy consumption in buildings. Legislation and planning and preparation specified energy labels for various rankings adds to basic understanding of energy consumption in the building sector and separate from that, with increased public understanding and buyers of housing in this regard, increase the value of the transaction between vendors and buyers will find an approach and a factor such as the rate of energy exchange will be as part of the main criteria. To put it briefly stated, if a building is energy efficient is desirable goal to achieve this goal should be taken the main steps together with the intricacy to perfection or achieve lower power consumption achieved in buildings sector [10]. In this way, give advice to builders and buyers' awareness regarding the importance of obtaining high quality modest fee could be one of the main stages. Despite the high costs spent for such projects in the first place, amount of energy savings and economic benefits in the long run it is effective and profitable. Reduce waste and energy consumption maintain and store these resources and generate revenue through the sale of the state [11].(Fig.5-6)

Figure 5. The theoretical model of factors influencing the reduction of energy consumption (source: writer)

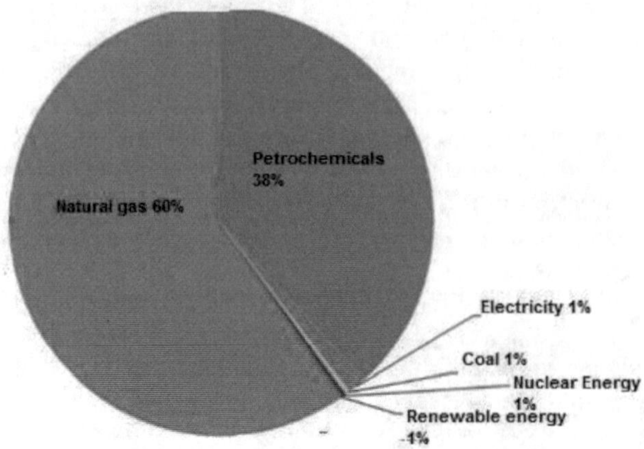

Figure 6. The primary and overall energy consumption in Iran [12]

Another most important thing in energy consumption is behavior of the residents and users. Which has a significant impact on energy consumption in buildings. Simple tasks, making culture to change daily habits, using smart windows with low emissions and switching to simple windows, sealing doors and windows and replacement of structural elements undesirable and also other passive strategies can play an important role. Behavior affecting energy use in buildings can be named the following options: 1) Turn on and off to time cooling and heating systems, 2) Use curtains suitable for different rooms, 3) Opening and closing time according to performance and type of pop-up windows, 4) The use of heating and cooling systems in accordance with spaces and special programming for different conditions throughout the seasons. Application of these strategies in a residential area need to awareness of the circumstances and the responsibility of individuals to resources and its own national interests to consider more sensitive to the environment and energy consumption. It was very important attracting the public trust, good policy and financial support and regulation tailored to the goals of reducing energy consumption and the use of

IOP Conf. Series: Earth and Environmental Science **159** (2018) 012030 doi:10.1088/1755-1315/159/1/012030

renewable energy systems. It is considered as the basic and fundamental and this second part of sector collaboration with the type of energy and use to required mechanical systems compared to basic needs and the primary need and feasibility choices in the strategy will be an architecture with high efficiency. [13] One of the basic strategies for saving energy is optimization of design. What an optimized architectural design of the form and geometry involved design or intelligent systems in the design, also sought save resources and costs. If adherence to specific criteria based on the design and geometry, without increasing costs or scarce facilities and additional manpower that can bring countless problems, the ultimate goal achieved to reduce energy consumption or energy-efficient building design. This method can be used for different buildings and provide its needs over the life cycle of buildings. Including the impact of optimization of design can be noted in the opening system and climate use of special materials in different areas and preserving the dominant form of architecture region. In this type of design can be considered material apart from the architect. In this case considered amount of heat transfer coefficient materials. This field of design methods based to reduce environmental damage and reduce energy consumption [14].

3. Renewable Energy

Oil consumption in Iran has a constant slope increasing and manufacturing of that has been high volatility. According to international experts about the energy crisis, in the not too distant future, Iran will be importing oil. Renewable energy from substances such waste can be used as a strategy in the country is part of the energy consumption. Renewable energy to say which types of energy the kind of energy source, unlike renewable energies (fossil) it is the nature re-emerged in a short period of time, or be renewed. In fact, the energy from the sun, wind, biomass, geothermal, hydro, wave and tidal, biogas and liquid biofuels are renewable energies. As well as fuels derived from municipal solid waste, hospital and household waste such as tires, plastics, waste oil and other energy sources similarly are the energy from waste. In Iran during recent years accommodate necessity clean energy in the energy mix of the country, is taken into consideration. But The risk appetite from investors Compared to the high cost of primary and The uncertainty of future investment return (Because of new technology), on The other hand The Ambiguity of people's welcoming of this technology and perhaps defeat in the first round, Brought down risk appetite for investors. However, according to experts in this field, all electricity needs in Iran supply through solar energy. While unused capacity in the country is such that, electricity production in Iran, including the establishment and maintenance of power plants and power generation, Tens of billions of dollars fuel and destroys the country's foreign exchange earnings in annual. There are huge oil and gas resources and cheap energy prices through subsidies keep in Iran, caused our country left behind in the advanced industrial nations of new energies. [15]

3.1. Energy situation in Iran

In most countries we have thermal regulations that enable us to control building's energy consumption. We also dispose of efficient techniques adapted to climate and to architectural culture. Building's envelops are designed to better insulate indoor climate from outdoor one by controlling heat and mass transfers between them. Though, this practices omit to consider the indirect effect that envelops have on building's energy consumption. Indeed, the radiative characteristics of the buildings envelop and geometrical arrangement of them in the city play an important role in the urban heat. this one modifies building's energy consumption, because of the increase of air and surface's temperatures, besides it leads to uncomfortable and unhealthy situations for inhabitants in summer. [16] Iran is among the countries that have abundant energy resources. Iran since 1913 has always been considered one of the major oil exporter. This country located between two source of energy in the world (the Caspian Sea and the Persian Gulf). This country in terms of the international situation is put in special circumstances. Iran by having 137.6 billion barrels of oil allocated 10.3% of global oil reserves. In the field of natural gas, with 29.61 trillion cubic meters, 15.8% of world gas reserves, is in the first place in the Middle East and second place of the world. Energies of oil and gas in Iran have the highest share in energy consumption. In the images 7 and 8 visible the share of renewable energy in the electricity production in the country and the status of renewable energy in the world. In 2010, approximately 1.48 percent of total electricity production comes from renewable energy. This value is much lower than amount of global (19 percent) [16]. Among renewable energy, the share of hydropower in the country is 1.45 percent more than other renewable energy. About the necessity of using renewable energy in Iran pointed to two cases. Oil consumption in Iran increased by a constant angle, but production has considerable fluctuations. Energy crisis is "of ice" in Iran. (image number) in the next few years will collide two charts and country instead of oil-producing will become a country consumer. (Iran Energy Balance Sheet, 2007) in this case the country in a state of limbo placed in the energy production and there is no return way. The only remedy for this situation is critical in the the use of renewable energy is in the long-term planning. According to the International Energy Agency in the 2012 Iran paid 82 billion government subsidy. The amount of subsidies paid by one-tenth of the total subsidy in the world. According to the statistics of the world's the largest donor subsidies for fossil fuel consumption. Even if half of this amount of money spent on setup and deployment of renewable energy systems, in the near future we can reduce pollution and preserve the national capital. This support, unlike the renewable is gradually dying out and the return on investment will be used to control in a short time [17].(Fig.7-8)

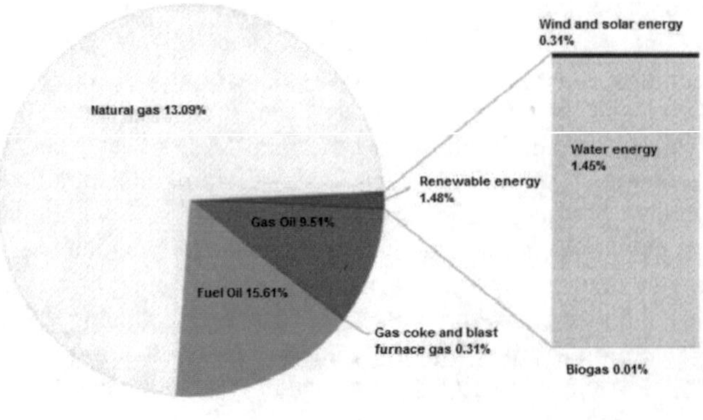

Figure 7. Graph of share of renewable sources in the electricity production in Iran in 2010 [18]

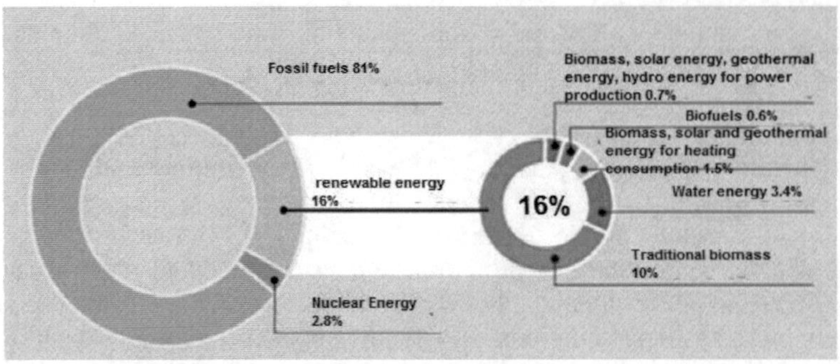

Figure 8. Renewable energy situation in the world in 2009 [18]

In the Figure 9, We can see Graph of oil production and consumption in Iran Oil consumption in Iran increased by a constant angle, but production has considerable fluctuations. Energy crisis is "of ice" in Iran. In the next few years will collide two charts and country instead of oil-producing will become a country consumer. Therefore, due to excessive consumption of energy in the country, needs to formulate a program for the use of other energy sources in the country [19]. Every living organism absorbs solar energy and keeps your storage as called biomass. Each year, through photosynthesis, the equivalent of several times the world's annual consumption of energy, solar energy saved in leaves, trunks and branches of trees [19]. Therefore, the biomass of renewable energy sources is unique to store solar energy. Today's world is the world of development, at the same time, a growing world progress and development in various fields and in the meantime created the proper disposal of solid waste and energy from waste plants is also considered one of the development. In recent years, especially electric energy production from waste has been considered and the law also provided programs and support for it has been on the agenda [19].(Fig.9-10)

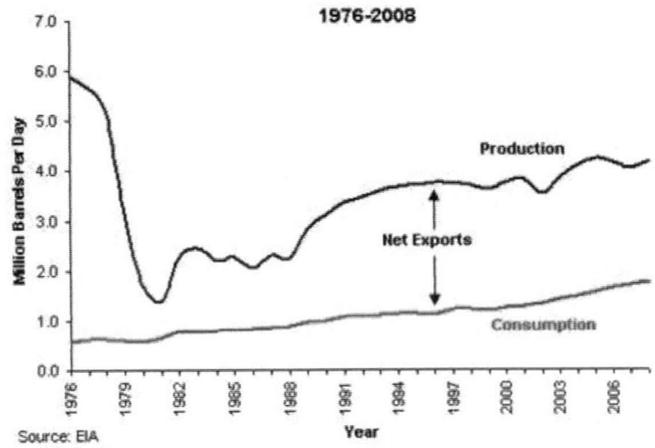

Figure 9. Graph of polymer production and consumption of petroleum products in Iran [18]

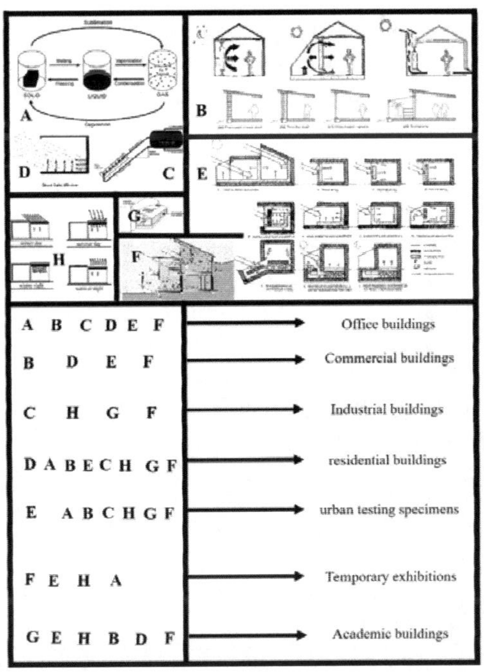

Figure 10. Hybrid Strategies Chart for High-Performance Buildings [Source:Writer]

IOP Conf. Series: Earth and Environmental Science **159** (2018) 012030 doi:10.1088/1755-1315/159/1/012030

4. Conclusion

The first step in designing is detailed understanding of climate energy architecture, in order to identify a suitable climate In an energy efficient design we should examined different parameters affecting the climate In the the building behavior, including behavior and thermal analysis. This parameters and tools is likely together and complement each other by factors such as thermal comfort. Some of these tools, such as: Psychometrics and environmental charts, graphs radiation - Ghosting can play an important role. In the developing countries such as: Iran uses one-dimensional strategies to reduce energy consumption is not responsive according to the specific circumstances of the country and on the other hand there are problems in various fields (pollution and reducing sources of energy). To achieve the objectives of reducing energy consumption in residential buildings in the order to reduce the use of nonrenewable energy sources and Annihilation of their emissions, we need to determine the strategy in two main sections, including preparatory measures and actions are final. These parts, including the mid-term strategies and directives to improve and facilitate the path. The initial or introductory sector includes measures of social, political, economic, organizations and agencies of the government. This includes strategies such as, informing the people and culture of in the use of heating and cooling systems and select the building based on energy standards. Another important factor in the introductory section is financial support and policy relevant organs of the goals and timing. In the second part with respect to the substrate examined practical actions and strategies and the quality of project objectives. For example, select the type of renewable energy systems according to their use and the combination of factors will have an impact system. In summary, in the developing countries does not work one-dimensional action in the field of energy efficiency and use of renewable energy systems. This cannot be achieved unless, by combining and using a combination of several systems and long-term strategies of the basic and fundamental and practical steps.

5. References

[1]Heidari F,Mahdavinejad M.(2017). High-performance Sonitopia (Sonic Utopia): Hyper intelligent Material-based Architectural Systems for Acoustic Energy Harvesting,2nd International Conference on Green Energy Technology, Earth and Environmental Science, Volume 83(1),012021

[2] Nasrollahi,F.(2010).CO2 Emission Inventory in Iran, Young Cities Research Project, Berlin ,university of Technology.

[3] Web site: http://www.wri.org/ World Resources Institute 2014, CAIT database

[4] Hollinger, Keith H. (2008), Trade Liberalization and The Environmental Study of Natta's Impact in Elpaso,Texas and Juarez ,MexicoVirginia Polytechnic Institute and state university,1-79.

[5] Khalil, s.Inam,z.(2006),Is Trade Good for Environment ? A Unit Root Co-Integration Analysis. Journal of Pakistan Development Review,45(4) ,1187-1196.

[6] Chaitanya,k.(2007).Rapid economic growth and Industriazation in India ,China and Brazil: At what Cos?.William Davidson Institute working paper,No.897.

[7] Nasrollahi, F.(2009).Energy Efficient Architecture for Tehran, IFHP World Congress 2009, Berlin , Germany.

[8] Web site://www.iea.org/ International Energy Agency

[9] Mahdavinejad M, Fallahtafti R. Optimisation of Building Shape and Orientation for Better Energy Efficient Architecture, International Journal of Energy Sector Management 2015; 9(4): 593 – 618.

[10] Nasrollahi,F.(2009).Climate and Energy Responsive housing in Continental Climates, university of TU Berlin,ISBN:978-3-7983-2144-1,278P.

[11] Mahdavinejad M, Javanroodi K. Natural Ventilation Performance of Ancient Wind Catchers, an Experimental and Analytical Study – Case Studies: One-Sided, Two-Sided and Four-Sided Wind Catchers, Int. J. Energy Technology and Policy 2014; 10(1): 36-60.

[12] Nasrollahi,F.(2009).Energy Efficient Architecture for Tehran, IFHP World Congress 2009, Berlin , Germany.
[13] Mahdavinejad M, Ghasempourabadi M, Ghaedi H. The role of form compositions in energy consumption of high-risebuildings(case study:Iran,Tehran),AdvMatRes2012;488-489:175– 181.
[14] Mohtashami N, Mahdavinejad M, Bemanian M. Contribution of City Prosperity to Decisions on Healthy Building Design: A case study of Tehran, Front. Arch. Res. 2016: 5: 319–331.
[15] Mahdavinejad M, Doroodgar A, Mashayekhi M. Utilization of Wind Power as a Renewable Energy in Asbads, Case of Iran, Sistan. Adv Mat Res 2012; 433: 1141-1145.
[16] Nasrollahi,F.(2010).CO2 Emission Inventory in Iran, Young Cities Research Project, Berlin ,university of Technology.
[17] Mahdavinejad,M.Bemanian,M.Khasksar, N.Abdolvardi,G.(2011).Choosing efficient types of smart windows in tropical region regarding to their advantage productivities, International Conference on intelligent building and management,Sydney.
[18] National Iranian Oil Company (2007). Iran Energy Balance Sheet.
[19]Mahdavinejad,M.Abedi,Mahmoud.(2012).Subsidiz_Reform Plan and Energy Efficiency in building Energy Consumption in case of Iran, SAVAP International (Society for the Advancement of Education through visionary Academicians/Researchers for Peaceful Globe),Academic Research International,(3),637.

IOP Conf. Series: Earth and Environmental Science **159** (2018) 012031 doi:10.1088/1755-1315/159/1/012031

Numerical Analysis on the Structure of Wall Implanted with Heat Pipes

Zhengyu Wang[1] and Xianghui Su[2]

1 Master Candidate, Nanjing University of Aeronautics and Astronautics, Nanjing, China
2 Associate Professor, Nanjing University of Aeronautics and Astronautics, Nanjing, China
E-mail: 1149688760@qq.com

Abstract. A type of passive solar energy utilization technology, the wall implanted with heat pipes, with a double vacuum glass panel be set to isolate the condenser section from the outdoor environment, is studied theoretically in this paper, to improve the utilization of solar energy on building walls with external thermal insulation in winter in northern China. A two-dimensional transient heat-transfer model is developed to analyse this instruction and then it is solved by using finite difference method (FDM). Numerical calculation on this construction indicates that, the temperature of the inner surface increases by 12.4 degrees with the glass panel in typical climate of Beijing, more than the figure that is 5.8 degrees at most without the structure. It can improve the indoor thermal environment effectively.

1. Introduction

Walls are the important part of building envelope, as well as the most part of the heat dissipation. Due to the bad climate in winter in northern China, and the thermal resistance of building walls can't reach the standard, it's necessary to use external thermal insulation for a comfortable indoor environment. However, it reduces the heat transfer of solar energy received by south wall while it reduces the indoor thermal loss [1].

To solve the problem above, this paper chooses a structure, wall implanted with heat pipes, and does some studies. Heat pipes is one of the most efficient heat transfer elements, and it can transfer a large number of heat without additional power consumption [2-5]. Do not influence the thermal resistance of walls, the heat pipes installed in walls can transfer the solar radiation outdoor to inside room and reduce the heating costs. Zhang et al. [6] and Tan et al. [7] studied the heat transfer situation of walls implanted with heat pipes. They simplified the structure to cylinders in walls, and did experiments in steady state. In fact, the real climate is transient and the heat transfer in this structure is complex. This paper put forward a new heat transfer model and calculation method, establishes a two-dimensional transient heat-transfer model of walls, and calculates the effects of the structure numerically.

2. Theoretical analysis

2.1. Wall structure

The object of study in this paper is traditional wall in northern China countryside, which is called 370 wall (355 mm bricks and 10mm cement mortar), with a 50 mm EPS as insulating layer. The structure wall implanted with pipes is installed in southern wall as shown in figure 1. The evaporator section of heat pipes is set in plastering on wall's external surface and contact with the ambient environment,

IOP Conf. Series: Earth and Environmental Science **159** (2018) 012031 doi:10.1088/1755-1315/159/1/012031

with a double vacuum glass panel on the surface. The condenser section of heat pipes is located in plastering at wall's internal surface. The adiabatic section is set in wall aslant.

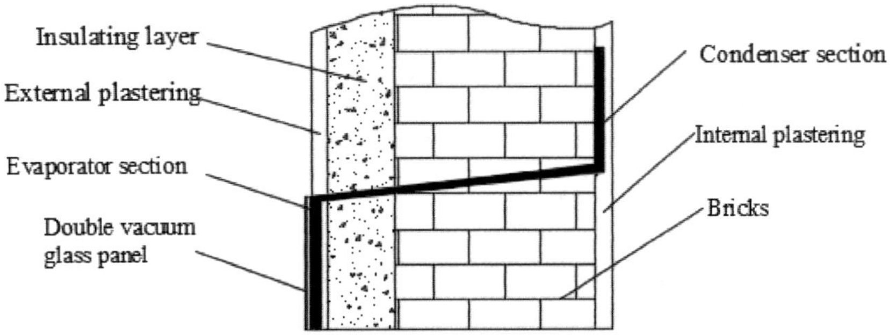

Figure 1. The structure of wall implanted with heat pipes.

2.2. Analysis of heat transfer in wall structure

In order to investigate the progress theoretically, the following assumptions are applied:

(1) Doors' and windows' effects to the wall's heat transfer are ignored.

(2) The wall's bottom surface is isolated.

When the ambient temperature is low and the heat pipes don't work, the progress in wall is a common two-dimensional transient heat transfer, and the heat pipes can be ignored.

When the ambient temperature increases and the heat pipes start to work, the progress in wall will be analyzed in next subsection.

2.3. Analysis of heat transfer in heat pipes

The heat pipes in this paper are gravity heat pipes which has no wicks, working medium flows driven by gravity and pressure. The following assumptions are applied:

(1) The heat transfer in heat pipes wastes no time.

(2) There is no heat transfer limitation.

(3) The temperature in different part of evaporator section is equal, so is the condenser section.

Heat pipes' working is stated blow: When the ambient temperature and solar radiation are low, the temperature of evaporator section is lower than the starting temperature of heat pipes, and the working medium will be heated but the heat pipes don't work. With the ambient temperature and solar radiation increasing, the temperature of evaporator section increases, the working medium boil and the heat pipes start to work. In this situation, energy is transferred from evaporator section to working medium, then from working medium to condenser section, and from condenser section to indoor environment. In this process, the temperature of condenser section is not stable and not easily get because the ambient environment temperature and solar radiation is transient, and the heat pipe is not at full capacity.

After heat pipes' starting, the energy balance equation of evaporator section is:

$$\alpha_m A_m I - A_m\left(T_e - T_{out}\right)/R_L - A_m\left(T_e - T_{in}\right)/R_{in} = \left(T_e - T_s\right)/\left(R_1 + R_2\right) \tag{1}$$

Where α_m is the absorption coefficient to the solar radiation (dimensionless); A_m is the area of evaporator section (m^2); I is the solar radiation (W·m^{-2}); T_e is the surface temperature of evaporator section (K); T_s is the saturation temperature of working medium (K); T_{in} and T_{out} are indoor temperature and ambient temperature, respectively (K); R_L represents the thermal resistance to ambient environment through the glass panel (m^2·K·W^{-1}); R_{in} represents the thermal resistance to indoor environment (m^2·K·W^{-1}); R_1 is the thermal resistance of evaporator section's shell (K·W^{-1}); R_2

is the thermal resistance of heat convection between evaporator section's shell and working medium ($K\cdot W^{-1}$).

In equation (1), the term $A_m(T_e - T_{out})/R_L$ is the heat loss to ambient environment, and R_L can be obtained by equation (2) as follows:

$$R_L = \delta_b/\lambda_b + 1/h_{out} \tag{2}$$

Where δ_b is the thickness of glass panel (m); λ_b is the thermal conductivity of glass panel ($W\cdot m^{-1}\cdot K^{-1}$); h_{out} is the convective heat transfer coefficient between ambient and glass panel ($W\cdot m^{-2}\cdot K^{-1}$).

In equation (1), the term $A_m(T_e - T_{in})/R_{in}$ is the heat loss to indoor environment through the wall, and R_{in} can be obtained by equation (3) as follows:

$$R_{in} = \sum_i \delta_i/\lambda_i + 1/h_{in} \tag{3}$$

Where δ_i is the thickness of each materials of wall (m); λ_b is the thermal conductivity of each materials of wall ($W\cdot m^{-1}\cdot K^{-1}$); h_{in} is the convective heat transfer coefficient between indoor environment and wall's internal surface ($W\cdot m^{-2}\cdot K^{-1}$).

Let the right term in equation (1) be Q_u. Q_u is the heat that is transferred to indoor environment through the heat pipes. R_1 can be obtained by

$$R_1 = 1/(2\pi l_e \lambda_{HP}) \ln d_o/d_i \tag{4}$$

Where d_o and d_i are external diameter and inner diameter of heat pipes (m); l_e is the length of evaporator section (m); λ_{HP} is the thermal conductivity of heat pipes' shell ($W\cdot m^{-1}\cdot K^{-1}$).

And R_2 can be obtained by

$$R_2 = 1/(\pi d_i l_e h_{i,e}) \tag{5}$$

Where $h_{i,e}$ is the convective heat transfer coefficient between evaporator section's shell and working medium ($W\cdot m^{-2}\cdot K^{-1}$), the equation [8] is:

$$h_{i,e} = 0.32 \rho_l^{0.65} \lambda_l^{0.3} c_{pl}^{0.7} g^{0.2} q_e^{0.4} (P_{sat}/P_a)^{0.23} / (\rho_v^{0.25} h_{f,g}^{0.4} \mu_l^{0.1}) \tag{6}$$

Where the footnote l and v represent liquid and vapour parameter of working medium; ρ is density ($kg\cdot m^{-3}$); c_p is specific heat ($kJ\cdot kg^{-1}\cdot K^{-1}$); q_e is the heat flux in evaporator section in radial direction ($W\cdot m^{-2}$); P_{sat} and P_a are saturation pressure of working medium and ambient pressure (Pa); $h_{f,g}$ is the latent heat of vaporization of working medium ($kJ\cdot kg^{-1}$); μ_l is dynamic viscosity ($Pa\cdot s$).

With these equations above, the surface temperature of evaporator section and Q_u can be acquired. Then, the condenser section is seen as volume heat source, and its value is

$$S_0 = Q_u / l_c d_o^2 \tag{7}$$

3. Mathematical model and solution

The two-dimensional transient heat-transfer equation with internal heat source is

$$\rho c \frac{\partial T}{\partial t} = \frac{\partial}{\partial x}\left(\lambda \frac{\partial T}{\partial x}\right) + \frac{\partial}{\partial y}\left(\lambda \frac{\partial T}{\partial y}\right) + S \tag{8}$$

The initial condition is the initial temperature distribution in the wall.

$$T(x,y)\Big|_{t=0} = T(x,y) \tag{9}$$

The external and internal boundary condition are third boundary condition as follows

$$\lambda \frac{\partial T(x,y,t)}{\partial x}\Big|_{x=0} = h_{out}(T(0,y,t) - T_z(t)) \tag{10}$$

$$\lambda \frac{\partial T(x,y,t)}{\partial x}\Big|_{x=L} = h_{in}(T(L,y,t) - T_{in}(t)) \tag{11}$$

Where T_z is outdoor colligate temperature calculated by equation (12).

$$T_z(t) = T_{out}(t) + \alpha I / h_{out} \tag{12}$$

Where α is the absorption coefficient to the solar radiation (dimensionless).
The above and bottom boundary condition are thermal insulating boundary conditions.
The model is solved with finite difference method (FDM). And the boundary condition is handled by replenishing equations of boundary node. At last the transient temperature distribution can be obtained.

4. Numerical analysis results

4.1. Physical parameter

The thickness of the wall is 0.435m, the height is 2.8m. And respectively each layer of the wall from outdoor to indoor is list in Table 1.

Table 1. The physical parameter of each layer of the wall

Materials	Thickness (mm)	Density ($kg \cdot m^{-3}$)	heat conductivity ($W \cdot m^{-1} \cdot K^{-1}$)	specific heat ($kJ \cdot kg^{-1} \cdot K^{-1}$))
External plastering	10	1900	0.93	1.0
thermal insulation material	50	18	0.04	2.5
Bricks	365	1400	0.8	1.1
Internal plastering	10	1900	0.93	1.0

The convective heat transfer coefficient of surface of internal wall and external wall are 8.7 $W \cdot m^{-2} \cdot K^{-1}$ and 23 $W \cdot m^{-2} \cdot K^{-1}$.
The heat pipes are copper-water heat pipes, the length of evaporator section and condenser section are both 1m, and the external diameter is 5mm while inner diameter is 3.5mm. Water's saturation temperature is 24℃ and the latent heat of vaporization is 2443$kJ \cdot kg^{-1}$.
The double vacuum glass panel's thickness is 18mm, and its thermal conductivity is 0.2 $W \cdot m^{-1} \cdot K^{-1}$.

4.2. Case calculation and analysis

This paper choose the buildings in Beijing countryside and its climate of temperature and solar radiation in 150 hours is in figure 2.
The variation tendency of inner surface temperature for 150 hours is shown in figure 3. And it can be easily seen that, the wall with heat pipes can lead to a higher inner surface temperature that is 14℃ at most and a temperature increase of 5.8℃.When using the double vacuum glass panel, the effects will be better, the inner surface temperature will be 22℃ at most and the temperature increase is 12.4℃.

The quantity of heat transferred by each heat pipe is shown in figure 4. It indicates that each heat pipe can transfer 1.6W at most.

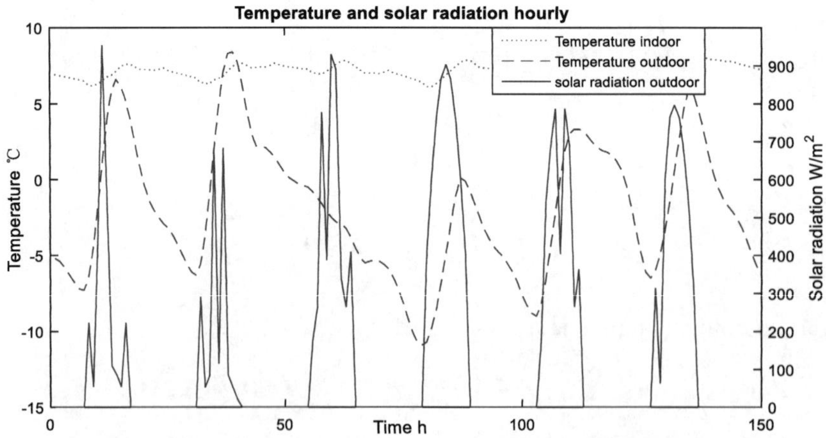

Figure 2. The indoor and outdoor temperature and solar radiation

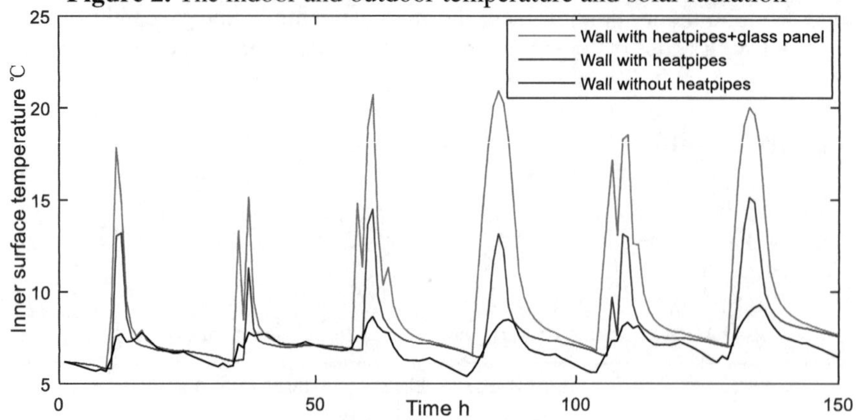

Figure 3. Inner surface temperature of the wall

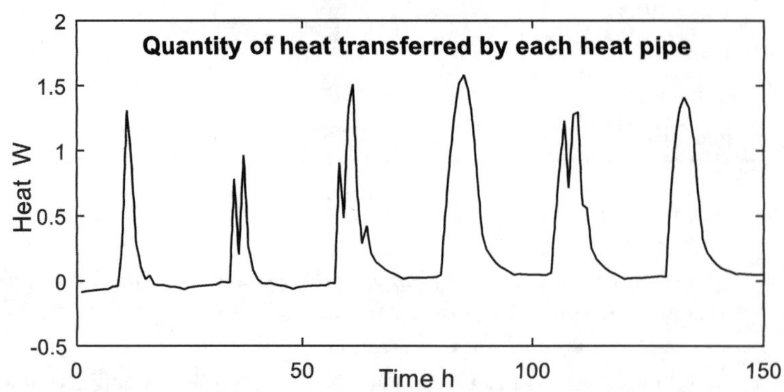

Figure 4. Quantity of heat transferred by each heat pipe.

5. Conclusion

In this paper, a type of passive solar energy utilization technology, the wall implanted with heat pipes is studied. And a two-dimensional transient heat-transfer model is developed to analyse the transient heat transfer in wall numerically. Then the model is solved in the typical climate in Beijing. The results shows that, this structure can improve the inner surface temperature for 12.4℃. As a result, it can signally improve the indoor environment. According to the results, the structure's effects is impacted by the area of heat pipes, and it will work better if combining with active solar heating method.

6. Reference

[1] H.E. Zirnhelt, R.C. Richman, The potential energy savings from residential passive solar design in Canada [J]. Energy & Buildings. 103 (2015) 224–237.

[2] Li A, Xu X, Sun Y. A study on pipe-embedded wall integrated with ground source-coupled heat exchanger for enhanced building energy efficiency in diverse climate regions [J]. Energy & Buildings. 121 (2016):139-151.

[3] Gedik E, Yılmaz M, Kurt H. Experimental investigation on the thermal performance of heat recovery system with gravity assisted heat pipe charged with R134a and R410A[J]. Applied Thermal Engineering. 99 (2016):334-342.

[4] Sun X, Zhang L, Liao S. Performance of a thermoelectric cooling system integrated with a gravity-assisted heat pipe for cooling electronics [J]. Applied Thermal Engineering. 116 (2017):433-444.

[5] Chnookaraju B, Kurmarao P S V, Sarada S N. Thermal Analysis of Gravity Effected Sintered Wick Heat Pipe [J]. Materials Today Proceedings. 2(4-5) (2015):2179-2187.

[6] Zhang Z, Sun Z, Duan C. A new type of passive solar energy utilization technology - the wall implanted with heat pipes [J]. Energy & Buildings. 84 (2014):111-116.

[7] Tan R, Zhang Z. Heat pipe structure on heat transfer and energy saving performance of the wall implanted with heat pipes during the heating season [J]. Applied Thermal Engineering. 102 (2016):633-640.

[8] D.A.Reay, P.A.Kew, Heat Pipes, fifthed.Butterworth-Heinemann, United Kingdom, 2006.

IOP Conf. Series: Earth and Environmental Science **159** (2018) 012032 doi:10.1088/1755-1315/159/1/012032

Oxidative Pyrolysis of Agricultural Residues in Gasification and Carbonization Processes

Xuan-Huynh Pham[1,2], Laurent Van De Steene[1,2], Bruno Piriou[1], Sylvain Salvador[3]

[1] CIRAD, UPR BioWooEB, F-34398 Montpellier, France.
[2] Department of Energy, University of Science and Technology of Hanoi, Vietnam Academy of Science and Technology, 18 Hoang Quoc Viet, Hanoi, Vietnam.
[3] RAPSODEE, CNRS UMR 5203, Mines-Albi, Campus Jarlard, 81013 Albi Cedex 09, France.
Email: huynh.pham_xuan@cirad.fr

Abstract. Oxidative pyrolysis, in which heat is provided by the partial oxidation, occurs in many fixed bed reactors of air staged gasification and advanced carbonization processes. In these reactors, an Oxidation Zone (OZ) propagates and separates the virgin biomass from the produced char, and self-sustains the process. To investigate the energy transfer efficiency and to characterize the OZ features, oxidative pyrolysis was performed in a 20 cm diameter fixed bed reactor with Wheat straw pellets. Temperatures and biomass bed height were measured continuously during the experiment. The OZ propagated at 0.41 cm/min and consumed about 11 % of the biomass. The role of bed compaction on the effective propagation velocity was highlighted. Char yield was measured to 28.6 % and char accounted for 39.56 % of the total energy content of the Wheat straw. Maximum temperature in the bed reached 780°C. This paper proposes a new insight into the OZ features when it propagates in a biomass fixed bed.

1. Introduction

Agricultural residues must be disposed in order to avoid contamination, pest growth, occupying large expanses of land and hindering agricultural work. In addition, agricultural residues can be exploited for recovering specific components or useful energy. Therefore, valorization of agricultural residues for energy purposes has multiple advantages: minimizing the environmental impacts, increasing economic benefits for the farmers, and reducing the reliance on fossil fuels [1], [2].

Holding a huge potential and the carbon-rich composition, agricultural residues are being considered as the major contributors of resources for energy and material productions. The global annual production of agricultural residues of the six most important crops (barley, maize, rice, soybean, sugar cane and wheat) in 227 countries and territories of the world is reported to be $3.7x10^{12}$ kg of dried matter. This residues production is proportional to crop yields [1], [3].

Thermo-chemical conversion, i.e. combustion, gasification and pyrolysis can convert agricultural residues into variety of useful energy types such as heat, electricity and materials like charcoal, syngas and bio-oil. Pyrolysis for charcoal production, carbonization, is one of the most widely used technologies. It is normally known as the thermal decomposition of organic components of biomass in an inert gas atmosphere, at moderate temperature. In such technology, the external energy source is needed to provide heat for the pyrolysis reactions. However, some current studies have shown that the partial oxidation of biomass and products, so called *oxidative pyrolysis*, is an effective way to provide

energy to the process [4], [5]. Oxidative pyrolysis allows the autothermal operation; and contributes to reduce complexity of the system and operating cost.
Here, we characterized the oxidation features in terms of propagation velocity, temperature, char yield, equivalence ratio and energy biomass conversion to pyrolysis products.

2. Material and equipment

2.1. Material

In this study, we used Wheat straw as a specific agricultural residue (Table 1). The pellet form was used to avoid the bridging problem occurring in stratified fixed bed reactors [6]–[8] and to ensure the homogeneity in the biomass bed properties.

Table 1. Wheat straw pellets properties

Density (kg/m^3)	Moisture (wt%)	Proximate analysis (wt%, d.b.)			Ultimate analysis (wt%, d.a.f.)				LHV, (MJ/kg)
		Volatile mater	Ash	Fixed carbon	C	H	N	O	
549 ±1.42	8.75 ±0.05	78.75 ±1.89	8.17 ±0.09	13.08 ±1.89	58.91 ±0.18	5.07 ±0.04	1.10 ±0.10	34.92 ±0.24	17.45 ±0.12

2.2. Equipment description

Oxidative pyrolysis was studied in a fixed bed reactor (Figure 1). It consists of a 310 type refractory steel tube (20cm I.D.; 160 cm L) surrounded by a 30 cm thick refractory wool insulation. Thermocouples are installed every 10 cm inside the bed, allowing continuous measurement of temperature. In addition, a laser beam device is placed above the biomass bed to continuously measure the bed height during experiments allowing the calculation of the bed compaction.

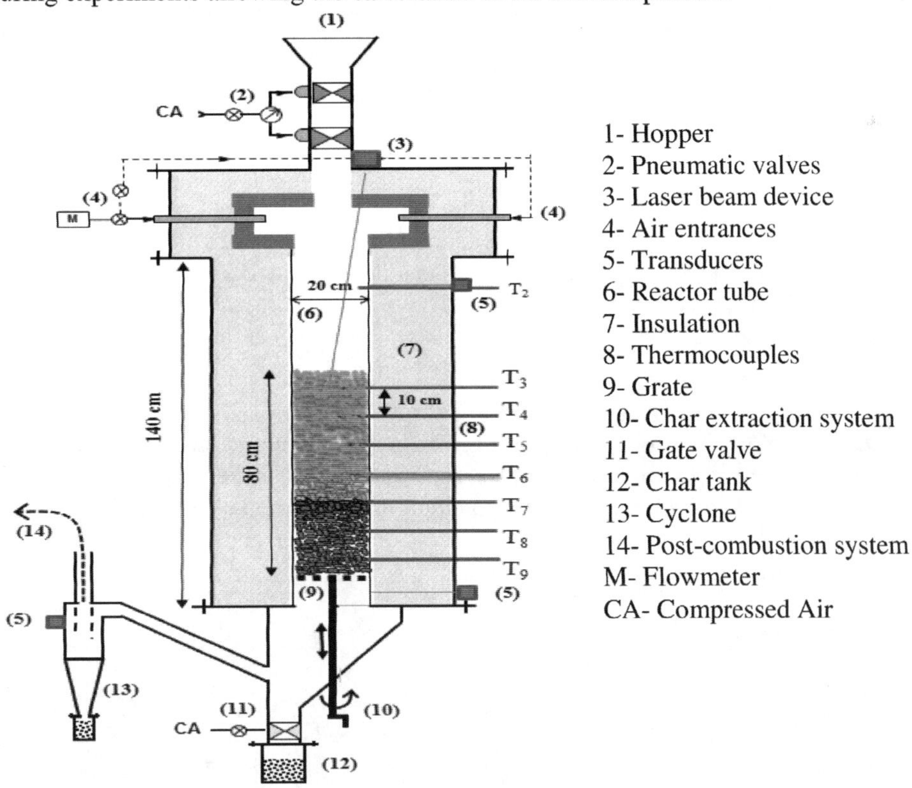

Figure 1. The fixed bed reactor

IOP Conf. Series: Earth and Environmental Science **159** (2018) 012032 doi:10.1088/1755-1315/159/1/012032

The experimental procedure consists in igniting the OZ at the bottom of a 70 cm high biomass bed and following its upward propagation (about 10 cm of charcoal was used for the ignition). For this purpose, air flow is supplied at the top of the reactor and passes down through the biomass bed. As an example, evolution of temperatures at the different thermocouples presented in figure 2a shows the propagation of the OZ through a Wheat straw pellets bed while figure 2b shows the bed height reduction during the test.

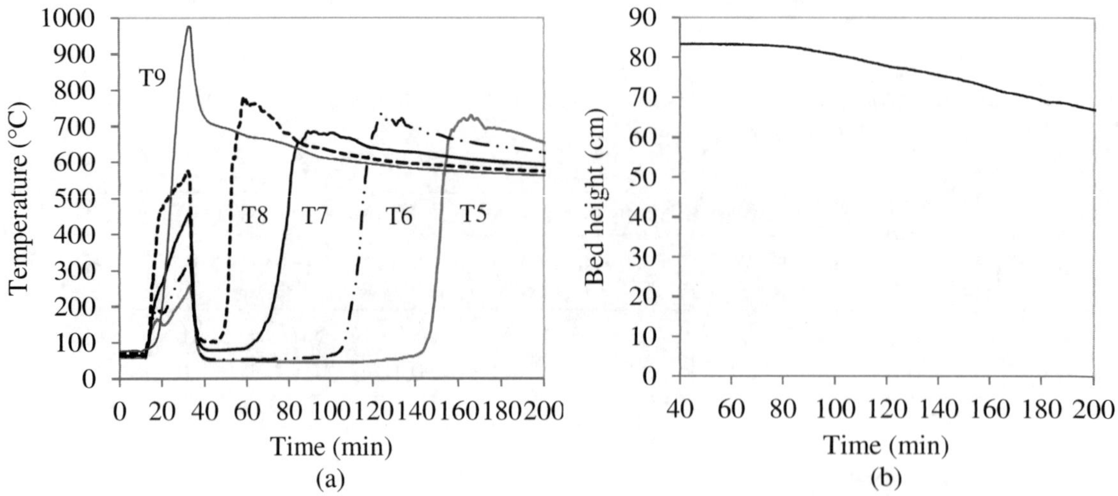

Figure 2. Typical experimental results with an air flux of 0.022 kg/m^{2}s:

(a) Temperature profile and (b) Bed height reduction

In Figure 2a, first the temperature in T9 reach 950-1000°C after about 30 mins following the ignition. Then, the air flux was reduced to the studied value of 0.022 kg/m^{2}s involving a decrease of the temperature tends to about 600 °C up to the end of the experiment. Then, the oxidation zone propagates upward from T9 position to the top of the bed, in T5 position. The passage is observed by the increase of each temperature measured by thermocouples located all along the biomass/char bed each 10 cm. Thus, the temperature profiles versus time can be used for determination of the OZ propagation velocity in the biomass bed.

The bed height remains constant up to 80 mins, then, continuously reduces up to the end of the test (Figure 2b). This fact reveals that the oxidation zone propagates in a stationary regime only after reaching T7. Thus, in further discussion, we focus on the zone behavior from T7 to T5, i.e. 25cm to 45 cm respectively from the bottom of the bed. The bed height measurement allows the determination of the bed compaction velocity, and thus the biomass consumption rate.

3. Method for characterizing the oxidation zone features

From these temperature and bed height measurements, a new method has been proposed to calculate the upward propagation velocity of the oxidation zone. In this reverse smoldering configuration, the OZ propagates upward while the bed compaction makes biomass going down into the oxidation zone. The bed compaction velocity needs to be taken into account when determining the biomass consumption rate (Figure 3).

IOP Conf. Series: Earth and Environmental Science **159** (2018) 012032 doi:10.1088/1755-1315/159/1/012032

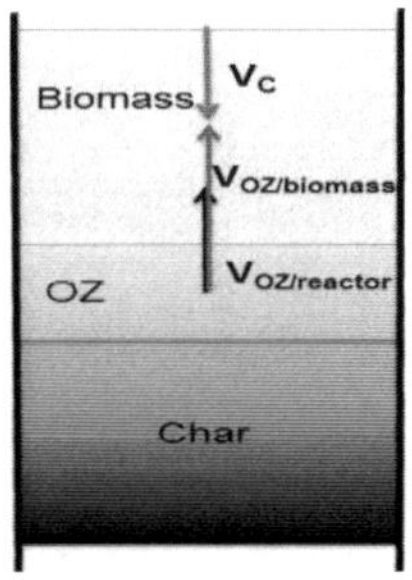

Figure 3. Simplified sketch illustrating OZ velocities and compaction velocity in fixed bed

We define $V_{OZ/reactor}$ (cm/min) as the apparent upward velocity. It is the OZ velocity related to the reactor, the one we observe from outside of the reactor. It is measured by the temperature evolution inside the reactor as described in next section. $V_{OZ/biomass}$ (cm/min) is defined as the velocity of the oxidation zone related to biomass, which is an effective velocity of propagation of the oxidation zone in the biomass. V_C (cm/min) is defined as the compaction velocity; it is measured by the laser during experiments and allows the calculation of Voz/biomass by following equation:

$$V_{OZ/biomass} = V_{OZ/reactor} + V_C \tag{1}$$

Here, we assume that the bed compaction only occurs within the oxidation zone. In other words, there is no compaction in the bed of biomass and in the bed of produced char.

3.1. Measurement of $V_{OZ/reactor}$

The position of the oxidation zone at the elevation of a given thermocouple Ti is indicated when dTi/dt is maximum. We define t^i_{OZ} as the time coresponding to the passage of the OZ at that elevation (Figure 4).

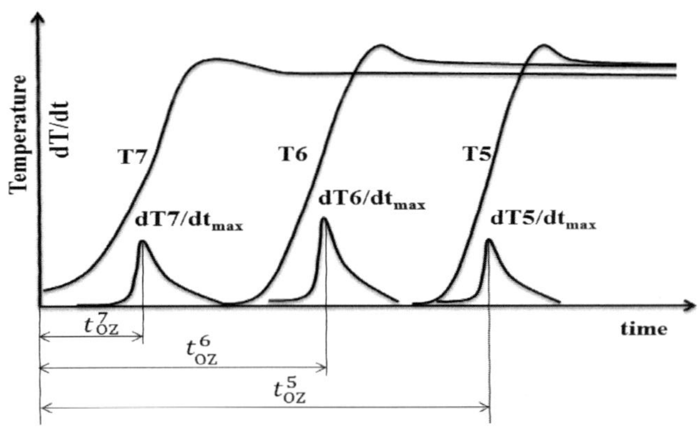

Figure 4. OZ passage time determination.

The position of the oxidation zone in the reactor versus time can be plotted (Figure 5). The upward propagation velocity of the OZ can be determined from the derivative of the position function. For this purpose, a fitting of the experimental data was performed (Figure 6).

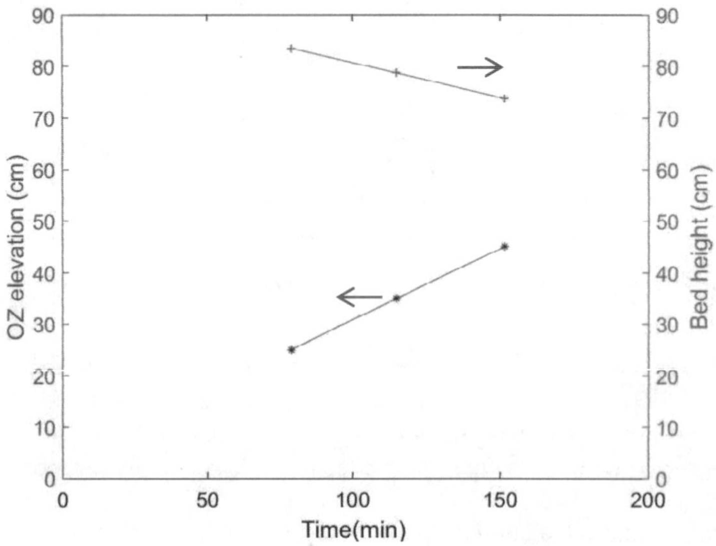

Figure 5. The evolution of the OZ propagation rate along a Wheat straw bed (air flux: 0.022 kg/m^{2}s).

As shown in Figure 5, the oxidation zone reached T7 position (Z=25cm) at around 80 mins, then propagated upward and reached T6 (Z=35cm) at 118 mins and T5 (Z=45cm) at 156 mins. The total bed height reduced from 82 cm when OZ arrived at T7 (25 cm) to around 74 cm when OZ arrived at T5 (45 cm).

3.2. Measurement of the biomass consumption and char production

The hypothesis that the compaction happens only within the oxidation zone allows the estimation of the apparent OZ char velocity ($V_{oz/char}$):

$$V_{OZ/char} = V_{OZ/reactor} \tag{2}$$

The biomass consumption rate then is calculated as following equation:

$$\dot{m}_{biomass} = V_{OZ/biomass} * A_{reactor} * \rho_{biomass} \tag{3}$$

Char production rate:

$$\dot{m}_{char} = V_{OZ/reactor} * A_{reactor} * \rho_{char} \tag{4}$$

ρ_{char} is the density of char which was measured about 236 kg/m^3.

3.3. Equivalence ratio (ER)

The ratio of air to biomass is written:

$$Air\!: biomass = \frac{Air\ mass\ flux}{Biomass\ consumption\ rate} \tag{5}$$

Thus, the ER is classically defined as follow:

$$ER = \frac{Air\!: biomass}{Air\!: biomass_{stoichiometry}} \tag{6}$$

Note that the Air: biomass stoichiometry is 5.38 calculated from the ultimate analysis (Table 1) of Wheat straw pellets.

4. Results and discussions

4.1. Propagation velocities of the oxidation zone

IOP Conf. Series: Earth and Environmental Science **159** (2018) 012032 doi:10.1088/1755-1315/159/1/012032

Propagation velocities include the apparent velocity ($V_{OZ/reactor}$) and the effective velocity ($V_{OZ/biomass}$). In figure 6, the apparent velocity of the oxidation zone was shown about 0.28 cm/min while the bed compaction velocity was found to be 0.13 cm/min. As a consequence, the effective velocity of the oxidation zone in the biomass bed was calculated to 0.41 cm/min. Bed compaction and propagation velocities were almost constant during the test period, which confirm that a stationary regime was reached in the zone from 25 cm to 45 cm in the reactor (from T7 and T5).

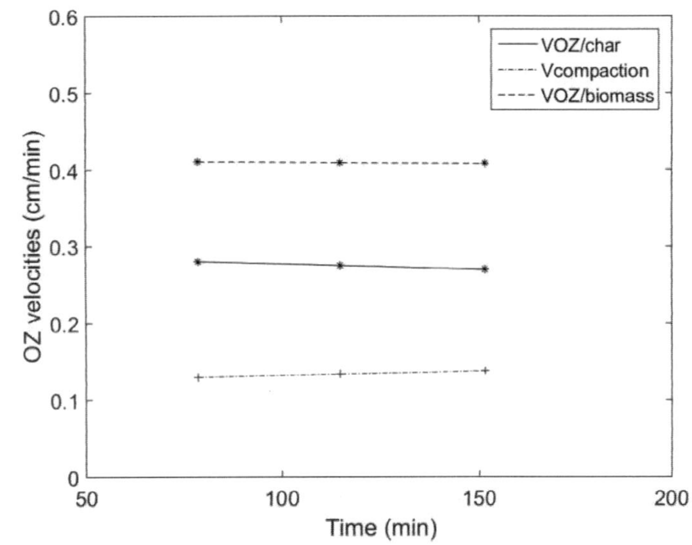

Figure 6. OZ propagation velocities with an air flux of 0.022 kg/m^{2}s.

These results show that in downdraft fixed bed configuration, the bed compaction accelerated the propagation of the oxidation zone. However, few studies [9], [10] have taken the bed compaction into account when calculating the propagation velocity of the oxidation zone.

4.2. Temperature

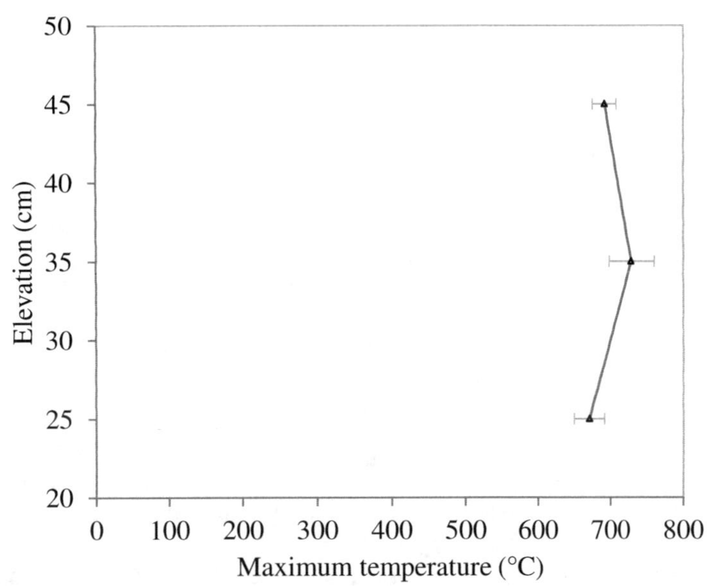

Figure 7. Maximum temperature profile with an air flux of 0.022 kg/m^{2}s.

The maximum temperature (Figure 7) in the bed of Wheat straw pellets oscillated between 650°C and 780°C and no hot spot temperature along the bed was observed. Temperature is a major influencing factor regarding pyrolysis products, yields and composition. It is reported that pyrolysis temperature is increased, the char yield decreased [11], [12].

4.3. Biomass consumption rate, char production rate and equivalence ratio

Propagation velocities allow the calculation of the biomass consumption rate and the char production rate (Figure 8). In our operating conditions, biomass consumption rate was calculated to 4.2 kg/h and the char production rate to 1.2 kg/h. The char yield defined as the ratio of char production rate over biomass consumption rate was 28.6 %. This result shows that char yield in the oxidative pyrolysis was in line with the typical slow pyrolysis processes [13].

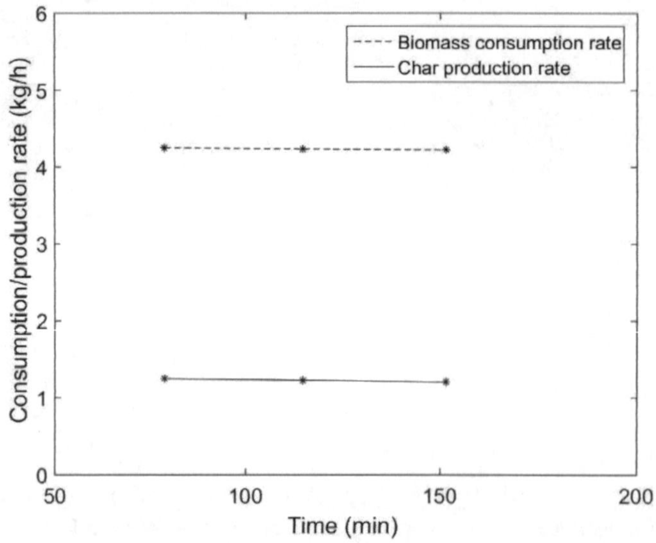

Figure 8. Biomass consumption rate and char production rate with an air flux of 0.022 kg/m^{2}s.

The energy transferred from biomass to the char was calculated from the measurements of char yield and the lower heating value (LHV) of charcoal and biomass.

The LHV was measured to 24.14 MJ/kg and 17.45 MJ/kg for charcoal and biomass, respectively.

$$E_E = \frac{\dot{m}_{char} * LHV_{char}}{\dot{m}_{biomass} * LHV_{biomass}} = 39.56\ \% \qquad (7)$$

Energy transfer to the char was about 39.56 %. The rest of energy in biomass was transferred to the other products such as permanent gases and liquid fuels.

Finally, the ER of our autothermal pyrolysis process was calculated to be 0.11. It means that the oxidation zone consumed 11 % biomass to provide heat for self-sustain the process.

5. Conclusions

The aim of this work was to investigate the oxidative autothermal pyrolysis in fixed beds of agricultural residues pellets. The OZ features in terms of propagation velocity, temperature, char yield and equivalence ratio were discussed. Regarding the air mass flux of 0.022 kg/m^{2}s, the oxidation zone propagates at an effective velocity of 0.41 cm/min in the biomass bed. The maximum temperature oscillated between 650°C and 780°C. The OZ consumes 11 % of the Wheat straw to provide energy for the autothermal process. Char yield and energy transferred from biomass to the char were measured to 28.6 % and 39.56 %, respectively.

These results suggest the direction for the valorisation agricultural residues for energy purpose by oxidative pyrolysis. The future work will focus on the characterization the oxidation zone features in fixed bed of various biomass types.

References

[1] Bentsen N S, Felby C and Thorsen B J 2014 Agricultural residue production and potentials for energy and materials services *Prog. Energy Combust. Sci.* **40** 59–73

[2] Malico I, Carrajola J, Gomes C P and Lima J C 2016 Biomass residues for energy production and habitat preservation. Case study in a montado area in Southwestern Europe *J. Clean. Prod.* **112** 3676–83

[3] Colantoni A, Evic N, Lord R, Retschitzegger S, Proto A R, Gallucci F and Monarca D 2016 Characterization of biochars produced from pyrolysis of pelletized agricultural residues *Renew. Sustain. Energy Rev.* **64** 187–94

[4] Anca-Couce A, Zobel N, Berger A and Behrendt F 2012 Smouldering of pine wood: Kinetics and reaction heats *Combust. Flame* **159** 1708–19

[5] Daouk E, Van de Steene L, Paviet F, Martin E, Valette J and Salvador S 2017 Oxidative pyrolysis of wood chips and of wood pellets in a downdraft continuous fixed bed reactor *Fuel* **196** 408–18

[6] Ummadisingu A, Sheth P and Babu B V 2010 Experimental Studies on Gasification of Pine Wood Shavings in a Downdraft Biomass Gasifier **31** 2–7

[7] Paulauskas R, Striūgas N and Misiulis E 2014 Experimental and theoretical investigation of wood pellet shrinkage during pyrolysis 1–11

[8] Guo F, Dong Y, Dong L and Guo C 2014 Effect of design and operating parameters on the gasification process of biomass in a downdraft fixed bed: An experimental study *Int. J. Hydrogen Energy* **39** 5625–33

[9] Pillier L, Idir M, Molet J, Matynia A and De Persis S 2015 Experimental study and modelling of NOx formation in high pressure counter-flow premixed CH4/air flames *Fuel* **150** 394–407

[10] Mahapatra S, Kumar S and Dasappa S 2016 Gasification of wood particles in a co-current packed bed: Experiments and model analysis *Fuel Process. Technol.* **145** 76–89

[11] Mahapatra S and Dasappa S 2014 Experiments and analysis of propagation front under gasification regimes in a packed bed *Fuel Process. Technol.* **121** 83–90

[12] Kumar M, Gupta R C and Sharma T 1992 Effects of carbonisation conditions on the yield and chemical composition of Acacia and Eucalyptus wood chars *Biomass and Bioenergy* **3** 411–7

[13] Katyal S, Thambimuthu K and Valix M 2003 Carbonisation of bagasse in a fixed bed reactor: Influence of process variables on char yield and characteristics *Renew. Energy* **28** 713–25

Acknowledgements

Authors would like to thank Mr. Jérémy Valette for his technical assistance and Dr. Jean-Michel Commandré for his advices and material supports for this study.
This work is funded by the Vietnamese International Education Development (VIED) and by the agreement with the CIRAD-France.

IOP Conf. Series: Earth and Environmental Science **159** (2018) 012033 doi:10.1088/1755-1315/159/1/012033

Rice Husk Gasification: from Industry to Laboratory

Hong Nam Nguyen[1], Laurent Van De Steene[2,3], Thi Thuy Hang Le[4], Duc Dung Le[5] and Minh Ha-Duong[6]

1 Energy Department, University of Science and Technology of Hanoi, Hanoi, Vietnam
2 CIRAD, UPR BioWooEB, F-34398 Montpellier, France
3 BioWooEB, Univ Montpellier, CIRAD, Montpellier, France
4 Institute of Energy Science, VAST, Vietnam
5 Hanoi University of Science and Technology (HUST), Hanoi, Vietnam
6 The international research center on environment and development, CNRS, France
Email: nguyen-hong.nam@usth.edu.vn

Abstract. Rice husk gasification (RHG) has been increasingly paid attention in rice-producing countries. Nevertheless, information related to this technology remains small and fragmented. In this paper, the status of RHG has been summarized, highlighting domestic and industrial applications, as well as a scientific review. Moreover, an experimental parametric study was performed to measure: a) the influence of temperature and heating rate on RH pyrolysis, and b) the role of temperature, partial pressure, and heating rate on RHG, under H_2O or CO_2 atmosphere. Regarding pyrolysis, an increase in the final pyrolysis temperature lowers the RH char yield but does not have much effect on the char morphology. Regarding gasification, an increase in the reaction temperature from 900 to 1000°C accelerates about 4.8 and 7 times the char conversion rate, under 0.2 atm of CO_2 or H_2O, respectively. The conversion rate decreases about 1.8 times when CO_2 partial pressure decreases from 1 to 0.2 atm, and about 1.6 times when H_2O partial pressure decreases from 0.4 to 0.1 atm. At 900°C and 0.2 atm, steam gasification was about 3.5 times faster than Boudouard reaction.

1. Introduction

Rice has been grown in more than 75 countries with a world paddy production of 747 million tons in 2016 [1]. This would result in approximately 149 million tons of rice husk (RH), considering a ratio of 0.2 ton of RH for each ton of paddy. While RH has been traditionally used in low-value applications, its potential as a feedstock to generate heat and electricity is attracting increasing attention.

Biomass gasification is a thermochemical conversion process that converts biomass into a CO and H_2 rich-gas called syngas. This latter can be used to produce heat, electricity or fuel transportation. Some different generic types of gasifiers have been developed and commercialized: fixed bed, fluidized, entrained flow and multi-stage gasifier [2]. Gasification involves a series of processes, namely drying, pyrolysis, volatiles oxidation/cracking, and char gasification. Char gasification is particularly important as it controls the syngas production, the complete carbon conversion and thus the efficiency of the whole process. The main reactions involved during char gasification are the following:

$$C + H_2O \rightarrow CO + H_2 \text{ (Steam gasification)} \quad (1)$$
$$C + CO_2 \rightarrow 2CO \text{ (Boudouard reaction)} \quad (2)$$

Currently, wood is principally used for gasification. However, regarding socio-economic issues, diversifying the feedstock is becoming the biggest challenge for this technology. RH gasifiers are increasing rapidly in number in rice-producing countries [2], [3]. Nevertheless, information about current applications and development trends have been small, fragmented and difficult to check. In this context, the first objective of this study was to establish an updated state of the art of rice husk gasification (RHG) to provide a new insight on both industrial and scientific issues. Concerning scientific studies, most of the authors focus on a pilot scale with the aim of evaluating and validating their own designs. Meanwhile, fundamental studies of RHG exist but in a very low number [4]–[7], highlighting a scientific need for further investigation in that field. Therefore, the second objective of this work was to conduct a fundamental parametric study on RHG to provide relevant data on the features of RH pyrolysis, steam gasification, and Boudouard reaction. Results and data produced could be useful to set up industrial or academic codes, necessary for the conception of new RH gasifiers.

2. State of the art of RHG

2.1. Existing technologies and applications

2.1.1. RH gasifier stoves for domestic applications

RH gasifier stoves (Figure 1) have been developed for decades in developing countries for cooking. Most of these gasifiers are based on "top-lit updraft" principle. Firstly, the fuel burns because of primary air supplied to the lower side of the stove. Secondly, the gases released are burned in a flame on the top of the stove, where the secondary air is injected. This design is known to generate less harmful emissions than traditional stoves. Many local manufacturers are positioned in this market with their own designs. Thus, the product efficiency varies greatly from one to another. RH gasifier stoves do not differ much from wood gasifier stoves. Considering the rice husk, small design changes are required to take into account of the specific features of this feedstock, such as high ash content and low density.

Regarding on-going development, several designs have been tested to combine cooking and biochar production [8]. The upscaling of gasifier stoves for more heat demanding applications (collective cooking or small industry) is an interesting development path for this technology. In these cases, the continuous operation of the gasifier is the main challenging issue.

2.1.2. RH gasifiers for heat production

Regarding existing commercial gasifiers, downdraft fixed-bed is the most widespread technology for heat application (Figure 2). The main reasons lie on its simplicity in design and affordability in price. These gasifiers are most common in rice mills to use RH generated by the milling process. It is noteworthy that rice mills can be self-sufficient in terms of energy by using RHG [3]. Heat produced from RH gasifiers can be used for drying, bricks/ceramic cooking, or any processes requiring steam.

Figure 1. Various designs of RH gasifier stoves in Vietnam (Picture taken by author)

Figure 2: A RH gasifier for heat generation

(Source: https://hayvip.com)

Figure 3: RH gasifier for power generation in Cambodia (Picture taken by author)

The technical constraints of these gasifiers are minor. When a RH gasifier is used for direct heat, there is no need to cool down the syngas, neither does the tar have to be removed. Therefore, these systems do not require expensive gas cleaning equipment. Thermal power of these gasifiers is generally small or medium to match with the local resource availability, and/or the heat required by end-users.

Considering the price of the product and the needs of users, the gasification technology for heat production is quite mature. There are possibilities to improve the efficiency of the system or the quality of the gas, but the main challenge concerns the improvement of the lifespan of the gasifier.

2.1.3. RH gasifiers for power generation

India, China, Myanmar, and Cambodia are the leading countries with dozens of installed gasifiers for power generation (Figure 3). Some other countries start to import and develop the technology, such as Indonesia, Bangladesh, Nepal, and the Philippines, but the number of installed units remains weak. A few power plants using RHG have also been identified in India, Thailand, Cambodia, and Myanmar.

Many international manufacturers produce biomass gasifiers for power generation, but as far as we know, only a few of them provide gasifiers that can really operate with RH. Table 1 summarizes the list of main international manufacturers of RH gasifiers by countries.

Table 1. Main international manufacturers by countries

Country	Manufacturer	Technology
China	Qingdao Kexin New Energy Technology Co. Ltd.	Downdraft fluidized bed
	Weifang Haitai Power Machinery Co., Ltd.	Downdraft fluidized bed
	Hefei Debo Bioenergy Science & Technology Co., Ltd.	Downdraft fixed-bed
	Zhengzhou Haoqili Machinery Co., Ltd.	Downdraft fluidized bed
	Shangqiu Haiqui Machinery Equipment Co., Ltd.	Downdraft fixed-bed
	Wuxi Teneng Power Machinery Co., Ltd.	Circulating fluidized bed
India	Agro Power Gasification Plant Pvt. Ltd.	Downdraft fixed-bed
	S. S. Fabrication	Downdraft fixed-bed
	Urja Gasifiers Pvt. Ltd.	Downdraft fixed-bed
	The Energy and Resources Institute (TERI)	Downdraft fixed-bed
	Ankur Scientific Energy Technologies Private Limited	Downdraft fixed-bed
Japan	Yanmar Co, Ltd.	Downdraft fixed-bed
USA	PRM Energy Systems, Inc.	Updraft fixed-bed
Singapore	Trillion Gasifier	Downdraft fixed-bed

RH gasifiers for power generation are generally of downdraft fixed-bed type. Gasifiers are coupled either with a gas engine or a dual-fuel engine. Such this coupling is very demanding regarding gas quality, as no tars and particles can be injected into the engine. Cyclones and scrubbers are the most common gas cleaning equipment to remove tars and particles. This solution is simple and cheap, but produces wastewater in large quantities, causing serious environmental issues [3].

Development orientations focus on introducing technologies other than fixed-bed to the market, in the objective to provide a better gas quality and to improve the size of the system.

2.2. Pilot and laboratory studies

The scientific literature relates to works in RHG both at the pilot and laboratory scale. Regarding studies on a pilot scale, the authors mainly performed experiments to evaluate the performance of their gasifier when using RH [9]–[13]. In most cases, the technologies remain classical (fixed or fluidized bed type) and none clearly highlights a specific innovation. Information about innovative designs is probably kept confidential for commercialization. Therefore, studies at the laboratory scale have made a greater scientific contribution by providing knowledge and deeper understanding regarding RHG.

Zhai et al. [4] conducted pyrolysis experiments of RH in a fixed-bed gasifier and showed that an increase in final pyrolysis temperature from 600 to 1400°C influences significantly porosity and surface area: the latter presenting a maximum value at 1000°C. Moreover, a char produced at high temperature is less reactive to H_2O during gasification. RH char gasification under H_2O and CO_2 atmosphere has been investigated by Bhat et al. [7] using either RH particles or fine powder. However, the results regarding reaction kinetics remain questionable when compared with the literature [14]. The effect of operating conditions on Boudouard reaction has been studied by few authors [5], [6]. Results showed that temperature strongly influences gasification kinetics. An increase from 750 to 850°C leads to a reduction of 9 times in the reaction rate [5], and 7.5 times when temperature increases from 850 to 950°C [6]. The partial pressure of reacting gas is shown to have a slight influence on conversion. The reaction rate increases 1.2 times when CO_2 partial pressure increases from 50 to 100% [5], and 1.5 times from 25 to 100% [6]. Meanwhile, low pyrolysis heating rate (HR) varying in the range of 5 – 20°C.min^{-1} has no effect on RH char gasification kinetics [5]. Regarding kinetic model, the modified random pore model is shown to be suitable to describe the RH char gasification with CO_2. Given the low number of articles found in the literature, there is actually no complete parametric study for RHG under CO_2 and H_2O. Meanwhile, results of existing studies focus mainly on a microscopic scale, making it difficult to apply to real particles behaviour in a gasifier. For this reason, we carried out an experimental parametric study of RHG under H_2O and CO_2 atmosphere at particle scale.

3. Experimental study

3.1. RH feedstock
The RH used for experiments was the popular "Te do", taken from the Red River Delta, Vietnam. Its composition is given in Table 2.

Table 2. Composition of "Te do" RH

M%	VM_{db}%	A_{db}%	FC_{db}%	HHV (MJ/kg)
10.0	64.6	16.4	19.0	15.1

M: Moisture (as received), VM: Volatile matter, A: Ash, FC: Fixed carbon, HHV: higher heating value, db: dry basis.

3.2. RH char preparation
For the purpose of the study, five different chars were produced by varying the HR (5, 20 and 1800°C.min^{-1}) and the final temperature T_{final} (600, 750 and 900°C). About 400 mg of RH were placed in an air-tight refractory steel box of 25 cm diameter and 20 cm height. The box was swept with N_2 to avoid oxidation and placed in the muffle furnace. Note that this system allowed us to produce a big quantity of char that will be used for the char gasification study. The T_{final} and HR of the furnace were controlled. For the char produced at high HR, i.e. 1800°C.min^{-1}, we used the macro-thermogravimetric reactor described below. Whatever the pyrolysis reactor used, the T_{final} was maintained for one hour.

3.3. Experimental setup
A new macro-thermogravimetric reactor was set up to experiment char gasification kinetics (Figure 4).

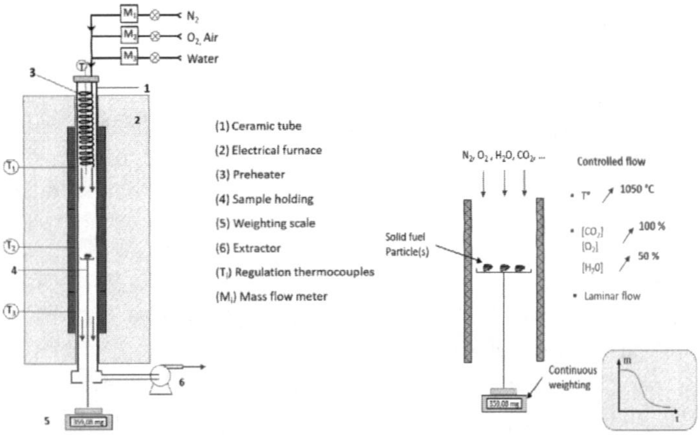

Figure 4. Macro-thermogravimetric reactor and its principle

It consists of a 111 cm long, 7.5 cm i.d. ceramic tube (1), inserted in an electrical furnace (2). The heating is ensured by three independently controlled heating zones, allowing a uniform temperature along the reactor. The reaction atmosphere is generated by a mixture of N_2 and a reacting gas (H_2O, O_2, air or CO_2) in selected proportions. Each gas is controlled by mass flowmeters (M_1, M_2, and M_3). The gas mixture is preheated in a 2 m long coiled tube (3) located in the upper part of the reactor.
The experiment consists of gasifying RH char particles at atmospheric pressure, in a well-controlled atmosphere in terms of temperature and partial pressure of reacting gas. Sample mass was measured and recorded continuously. For each experiment, the reactor was first heated to the desired operating temperature. Then the sample holding (4) containing char particles was lifted from the bottom of the reactor to the desired position and maintained under N_2 to release all remaining VM. When a constant mass was achieved, gas flows were established. As gasification took place, the mass of the char progressively decreased until constant mass - that corresponded to the ash content - was obtained.
The conversion X during gasification was calculated as follows:

$$X = \frac{m_i - m}{m_i - m_{ash}} \quad (3)$$

where m_i, m, and m_{ash} are respectively the initial mass, the mass at time t and the mass of ash.
The gas flow rate of 5Nl.min^{-1} and the samples of 100 mg were chosen for all experiments. We previously checked that these values allowed gasification to take place in a chemical regime. A deviation of less than 10% in measurements was observed, which is acceptable considering the HR heterogeneity and equipment accuracy [15]. All the data below are an average of at least 2 repeatability experiments.

3.4. Pyrolysis of RH

3.4.1. Influence of temperature

Pyrolysis experiments were performed (Table 3) considering 3 T_{final} of 600, 750 and 900°C and the same HR of 20°C.min^{-1} for this study. RH char yield slightly decreases with an increase in T_{final}, from 39.5% at 600°C to 36.6% at 900°C. As a consequence, the ash content in the char is increasing accordingly.

Table 3. Influence of final temperature Tfinal (HR of 20°C.min-1)

T_{final} (°C)	VM_{db}%	A_{db}%	FC_{db}%	HHV (MJ/kg)	Yield (%)	Energy transferred in char (%)
600	7.3	37.8	54.9	19.0	39.5	49.7
750	5.5	38.3	56.2	19.2	38.1	48.4
900	2.9	39.9	57.2	19.3	36.6	46.8

T_{final} has an impact on VM in the char. VM was measured to 7.3% and 2.9% for T_{final} of 600 and 900°C, respectively. In contrarily, T_{final} doesn't affect much higher heating value (HHV) of the char. Besides, from the measurement of char yield and HHV, the ratio of energy transferred from the RH to the char have been calculated. It can be seen that about 47% of energy was transferred in the char after pyrolysis at 900°C. This value slightly increases at lower T_{final} due to an increase in the char yield.

3.4.2. Influence of heating rate

Pyrolysis experiments were performed at HR of 5, 20 and 1800°C.min^{-1}. A small change in HR did not impact char yield. In contrarily, when HR is highly increased, the same effect as that of T_{final} was observed. Char yield, HHV, and energy transferred in the char significantly decreases (Table 4).

Table 4. Influence of heating rate (T_{final}: 900°C)

HR (°C.min^{-1})	VM_{db}%	A_{db}%	FC_{db}%	HHV (MJ/kg)	Yield (%)	Energy transferred in char (%)
5	2.6	39.2	58.2	19.6	36.9	47.9
20	2.9	39.9	57.2	19.5	36.6	47.2
1800	1.9	46.5	51.6	16.3	29.7	32.0

3.4.3. Morphology of RH and RH chars

The SEM of the surface of RH particle and its chars (Figure 5) showed a high degree of roughness. A large number of serrated peaks are placed in lines and parallel to each other. Some pointed thorns are found on the surface and these are easy to be broken by thermal effects. There is no obvious difference between chars produced at different operating conditions. However, the surface of RH char produced at 900°C, 1800°C.min^{-1} seems to be more damaged, expressed through the presence of some cracks and a few small pores. This may presumably due to the sudden change of temperature during pyrolysis.

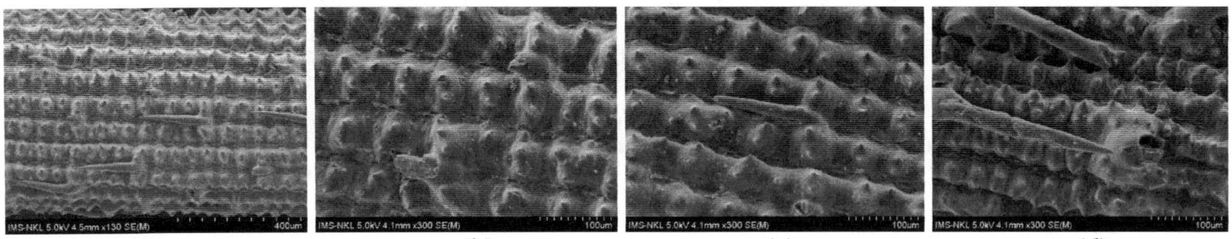

(a) (b) (c) (d)

Figure 5. SEM images of surface of RH and its chars. (a): RH, (b): 600°C, 20.min^{-1} char, (c): 900°C, 20.min^{-1} char and (d): 900°C, 1800°C.min^{-1} char

3.5. Gasification of RH char particle

3.5.1. Influence of reaction temperature

The influence of temperature on conversion during char-H_2O and char-CO_2 gasification was investigated by varying temperature from 850 to 1000°C (Figure 6) with a partial pressure of 0.2 atm in N_2.

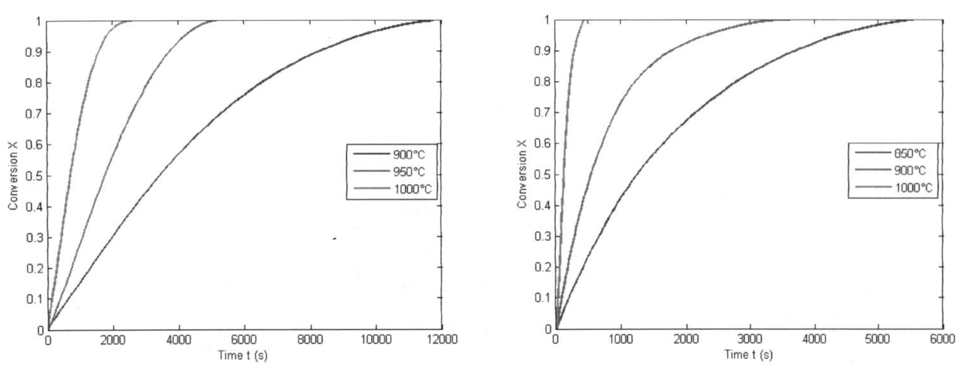

(a) (b)

Figure 6. Influence of reaction temperature on char conversion (a) under CO_2 and (b) under H_2O atmosphere (Partial pressure of the reacting gas: 0.2 atm)

Under CO_2 atmosphere, gasification was completed after 12000, 5000 and 2500s at the temperature of 900, 950 and 1000°C, respectively. Thus, a 100°C increase in gasification temperature increases 4.8 times the reactivity of CO_2 gasification. Under H_2O atmosphere, gasification was completed after 5500, 3500 and 500s at a temperature of 850, 900 and 1000°C, respectively. Thus, a 100°C increase in gasification temperature increases 7 times the reactivity of H_2O gasification. The influence of temperature on CO_2 or steam gasification of RH char is considerably different for the one of wood char. Indeed, a ratio of only 2 to 3 was reported for wood char gasification when increasing the temperature from 900 to 1000°C under the same operating conditions of CO_2 or H_2O [16, 17].

3.5.2. Influence of partial pressure

The influence of partial pressure was studied by varying CO_2 one from 0.2 to 1 atm, and H_2O one from 0.1 to 0.4 at 900°C (Figure 7). An increase in the partial pressure increases the conversion rate of RH char gasification. Under CO_2 atmosphere, gasification was completed after 12000, 10500 and 6500s for the partial pressure of 0.2, 0.4 and 1 atm, respectively. The gasification rate was thus about times faster with 1 atm of CO_2 than with 0.2 atm of CO_2. Under H_2O atmosphere, gasification was completed after 4500, 3500 and 2800s for the partial pressure of 0.1, 0.2 and 0.4 atm, respectively. Gasification was thus about 1.6 times faster with 0.4 atm of H_2O than with 0.1 atm of H_2O.

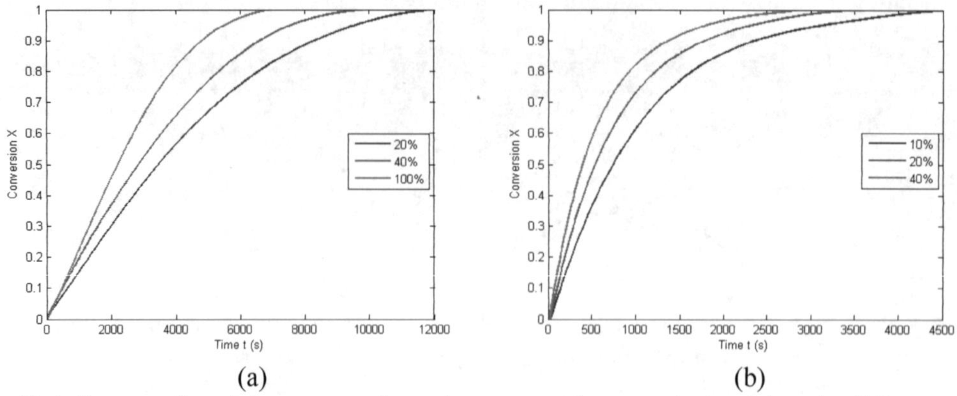

Figure 7. Influence of partial pressure of reacting gas on char conversion (a) under CO_2 atmosphere and (b) under H_2O atmosphere (Temperature of reacting gas: 900°C)

Regarding gasifier operating conditions, the impact of partial pressure on kinetics is much lower than that of temperature. Comparing the role of reacting gas nature at 900°C, steam gasification of RH char was about 3.5 times faster than under CO_2 atmosphere. These results are similar compared to wood char where gasification with CO_2 is 2 to 5 times slower than with H_2O under the same conditions [17], [18].

3.5.3. Influence of heating rate

Pyrolysis HR influences the gasification conversion rate of wood char significantly [15]. No effect on RH char conversion has been reported with chars produced at low HR of 5, 10 and 20°C.min^{-1} [5], but this result cannot be extrapolated to high HR. Therefore, in our experiments, we compared the steam gasification conversion rate of chars produced at very different HR: 20 and 1800°C.min^{-1} (Figure 8).
Results showed that a high pyrolysis heating rate accelerates the conversion rate of rice husk char gasification. For a char produced at 20°Cmin^{-1}, gasification was complete after 3500s, whereas for a char produced at 1800°Cmin^{-1}, gasification was complete after 2300s. The gasification rate was thus about 1.4 times faster with a char produced at 1800°Cmin^{-1} than with a char produced at 20°Cmin^{-1}. The effect of heating rate on rice husk char gasification seems to be less important compared to results obtained with wood char gasification, where a much clearer difference in reaction rates was found between chars produced at different heating rates of 2.6°Cmin^{-1}, 12°Cmin^{-1} and 900°Cmin^{-1} [15].

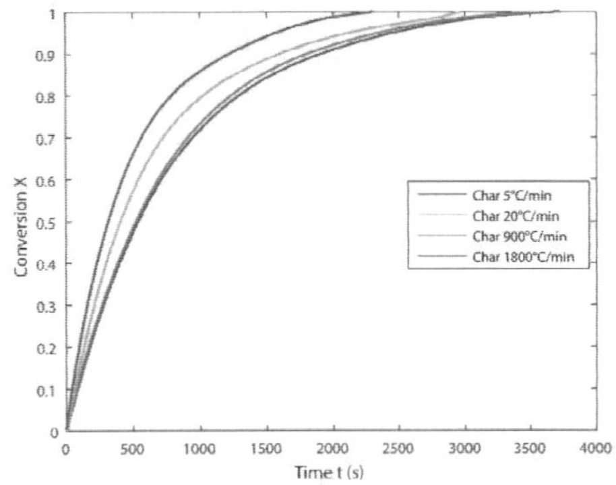

Figure 8. Influence of HR on char conversion (H_2O partial pressure: 0.2 atm, temperature: 900°C)

4. Conclusion

An updated state of the art of RHG technologies was established, from industrial and domestic applications to pilot and laboratory studies. Recent trends and ideas for the development of RH gasifiers at different scales have also been summarised. RH char gasification characteristics at particle scale under CO_2 or H_2O atmosphere were also investigated through a complete parametric study. An increase in T_{final} decreases char yield, remaining VM and energy transferred to the char but does not have much effect on HHV and char morphology. The same effect was shown with a significant increase in HR, except for the fact that we observed small structure damages at the char surface. The temperature was the main factor affecting conversion rate of RH char. An increase from 900 to 1000°C in 0.2 atm accelerates about 4.8 and 7 times the reactivity, under CO_2 or H_2O atmosphere, respectively. A change in the partial pressure of gasification agents has a lesser influence compared to the temperature, but also plays a role in char conversion. The conversion rate decreases in a ratio of about 1.8 when CO_2 partial pressure decreases from 1 to 0.2 atm, and about 1.6 when H_2O partial pressure decreases from 0.4 to 0.1 atm. A significant increase in HR from 20°Cmin^{-1} to 1800°Cmin^{-1} enhanced 1.4 times rice husk gasification kinetics. RHG with H_2O at 900°C, 0.2 atm is about 3.5 times faster than CO_2 atmosphere, which is quite similar to wood gasification. To sum up, this parametric study could provide relevant data to support modelling, design or optimisation of RH gasifier.

5. References

[1] FAO. Rice market monitor. http://www.fao.org/ (2016).
[2] Abdul Salam P, Kumar S, Siriwardhana M. *The status of biomass gasification in Thailand and Cambodia*. Asian institute of technology, 2010.
[3] Nguyen HN, Ha-Duong M, Van de Steene L. A critical look at the rice husk gasification in Cambodia: Technology and Sustainability. *J Sci Technol*; 53.
[4] Zhai M, Xu Y, Guo L, et al. Characteristics of pore structure of rice husk char during high-temperature steam gasification. *Fuel* 2016; 185: 622–629.
[5] Alvarez J, Lopez G, Amutio M, et al. Kinetic Study of Carbon Dioxide Gasification of Rice Husk Fast Pyrolysis Char. *Energy Fuels* 2015; 29: 3198–3207.
[6] Gao X, Zhang Y, Li B, et al. Determination of the intrinsic reactivities for carbon dioxide gasification of rice husk chars through using random pore model. *Bioresour Technol* 2016; 218: 1073–1081.
[7] Bhat A, Ram Bheemarasetti JV, Rajeswara Rao T. Kinetics of rice husk char gasification. *Energy Convers Manag* 2001; 42: 2061–2069.
[8] Nsamba HK, Hale SE, Cornelissen G, et al. Improved Gasification of Rice Husks for Optimized Biochar Production in a Top Lit Updraft Gasifier. *J Sustain Bioenergy Syst* 2014; 04: 225.
[9] Htut Y, Win MM, Myat Khine M. Experimental Investigation of the Pilot Scale Downdraft Gasifier. *IJRCMCE* 2015; 2: 13–18.

[10] Karmakar M, Mandal J, Haldar S, et al. Investigation of fuel gas generation in a pilot scale fluidized bed autothermal gasifier using rice husk. *Fuel* 2013; 111: 584–591.

[11] Ma Z, Zhang Y, Zhang Q, et al. Design and experimental study of a two-stage pilot scale biomass fixed bed gasifier. In: *2013 International Conference on Materials for Renewable Energy and Environment*. 2013, pp. 287–290.

[12] Jairo Ramírez Behainne J, Martínez J. Performance analysis of an air-blown pilot fluidized bed gasifier for rice husk. *Energy Sustain Dev*; 18. Epub ahead of print 1 January 2013. DOI: 10.1016/j.esd.2013.11.008.

[13] Ma Z, Zhang Y, Zhang Q, et al. Design and experimental investigation of a 190 kWe biomass fixed bed gasification and polygeneration pilot plant using a double air stage downdraft approach. *Energy* 2012; 46: 140–147.

[14] Di Blasi C. Combustion and gasification rates of lignocellulosic chars. *Prog Energy Combust Sci* 2009; 35: 121–140.

[15] Mermoud F, Salvador S, Van de Steene L, et al. Influence of the pyrolysis heating rate on the steam gasification rate of large wood char particles. *Fuel* 2006; 85: 1473–1482.

[16] Mermoud F, Golfier F, Salvador S, et al. Experimental and numerical study of steam gasification of a single charcoal particle. *Combust Flame* 2006; 145: 59–79.

[17] Van de steene L, Tagutchou JP, Escudero Sanz FJ, et al. Gasification of woodchip particles: Experimental and numerical study of char–H2O, char–CO2, and char–O2 reactions. *Chem Eng Sci* 2011; 66: 4499–4509.

[18] Harris DJ, Smith IW. Intrinsic reactivity of petroleum coke and brown coal char to carbon dioxide, steam and oxygen. *Symp Int Combust* 1991; 23: 1185–1190.

IOP Conf. Series: Earth and Environmental Science **159** (2018) 012034 doi:10.1088/1755-1315/159/1/012034

Impact of Co-firing Straw for Power Generation to Air Quality: A Case Study in Two Coal Power Plants in Vietnam

A H Truong[1,3] and M Ha-Duong[2]

1 Researcher, Centre for Research and Technology Transfer, Vietnam Academy of Science and Technology, Hanoi, Vietnam
2 Director of research, CIRED - CNRS, Paris, France
3 PhD candidate, University of Science and Technology of Hanoi, Hanoi, Vietnam
E-mail: truonganha87@gmail.com

Abstract. Open field burning of rice straw regularly contributes to severe air quality issues affecting millions of inhabitants in the city of Ha Noi. We examine how much replacing open field burning by co-firing mitigates local air pollutants and greenhouse gases emissions. We select two coal power plants located in the North of Vietnam as specific examples. Our findings show that co-firing straw in these plants at 5% mixing ratio on heat basis can reduce greenhouse gas emission as well as air pollutant emissions (SO_2, PM10 and NO_x) from 3% up to 13%. We examined the social value of these emission reductions using external costs factors. The health benefits of improving air quality by disposing of straw at a large coal power plant instead of open field burning are over ten million USD per year. This is the same order of magnitude as the technical costs of co-firing. Greenhouse gas emissions reduction benefits appear smaller.

1. Introduction

Vietnam's major crop is rice with annual production of 45 Mt in 2015 [1]. Rice straw and rice husk are the by-product of rice production. Residue to product ratio of rice straw is 1:1, which means that to produce 1 t of rice, there will be 1 t of straw [2], and implies that about 45 Mt of straw is produced every year in Vietnam. In the past, straw was used as cooking fuel, cattle feed or composted [3, 4]. Nowadays, with the development in rural areas, straw is less or no longer used for such purposes. It becomes a waste to be disposed of to free the land for the next crop.

The simplest and cheapest way for the farmer to manage the large amount of straw produced after harvesting is to burn it right in the paddy field. According to Nguyen (2012), 60 to 90% of the rice straw in the rice cultivation areas around Hanoi was burned in open fields, and greenhouse gases (GHG) emissions from open field burning of rice straw in Red River Delta caused an environmental damage equivalent to 19 to 200 million USD per year, depending on the social value of CO_2 [5]. Open field burning is an uncontrolled combustion process. It releases significant amounts of air pollutants beyond CO_2, such as CO, SO_2, NO_x and particulate matter (PM). These pollutants have negative impacts to local air quality and human health.

In our study, we look at co-firing technology as an alternative to uncontrolled in-field burning for rice straw disposal.

Co-firing generally means the combustion of two different fuel at the same time. Co-firing biomass such as wood chips, pellets, rice husk or straw is a cost-effective way to increase the renewable energy content of the power generation sector, thus mitigate greenhouse gas emissions. Specifically in this text, co-firing will refer to burning rice straw along with coal in a coal-fired power plant. For co-firing, straw must be collected and transported from fields to the plant. At low mixing ratio (<10%), biomass can simply be ground to fine particles and mixed with coal to burn in the furnace without a costly retrofit of the plant. Burning one kilogram of straw does not generate as much heat as burning one kilogram of coal, but still the heat content of straw allows using less coal.

Co-firing straw with coal in coal power plants has many effects on emissions:

- The amount of straw burnt remains the same, but its energy is not wasted. It can partially substitute coal for electricity generation. Net greenhouse gases emissions are reduced by burning less coal.
- Co-firing reduces the amount of straw burned in open fields during harvesting season. A diffuse pollution problem is replaced by a point pollution problem. Then, pollutants can be treated by the plant's emission control systems.
- Typically, straw contains much less sulphur than the coal it replaces, hence reducing the SO_2 emissions.
- As a side benefit, a fraction of biomass can improve the combustion conditions at the plant and lower the quantity of some other pollutants generated from the coal itself.

In this study, we conduct a quantitative assessment of how co-firing rice straw improves local air quality as well as its CO_2 emissions mitigation. We add up the emissions at the power plant with the emissions in the fields and compute the emission reduction associated with co-firing.

2. Method

We consider direct co-firing at a biomass-mixing ratio of 5% on heat basis in two coal power plants. The old Ninh Binh coal power plant has a capacity of 100 MW. The new Mong Duong 1 coal power plant is ten times bigger with an installed capacity of 1080 MW. Both are located in the north of Vietnam, within the Red River Delta where open field burning contributes to serious air quality issues affecting the lives of millions of people every year.

We estimate the emissions of four pollutants: particulate matter (PM10), SO_2, NO_x and greenhouse gases (GHG).

The system boundary for study includes three segments: (1) straw/coal combustion in plants, (2) straw/coal transportation and (3) straw burned in the open field. We assumed that the rice production and coal mining remain the same with or without co-firing. Table list the input parameters for emission calculation.

Table 1. Input parameters for emission calculation

	Mong Duong 1	Ninh Binh
Coal consumption (Mt/y)	2.71	0.43
Coal transport distance (km)	0	200
Straw required for co-firing (kt/y)	226	40
Transportation activity (Mt·km)	19.27	0.53
Straw collection radius (km)	68	14
ESP efficiency	99.6%	99.2%
Desulphurization efficiency	98.2%	N/A

In the baseline, there is no straw co-fired in the two power plants. Emissions come only from coal combustion, coal transportation and open field burning of the straw produced in the provinces around the plant.

For the combustion activities, emission is calculated by Equation (1) with the emission factor of each pollutant for coal and straw shown in Table . Emission control systems in the power plants can filter some pollutants. Thus, we consider the efficiency (η) of such system in our calculation. For the pollutants emitted without any control measure, the efficiency is zero.

$$Emission_{pollutant} = \sum_{fuel} Quantity_{fuel} \times Emission\ Factor_{fuel,pollutant} \times (1 - \eta_{control\ system}) \quad (1)$$

Emission for transportation activities is calculated by Equation (2) using emission factors listed in Table 2, and the activity levels (in t·km) estimated as follows. Coal is supplied to Mong Duong 1 by conveyor belt from the coalmine nearby. Since the transport distance is small, we assume that the emission from coal transportation by conveyor belt is negligible. Ninh Binh gets its coal supply by barge through 200 km distance. We assume that straw is delivered to the plants by 20 t unit load trucks.

$$Emission_{pollutant,transport\ mode} = Transport\ activities \times Emission\ factor_{pollutant,transport\ mode} \quad (2)$$

We converted the emission reduction quantities into monetary values using external costs factors. We estimate the benefit from CO_2 emission reduction by using a carbon price derived by international comparisons. From the health benefits of local air quality improvement, we used a range of three studies [6]–[8] as shown in Table 3 in order to deal with the deep uncertainty regarding these cost factors.

The Appendix tabulates the numerical values of the main engineering and economic parameters. Complete data and code of the system model developed by the authors [9] is available for download on Github.

Table 2. Emission factors

		CO_2	SO_2	PM10	NO_x
Fuel combustion (kg/t of fuel)	*Coal-Mong Duong 1* [14]	1 877	11.5	43.8	18
	Coal-Ninh Binh [14]	2 081	11.5	26.1	18
	Straw[15]	1 004	0.18	9.1	2.28
Transport mode (g/t/km)	*Conveyor belt*	0	0	0	0
	Road [16, 17]	110	$1.5\ 10^{-4}$	$2\ 10^{-4}$	0.134
	Barge [18]	71	$1.6\ 10^{-2}$	$2.6\ 10^{-3}$	0.4

Table 3. Social cost of air pollutant emissions (USD/t of emitted pollutant)

Pollutant	External cost factors [6]	Health damage cost [7]	Health damage cost [8]
SO_2	3 680	3 767	1 134
PM10	2 625	5 883	2 496
NO_x	2 438	286	1 398

3. Results

Figure 1 summarizes results on emissions estimation. There are four dimensions: at each plant, we represent the emission of each pollutant, from three activities, comparing two scenarios: baseline and co-firing. These dimensions are arranged as follows. The left panel shows emission from Mong Duong 1 case, the right panel shows emissions from Ninh Binh case. Vertically, each of the four stacked bar pairs corresponds to one pollutant. Within the pair, the top bar represents emissions with co-firing, the bottom bar the baseline emissions. In all of the eight pairs, the emissions are lower with co-firing than the baseline. This result shows that co-firing reduces emissions of all pollutants considered, in both plants.

The horizontal axes represent the quantity of pollution emitted in one year. In absolute terms, Mong Duong 1 is larger and pollute more, so its scale at the top goes to 60 kt/y compared to Ninh Binh's 9 kt/y. Relative to the quantity of electricity produced, Mong Duong 1 is less polluting, but that is besides the discussion here.

Local air pollutants emissions (SO_2, PM10 and NO_x) are plotted against the top axis and GHG emission is plotted against the bottom axis. These scales differ since GHG emission is higher than local air pollutants emission by three order of magnitude. GHG emission reduction will be much higher than NO_x and PM10 emission reduction in absolute mass.

Total length of the bar represents the total emissions, adding up the three segments of the system. Co-firing straw with coal at 5% at Mong Duong 1 could save 249 423 tCO_2eq per year (see Table). This number is 44 788 tCO_2eq per year for Ninh Binh. In percentage compared to baseline, PM10 has the highest relative emission reduction. Although it has the smallest reduction in t per year, SO_2 reduction is the second highest in term of percentage (Table).

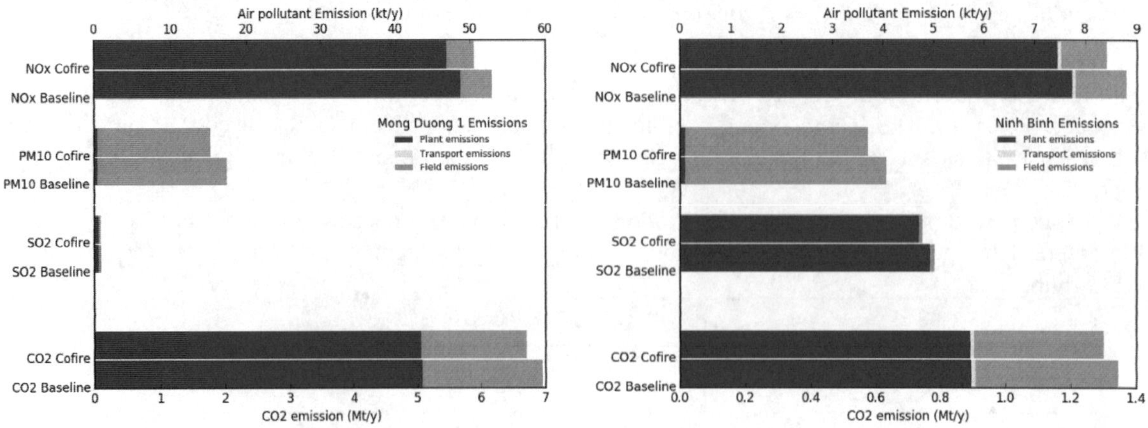

Figure 1. Co-firing reduces emissions compared to baseline in the two power plants. Mong Duong 1 is left, Ninh Binh is right. Ninh Binh plant emits less CO_2 than Mong Duong 1 plant because it is ten times smaller. However, its SO_2 emissions are larger because it lacks a desulphurization system. The PM10 emissions mostly come from open-field burning because plants have dust filters and precipitators.

Comparing left and right plots on Figure 1 shows that the benefits of co-firing depend mainly on the characteristics of the plant: size, existence of pollutant controls. When the plant has pollutants control technologies, the related pollutant emission from combustion activity in the plant is reduced significantly. With both ESP and desulfurization system, PM10 and SO_2 emission from Mong Duong 1 power plant is negligible compare to CO_2 and NO_x emission. For Ninh Binh, ESP system also filtered PM10 emission significantly. Combustion in power plant accounts for the majority of NO_x emissions in both plants since there is no NO_x control measure installed.

Since there is no emission control measure in both power plants, the combustion of straw in co-firing only have very slight effect on emission reduction of GHG and NO_x due to lower emission factor of straw compare to coal. Instead, emission reduction mostly comes from the amount of coal saved from burning in the plant by co-firing for these pollutants. Emission reductions of SO_2 and PM10 in Mong Duong 1 (and only PM10 in case of Ninh Binh) reflect the amount of coal consumption reduced in plants and the effect of emission control technologies deployed in the plants.

Table 4 summarizes emission reductions of each pollutant by activities. For both studied plants, emission from transportation is negligible and not visible in Figure 1, except for GHG emissions from Ninh Binh. However, we find that for Mong Duong 1, biomass transportation increases GHG and air pollutants compare to baseline case. Mong Duong lie next to coalmine and coal is delivered by conveyor belt, coal transportation activity is assumed zero. While coal is supplied from nearby mine, biomass is collected and transported with collection radius of 68 km. With capacity ten times larger than Ninh Binh, biomass transportation activity is 19.27 million t·km which add 2 172 tCO_2eq per year to the total emission. For Ninh Binh, straw is collected and transported locally while coal is transported from coalmine 200 km away. Thus, the transportation activity in co-firing case emits 546 tCO_2eq less than baseline case.

Emissions from plant's stack is much larger than emission from field for the pollutants that are emitted without control measure (GHG and NO_x in case of Mong Duong 1; GHG, NO_x and SO_2 in case of

Ninh Binh). PM10 and GHG account for the largest part of emissions from open field combustion of rice straw.

Combustion activity in power plant contribute most to emission reduction of the pollutants that not controlled by any measure. For example, NO_x emission reduction is largest from plant combustion in both plant. SO_2 emission reductions also mostly comes from plant combustion activity in Ninh Binh case. This reflects the effect of replacing part of coal in the plants by straw, which has emission factor of SO_2 60 times lower than that of coal and emission factor of NO_x 8 times lower.

Table 4. Emission reductions by activities (t/y)

Plant	Activity	GHG	SO_2	PM10	NO_x
Mong Duong 1	*Plant combustion*	25 599	27	15	1 899
	Transportation	-2 172	0	0	-3
	Field combustion	225 996	41	2049	513
Ninh Binh	*Plant combustion*	4 416	237	2	292
	Transportation	546	0.1	0.2	3
	Field combustion	39 827	7	361	90

The benefit of GHG emission reduction is estimated using social carbon value. Carbon pricing policy for electricity production is not yet in place for Vietnam. We looked at "State and trends of carbon pricing" [10], a review of existing carbon pricing instruments in other countries, to choose a social carbon value for our analysis. Kossoy et al.'s [10] observe a wide range of carbon prices from 1 to 130 USD/tCO_2e. However, 99 % of emission are priced less than 30 USD/tCO_2e and 85% are priced less than 10 USD/tCO_2e. We have plotted the carbon prices of all countries listed in the mentioned report against their Gross Domestic Production (GDP) per capita to see in which range that carbon value for Vietnam could fall in (see Figure 2). Vietnam GDP per capita is much lower than the lowest income countries that present in the list, which is China. China has carbon tax rate at 5 USD/tCO_2e and has the 2014 GDP per capita at 7 590 USD while the GDP per capita for Vietnam is only 2 052 USD. For this reason, we choose the lower bound of the observed range. The social carbon value used to estimate the benefit from GHG emission reduction in this study is 1 USD/tCO_2eq.

Considering a social carbon value at 1 USD/tCO_2eq, the environmental benefits from greenhouse gas emission reduction of co-firing for Mong Duong 1 is 250 kUSD/year and for Ninh Binh is 45 kUSD/year.

Table 5. Emission reductions and associated benefits for two plants

	Mong Duong 1			Ninh Binh		
Pollutant	Emission reduction (t/y)	Emission reduction (%)	Benefit (kUSD/y)	Emission reductions (t/y)	Emission reduction (%)	Benefits (kUSD/y)
GHG	249 423	3.7%	249	44 788	3.4%	45
SO_2	68	8%	77-256	244	4.9%	277 - 921
PM10	2 063	13.4%	5 149-12 141	362	9.8%	904-2 133
NO_x	2 409	4.6%	689 – 5 873	382	4.4%	109-931

IOP Conf. Series: Earth and Environmental Science **159** (2018) 012034 doi:10.1088/1755-1315/159/1/012034

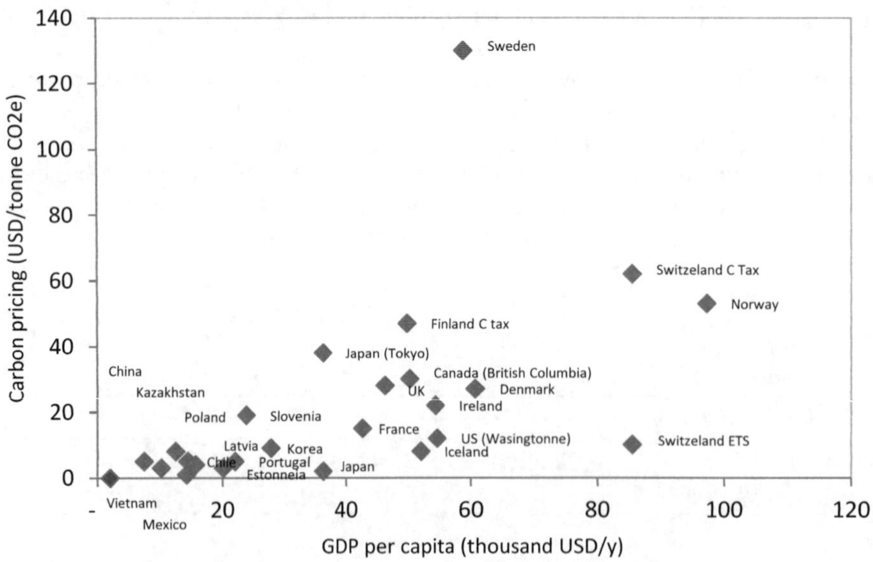

Figure 2. Carbon pricing policy against countries' GDP. Carbon pricing policy for electricity production is not yet in place for Vietnam. Vietnam GDP per capita is much lower than the lowest income countries that have carbon pricing policy, which is China.

We use health damage cost specified for each air pollutant to monetarize the impact of emission reduction to public health. We refer to the health damage costs used to estimate health impact in other neighbour countries such as Thailand and China in previous studies. Although these values vary between countries, the health impact from PM10 emissions remain the highest.

Health benefit from reducing air pollution emission is 8.6 - 13.3 million USD per year for Mong Duong 1 and 1.7 - 3.2 million USD per year for Ninh Binh. The majority of health benefit comes from PM10 emission reduction. This is because PM10 has the highest social cost as seen in

Table 3.

Benefit of air pollutant mitigation does not proportionate to the size of the coal power plant. SO_2 emission reduction in Mong Duong 1 is much smaller compare to Ninh Binh despite that it is 10 times larger in capacity. As a result, the benefit from SO_2 emission reduction is higher in Ninh Binh case. This is because with high efficiency desulphurization system, the emission from coal combustion is significantly reduced even without co-firing. Thus, for co-firing scenario, the substitution of coal by straw has less impact compare to Ninh Binh case.

4. Discussion

Co-firing rice straw with coal in coal power plants reduces local air pollutants emissions by two effects. First, the substitution of coal by straw lowers the air-borne pollutants emission at the plant because straw has lower emission factors than coal. Second, co-firing reduces the amount of straw burned in the open field. Instead, straw will be burned in coal boiler and the emission will be reduced by plant's filtering systems such as electrostatic precipitation, desulfurization system.

Co-firing straw offers a measure to reduce greenhouse gas and air pollutant emissions. Air pollutant emission reductions depend control technologies deployed by the power plants. In the baseline case, the amount of straw required for co-firing is open burned without any emission control measure. In the co-firing case, when straw is combusted in the plant's boiler, pollutants such as PM10 and SO_2 are filtered. The technologies for air pollutant control in power plants (such as electrostatic precipitator, $DeNO_x$ and $DeSO_2$) are commercially available. This is not the case for CO_2 emission control technology. Carbon capture and storage (CCS) is the technology of capturing the CO_2 produced by a power plant in order to inject it into deep geological formations. CCS is not yet considered in the latest

National Power Development Plan of Vietnam [11]. Cost of technology is one of the main barrier since it would increase the cost of electricity produced by coal power plants to a point where they may be uncompetitive with other generation technologies [12]. In the long run however, biomass energy with CCS has been proposed as a technologically plausible way to extract CO_2 from the atmosphere. Co-firing is a step in this direction.
Co-firing rice straw for heat and power generation already exist in other countries such as China and the US. It is known to work, however there are technical and economic difficulties. Rice straw is a scattered resource, which requires work for collection and transportation. Mobilizing large amount of straw for co-firing in large coal power plants imposes additional cost to the plants. Technical costs for retrofitting coal power plants for co-firing includes investment cost, fuel cost and operation and maintenance cost. The costs for retrofitting coal power plants for co-firing with co-feed, the simplest and cheapest co-firing technology, ranges from 300 to 700 USD/kW [13]. Operation and maintenance cost is about 2.5 to 3.5% of capital cost. Fuel cost depends greatly on the origin, type and composition of biomass feedstock. Since the costs of co-firing are concentrated and private, while its benefits are diffuse and public, co-firing raises interesting political economy problems.
There are alternatives to open field burning for disposal of rice straw. Making rice straw pellets is one option. Rice straw pellets are a better fuel and technically more suitable for co-firing than raw rice straw since pellets have a higher heating value, higher grind ability and are easier to transport and to store. However, it is difficult to produce pellet form rice straw, compared to other types of biomass. Additives are usually required. Costs of pelletizing imply that using raw straw is cheaper for power plants co-firing at a large scale. Pellets appear more economical for small-scale applications such as small boiler for heat production or cook stove.
A central result of our numerical analysis is that benefit from GHG emission reduction are small compared to health benefits. Lacking estimates in the Vietnamese context, we used a range of international studies for external costs factors of local air pollution. Regarding the value of GHG emission mitigation, we believe the 1 $/t$CO_2$ carbon value used is appropriate in a low middle-income country like Vietnam. This value could increase if there were effective international clean development mechanisms. There is a Joint Crediting Mechanism (JCM), a bilateral carbon credits trading scheme between Vietnam and Japan. GHG benefits of co-firing could also increase if Vietnam implements a carbon market –it is expected that the legal framework for carbon market should be set up by 2020– or renewable energy portfolio standards in the power sector.

5. Conclusion

Co-firing reduces GHG emissions by replacing coal, and it reduces pollutant emissions detrimental to local air quality caused by open-field burning. Our study quantified these benefits in the context of Vietnam today. We find that the local air improvement is more important than GHG mitigation in terms of social value. Co-firing straw has large positive impact on public health: we estimate the social value of avoided air pollution at –around ten million USD per year for the large plant and –two million USD per year for the smaller plant. The largest part of these benefits comes from reducing the PM10 emissions. Co-firing straw with coal in power plants would improve Hanoi air quality.
In our future work, we will extend our research to include all the operating coal power plants in the North of Vietnam to assess fully the impact of co-firing to air pollutant and greenhouse gas mitigation in Red River Delta region.

7. References

[1] General Statistic Office 2017. *Statistical Handbook of Vietnam 2016*.

[2] Arvo L, Nguyen DC 2013. *Development of biomass fuel chains in Vietnam*. Technical Research Centre of Finland.

[3] Hoang AL, Nguyen TTH, Le TL 2013. Estimated gas emission from burning rice straw in open fields in Thai Binh province. *Sci J Vietnam Natl Univ*; **29**: 26–33.

[4] Dinh MC, Hoang AL, Hoang XC 2016. Calculation of Gas Emission from Rice Straw Open Burning in Ninh Binh Province for 2010 - 2015 Period and Proposal of Mitigation Solutions. *VNU J Sci Earth Environ Sci*; **32**: 70–76.

[5] Nguyen MD 2012. Estimation of air pollutant emission from rice straw combustion in the open air in Red River Delta. *J Sci Dev*; **10**: 190–198.

[6] Zhang Q, Weili T, Yumei W, et al 2007. External costs from electricity generation of China up to 2030 in energy and abatement scenarios. *Energy Policy*; **35**: 4295–4304.

[7] Sakulniyomporn S, Kubaha K, Chullabodhi C 2011. External costs of fossil electricity generation: Health-based assessment in Thailand. *Renew Sustain Energy Rev*; **15**: 3470–3479.

[8] Hainoun A, Almoustafa A, Seif Aldin M 2010. Estimating the health damage costs of syrian electricity generation system using impact pathway approach. *Energy*; **35**: 628–638.

[9] Truong AH 2015. *Feasibility and sustainability of co-firing biomass in coal power plants in Vietnam*. Master of Science Thesis, University of Science and Technology of Hanoi.

[10] Kossoy A, Peszko G, Oppermann K, et al 2015. *State and trends of carbon pricing*. Washington DC: World Bank and Ecofys http://www.worldbank.org/content/dam/Worldbank/document/Climate/State-and-Trend-Report-2015.pdf (September 2015).

[11] Ha-Duong M, Nguyen-Trinh HA 2017. Two scenarios for carbon capture and storage in Vietnam. *Energy Policy*; **110**: 559–569.

[12] Asian Development Bank. *Prospect for Carbon Capture and Storage In Southeast Asia*. Asian Development Bankhttp://www.adb.org/publications/prospects-carbon-capture-and-storage-southeast-asia (2012).

[13] IRENA 2013. *Biomass Co-firing-Technology Brief*. IRENA and IEA-ETSAP.

[14] IPCC 2006. *2006 IPCC Guidelines for National Greenhouse Gas Inventories*http://www.ipcc-nggip.iges.or.jp/public/2006gl/index.html.

[15] Shafie SM, Mahlia TMI, Masjuki HH 2013. Life cycle assessment of rice straw co-firing with coal power generation in Malaysia. *Energy*; **57**: 284–294.

[16] Binh NT, Tuan VA 2016. Greenhouse Gas Emission from Freight Transport-Accounting for the Rice Supply Chain in Vietnam. *Procedia CIRP*; **40**: 46–49.

[17] Department for Environment, Food and Rural Area for Environment 2017. Emission factors for transport - Defra, UKhttp://naei.defra.gov.uk/data/ef-transport (accessed 10 April 2017).

[18] Van Dingenen R, Muntean M, Janssens-Maenhout G, et al 2016. *Analysis of Air Pollutant Emission Scenarios for the Danube region*. Technical Report EUR 28521 EN, Joint Research Center http://publications.jrc.ec.europa.eu/repository/bitstream/JRC104828/jrc104828_scenario_analysis_for_the_danube_region_final_w_identifiers.pdf (accessed 18 June 2017).

Pyrolysis and Char Oxidation of Biomass and Coal Blends: Kinetic Study Using Thermogravimetric Analysis

Hoang Anh Tran[1], Laurent Van de Steene[1,2] and Duc Dung Le[3]

1 University of Science and Technology of Hanoi (USTH), Vietnam Academy of Science and Technology, Hanoi, Vietnam
2 CIRAD, UPR BioWooEB, Hanoï, Vietnam & BioWooEB, Univ Montpellier, CIRAD, Montpellier, France
3 Hanoi University of Science and Technology (HUST), Hanoi, Vietnam
E-mail: tran-hoang.anh @usth.edu.vn

Abstract. Vietnam as an agricultural country has a high potential in biomass, especially agricultural and forestry wastes. This resource offers a promising way to develop co – combustion technology of biomass and coal in Vietnam and thus tackle the environmental issues. A fundamental research was established to study combustion of Vietnamese rice straw and coal. Blends was made by mixing manually 50% wt. of rice straw and 50% wt. of coal. Pyrolysis in inert gas, char oxidation in oxidizing gases (10 % O_2 in N_2) and combustion process in oxidizing gases (10 % O_2 in N_2) were implemented. A particular attention was paid to the behaviour of rice straw/coal blends during thermochemical transformation. Concerning pyrolysis process, the char yield of the mixture is equivalent to weighted sum of the char yield of two solid fuels. During the combustion of the blends, we observed a coupling between pyrolysis and char oxidation. The exothermic oxidation reactions could overheat the sample itself and accelerate the pyrolysis and char oxidation kinetics.

1. Introduction

In Vietnam an abundant fossil fuel, coal is used mostly as a principal energy resource to produce heat and electricity. In 2015, electricity production coming from coal fire power accounted for 33.5% and was classified in the second position after hydropower at 38% [1]. The latest "National Power Development Plan" indicated that Vietnam will continue developing coal as the main energy source for electricity production. Coal should contribute to 53.2% of the total power generation by 2030. With the increasing number of coal power plants, Vietnam will have to face to many environmental and socio-economic challenges from coal import to pollution and human health issues.

In parallel, biomass is a promising source of energy to be exploited in order to support national energy security and reduce the dependency to fossil fuels. The largest biomass sources for energy production in Vietnam are crop residues and wood. The crop residues can be classified in two categories – agricultural by-products (rice straw, corn leaves and cobs etc.) and agro-industry by-products (rice husk, bagasse, peanut shells, coffee husks, etc.). The wood fuel consists of sawdust, wood chips and bark [2]. Despite of the low heating value of biomass compared to coal, it remains a promising resource because of its large quantity available [3].

In this context, co-combustion of biomass and coal is an opportunity to take advantage of the domestic green resources and to limit the use of fossil fuel in coal power plants. In order to overcome this issues and to develop this technology in Vietnam, a better expertise and understanding of behaviour of biomass/coal blending during combustion is required. In this study, rice straw and anthracite coal were selected because of their large availability in Vietnam.
In general, the combustion of solid fuel can be decomposed into several steps: (1) drying; (2) pyrolysis process, that consists to release volatile gases; (3) oxidation of volatile matters near the particle; (4) char oxidation that converts the solid carbon into CO_2 [4]. In this work, we concentrated separately on the pyrolysis step in inert environment, on the char combustion and on the whole combustion process considering as fuel successively: rice straw, coal and their mixture.
Several research was investigated to study the kinetic of different nature of coal and biomass. Ruiling Du at al. experimented the pyrolysis of Shenfu coal of China. They determined three zones in which the kinetic mechanism changes was accompanied by a radical change in mass loss. The temperature range of each zone is influenced by the heating rate. In the first one, at the lowest temperature range (105 °C – 386 °C at the heating rate of 25 °C/min), the absorbed gas and the water are vaporized. In the second zone (386 °C – 495 °C at the heating rate of 25 °C/min), the volatile matters are released. The coal is transformed into a semi coke. Finally, this semi coked coal are finally coked (495 °C – 900 °C at heating rate of 25 °C/min). The influence of heating rate was observed more remarkable in the second and the third zone than the first one. [5] [6]. Garima Mishra et al. investigated the pyrolysis of rice straw and found that there are three stages like coal: moisture release (until 150 °C), active pyrolysis (150 °C – 350 °C) and passive pyrolysis (350 °C – 700 °C) [7]. In parallel, the oxidation of these solid fuels was also examined by several research groups [3] [8] [9] [10].
The objective of this work is to study the kinetic of rice straw/coal mixture and to understand the coupling effect occurred in the pyrolysis, char oxidation and the overall combustion process.

2. Materials and experiments

2.1. Materials

In this study, an anthracite coal (Coal) was taken from Ninh Binh coal-fired power plant, situated in the North of Vietnam. This coal is classed under the "low rank coal" category because of its high ash content and its low burning efficiency. The particle size was in the range 45 – 75 μm. The rice straw "Rơm nếp" (RS) was taken in the rural area of Hanoi during harvest season. After being dried under the sunlight for a half day, the sample was milled and sieved to obtain a particle size lower than 200 μm. The proximate analysis, as well as the higher heating value of two materials are shown in the Table 1. The coal has significantly different properties compared to the rice straw because of its higher ash content and lower volatile matters. Characterization tests were carried out at least three times to ensure the measurement's repeatability.

Table 1. Proximate analysis of coal and rice straw – dry basis

	FC (%)	A (%)	V (%)	M (%)	HHV (MJ/kg)
Coal	61.39	30.37	8.24	2.51	23.1
Rice straw	15.80	15.48	68.72	9.31	12

2.2. Experiments

The thermo-gravimetric analysis using a macro TG reactor set up at USTH was investigated in non-isothermal operation mode (Figure 1). The temperature and the mass of the sample were recorded continuously. The temperature surrounding the sample is measured in the position of 2 cm above the crucible. The equipment consisted of a reactor (1), an electrical furnace (2), three thermocouples (T_i) and a micro balance (Precisa ES 320A) (5). The maximum heating rate allowed by the system was 5 °C/min. The different gases are mixed before entering the reactor and controlled by the mass flow meter – controllers (M_i) with a range of 0 – 10 NL/min. The oxygen concentration can be controlled from 0 to 100% by adjusting the mass flow rates of N_2 and O_2. These gases were preheated (3) in a 2-meter-long coiled tube located in the upper part of the reactor.

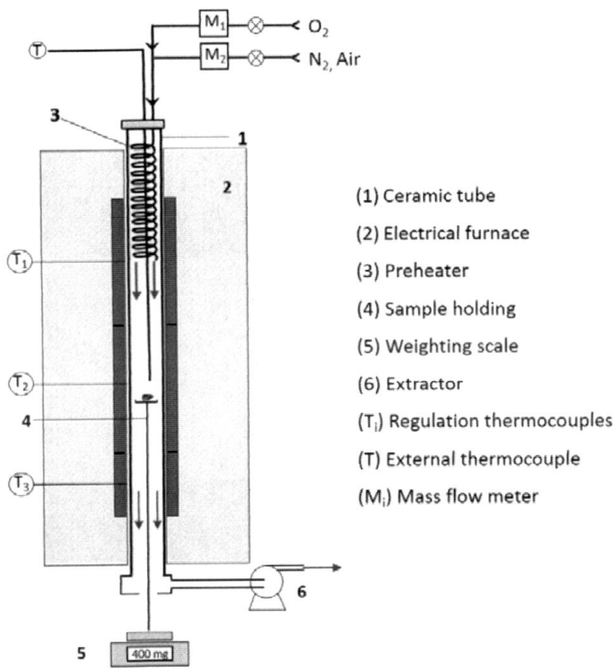

Figure 1. Macro thermo-gravimetric

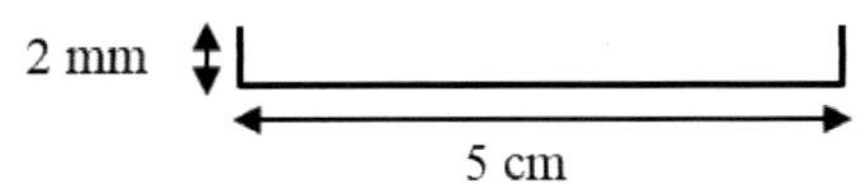

Figure 2. Geometrical view of the alumina crucible

The sample, i.e. rice straw, coal or their blends was weighted to about 400 mg ± 2 mg and was carefully spread on the 5 cm x 2 mm alumina crucible (Figure 2). When operating with the mixture of rice straw and coal, the blends was prepared manually in the proportion of 50 % wt. – 50 % wt. The total gases flow rate was set to 5 NL/min. For combustion process, the concentration of oxygen in nitrogen was fixed to 10% and assured by adjusting the mass flow rate of O_2 and N_2. Before starting experiments, the gas mixture was circulating inside the reactor for at least 20 minutes to flow out the eventual residual gases. The sample was positioned in the center of the reactor. The reactor was then heated from ambient temperature to the desired temperature with a heating rate of 5°C/min.

Three series of experiments were preformed independently: pyrolysis, char oxidation and combustion. Regarding the pyrolysis, the sample was heated to 950°C in 100% N_2. Concerning the char oxidation experiments in the atmosphere of 10% O_2, the char of rice straw and coal were produced by pyrolyzing the raw material during 7 minutes at 900 °C in a muffle furnace. The char sample was then heated to 900°C. The combustion process experiments were performed in the same experimental condition as char oxidation but using the raw samples.

Some preliminary tests were done to identify the isothermal zone in the reactor and thus the best position for the sample and to check the accuracy of the equipment at high temperature. For the identification of the isothermal zone, a temperature profile was measured along the reactor at different temperature (700°C, 800°C and 900°C). The tests showing that sample could be placed between 41 cm – 57 cm from the bottom of the reactor. For all following experiments, the sample was put at the position of 41 cm from the bottom. To check the accuracy of the system, some "blank tests" were carried out without sample. A deviation less than 2 mg per hour was found. It was accepted compared to the mass of the sample.

3. Results and discussion

For the whole study, the mass loss is presented (Equation 1) by its normalized mass (m*) based on the initial mass of dry sample $m_{i\,(dry)}$.

$$m^* = m/m_{i(dry)} \quad (1)$$

3.1. Pyrolysis process of coal, rice straw and their mixture

The Figure 3 and Figure 4 show respectively normalized mass loss (TG curve) and derivative mass loss (DTG curve) of rice straw, coal and their mixture during pyrolysis process under N_2 atmosphere. Because of its low volatile matters, the coal DTG was much lower than the one of rice straw and the mixture. Note that the derivative mass of coal in the Figure 4 has been multiplied by 10 to facilitate the reading.

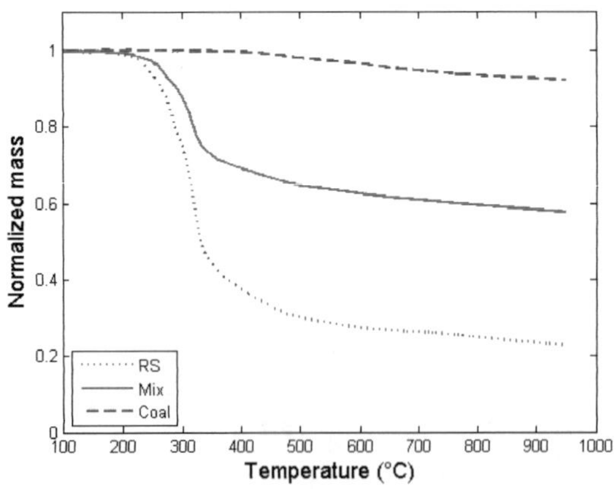

Figure 3. Normalized mass loss versus temperature during pyrolysis process of rice straw, coal and mixture – dry basis.

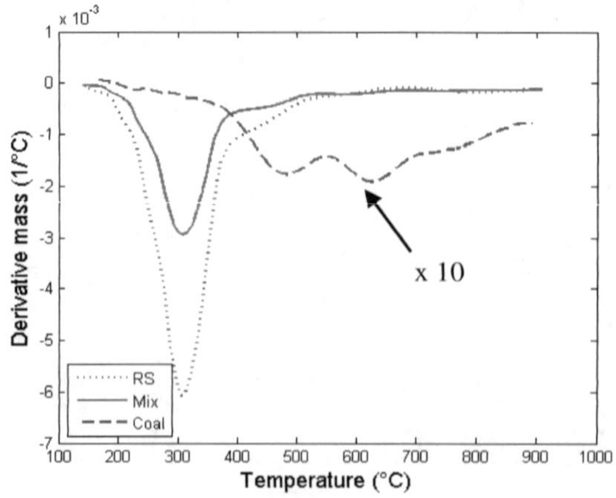

Figure 4. Derivative mass loss versus temperature during pyrolysis process of rice straw, coal and mixture.

The char yield of rice straw (22.8 %) is significantly lower than coal's one (92.3 %). The reason is obviously the very high difference in volatile matters content. The char yield of rice straw/coal mixture is 57%. This char yield of the mixture is equivalent to the weighted sum of the char yield of the two constitutive solid fuels. The maximum mass loss is measured at low temperature for rice straw ($T_{DTG\ max}$ = 308°C) and for mixture ($T_{DTG\ max}$ = 307°C), and at higher temperature for coal ($T_{DTG\ max}$ = 480 and 627 °C). The pyrolysis of rice straw and mixture occurs in a narrow temperature range compared to the pyrolysis of coal. Moreover, the rice straw pyrolysis starts at a lower temperature than the blends pyrolysis. It seems that the presence of coal in the mixture delays the reaction. The temperature $T_{DTG\ max}$ and the derivative mass loss of each sample are summarized in the Table 2.

3.2. Oxidation process of char of coal, rice straw and their mixture

The Figure 5 and Figure 6 show respectively normalized mass loss (TG curve) and derivative mass loss (DTG curve) of char of rice straw, coal and their mixture during oxidation process under oxidizing atmosphere (10% O_2 in N_2).

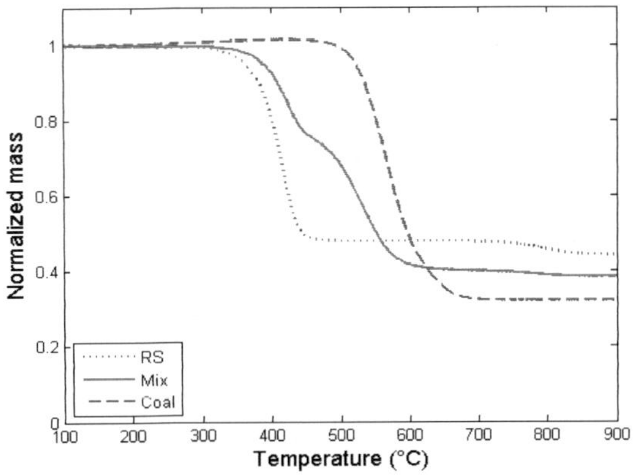

Figure 5. Normalized mass loss versus temperature during oxidation process of char of rice straw, coal and mixture – dry basis.

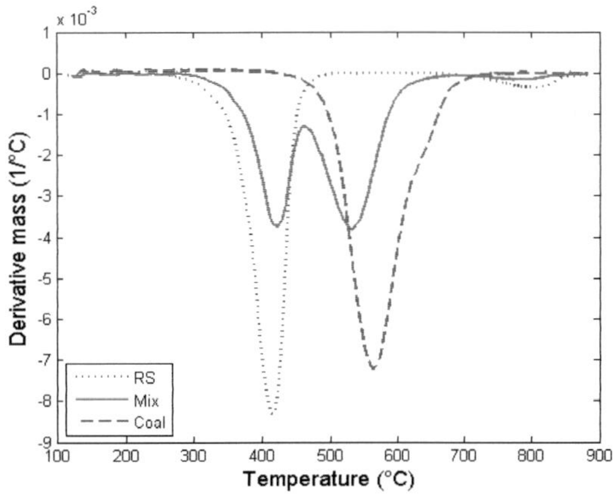

Figure 6. Derivative mass loss versus temperature during oxidation process of char of rice straw, coal and mixture.

The oxidation of char from individual feedstock occurs in a narrow range. The mass loss is maximum at the temperature $T_{DTG\ max}$ = 416°C for char of rice straw and $T_{DTG\ max}$ = 564°C for char coal. Their char's mixture oxidation shows two peaks: the first one corresponds to the oxidation of rice straw char and the second one is related to the oxidation of char coal. When mixing char of rice straw and coal, the first char oxidation peak appears at $T_{max\ DTG}$ = 421 °C, a slightly higher temperature than for rice straw char alone. The second char oxidation peak is measured at T_{max} = 532 °C, a lower temperature than in case of char coal alone. The intensity of the peak is -0.0083 $°C^{-1}$ and -0.0072 C^{-1} for rice straw and coal respectively. For the char mixture, DTG_{max} of the first peak is -0.0037 C^{-1} that represents a half of the char rice straw one. DTG_{max} of the second peak is -0.0038 C^{-1} that is equivalent to a half of the char coal. The temperature value $T_{DTG\ max}$ and the derivative mass loss of each sample are summarized in the Table 2

3.3. Combustion process of coal, rice straw and their mixture

The Figure 7 and Figure 8 show respectively normalized mass loss (TG curve) and derivative mass loss (DTG curve) of rice straw, coal and their mixture during combustion process under oxidizing atmosphere (10% O_2 and 90% N_2).

Figure 8 shows that for rice straw, two zones are identified during combustion. The first zone corresponds to the pyrolysis step and the second one is identified to the oxidation of the produced char [11]. In this case, the maximum mass loss occurs at $T_{DTG\ max}$ = 283 °C, tightly lower than temperature in inert atmosphere ($T_{DTG\ max}$ = 308 °C). The same behavior is observed in char oxidation zone, the maximum mass loss during rice straw combustion occurs at lower temperature ($T_{DTG\ max}$ = 391 °C) than in case of char oxidation only ($T_{DTG\ max}$ = 416 °C). Concerning coal combustion, only one peak related to the char oxidation is measured at $T_{DTG\ max}$ = 556 °C because of low volatile content. A small gain in mass (normalized mass > 1) is obtained because of some absorption phenomenon at low temperature [12]. Regarding the rice straw/coal mixture, the same observations are the same than those explained for the char oxidation. The first maximum mass loss is measured at $T_{DTG\ max}$ = 294 °C and the second maximum mass loss at a temperature $T_{DTG\ max}$ = 549 °C. All results are centralized in Table 2. These results show that there is a coupling between pyrolysis and char oxidation when mixing two solid fuels. The oxidation of volatile matters at the end of pyrolysis process overheats the sample and accelerates the oxidation of char.

IOP Conf. Series: Earth and Environmental Science **159** (2018) 012035 doi:10.1088/1755-1315/159/1/012035

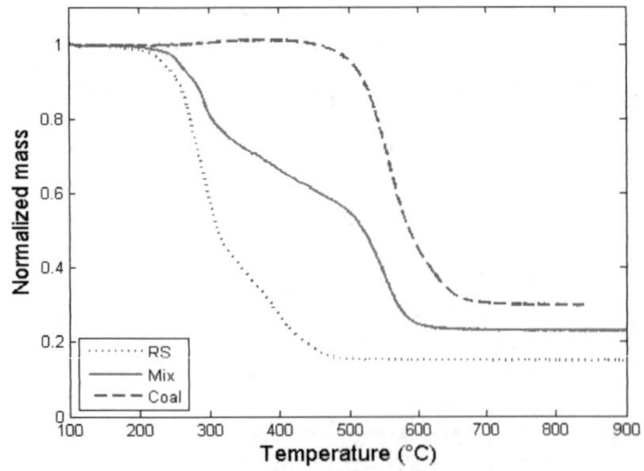

Figure 7. Normalized mass loss versus temperature during combustion process of rice straw, coal and mixture – dry basis.

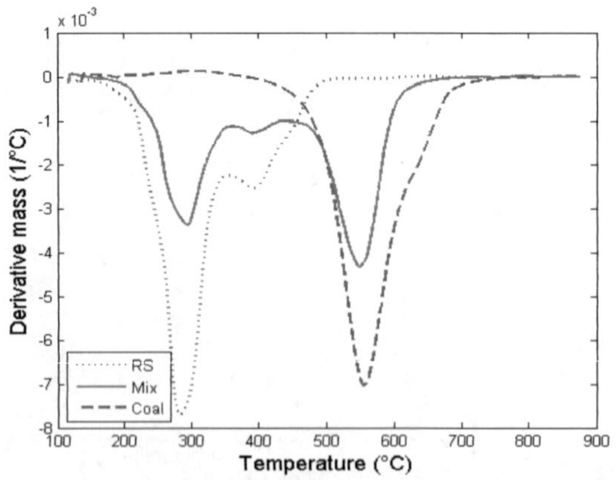

Figure 8. Derivative mass loss versus temperature during combustion process of rice straw, coal and mixture.

Table 2. The temperature at maximum mass loss (a) and the maximum DTG (b) of each sample during pyrolysis, oxidation of char and combustion process

(a)	Peak temperature (°C)					
	Coal		RS		Mix	
	Peak 1	Peak 2	Peak 1	Peak 2	Peak 1	Peak 2
Pyrolysis	480	627	308		307	
Oxidation	564		416		421	532
Combustion	556		283	391	294	549

(b)	Mass loss rate maximum (1/°C)					
	Coal		RS		Mix	
	Peak 1	Peak 2	Peak 1	Peak 2	Peak 1	Peak 2
Pyrolysis	$-1{,}77*10^{-4}$	$-1{,}92*10^{-4}$	-0,0061		-0,0029	
Oxidation	-0,0072		-0,0083		-0,0037	-0,0038
Combustion	-0,007		-0,0077	-0,0025	-0,0034	-0,0043

4. Conclusion

The combustion of rice straw, coal and their blends has been experimentally investigated in a new macro TG reactor. Pyrolysis and char oxidation have been studied firstly independently and then the coupled in combustion process was investigated. For pyrolysis experiments, the char yield of the mixture is equivalent to the weighted sum of the two char yield from rice straw and coal. Regarding the combustion process, there is a coupling phenomenon between pyrolysis and char oxidation when mixing two solid fuels. The oxidation reactions near or inside the particles could overheat the sample. Thus, the temperature inside particles may be higher than the temperature measured around the particle. Consequently, this phenomenon could accelerate the char oxidation at lower temperature. Further research could be carried out to get a better understanding of the interaction and the synergetic effects between coal and rice straw during combustion process. These results could be very useful for further research aiming for industrial application in developing a kinetic model of the combustion.

5. References

[1] EVN 2017 *Vietnam electricity annual report 2016* EVN.
[2] Arvo Leinonen NDC. Development of biomass fuel chains in Vietnamhttp://www.vtt.fi/inf/pdf/technology/2013/T134.pdf (accessed 12 October 2017).
[3] Calvo LF, Otero M, Jenkins BM, et al. 2004 Heating process characteristics and kinetics of rice straw in different atmospheres. *Fuel Process Technol* **85** 279–291.
[4] Boiko EA, Pachkovskii SV. 2004 A kinetic model of thermochemical transformation of solid organic fuels. *Russ J Appl Chem* **77** 1547–1555.
[5] Du R, Wu K, Zhang L, et al. 2016 Thermal behavior and kinetic study on the pyrolysis of Shenfu coal by sectioning method. *J Therm Anal Calorim* **125** 959–966.
[6] Wang H, Zhang J, Wang G, et al. 2016 Characteristics and kinetic analysis of co-combustion of brown coal and anthracite. *J Therm Anal Calorim* **126** 447–454.
[7] Mishra G, Bhaskar T. 2014 Non isothermal model free kinetics for pyrolysis of rice straw. *Bioresour Technol* **169** 614–621.
[8] Contreras ML, García-Frutos FJ, Bahillo A. 2016 Study of the thermal behaviour of coal/biomass blends during oxy-fuel combustion by thermogravimetric analysis. *J Therm Anal Calorim* **123** 1643–1655.
[9] Yu Z, Fang S, Lin Y, et al. 2015 Investigation of Rice Straw Combustion by Using Thermogravimetric Analysis. *Energy Procedia* **75** 144–149.
[10] Wang P, Hedges SW, Casleton K, et al. 2012 Thermal Behavior of Coal and Biomass Blends in Inert and Oxidizing Gaseous Environments. *Int J Clean Coal Energy* 01: 35.
[11] Yu Z, Fang S, Lin Y, et al. 2015 Investigation of Rice Straw Combustion by Using Thermogravimetric Analysis. *Energy Procedia* **75** 144–149.
[12] Wang G, Zhang J, Shao J, et al. 2014 Thermogravimetric Analysis of Coal Char Combustion Kinetics. *J Iron Steel Res Int* **21** 897–904.

IOP Conf. Series: Earth and Environmental Science **159** (2018) 012036 doi:10.1088/1755-1315/159/1/012036

Pyrolysis of Corn Residues: Kinetic Analysis using Discrete Distributed Activation Energy Model

T Onsree[1], P Sittisun[1], R Sasaki[2] and N Tippayawong[1]*

1 Department of Mechanical Engineering, Chiang Mai University, Chiang Mai, 50200, Thailand
2 Department of Mechanical Engineering, Nagaoka University of Technology, Niigata, 940-2188, Japan
E-mail: n.tippayawong@yahoo.com

Abstract. Agro-residues are usually generated in large amount. These biomass materials may be utilized as renewable energy source via thermochemical conversion processes. Knowledge on kinetic of thermal decomposition behavior of biomass is important in reactor design and thermal process optimization. In this work, thermogravimetric (TG) data (35 to 900°C, at heating rates of 5 to 50°C/min) from pyrolysis of corn residues was analyzed using discrete distributed activation energy model (DAEM). Prediction accuracy of the simulated data was verified by comparing experimental with calculated data of conversion at selected heating rates. Kinetic parameters including activation energy and pre-exponential factor obtained from this method were found to give correct conversion curves for different heating rates simultaneously with correlation coefficients greater than 0.99, within the ranges considered. According to the discrete DAEM, variation in activation energy with conversion implied changes in mechanism during pyrolysis process. The activation energies were observed to be in the range of 100 – 120 and 80 – 180 kJ/mol for the macro-TG and typical TG analysers, respectively.

1. Introduction

Corn is a major economic crop in the world. It is widely planted for human food and animal feed. After harvesting, corn residues are usually left as waste. Some might be disposed of by burning openly in the fields, leading to high emissions of gaseous and particulate pollutants to atmospheric environment [1]. These agro-residues should be utilized for renewable energy, in which potentially 65,000 TJ per year was estimated for Thailand [2], [3].

Pyrolysis is a thermochemical process which converts raw biomass into valuable fuels. It is a promising method in utilization of biomass for renewable energy. By pyrolysis, the biomass material can be broken down to generate volatile gases, liquid products and solid char [4]. These products can be used as fuel or a source of useful chemicals [5]. It is therefore useful to understand the pyrolytic mechanism and its transformation process for subsequent generation of fuels, chemicals, and materials. Kinetic analysis is essential for engineering applications in views of design, building up, control and optimization of the thermal process. Kinetic parameters of reaction are necessary for characterizing and accurately predicting the thermal behaviours.

Kinetics of pyrolysis is usually investigated by a thermogravimetric (TG) analyser, which is among the most common and powerful instruments to study thermal behaviours of a few milligram of sample in the form of powder. It has been shown that pyrolysis is dependent on the biomass main components i.e. hemicellulose, cellulose, and lignin [6]. Thermal decomposition is strictly controlled by chemical kinetics. However, at a commercial scale of pyrolysis plant, big particle size and large amount of biomass are used, the pyrolytic reaction is limited by transport phenomena [7].

In TG and macro-TG analysis experiments, the pyrolytic behaviours (mass loss and mass loss rate) are measured as functions of temperature and time in an inert atmosphere [8]. By using the experiments data, many mathematical methods such as model-fitting and model-free (iso-conversional model) were used [9–12]. However, these methods are still rather limited in applying to the kinetics of biomass pyrolysis because of the high heterogeneousity such as catalysis of alkalis and secondary reactions of the pyrolysis process [13]. Distributed activation energy model (DAEM) is similar to model-fitting and shows good prediction for whole conversion range. The DAEM assumes that infinite irreversible independent parallel reactions occur simultaneously during pyrolysis. Such reactions share a constant pre-exponential factor (A) but the activation energy (E) is further assumed as a continuous distribution such as Gaussian distribution [14]–[16]. The infinite reactions are simplified by discretizing it to many but finite reactions as n parallel first-order reactions, known as discrete DAEM. Each of these reactions has its own activation energy and pre-exponential factor. Among these methods, the discrete DAEM is straightforward in making calculation and giving accurate prediction [17]–[20].

In the present study, pyrolysis of corn residues was investigated. The biomass residues in powdered form were tested in a typical TG analyser, while those in pelletized form were conducted in a macro-TG analyser. The kinetic behaviours of corn residue pyrolysis were evaluated using the discrete DAEM and the accuracy of predictions was examined.

2. Materials and methods

2.1. Experimental

In this study, corn residues were obtained from a local corn field in Chiang Mai, Thailand. The raw material was dried naturally. One part of the sample was then crushed and sieved with 120 mesh. The other part was densified in a pelleting machine to become 8 mm in diameter and 25 mm in length. Both samples were then stored in an airtight plastic bag at room temperature prior to experimentation.

The powdered sample was pyrolyzed in a Mettler Toledo TG analyser model 851e with 0.1 μg resolution. An initial weight of about 10 mg was placed in an alumina crucible and then heated from room to the final temperature of 900°C at three different heating rates of 10, 30, and 50°C/min under high purity nitrogen with a flow rate of 20 mL/min. The pelletized sample with 40 g each was pyrolyzed in a macro TG analysis setup with 0.01 mg resolution, shown in Figure 1. High purity nitrogen at a flow rate of 5 L/min was used as a carrier gas. The pellet was loaded in a basket and then heated from room to the final temperature of 900°C at three different heating rates of 5, 10, and 15°C/min. The data of weight loss and temperature evolution were monitored and recorded continuously. Experiments were repeated in triplicate for all conditions.

2.2. Kinetic analysis

2.2.1. Distributed activation energy model

Kinetic of solid-state pyrolysis in non-isothermal conditions at linear heating rate can be explained using Arrhenius equation [13]. The differential form of the reaction rate (r) can be expressed as follows:

$$\mathrm{r} = \frac{d\alpha}{dT} = -\frac{A}{\phi}\exp\left(-\frac{E}{RT}\right)\mathrm{f}(\alpha) \tag{1}$$

where α is the conversion, T is the absolute temperature (K), A is the pre-exponential factor (1/s), E is the activation energy (kJ/mol), R is the universal gas constant (8.314 J/mol·K), ϕ is the linear heating rate (K/s), f(α) is a reaction model and its independent variable, α, which can be defined as:

$$\alpha = \frac{m_0 - m(t)}{m_0 - m_f} = \frac{v(t)}{v_0} \tag{2}$$

IOP Conf. Series: Earth and Environmental Science **159** (2018) 012036 doi:10.1088/1755-1315/159/1/012036

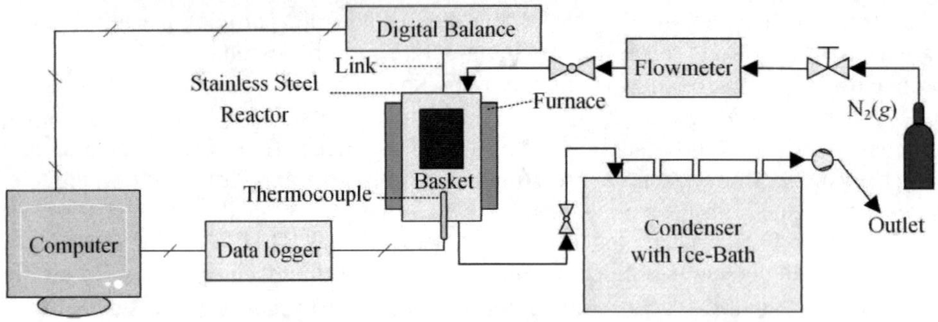

Figure 1. Schematic of a macro-TGA experiment.

where m_0 is the initial mass, $m(t)$ is the instant mass at any time t, and m_f is the final mass of the sample remaining after the pyrolysis reaction. $v(t)$ and v_0 are the mass of volatiles at any time and the initial mass of volatiles. Integral form of equation (1) is given as:

$$g(\alpha)=\int_0^{\alpha}\frac{d\alpha}{f(\alpha)}=-\frac{A}{\phi}\int_{T_0}^{T}\exp\left(-\frac{E}{RT}\right)dT=-\frac{A}{\phi}\psi(E,T) \tag{3}$$

which $\psi(E,T)$ is a temperature integral without an analytical solution**.** Replacing f(α) by first-order reaction, f**(**α**)** = 1**-**α, equation (3) can be rewritten in the following form:

$$1-\alpha=\exp\left[-\frac{A}{\phi}\psi(E,T)\right] \tag{4}$$

DAEM assumes that infinite irreversible independent parallel reactions occur simultaneously during pyrolysis [21]. All the reactions share a constant pre-exponential factor, and the activation energy is further assumed as a continuous distribution [22]. According to equation (4), the conversion function of DAEM over the entire temperature range becomes:

$$x=1-\alpha=\int_0^{\infty}\exp\left[-\frac{A}{\phi}\psi(E,T)\right]f(E)dE \tag{5}$$

where x represents the remaining conversion, $f(E)$ is the probability density function (PDF) of the activation energy distribution. Several PDFs such as Gaussian, Weibull, Gamma and others have been reported [23].

2.2.2. Discrete DAEM

The infinite reaction is simplified by discretizing it to finite reactions as n parallel first-order reactions. Each of these reactions has its own activation energy and pre-exponential factor [24]. For the pyrolysis of biomass under non-isothermal conditions, biomass decomposes via several such first-order reactions:

$$\frac{dm_i}{dT}=-\frac{A_i}{\phi_j}\exp\left(\frac{-E_i}{RT}\right)m_i \tag{6}$$

where m_i, A_i, and E_i are the instantaneous mass, pre-exponential factor, and activation energy of i-th reaction, respectively. The integral form of equation (6) is:

IOP Conf. Series: Earth and Environmental Science **159** (2018) 012036 doi:10.1088/1755-1315/159/1/012036

$$M_i = M_{i,0}\exp\left[-\frac{A_i}{\phi_j}\psi(E_i,T)\right] \tag{7}$$

where M_i is the mass of i-th reaction, $M_{i,0}$ is the initial value of M_i. From equations (6) and (7), the discrete DAEM can be expressed as:

$$x = \frac{\sum_{i=1}^{n} M_i}{\sum_{i=1}^{n} M_{i,0}} - \sum_{i=1}^{n} w_{char,i} = \sum_{i=1}^{n} f_{i,0}\exp\left[-\frac{A_i}{\phi_j}\psi(E_i,T_i)\right] \tag{8}$$

where $f_{i,0}$ is the initial mass fraction of i-th reaction, T_i and $w_{char,i}$ are the corresponding temperature and char mass of i-th reaction, respectively. Subscripts run from 1 to n for i and 1 to 2 which j = 1 and 2 means ϕ = 5 and 10°C/min, respectively. If the values of E_i and A_i are known, the unknown, $f_{i,0}$, in equation (8) will be solved by using linearly equation (9) as:

$$x = \begin{bmatrix} \Psi_1(E_0,T_0) & \Psi_2(E_0,T_0) & \dots & \Psi_n(E_0,T_0) & 1 \\ \Psi_1(E_1,T_1) & \Psi_2(E_1,T_1) & \dots & \Psi_n(E_1,T_1) & 1 \\ \vdots & \vdots & & \ddots & \\ \Psi_1(E_n,T_n) & \Psi_2(E_n,T_n) & \dots & \Psi_n(E_n,T_n) & 1 \end{bmatrix} \times \begin{bmatrix} f_{1,0} \\ f_{2,0} \\ \vdots \\ f_{n,0} \end{bmatrix} \tag{9}$$

where T_0 is the initial temperature of biomass and $\Psi(E_i,T_i)$ is the double exponential term:

$$\Psi(E_i,T_i) = \exp\left[-\frac{A_i}{\phi_j}\psi(E_i,T_i)\right] \tag{10}$$

To calculate the initial mass fractions decomposing in each reaction i, a set of reactions must be generated first with a known value of E and A by assuming that there will be a single reaction dominating at the same x under two different heating rates (e.g. ϕ_1 and ϕ_2). And one characteristic x represents the only reaction taking place at this x. Then, $f_{i,0}(\phi_1,T_1) = f_{i,0}(\phi_2,T_2)$ at two particular temperature, T_1 and T_2 so that the E_i can be calculated by:

$$\Psi_i(E_1,T_1) = \Psi_i(E_2,T_2) \tag{11}$$

To determine the pre-exponential factor, A_i, once E_i already known from equation (11) is used with the assumption of $\Psi_i = e^1 \approx 36.8\%$ which a value of $\Psi_i = e^1$ means that the conversion of a single first-order reaction would reach a maximum in the decomposition rate [17]. This allows A_i to be calculated from the double exponential term as equation (12) below:

$$\ln(\Psi_i) = -1 = \frac{A_i}{\phi_1}\left[T_0\exp\left(-\frac{E_i}{RT_0}\right) - \frac{E_i}{R}\int_{u_0}^{\infty}\frac{\exp(-u)}{u}du - T_2\exp\left(-\frac{E_i}{RT_2}\right) + \frac{E_i}{R}\int_{u_2}^{\infty}\frac{\exp(-u)}{u}du\right] \tag{12}$$

where $u_i = E_i/(RT_i)$. When A_i and E_i are known, the initial mass fraction, $f_{i,0}$, in equation (9) can, then, be solved by non-negative linear least squares method which is the *lsqnonneg* algorithm in Matlab software version 7.11.0; R2010b and the results must be guaranteed via the following constraints.

$$\begin{cases} f_{i,0} \in [1] \\ \sum_{i=1}^{n} f_{i,0} = 1 \end{cases} \tag{13}$$

3. Results and discussion

3.1. Thermal degradation characteristics

TG and derivative TG curves are shown in Figures 2a and b for the macro- and typical TG analysers. Both sets of curves showed similar patterns. Higher heating rates appeared to delay devolatilization to higher temperatures. For the same heating rate, temperature lagging effect was more apparent for pyrolysis in the macro TG than in the typical TG analysers. This was pointed out by Becidan et al. [7] that for the same degree of conversion, the macro-TG analyser took longer time, compared to typical TG analyser. It was also reported that the pyrolysis of wood occurred faster in micro TG than in macro-TG analysers because of the heat- and mass-transfer limitations of the larger samples used in the macro-TG analyser causing a lag in the temperature evolution of the biomass samples [25].

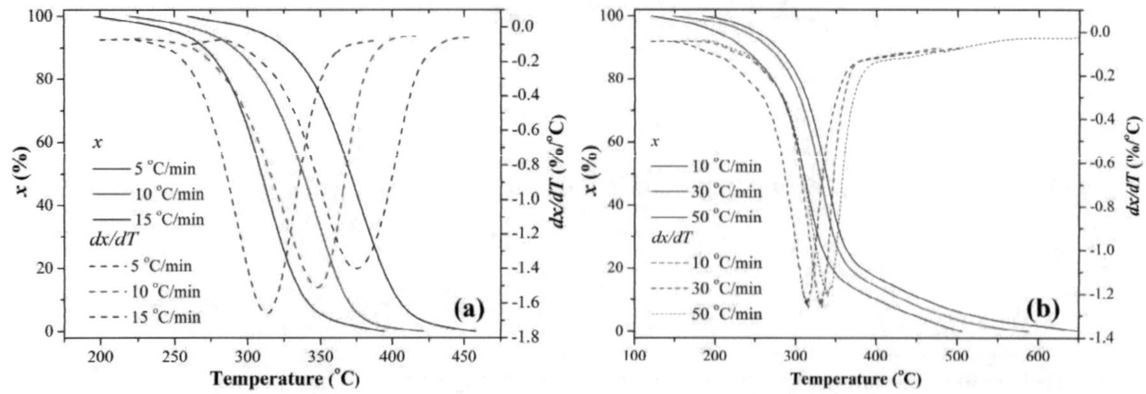

Figure 2. Remaining conversion and reaction rate for devolatilization of corn residues at different heating rates in (a) the macro-TG and (b) typical TG analysers.

3.2. Discrete DAEM kinetics

3.2.1. Modelling and simulation results

According to the discrete DAEM, each remaining conversion represents one first-order reaction which has a characteristic activation energy and pre-exponential factor. In this work, 39 parallel first-order reactions were assumed. From the discrete DAEM, three decomposition curves were obtained at 10, 30 and 50°C/min for the typical TG data and other three curves were obtained at 5, 10 and 15°C/min for the modified macro-TG data. Effect of temperature lag or delayed conversion on pyrolysis was taken into account when applying the discrete DAEM to the macro-TG data by shifting experimental TG curves to lower temperatures. The kinetic parameters obtained from the discrete DAEM were used in predicting pyrolysis behaviour of corn residue pellets. Results of simulated data from the pyrolysis of corn residues were subsequently compared with the corresponding experimental data, as shown in Figures 3a and b. It can be seen that the simulated data were closely similar to the experimental data. Correlation coefficients from regression analyses between the simulated and experimental data were evaluated to be greater than 0.99. The model was shown to be greatly accurate and effective in predicting pyrolysis behavior of corn residues.

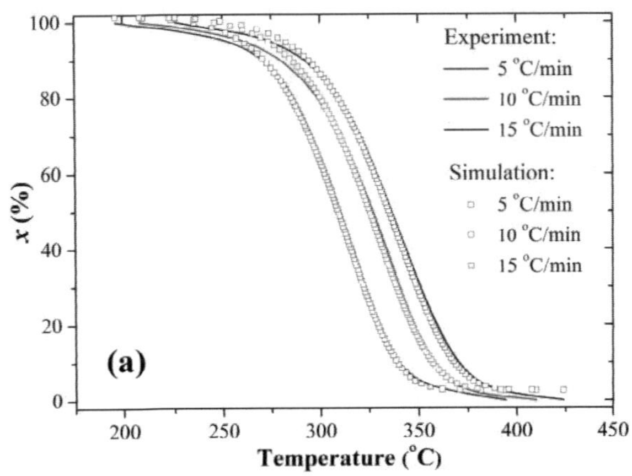

Figure 3. Comparison of simulated and experimental date at three different heating rates for (a) macro-TG and (b) TG analysers.

3.2.2. Kinetic parameters

Figures 4a and b show the kinetic parameters (activation energy, pre-exponential factor and initial mass fraction) derived from the macro-TG and typical TG experimental data. The activation energy was found to vary with conversion. For pyrolysis in the macro-TGA, as x increased, E was initially decreased from about 220 to 110 kJ/mol, changed slightly, and then increased to 190 kJ/mol towards the end of devolatilization step. The opposite trend was observed for pyrolysis in the typical TGA, whose E was varied between 40 to 180 kJ/mol. Variations of pre-exponential factor with conversion were similar to those of the activation energy. It should be noted that variation of activation energy with conversion can indicate change in decomposition mechanisms [20]. Although 39 reactions were assumed for the discrete DAEM, only several reactions contributed to the pyrolysis process, as can be seen from significant initial mass fractions shown in Figures 4a and b.

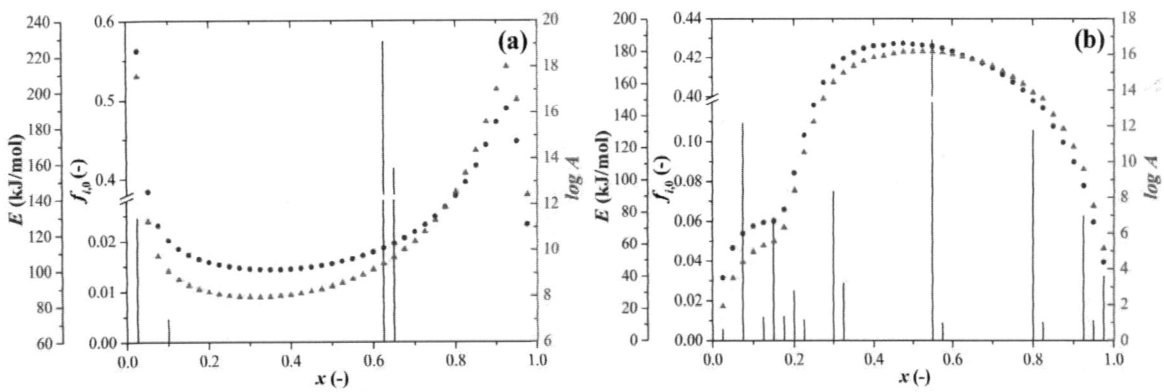

Figure 4. Discrete DAEM kinetic parameters (activation energy, pre-exponential factor and initial mass fraction) from (a) macro-TG and (b)TG analysers.

4. Concluding remark

Pyrolysis of corn residues in typical and macro-TG analysers was investigated. The thermal degradation behaviours were found to differ between the chemical kinetically controlled and the physically influenced reactions in typical and the macro-TG analysers, respectively. For a given conversion, effect of temperature lag was observed in the macro-TG cases, resulting from mass and heat transport limitation. Pyrolysis kinetics were also evaluated using the discrete DAEM. Simplification of the prediction process of thermal degradation of corn residues and good fits to the experimental data were obtained. The kinetic parameters derived from the discrete DAEM were found to simulate correctly the whole conversion curves simultaneously for different heating rates with high correlation coefficients.

5. References

[1] Tippayawong N, Pengchai P and Lee A 2006 *Inter. J. Environ. Sci. Technol.* **3** 359-69
[2] Tippayawong N, Chaichana C, Promwungkwa A and Rerkkriangkrai P 2011 *Energy Sustain. Dev.* **15** 69-72
[3] Wongsiriamnuay T and Tippayawong N 2015 *Biosys. Eng.* **139** 111–20
[4] Ceylan S and Topçu Y 2014 *Bioresour. Technol.* **156** 182–8
[5] Damartzis T, Vamvuka D, Sfakiotakis S and Zabaniotou A 2011 *Bioresour. Technol.* **102** 6230–8
[6] Meng A, Chen S, Zhou H, Long Y, Zhang Y and Li Q 2015 *Fuel* **157** 1–8
[7] Becidan M, Skreiberg O and Hustad J E 2007 *Fuel* **86** 2754–60
[8] White J E, Catallo W J and Legendre B L 2011 *J. Anal. Appl. Pyrolysis* **91** 1–33
[9] Šimon P 2004 *J. Therm. Anal. Calorimetry* **76** 123-9
[10] Sbirrazzuoli N, Vincent L, Mija A and Guigo N 2009 *Chemometrics Intel. Lab. Sys.* **96** 219–26
[11] Lin Y, Ma X, Ning X and Yu Z 2015 *Energy Convers. Manag.* **89** 727–34
[12] Jayaraman K and Gökalp I 2015 *Energy Convers. Manag.* **89** 83–91
[13] Vyazovkin S, Burnham A K, Criado J M, Pérez-Maqueda L A, Popescu C and Sbirrazzuoli N 2011 *Thermochim. Acta* **520** 1–19
[14] Wang J, Lian W, Li P, Zhang Z, Yang J, Hao X, Huang W and Guan G 2017 *Fuel* **207** 126–35
[15] Wang S, Lin H, Ru B, Sun W, Wang Y and Luo Z 2014 *J. Anal. Appl. Pyrolysis* **108** 78–85
[16] Cai J, Wu W and Liu R 2013 *Bioresour. Technol.* **132** 423–6
[17] Scott S A, Dennis J S, Davidson J F and Hayhurst A N 2006 *Chem. Eng. Sci.* **61** 2339–48
[18] Caballero J A and Conesa J A 2011 *Thermochim. Acta* **525** 40–9
[19] Cao H, Xin Y, Wang D and Yuan Q 2014 *Bioresour. Technol.* **172** 219–25
[20] Chen Z, Zhu Q, Wang X, Xiao B and Liu S 2015 *Energy Convers. Manag.* **105** 251–9
[21] Cai J and Liu R 2008 *Bioresour. Technol.* **99** 2795–9
[22] Wang S, Lin H, Ru B, Dai G, Wang X, Xiao G and Luo Z 2016 *Fuel* **185** 763–71
[23] Cai J, Wu W and Liu R 2014 *Renew. Sustain. Energy Rev.* **36** 236–46
[24] Zhou B, Zhou J and Zhang Q 2017 *Bioresour. Technol.* **241** 113–9
[25] Skreiberg A, Skreiberg O, Sandquist J and Sorum L 2011 Fuel **90** 2182-97

Acknowledgments

We wish to acknowledge supports from the Research and Researcher for Industry PhD program of the Thailand Research Fund (contract no. PHD57I0059), and Chiang Mai University.

IOP Conf. Series: Earth and Environmental Science **159** (2018) 012037 doi:10.1088/1755-1315/159/1/012037

Bio-oil Production from Ablative Pyrolysis of Corncob Pellets in a Rotating Blade Reactor

N Khuenkaeo[1] and N Tippayawong[1]

1 Department of Mechanical Engineering, Chiang Mai University, Chiang Mai, 50200, Thailand
E-mail: n.tippayawong@yahoo.com

Abstract. Fast pyrolysis serves as an alternative and eco-friendly method to convert biomass into valuable bio-oil and char. Bio-oil is a potential source for generating renewable biofuels. In this study, corncob pellets were pyrolysed in a laboratory rotating blade ablative pyrolysis reactor. Effects of plate temperature (450 – 550°C) and nitrogen flowrate (5 – 15 L/min) were investigated on pyrolysis product yields at a fixed rotating speed of 6 Hz. Temperature and inert gas flowrate were shown to affect product distribution and yields. Maximum bio-oil yields of almost 50% w/w on dry basis were obtained at 550°C and flowrate of 5 L/min. Composition of the bio-oil contained predominantly oxygenated hydrocarbons. From analysis of chemical and physical properties, the bio-oil derived from corncob pellets showed good fuel quality, comparable to those bio-oils reported in the literature.

1. Introduction

Corn is an important industrial crop in the northern region of Thailand. Large amount of corn residues such as corncobs are generated after harvest. Some of these residues have been disposed of improperly such as burning openly in farms. This practice may lead to wildfires, accidental fires to agricultural lands, causing local atmospheric pollution [1]. These biomass residues may be utilized for renewable energy. The use of biomass for energy and chemicals has tremendous potential to reduce the problems caused by our dependence on fossil fuels [2]. A simple physical upgrade of corncobs by densification has been proposed [3]. The ago-residues can be further improved to produce liquid biofuel via fast pyrolysis.

Fast pyrolysis is an effective thermochemical process for biomass conversion, in which biomass is rapidly heated in the absence of oxygen, and operated at reaction temperatures between 300 and 700°C [4], generating a majority of condensable vapours, solid char, and permanent gases. After quenching and rapid condensation of the vapours, a dark brown liquid called bio-oil is formed [5]. The bio-oil can either be combusted in boilers, engines and turbines to generate heat and power or upgraded to produce transportation fuels and chemical stocks [6], [7].

There are many types of fast pyrolysis reactor, including bubbling fluidized bed, circulating fluidized bed, free fall, auger, and ablative reactors. For most pyrolysis reactor configurations, the biomass feedstock needs to be processed into small sizes of around 2 mm or less to minimize heat transfer resistances throughout the material [8]. Fortunately, the ablative pyrolysis reactor provides an opportunity to use large particles of biomass which can potentially offer saving on processing costs [9, 10]. In ablative pyrolysis, the biomass is directly in contact with hot surface. There is a steep temperature gradient at the biomass surface, undergoing severe decomposition or sublimation and leading to the formation of a thin layer of reacting solid, shown schematically in Figure 1. The reacting layer moves towards the inside of the cold biomass. The conversion takes place only at the surface rather than the entire biomass particle, and the reaction rates are not limited by heat transfer throughout the whole particle [11].

In this work, a laboratory scale rotating blade ablative reactor was designed, developed and tested with corncob pellets. The goal was to experimentally verify if high yield of bio-oil can be produced. Operating factors that affect bio-oil yield were also parametrically evaluated.

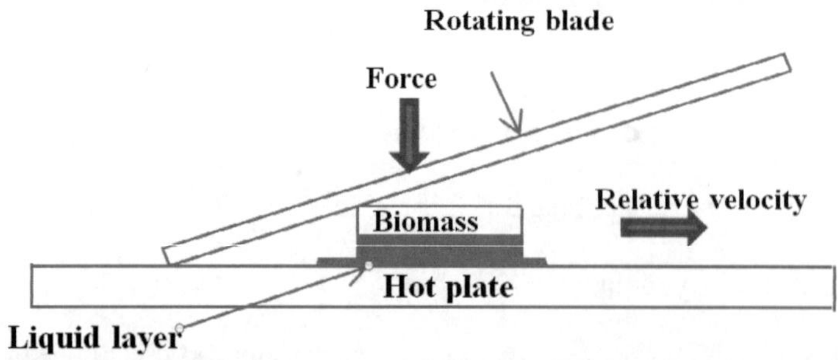

Figure 1. Basic principle of contact ablative pyrolysis mechanism.

2. Materials and Methods

2.1. Feedstock

Corncob pellets as feedstock in the experiment were supplied from Mae Cham agricultural cooperative in Chiang Mai, Thailand. The pellets were about 8–10 mm in size and kept in airtight zipper bags prior to use. The composition and energy content of the feedstock are given in Table 1 below.

Table 1. Composition of corncob pellet feedstock.

Proximate analysis (% w/w)				
Moisture	Ash	Volatile	Fixed carbon	
10.7	5.5	67.3	16.5	
Ultimate analysis (% w/w)[a] and energy content				
C	H	O	N	HHV(MJ/kg)[b]
49.1	6.3	44.1	0.4	20.1

[a] dry basis
[b] estimated from HHV (MJ/kg) = (C% x 0.3578) + (H% x 1.1356) + (N% x 0.0594) – (O% x 0.0854) – 0.974 [12]

2.2. Experimental setup

A rotating blade ablative reactor was designed and developed for fast pyrolysis of biomass. The ablative reactor was made of steel, cylindrical in shape with 260 mm in diameter, 80 mm high and 10 mm thick. The rotating blade had two propellers which were at 53° angle with a horizon, connected to axle shaft of motor. The rotational speed was adjustable by an inverter. The top of the reactor was a movable cover with static seal by graphite gasket placed between the cover and the reactor chamber. The hot plate formed the bottom of the reactor, directly heated by liquefied petroleum gas flame. Temperature of the hot plate was measured using a type K thermocouple and controlled manually via a gas regulator.

An experimental setup (shown in Figure 2) was composed mainly of a biomass hopper, nitrogen feeding, the ablative reactor, condenser and cooling water, bio-oil collection, and pyro-gas collection. The biomass hopper of 1.5 L capacity was made of brass steel material. It was connected with a nitrogen feeding to the ablative reactor. Downstream of the reactor was a counter-current, double-pipe glass condenser connected with oil and gas collection systems. Installation of a vacuum pump was connected to the system for driving pyrolysis vapours out of the reactor chamber.

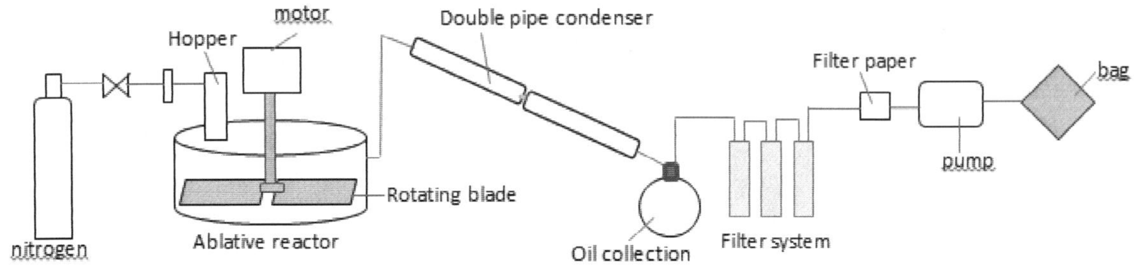

Figure 2. Schematic diagram of rotating blade ablative reactor system.

2.3. Experimental procedure

Prior to the start of each experiment, preheating was initiated and nitrogen was fed to the reactor to purge any oxygen or air inside. The biomass feedstock was loaded to the hopper, with the bottom valve closed and upper vent of the hopper open while feeding nitrogen in order to sweep any oxygen or air in the hopper. The required nitrogen flowrate was adjusted. Once the hot plate reached the set temperature, the rotating blade was set in motion and the biomass pellet was released from the hopper. Ablative contact between the hot plate with the biomass initiated fast pyrolysis. Volatile vapours were released and carried quickly by the nitrogen out of the reactor chamber to the cooling water condenser, where bio-oil was condensed. The remaining incondensable vapours flowed through the gas conditioning unit and collected in a gas bag. Each experimental run lasted for about 5-10 min in which no further release of volatile vapours was observed. The remaining solid char was then collected and kept in a closed container while all bio-oil samples were stored in a refrigerator for further analysis.
Effects of the hot plate temperature (450 – 550°C) and nitrogen flowrate (5 – 15 L/min) on pyrolysis product yields were investigated at a fixed rotating speed of 6 Hz. Each experimental condition was carried out and repeated at least in triplicate. The aim was to find out the optimum condition for obtaining highest bio-oil yield.

2.4. Product analysis

Pyrolysis product distribution and yields were determined. Each yield was relative to the initial loading of biomass. Elemental composition of the bio-oil and solid char were evaluated using a Thermo Scientific, CHNS/O analyser model Flash 2000. Gas chromatography and mass spectrometry (GC/MS) was used to determine chemical composition of the bio-oil. Pure acetone was used as solvent. The GC/MS analysis was performed using an Agilent Technologies model 7890A GC equipped with an Agilent Technologies 19091s-433 column (30 m x 0.25 mm x 0.25 μm) and an Agilent Technologies MS model 5975C. The carrier gas was helium at a flowrate of 1 L/min. The column temperature was initially held at 40°C for 5 min, then gradually increased to 280°C at 10°C/min, and finally maintained at 280°C for 10 min. The injector and detector temperatures were set at 250 and 280°C, respectively.

3. Results and discussion

3.1. Effects of nitrogen flowrate and hot plate temperature

Figure 3 shows mass balances, product distribution and yields of the ablative pyrolysis experiments. Bio-oil, solid char and incondensable gases were obtained. It was generally accepted that in fast pyrolysis, bio-oil is dominant, while char and gases are by-products. All cases of accumulative mass were found to be above 96% which were in acceptable level. Incremental mass loss may have related to losses during collection of liquid and solid samples, as well as undetectable gases.
With regards to product yield, nitrogen flowrate and hot plate temperature appeared to have significant effect. It was observed that for a given flowrate, bio-oil yield was increased with increasing hot plate temperature. This was expected, since higher plate temperature resulted in higher temperature gradient

and higher heating rate to the biomass sample, hence greater decomposition of the biomass and more volatile release may have occurred [11]. At the lowest temperature considered here, char yields were higher than bio-oil yield. This was possibly due to low heating rate, leading the reaction to become intermediate or slow pyrolysis. For a fixed temperature, increase in flowrate was found to decrease the yield of bio-oil. It was probable that higher flowrate may have reduced overall heat transfer to the biomass due to convection. The highest yield of the bio-oil obtained here was about 50% w/w on dry basis. This was smaller than those reported in the literature [11] in which 70% was typically obtained for similar configuration of reactor. It should be noted that at high flowrate, gas yields were increased, at the expense of both bio-oil and char yields. This was probably due to combination between secondary char decomposition and secondary thermal cracking of pyrolysis vapours [13].

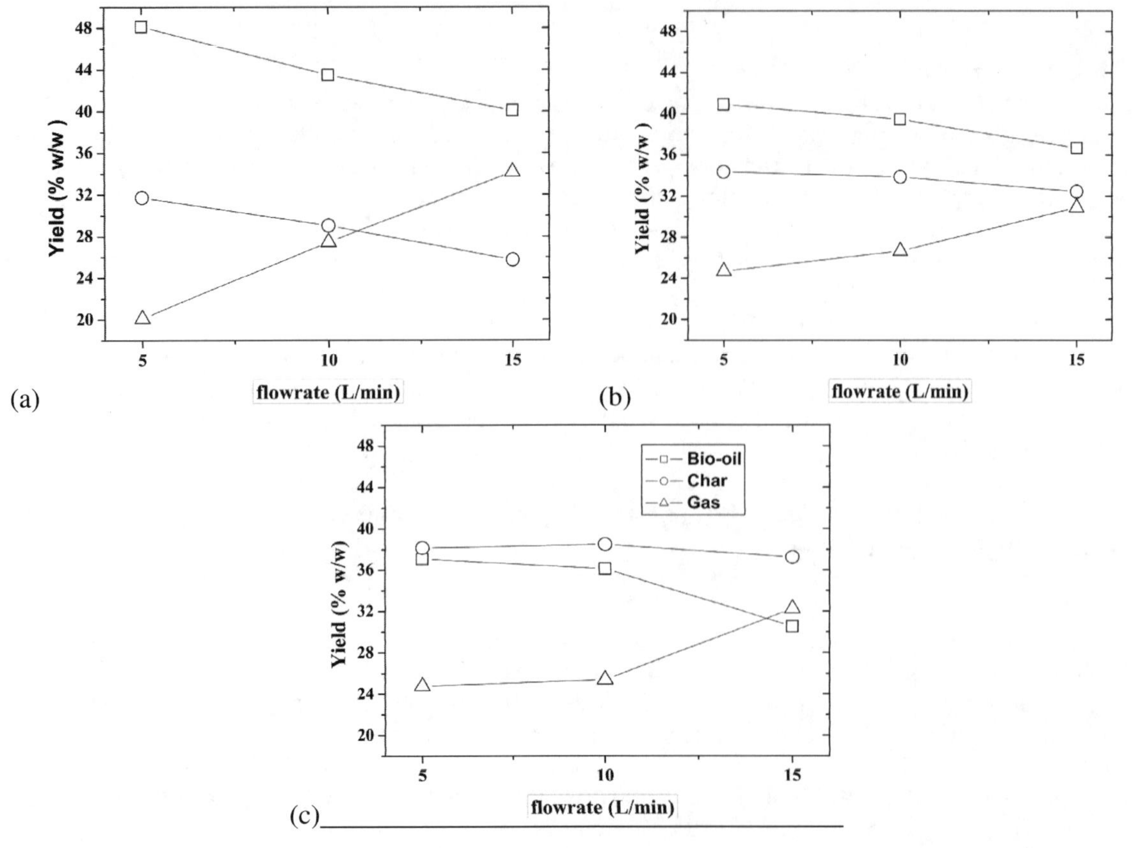

Figure 3. Effects of nitrogen flowrate and hot plate temperature on product distribution and yields, (a) for 550°C, (b) 500°C, (c) 450°C.

3.2. Characteristic of the bio-oil

Ultimate analysis and heating value of the bio-oil were determined and compared with literature data for bio-oils produced from ablative and fluidized bed pyrolysis of various agro-residues, shown in Table 2. It was found that the percentages of carbon, hydrogen and oxygen and the heating value were in similar range to other bio-oils. However, H/C and O/C atomic ratios were larger. This indicated that the bio-oil from this work had lower aromaticity and lower degree of deoxygenation, compared to other bio-oils. Taking into account the bio-oil heating values and their yields, the energy recovered in the form of liquid fuel was accounted for about 45% of the biomass energy input.

Chemical compounds of the bio-oil from the ablative pyrolysis process were also identified using GC/MS. The results are shown in Figures 4 a and b. It was observed that the bio-oils were composed of complex mixture of oxygenated hydrocarbons. The compounds observed in this investigation consisted of long chain products such as phenols, aromatics, esters, aldehydes, alcohols, organic acids and ketones.

Table 2. Comparison of elemental composition of bio-oils from this work with those in the literature.

Ref	This work	[9]	[14]
biomass	corncob pellet	pines	rice husk
reactor	ablative	ablative	fluidized bed
C	33.22	37.53	35.95
H	10.64	8.38	9.81
O	55.86	54.05	53.94
N	0.28	0.03	0.59
H/C	0.32	0.22	0.27
O/C	1.68	1.44	1.50
HHV (kJ/kg)	18.35	17.38	18.52

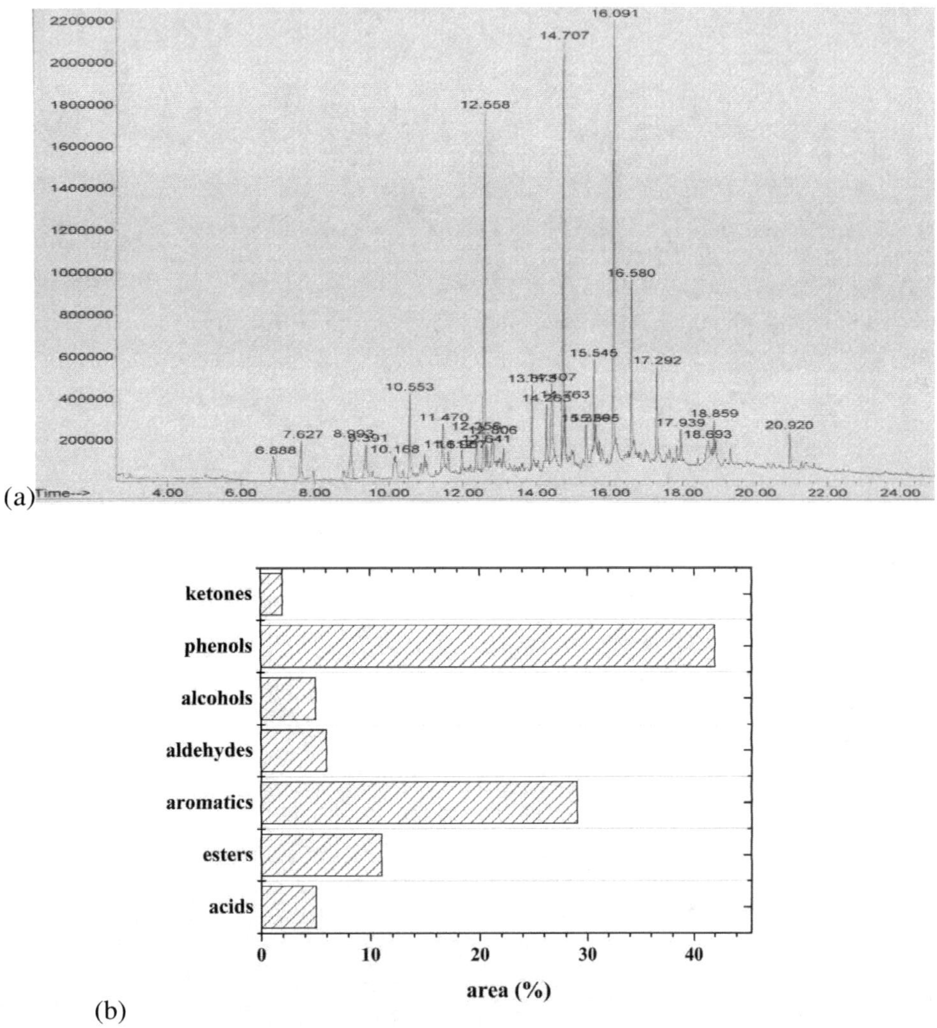

Figure 4. GC/MS analysis result and organic compounds of the bio-oil.

IOP Conf. Series: Earth and Environmental Science **159** (2018) 012037 doi:10.1088/1755-1315/159/1/012037

4. Conclusion

Agro-residues have great potential to produce bio-liquid fuels. Bio-oil from fast pyrolysis of agro-residues is of great interest in providing alternative source of renewable energy and biofuel. In this work, fast pyrolysis of corncob pellet in a rotating blade ablative reactor was reported. Pyrolysis products were found to be strongly affected by the hot plate temperature and sweeping gas flowrate. Optimum condition was identified at 550°C and flowrate of 5 L/min. High yields of bio-oil (50% w/w dry basis) was realized. The GC/MS analysis showed that the ablative pyrolysis of corncob pellets generated bio-oil with various phenolic compounds, and aromatics. In the present work, it was shown that corn residues were promising renewable sources for bio-oil production.

5. Reference

[1] Tippayawong N, Pengchai P and Lee A 2006 *Inter. J. Environ. Sci. Technol.* **3** 359-69
[2] Raguskas A J, Williams C K, Davidson B H, Britovsek G, Cairney J, et al. 2006 *Science* **311** 484-9
[3] Wongsiriamnuay T and Tippayawong N 2015 *Biosys. Eng.* **139** 111–20
[4] Demirbas M F, Balat M and Balat H 2011 *Energy Convers. Manag.* **52** 1815-28
[5] Guillain G, Fairouz K, Mar S R, Monique F and Lede J 2009 *Bioresour. Technol.* **100** 6069-75
[6] Visput T P, Zhang H, Sanna A, Xiao R, Huber G W 2010 *Science* **330** 1222-7
[7] Mortensen P M, Grunwaldt J D, Jensen P A, Knudsen K and Jensen A D 2011 *Appl. Catal.* A **407** 1-19
[8] Bridgwater A V 2003 *Chem. Eng. J.* **91** 87-102
[9] Luo G, Chandler D S, Anjos L C A, Eng R J, Jia P and Resende F L P 2017 *Fuel* **194** 229-38
[10] Apfelbacher A, Conrad S and Schulzke T 2014 *Environ. Prog. Sustain. Energy* **33** 669-75
[11] Lede J 2003 *J. Anal. Appl. Pyrolysis* **70** 601-18
[12] Leo G and Resende F L 2014 *J. Anal. Appl. Pyrolysis* **110** 100-7
[13] Pattiya A 2011 *Bioresour. Technol.* **102** 1959-67
[14] Cai W and Liu R 2016 *Fuel* **182** 677-86

Acknowledgment

We wish to acknowledge the financial support from Chiang Mai University.

IOP Conf. Series: Earth and Environmental Science **159** (2018) 012038 doi:10.1088/1755-1315/159/1/012038

Technical Problems of Wood Chips Utilization

I Czupy[1], F Szűcs[2] and A Vágvölgyi[3]

1 Associate Professor, University of Sopron, Sopron, Hungary
2 Sales Manager, Vértesi Erdő Ltd, Tatabánya, Hungary
3 Lecturer, University of Sopron, Sopron, Hungary
E-mail: czupy.imre@uni-sopron.hu

Abstract. The Hungarian renewable energy potential based on biomass. Wood biomass constitutes the large part of the raw material of operating heating plants. A significant part of raw material comes from forestry as firewood and harvesting losses. A lower part stems from short rotation energy plantation (currently only a few thousand hectare energy plantation are in Hungary), and attempts were made to utilize biomass production of non-conventional areas, as well (e.g. biomass from abandoned land, or located next to line-based facilities). The wood based biomass is utilized by the heating plants mainly as wood chips. This technology proposes more challenges: moisture content decisively influences the fuel value; optimization of moisture content depends on storage; combustion technique respect is important the fraction distribution etc. Our research based on the examination of these factors and the demonstration of the problems of wood chip management.

1. Introduction

The energy demand has shown a slow growth in Hungary. The country's dependency on energy import is quite high: in 2016th year, the domestic energy production was 424 436 TJ, while the import was 804 160 TJ, which means that the value of energy import is more than 65%. The share of the renewable energy consumption is around 9% and the solid biomass is around 82% (Figure 1) [1].

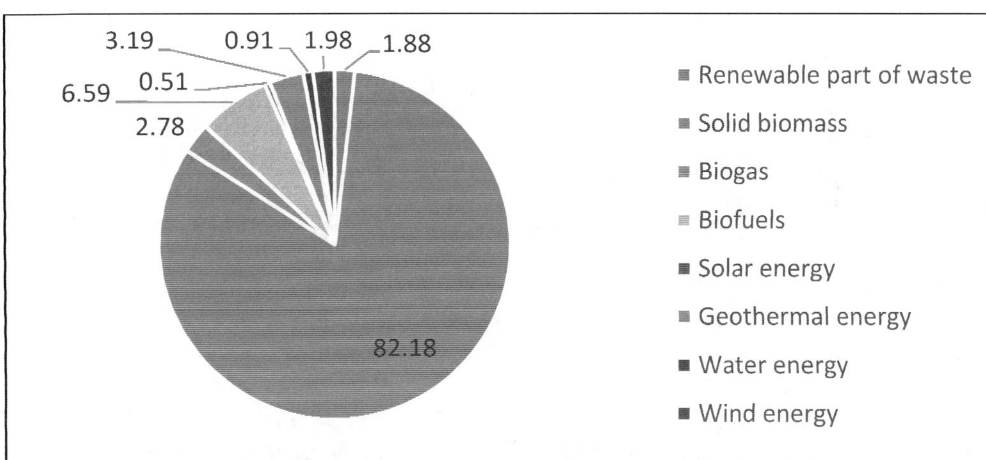

Figure 1. Primary renewable energy utilization in Hungary (%, in 2014) [1].

The Hungary's policy on renewable energy based on the EU strategies. The European Union aims to achieve three main goals by 2020:

- reducing the greenhouse gas emissions by 20%;
- reducing the energy consumption by at least 20%;
- achieving a 20% share of renewable energy in the global energy mix.

Hungary expected to reach at least 13%. In 2010th year the Hungarian Parliament adopted Hungary's Renewable Energy Utilization Action Plan extending the minimum required value of the share of the renewable energy in the global mix from 13% to 14.65%.

The main aim is to increase the use of renewable energy sources in Hungary to 186.3 PJ by 2020, including biomass with the target of 130.8 PJ [2].

Hungary's biomass resource estimated at 350-360 tonnes, representing two thirds of the renewable energy sources. Only one sixth of this resource currently utilized. Dendromass, i.e. wood-based biomass, constitutes a large part of biomass [3].

According to *Molnár* et al. [4], wood used for energy purposes may originate from four sources:

- the firewood of the traditional forest management (public or private sector);
- harvesting losses produced during tree utilization;
- wood industry by-products;
- timber of energy tree crops that is primarily utilized as wood chips (Figure 2).

Figure 2. Dendromass sources (Photo *F Szűcs*).

The benefits of dendromass utilization:

- sustainability;
- stability of the generation capacities;
- low production costs;
- easy manageability and industrial use;
- high calorific value (on average 13 MJ/kg, in case of water content of 25-30%);
- low ash and sulphur content;
- increase in rural employment.

The available quantity of dendromass examined several times. According to Molnár et al. [4], in Hungary ca. 3.5 million tonnes per year provided for energy purposes. The data of National Food Chain Safety Office showed that in recent years the amount of woodcut was 7.5 million gross m3 per year on average, which is 60% of the current growth (Figure 3).

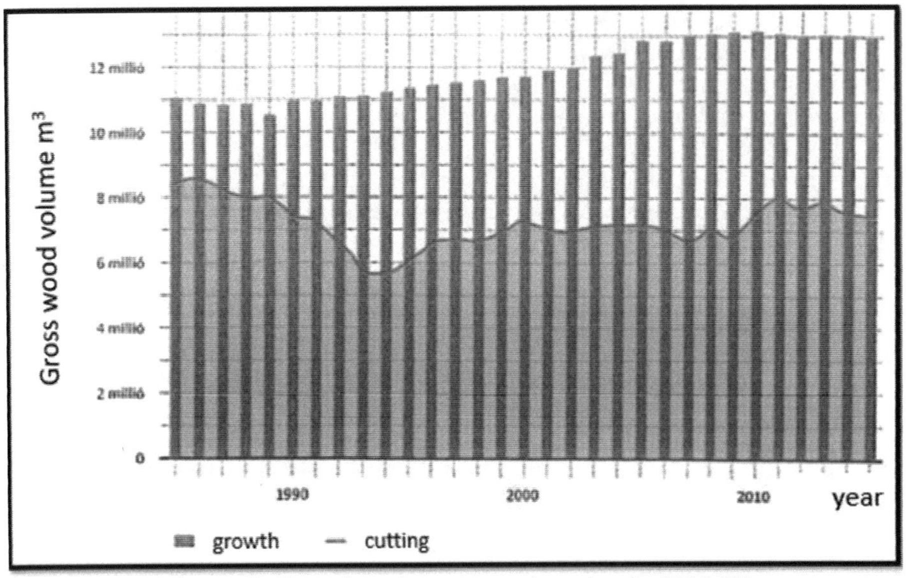

Figure 3. Growth and gross wood cutting in 2015 [5].

2. Utilization and problems of dendromass at Tatabánya power plant

The natural gas-fired power plant was converted to a wood chip-fired one in 2015-2016. The net value of the investment was HUF 6.2 billion. The total capacity of the fluidized bed technology is 94 MW. A 20 MW hot water boiler and two 37 MW steam boilers produce renewable energy. The power plant provides heat and hot water to 23 500 households and 2000 institutions. The plant utilizes 100 000 tonnes of wood chips as fuel per year on average. Dendromass utilization can reduce greenhouse gas emission by 67 000 tonnes per year. 1 m3 natural gas can be replaced with 2.5 kg wood chips.

In order to make the power plant sustainable, the adequate supply of raw materials has to be provided. Between 01.11.2016 and 04.30.2017 the quantity of wood chips was 69 636.52 lutro tonnes delivered by 2798 trucks to the power plant; the average weight of the trucks was 24.89 lutro tonnes/truck as illustrated in Figure 4.

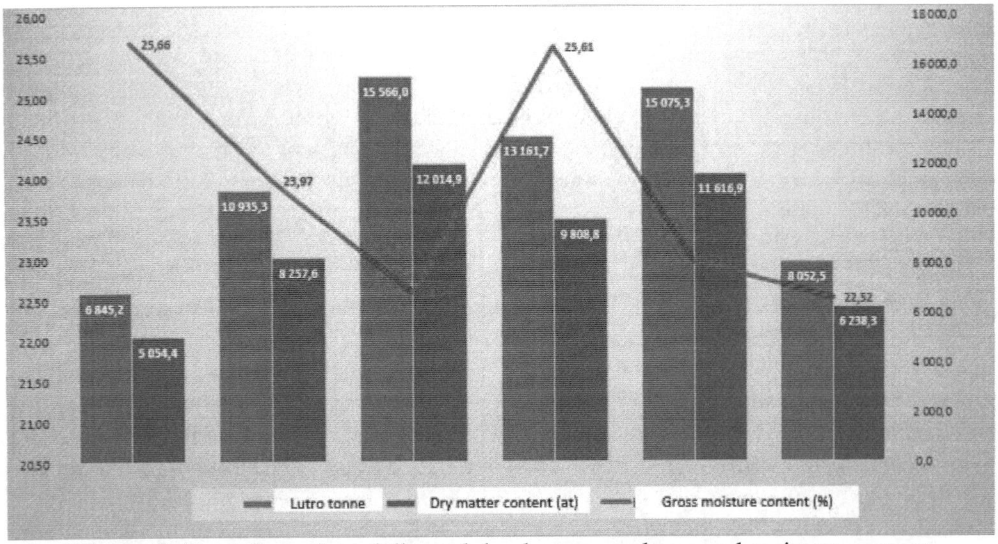

Figure 4. Tatabánya plant delivered dendromass volume and moisture content.

The basic condition of firewood supply is biomass integration. There are 12 members of the integration. In 2005 Vértes Erdő Ltd. launched its biomass program which provides renewable fuel for

IOP Conf. Series: Earth and Environmental Science **159** (2018) 012038 doi:10.1088/1755-1315/159/1/012038

Tatabánya Power Plant and Dunaújváros Power Plant in 2017. Thanks to biomass integration, the quantity of wood-based energy fuel marketed by the national forestry is 150 000 tonnes. The integration has currently 27 members, including private and public forestry sectors, out of which 12 are involved in the delivery to the Tatabánya Power Plant. The integration has long-term business benefits for both Vértes Erdő Ltd. and its partners. The supply of materials may come from several sources, e.g. energy tree crops.
The article demonstrates the qualities of acacia as well as the operation. The crop is three years old (2nd sprouting), the average number of sprouts is 5/per stem, the size of the area is 13.28, the production quantity is 609.52 tonnes, the yield is 45.89 t/ha, the average gross moisture content is 27.26%, and the dry matter content is 443.9 atro tonnes.
Figure 5 illustrates the machines and fuel consumption of energy plantation harvesting.

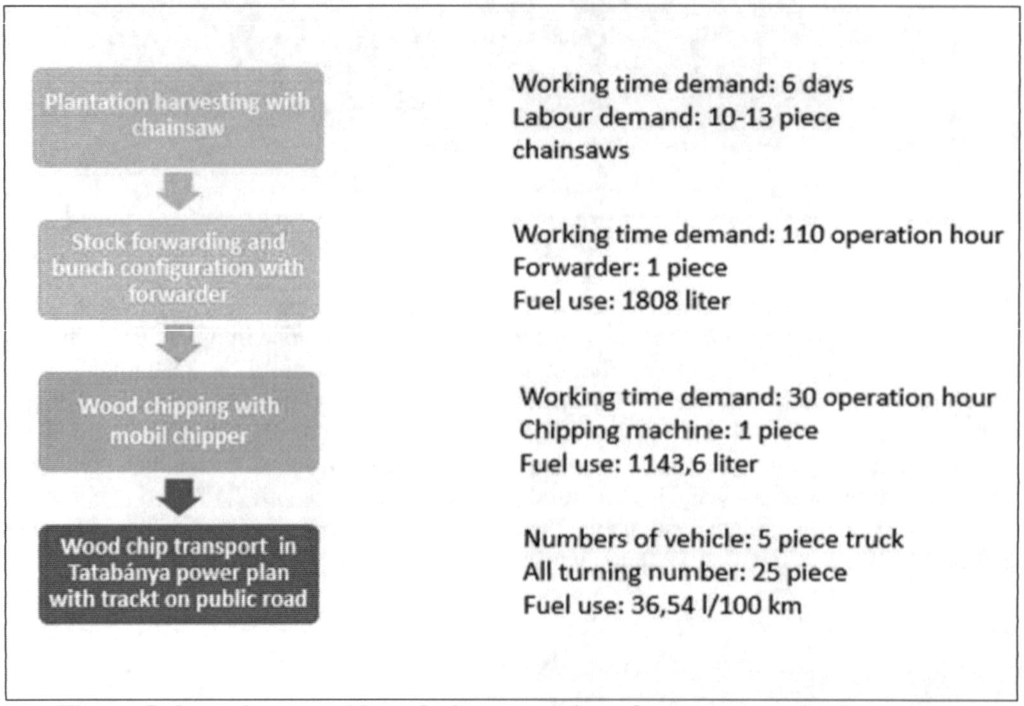

Figure 5. Operations, machines, fuel consumption of energy plantation harvesting.

The adequate water content (under 30%) can be guaranteed only by the management of wood chips.
The storage conditions (the size of chip, the size of the pile, the storage method, the aeration, etc.) affects the quality of wood chip. The moisture content of wood chips is a significant factor, which affects the calorific value and whose quantity has a relation to the temperature of the pile. At the beginning of the storage the moisture content of wood chips is high (ca. 55% in case of poplar chips), which results in a sudden temperature increase in the first phase of the storage. Due to the temperature increase, thermophilic and mesophilic fungi proliferate causing damage to wood and further increasing the temperature (a maximum of around 60% in case of poplar chips) [5].
The fungi can significantly impair the energy content and the quality of wood by the dissimilation of lignin; therefore, despite the decrease of the moisture content that improves the calorific value, wood will not be more beneficial in terms of energy utilization [6]. The fungi damage wood and may be harmful to human health during the storage. These problems cause dry matter and energy losses and have impact on the quality.
Austrian standard ÖNORM M 713 is applied for size classification of wood chips, which distinguishes fine-grade (G30), medium (G50) and coarse wood chips (G100). The most common storage method is G50 due to the simplicity of its utilization. However, research has shown that in case of poplar chips, mainly this size of wood chip generates the above mentioned problems [4].

It is expedient to store wood chips in piles for a short time (maximum 3 months) and to turn the pile roughly every two weeks.

The basic conditions of fuel supply were the increase of delivery capacity and the establishment of a logistics centre. For this purpose, Vértes Erdő Zrt. purchased delivery devices, increased storage capacity and established a logistics centre in Gánt. In the centre, there is a 4000 m2 depot which allows to meet the demand for safe fuel at Tatabánya Power Plant, even under unfavourable weather conditions. The logistics centre is 30 km far from the plant. The conditions of the production and management of wood chips, and of the water content measurement are given; moreover, a 60 tonnes weighbridge is also available.

Quality problems occurred mainly due to the fraction size of the wood chips. The heating system is technologically unsuitable for selecting wood chips that are longer than 120 mm. These separated with Doppstadt SM 620 plus type mobile star riddle (Figure 6). The output of the grading equipment is 1 ton/minute providing safe fuel to the heating plant. The selected oversized wood resized and utilized for energy purposes. Over the period under examination, the share of oversized wood chips covers the 2.2% of the total quantity.

Figure 6. Doppstadt SM 620 plus type mobile star riddle (Photo *F Szűcs*).

3. Conclusion

Hungary currently imports 65% of its energy needs. Renewable energy sources, including biomass utilization, play an essential role in reducing dependency on import. Dendromass, i.e. wood-based biomass, constitutes a significant part of biomass.

The authors of the article demonstrated the experience gained in the first heating season of the dendromass heating system, with a total output of 94 MW, established in 2015-2016 in Tatabánya, considering fuel supply and quality of fuel (wood chip). Creating an effective logistics system and guaranteeing the quality of wood chips are key factors: moisture content of less than 30% and the standard size of fraction.

4. References

[1] www.mekh.hu *(Electronic Materials).*

[2] Vágvölgyi A 2013 A fás szárú energetikai ültetvények helyzete Magyarországon napjainkig, üzemeltetésük, hasznosításuk alternatívái (Sopron) Doctoral thesis p 195.

[3] Czupy I, Vágvölgyi A and Horváth B 2012 The Biomass Production and its Technical Backgorund in Hungary. *Proc. Int. Conf. on Forestry Mechanization: "Forest Engineering: Concern, Knowledge and Accountability in Today's Environment"* ed T Pentek, T Prosinsky and M Sporcic (Zagreb: University of Zagreb) DOI 10.13140/2.1.4711.5207.
[4] Molnár S, Pásztory Z and Komán Sz 2013 A faenergetika minőségi fejlesztésének szakmai megalapozása (mire elég a magyar dendromassza?!) *(Electronic Materials*: www.fataj.hu).
[5] Horváth Zs, Marosvölgyi B, Idler C, Pecenka R and Lenz H 2012 Storage problems of poplar chips from short rotation plantations with special emphasis on fungal development *Acta Silv. Lign. Hung.* **8** pp 123-132.
[6] Barkóczy Zs 2009 A dendromassza alapú decentralizált energiatermelés alapanyagbázisának tervezése (Sopron). Doctoral thesis Reference lists.

Acknowledgments
The described article was carried out as part of the „Sustainable Raw Material Management Thematic Network – RING 2017", EFOP-3.6.2-16-2017-00010 project in the framework of the Széchenyi2020 Program. The realization of this project is supported by the European Union, co-financed by the European Social Fund.

IOP Conf. Series: Earth and Environmental Science **159** (2018) 012039 doi:10.1088/1755-1315/159/1/012039

Modelling of Aluminium-fuelled Power Plant with Gas Turbine

M S Vlaskin, M V Gaykovich[1] and A O Dudoladov

1 Joint Institute for High Temperatures of the Russian Academy of Sciences, Izhorskaya 13 Bldg 2, Moscow 125412, Russia
E-mail: advert.geom@gmail.com

Abstract. Presented work is devoted to the modeling of aluminum-fuelled power plant based on aluminum-water reactor, steam-hydrogen turbine, condenser and air-hydrogen fuel cell battery. Parameters of aluminum-water reactor were the following: pressure – 15 MPa, temperature – 600 K, steam to hydrogen mass ratio – 40. It was shown that without gas turbine the plant's electrical and total efficiencies are 12 % and 72 % respectively. If gas turbine with internal efficiency of 80 % is used the electrical efficiency of plant increases up to 25 ÷ 30 %. Total efficiency of power plant in this case can reach 80 %.

1. Introduction

Aluminum has been proposed for the role of perspective non-organic energy storage mater by a number of researchers [1]-[5]. Potentially high level of aluminum integration in future energy economy is bounded with high content of aluminum in the earth's crust, safety and moderate cost of aluminum storage and transportation [6], as well as possible regeneration of such energy carrier.

One of promising ways to convert the chemical energy of aluminum into useful forms of energy is aluminum oxidation in water or water steam with heat and hydrogen production [7]. When aluminum reacts with water, 15-16 MJ of heat, 0.111 kg of H_2 and more than 1.9 kg of aluminum oxide are produced per a kg of aluminum. Oxidation products (aluminum oxide or hydroxide) can be returned to the cycle of aluminum production or used for the production of adsorbents, catalysts, insulators, high-purity aluminas and other [8]-[10].

Aluminum oxidation in water under standard conditions is impossible due to formation on its surface of an oxide film. The search of ways of aluminum surface activation or oxide film destruction for fast and complete aluminum oxidation is an actual tusk today [7]. A number of methods of aluminum activation have been already proposed. These methods include the mechanochemical preparation of highly reactive aluminum powder alloyed with gallams of various compositions including Ga, In, Sn, Zn and other additions [11]-[12]. Enhanced aluminum oxidation in water is observed after aluminum co-milling with graphite [13] and silicon [14]. Aluminum particle surface is activated by milling with different metal oxides such as Al_2O_3, TiO_2, Co_3O_4, Cr_2O_3, Fe_2O_3, Mn_2O_3, NiO, CuO, ZnO, MoO_3, Bi_2O_3 and other [15]-[17]. Milling with salts was also proposed as perspective way of aluminum activation [18]-[20]; new aluminum surface created by milling is covered by salt during the storage of aluminum preventing its oxidation, such surface is revealed by salt dissolution in aqueous medium. Aluminum oxidation can be accelerated in alkaline aqueous solutions [21]. There is also incalculable quantity of methods which assume the use of additional chemicals that help to realize fast and complete aluminum oxidation in aqueous media.

Recently developed method of aluminum micron powder oxidation in high-temperature steam [22]-[24] allowed applying pure (without alkali and any other chemical activators) water as oxidant.

Kinetics of aluminum micron powders oxidation in high-temperature boiling water was studied in [25-26]. It was established that aluminum powders with average particle size from 4 to 70 μm, which are produced in industrial scale, were intensively oxidized within the special reactor under about 300 °C and 10 MPa water steam; the reaction time was several tens of seconds.

Further development of power plant based on "aluminum-water" reactor as high-pressure steam-hydrogen generator was based on the results of kinetic experiments and a number of designing investigations. One of such calculations was presented in [23]. It was devoted for thermodynamics of nonstop reactor operation. Reactor's thermo- and gas-dynamic parameters estimation and optimization was carried out in that work, an optimum parameter field (composed of reactor temperature, pressure, volume and others) in the view of thermodynamic effectiveness was determined.

Based on the results of [22]-[26] and other designing investigations an experimental co-generation aluminum-fuelled power plant was developed [27]. It consumes aluminum micron powder as primary fuel and pure water as primary oxidant. Hydrogen, which is produced within aluminum-water reactor, is used as secondary fuel for electrical energy generation via air-hydrogen fuel cell battery. Nominal hydrogen generation rate of the plant is 10 m^3/hour. Developed and created power plant outputs useful electrical energy and heat, and also it can produce hydrogen as end product. From 1 kg of aluminum the power plant produces 1 kWh of electrical energy and 5÷7 kWh of heat. Plant's electrical and total efficiencies are 12 % and 72 % respectively.

Main reason of low electrical efficiency of created aluminum-fueled power plant is that only chemical energy of hydrogen is converted into useful electrical energy and the heat of aluminum-water reaction doesn't used. At the same time the heat of aluminum-water reaction per a kg of Al is about 15-16 MJ and it is comparable with the chemical energy of 0.111 g of hydrogen produced from a kg of Al. So it is profitable to convert the heat of aluminum-water reaction into useful electrical energy as well.

One of possible way to convert the heat of aluminum-water reaction into useful electrical energy is the use of gas turbine after aluminum-water reactor and before a fuel cell. In this case steam-hydrogen mixture generated in aluminum-water reactor expands in steam-hydrogen turbine, then water is condensed in a condenser and then hydrogen is consumed by air-hydrogen fuel cell battery (figure 1).

Presented work is devoted to the modeling of aluminum-fuelled power plant with gas turbine shown in figure 1. Parameters of aluminum-water reactor are imported from our previous research [27]. The main aims of presented work are the development of calculation method and calculation of thermodynamic parameters of proposed plant.

2. Modeling methods

2.1. Input data for calculation

In this paper, we will consider the calculation of scheme of power plant based on the reactor of hydrothermal oxidation of aluminum described in our previous research work [22-27]. To oxidize micron fractions of aluminum powder with high rates and completeness of reaction in such reactor, a high temperature of 570 to 630 K and a pressure of 10 to 20 MPa is created. Aluminum is fed to the reactor in an aqueous suspension by means of a high-pressure pump. Aluminum hydroxide and hydrogen formed within the reactor are withdrawn from the bottom and the top of reactor correspondingly. During the continuous operation of reactor, the temperature and pressure are kept constant in it.

The composition of the steam-hydrogen mixture at the outlet from the reactor depends on the water excess ratio with respect to aluminum in the suspension entering the reactor [23]. The ratio of the mass of steam to the mass of hydrogen can be from 30 to 40 [24]. In this case, the thermal energy of the steam-hydrogen stream reaches 60 ÷ 80 % of the thermal energy released in the reactor (15-16 MJ per kg of Al), and taking into account the chemical energy of hydrogen, the steam-hydrogen mixture contains from 80 to 90 % of the chemical energy of aluminum (30.97 MJ/kg).

In present work we will consider the following parameters of aluminum-water reactor: pressure – 15 MPa, temperature – 600 K, steam to hydrogen mass ratio – 40. We assume that hydrogen output rate is 1 g/s that corresponds to aluminum consumption rate of 9 g/s (in case of full oxidation of aluminum). Taking into account the calorific value of aluminum (30.97 kJ/g), the input heat power is 278.7 kW.

IOP Conf. Series: Earth and Environmental Science **159** (2018) 012039 doi:10.1088/1755-1315/159/1/012039

2.2. Modelling of steam-hydrogen turbine

The main for calculation element of the scheme shown in figure 1 is steam-hydrogen turbine (or steam-gas turbine in common case). The process of expansion of the steam-hydrogen mixture in the steam-hydrogen turbine includes both the expansion of both components and the process of heat exchange between them. To calculate the process of steam-hydrogen expansion we will use two databases: Industrial Formulation for the Thermodynamic properties of Water and Steam from Inernational Association for the Properties of Water and Steam (IAPWS-1997) and IVTAN THERMO [28] for other individual substances (like H_2).

The input variable parameters in the modeling of steam-hydrogen turbine are total pressure P_0 and temperature T_0 of steam-hydrogen mixture at the input of turbine, contents of steam and hydrogen in steam-hydrogen mixture – $\frac{m_{H_2O}}{m_{H_2O}+m_{H_2}}$ and $\frac{m_{H_2}}{m_{H_2O}+m_{H_2}}$ respectively, pressure of the mixture at the output of turbine – *P*, and the internal efficiency of turbine - η. The parameter η characterizes the degree of irreversibility of the expansion process and depends mainly on the type of turbine and the flow characteristics of the working media. Parameters that are determined during the calculation are the initial and final partial pressures of steam and hydrogen, density of steam, final temperature, power of turbine and other parameters.

The process of isentropic expansion of steam-hydrogen mixture ($\eta = 1$) is conveniently represented as successive isentropic processes of expansion of its components from its initial partial pressures ($P^0_{H_2O}$ and $P^0_{H_2}$) and temperature T_0 to final partial pressures (P_{H_2O} and P_{H_2}) and the process of heat exchange between the components of the mixture without changing their pressures. The determination of the parameters of steam-hydrogen mixture after the isentropic expansion is flown from the solving of the following system of equations:

$$\begin{cases} P = P_{H_2O} + P_{H_2} \\ \upsilon = \frac{m_{H_2}}{m_{H_2O}} \times \frac{1}{M_{H_2}} \times \frac{RT}{P-P_{H_2O}} \\ \upsilon = \upsilon(P_{H_2O}, T) \\ h_{H_2O} = f(P_{H_2O}, T, \upsilon) \\ T^S_{H_2} = T_0 \times \left(\frac{P_{H_2}}{P^0_{H_2}}\right)^{\frac{k-1}{k}} \\ h^S_{H_2O} = \psi(P_{H_2O}) \\ h_{H_2} = C_{P_{H_2}} \times T \\ m_{H_2O} \times \left(h_{H_2O} - h^S_{H_2O}\right) = m_{H_2} \times C_{P_{H_2}} \times (T^S_{H_2} - T) \end{cases} \quad (1)$$

where $\upsilon(P_{H_2O}, T)$ – the dependence of the specific volume on pressure and temperature (it is known from IAPWS-1997); $h_{H_2O}=f(P_{H_2O}, T, \upsilon)$ – the dependence of the specific enthalpy of steam on pressure, temperature and specific volume (it is known from IAPWS-1997); $h^S_{H_2O}$ – the specific enthalpy of steam after isentropic expansion of steam from $P^0_{H_2O}$, T_0 to P_{H_2O}; $T^S_{H_2}$ – hydrogen temperature after isentropic expansion from $P^0_{H_2}$, T_0 to P_{H_2}; k – isoentropic index of hydrogen (the dependence of k on T is not taken into account); $h^S_{H_2O} = \psi(P_{H_2O})$ – the dependence of the specific enthalpy of steam on pressure and specific entropy, that in the case of known initial entropy is a function of one variable parameter (it is known from IAPWS-1997); h_{H_2O} – final specific enthalpy of steam; h_{H_2}, Cp_{H_2} – final specific enthalpy and heat capacity of hydrogen.

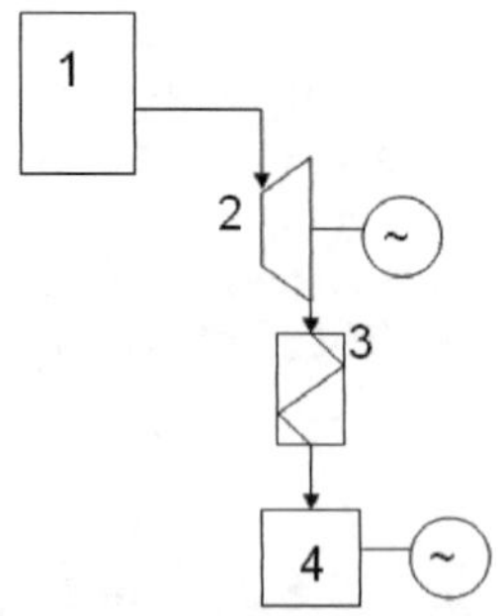

Figure 1. Scheme of aluminum-fuelled power plant with gas turbine: *1* – aluminum-water reactor, *2* – steam-hydrogen turbine, *3* – condenser, *4* – air-hydrogen fuel cell battery.

The initial partial pressures of steam $P^0_{H_2O}$and hydrogen $P^0_{H_2}$, as well as the specific volume of steam υ_0, are determined using known input calculation parameters: P_0, T_0 and $\frac{m_{H_2O}}{m_{H_2}}$. Solving the system of equations (1), we determine the final temperature (T), the partial pressures of steam and hydrogen (P_{H_2O} and P_{H_2}), the specific volume of steam (υ), the specific enthalpies of steam (h_{H_2O}) and hydrogen (h_{H_2}) after isentropic expansion of the steam-hydrogen mixture.
The performance of isentropic expansion of steam-hydrogen mixture is:

$$L = (m_{H_2O} + m_{H_2}) \times (h_0 - h) \tag{2}$$

where h_0 is the specific enthalpy of the steam-hydrogen mixture before expansion, and h is the specific enthalpy of the steam-hydrogen mixture after expansion. The initial and final specific enthalpies of the steam-hydrogen mixture are composed of the corresponding enthalpies of steam and hydrogen:

$$h_0 = \frac{m_{H_2O} \times h^0_{H_2O} + m_{H_2} \times h^0_{H_2}}{m_{H_2O} + m_{H_2}} \tag{3}$$

$$h = \frac{m_{H_2O} \times h_{H_2O} + m_{H_2} \times h_{H_2}}{m_{H_2O} + m_{H_2}} \tag{4}$$

where $h^0_{H_2O}$, $h^0_{H_2}$ - specific enthalpies of steam and hydrogen before expansion, h_{H_2O}, h_{H_2} - after expansion. By substituting the values of specific enthalpies found in equation (1) into equations (2)-(4) we find the performance of steam-hydrogen mixture in isentropic expansion.
In case of irreversible adiabatic expansion of steam-hydrogen mixture (η<1), the enthalpy of steam-hydrogen mixture after the expansion (h') will be greater than the final enthalpy of the isentropic process, and the performance of turbine will decrease in comparison with L by:

$$Q = (m_{H_2O} + m_{H_2}) \times (h' - h) = (m_{H_2O} + m_{H_2}) \times (h^0 - h) \times (1 - \eta) \tag{5}$$

The performance of the process of irreversible expansion is equal to:

$$L' = (m_{H_2O} + m_{H_2}) \times (h^0 - h) \times \eta \tag{6}$$

Irreversible expansion of steam-hydrogen mixture can be represented in the form of two successive processes: isentropic expansion of steam-hydrogen mixture and heat input (ΔQ) to the mixture. As the temperature of the mixture increases as a result of the heat input, the ratio of the partial pressures of steam and hydrogen generally changes. Heating of steam-hydrogen mixture at constant pressure is described by the equation:

$$\Delta Q = m_{H_2O} \times (h'_{H_2O} - h_{H_2O}) + m_{H_2} \times C_{P_{H_2}} \times (T' - T) = (m_{H_2O} + m_{H_2}) \times (h^0 - h) \times (1 - \eta) \tag{7}$$

where h'_{H_2O} is the specific enthalpy of steam, T' is the final temperature of the mixture after irreversible expansion.

The determination of the parameters of stem-hydrogen mixture after the irreversible adiabatic expansion is realized through the following system of equations:

$$\begin{cases} P = P'_{H_2O} + P'_{H_2} \\ \upsilon' = \frac{m_{H_2}}{m_{H_2O}} \times \frac{1}{M_{H_2}} \times \frac{RT'}{P-P'_{H_2O}} \\ \upsilon' = \upsilon\left(P'_{H_2O}, T'\right) \\ h'_{H_2O} = f(P'_{H_2O}, T', \upsilon') \\ h'_{H_2} = C_{P_{H_2}} \times T' \\ m_{H_2O}(h'_{H_2O} - h_{H_2O}) + m_{H_2}C_{P_{H_2}}(T' - T) = \left(m_{H_2O} + m_{H_2}\right)(h^0 - h)(1-\eta) \end{cases} \quad (8)$$

From system of equations (8) we find the temperature of steam-hydrogen mixture after irreversible expansion (T'), the final partial pressures of steam and hydrogen (P'_{H_2O} and P'_{H_2}), the specific volume of steam (υ'), the specific enthalpies of steam (h'_{H_2O}) and hydrogen (h'_{H_2}). Thus, the modeling of stem-hydrogen turbine is a successive solution of the systems of equations (1) and (8). It determines the final temperature of steam-hydrogen mixture after adiabatic expansion, the partial pressures of steam and hydrogen, the specific volume of steam and the specific enthalpies of steam and hydrogen.

3. Results and discussion

Table 1 and table 2 shows the power of steam-hydrogen turbine, the final temperature T, the final partial pressures of steam and hydrogen (P_{H_2O} and P_{H_2}), and the dryness of steam (X) and the thermal power of condenser (from steam condensation), depending on the final expansion pressure P of steam-hydrogen mixture. In table 1 the internal efficiency was $\eta = 1$, i.e. isentropic expansion. In table 2 $\eta = 0.8$.

Increasing of final pressure leads to decrease of work of steam-hydrogen turbine. On the other hand, the temperature of condenser increases with practically constant value of heat produced in the condenser. From the comparison of table 1 and table 2 it follows that with an increase in the irreversibility of the process of expanding, the performance of turbine decreases, the dryness of steam (X) at the outlet from it increases and, accordingly, the heat of condensation increases.

The development of turbine for steam-hydrogen mixture should take into account the degree of dryness of the steam, which affects the wear rate (erosion) of the working parts of turbine. When steam-hydrogen mixture is expanded in turbine with $\eta = 0.8$, the dryness of steam at the outlet of turbine changes from 0.75 to 0.8 for final pressures of 0.1 and 1 MPa respectively. It should be noted that such dryness may not be acceptable for some standard turbines.

The efficiency of a fuel cell depends on the specific power per a surface unit of its electrodes. Usually, with power increasing, the efficiency becomes less. So, if we use fuel cells with low power density, we have high efficiency (it can be up to 70-80 %). In this case, it is necessary to use fuel cell batteries with big total area of the active surface of electrodes. But, due to the relatively high cost of installed capacity of fuel cell batteries (over $ 2000), this approach becomes economically inexpedient. The operating mode of fuel cell is usually selected near the maximum power point. The efficiency of modern air-hydrogen fuel cells in operating mode lies in the range of 35-55 %.

If hydrogen output rate is 1 g/s and efficiency of fuel cell is about 50 %, we get about 50 kW of electricity and 50 kW of heat from fuel cell. Total power of considered plant is composed from power of steam-hydrogen turbine (that is in the range of 20-30 kW) and power of fuel cell (50 kW). The efficiency of plant represents the ratio, where the numerator is the sum of turbine power and fuel cell power, and the denominator is the input heat power of aluminum (with aluminum consumption rate of 9 g/s) – 278.7 kW. Neglecting mechanical and electrical losses in turbine, the overall electrical efficiency of the plant is about 25 ÷ 30 %. Therefore, the use of only gas turbine after aluminum-water reactor can potentially increase the efficiency of aluminum-fuelled power plant more than twice (from 12 % [27]). Total efficiency of power plant in this case reaches 70 ÷ 80 %.

IOP Conf. Series: Earth and Environmental Science **159** (2018) 012039 doi:10.1088/1755-1315/159/1/012039

Table 1. Power of steam-hydrogen turbine and heat power of condenser depending on final pressure of steam-hydrogen mixture (after expansion). η=1.

P, MPa	Turbine power, kW	T, K	P_{H2O}, MPa	P_{H2}, MPa	X	Condenser heat power, kW
0.1	36.920	364.79	0.075	0.025	0.67	61.252
0.2	33.039	384.52	0.150	0.050	0.69	61.519
0.3	30.667	397.23	0.226	0.074	0.70	61.550
0.4	28.928	406.84	0.301	0.099	0.71	61.510
0.5	27.487	414.67	0.377	0.123	0.72	61.496
0.6	26.301	421.32	0.453	0.147	0.72	61.441
0.7	25.219	427.14	0.529	0.171	0.73	61.435
0.8	24.317	432.32	0.605	0.195	0.74	61.363
0.9	23.548	437	0.681	0.219	0.74	61.251
1	22.800	441.29	0.757	0.243	0.74	61.188

Table 2. Power of steam-hydrogen turbine and heat power of condenser depending on final pressure of steam-hydrogen mixture (after expansion). η=0.8.

P, MPa	Turbine power, kW	T, K	P_{H2O}, MPa	P_{H2}, MPa	X	Condenser heat power, kW
0.1	29.536	365.52	0.077	0.023	0.75	68.525
0.2	26.431	385.25	0.154	0.046	0.76	68.029
0.3	24.533	397.96	0.231	0.069	0.77	67.611
0.4	23.142	407.56	0.308	0.092	0.78	67.170
0.5	21.989	415.39	0.385	0.115	0.78	66.925
0.6	21.041	422.03	0.462	0.138	0.79	66.600
0.7	20.175	427.84	0.539	0.161	0.79	66.368
0.8	19.454	433.02	0.616	0.184	0.79	66.162
0.9	18.838	437.69	0.693	0.207	0.80	65.856
1	18.240	441.97	0.770	0.230	0.80	65.622

4. Conclusion

Presented work is devoted to the modeling of aluminum-fuelled power plant based on aluminum-water reactor, steam-hydrogen turbine, condenser and air-hydrogen fuel cell battery. Parameters of aluminum-water reactor were the following: pressure – 15 MPa, temperature – 600 K, steam to hydrogen mass ratio – 40. It was shown that without gas turbine the plant's electrical and total efficiencies are 12 % and 72 % respectively. The results of the study confirmed that the efficiency of aluminum-fuelled power plant can be increased by using the gas turbine for steam-hydrogen mixture generated in aluminum-water reactor. It was shown that turbine with internal efficiency of 80 % allow increasing the overall electrical efficiency of the plant up to 25 ÷ 30 %. At the same time the scheme of plant considered in presented work is a simple and non-optimized example, and it is certainly can be improved by such standard operations like regeneration of heat or heat transfer to secondary contour. So this work may be continued and the model developed for calculation of the parameters of steam-hydrogen turbine can be used for that.

5. References

[1] Hiraki T, Yamauchi S, Iida M, Uesugi H and Akiyama T 2007 Process for recycling waste aluminum with generation of high-pressure hydrogen *Environmental Science & Technology* **41** 4454-7

[2] Shkolnikov E, Vlaskin M, Iljukhin A, Zhuk A and Sheindlin A 2008 2 W power source based on air-hydrogen polymer electrolyte membrane fuel cells and water-aluminum hydrogen micro-generator *Journal of Power Sources* **185** 967-72

[3] Woodall J M, Ziebarth J T, Allen C R, Sherman D M, Jeon J and Choi G 2008 *Recent Results on Splitting Water with Aluminum Alloys*: John Wiley & Sons, Inc.)

[4] Yang S and Knickle H 2002 Design and analysis of aluminum/air battery system for electric vehicles *Journal of Power Sources* **112** 162-73

[5] Li Q and Bjerrum N J 2002 Aluminum as anode for energy storage and conversion: a review *Journal of Power Sources* **110** 1-10

[6] Partington E C 1992 Strategic inventory management of renewable energy resources. Aluminium as an energy storage medium and market commodity *International Journal of Production Economics* **26** 211-6

[7] Shkolnikov E I, Zhuk A Z and Vlaskin M S 2011 Aluminum as energy carrier: Feasibility analysis and current technologies overview *Renewable and Sustainable Energy Reviews* **15** 4611-23

[8] Vlaskin M S, Grigorenko A V, Zhuk A Z, Lisitsyn A V, Sheindlin A E and Shkol'nikov E I 2016 Synthesis of high-purity α-Al2O3 from boehmite obtained by hydrothermal oxidation of aluminum *High Temperature* **54** 322-9

[9] Zhuk A Z, Vlaskin M S, Grigorenko A V, Kislenko S A and Shkolnikov E I 2016 Synthesis of high-purity α-Al2O3 from boehmite by high temperature vacuum treatment *Journal of Ceramic Processing Research* **17** 910-8

[10] Lisitsyn A V, Dombrovsky L A, Mendeleyev V Y, Grigorenko A V, Vlaskin M S and Zhuk A Z 2016 Near-infrared optical properties of a porous alumina ceramics produced by hydrothermal oxidation of aluminum *Infrared Physics & Technology* **77** 162-70

[11] Ilyukhina A V, Kravchenko O V, Bulychev B M and Shkolnikov E I 2010 Mechanochemical activation of aluminum with gallams for hydrogen evolution from water *International Journal of Hydrogen Energy* **35** 1905-10

[12] Dudoladov A O, Buryakovskaya O A, Vlaskin M S, Zhuk A Z and Shkolnikov E I 2016 Generation of hydrogen by aluminium oxidation in aqueous solutions at low temperatures *International Journal of Hydrogen Energy* **41** 2230-7

[13] Streletskii A N, Kolbanev I V, Borunova A B and Butyagin P Y 2004 Mechanochemically activated aluminium: Preparation, structure, and chemical properties *Journal of Materials Science* **39** 5175-9

[14] Soler L, Macanás J, Muñoz M and Casado J 2007 Aluminum and aluminum alloys as sources of hydrogen for fuel cell applications *Journal of Power Sources* **169** 144-9

[15] Deng Z-Y, Tang Y-B, Zhu L-L, Sakka Y and Ye J 2010 Effect of different modification agents on hydrogen-generation by the reaction of Al with water *International Journal of Hydrogen Energy* **35** 9561-8

[16] Wang H-W, Chung H-W, Teng H-T and Cao G 2011 Generation of hydrogen from aluminum and water – Effect of metal oxide nanocrystals and water quality *International Journal of Hydrogen Energy* **36** 15136-44

[17] Dupiano P, Stamatis D and Dreizin E L 2011 Hydrogen production by reacting water with mechanically milled composite aluminum-metal oxide powders *International Journal of Hydrogen Energy* **36** 4781-91

[18] Alinejad B and Mahmoodi K 2009 A novel method for generating hydrogen by hydrolysis of highly activated aluminum nanoparticles in pure water *International Journal of Hydrogen Energy* **34** 7934-8

[19] Mahmoodi K and Alinejad B 2010 Enhancement of hydrogen generation rate in reaction of aluminum with water *International Journal of Hydrogen Energy* **35** 5227-32

[20] Czech E and Troczynski T 2010 Hydrogen generation through massive corrosion of deformed aluminum in water *International Journal of Hydrogen Energy* **35** 1029-37

[21] Ambaryan G N, Vlaskin M S, Dudoladov A O, Meshkov E A, Zhuk A Z and Shkolnikov E I 2016 Hydrogen generation by oxidation of coarse aluminum in low content alkali aqueous solution under intensive mixing *International Journal of Hydrogen Energy* **41** 17216-24

[22] Bersh A V, Lisitsyn A V, Sorokovikov A I, Vlaskin M S, Mazalov Y A and Shkol'nikov E I 2010 Study of the processes of steam-hydrogen mixture generation in a reactor for hydrothermal aluminum oxidation for power units *High Temperature* **48** 866-73

[23] Vlaskin M S, Shkolnikov E I, Lisicyn A V, Bersh A V and Zhuk A Z 2010 Computational and experimental investigation on thermodynamics of the reactor of aluminum oxidation in saturated wet steam *International Journal of Hydrogen Energy* **35** 1888-94

[24] Vlaskin M S, Shkol'nikov E I, Lisitsyn A V and Bersh A V 2010 Thermodynamic calculation of the parameters of a reactor for oxidizing aluminum in wet saturated steam *Thermal Engineering* **57** 794-801

[25] Vlaskin M S, Shkolnikov E I and Bersh A V 2011 Oxidation kinetics of micron-sized aluminum powder in high-temperature boiling water *International Journal of Hydrogen Energy* **36** 6484-95

[26] Shkolnikov E I, Shaitura N S and Vlaskin M S 2013 Structural properties of boehmite produced by hydrothermal oxidation of aluminum *The Journal of Supercritical Fluids* **73** 10-7

[27] Vlaskin M S, Shkolnikov E I, Bersh A V, Zhuk A Z, Lisicyn A V, Sorokovikov A I and Pankina Y V 2011 An experimental aluminum-fueled power plant *Journal of Power Sources* **196** 8828-35

[28] Belov G, Iorish V and Yungman V 2000 Simulation of equilibrium states of thermodynamic systems using IVTANTERMO for Windows *High Temperature* **38** 191-6

IOP Conf. Series: Earth and Environmental Science **159** (2018) 012040 doi:10.1088/1755-1315/159/1/012040

Techno-economic Assessment of Turbine Inlet Air Cooling for Cogeneration Plant in Malaysia Climate

Saw Chun Lin[1], Choong Chee Guan[1], Azmi Naroh[1], Danny Tam Hong Khai[2] and Mohd Kamarul Yusri Yusof[2]

1 Centre of Air Conditioning and Refrigeration, Politeknik Ungku Omar Ipoh, Perak, Malaysia
2 Shinryo (Malaysia) Sdn. Bhd. Kuala Lumpur, Malaysia
E-mail: sawchunlin@puo.edu.my

Abstract. The high ambient and humid Malaysia has affected the performance of gas turbine generators. The power augmented from the gas turbine generator can be enhanced by implementing turbine inlet air cooling (TIAC) technologies. In this paper, techno-economic analysis of combined heat and power (CHP) plant using an aero-derivative gas turbine (GE LM6000PD) is presented based on annual performance simulation. There are three cases of comparison that is Non TIAC, TIAC with steam absorption chiller (SAC) and TIAC with electric compression chiller (ECC) is simulated using GT Pro software. The results show that the performance of TIAC with SAC yielding higher annual net electricity output compared Non TIAC and TIAC with ECC but incur highest capital investment and capital outflow. Thus, TIAC with ECC is the best choice for CHP plant due to highest NPV, IRR, fastest PBT and highest net revenue compared to TIAC with SAC and Non TIAC.

1. Introduction

Cogeneration or combined heat and power (CHP) system produces electricity power using turbine unit and the waste gas is utilised to generate steam via heat recovery steam generator (HRSG). The technical advantage of cogeneration systems is the ability to improve the fuel consumption in production of electrical and thermal energy. Hence, less fuel is required to produce a specific amount of electrical and thermal energy in a single cogeneration cycle compared to conventional technologies. Adzuieen et al. [1] evaluated the heat recovery steam generator (HRSG) for gas fired district cooling plant DCP) Universiti Technologi Malaysia in Tronoh Perak. The exergy of destructive (energy losses to environment and irreversibility within the system boundaries) accounted 36% of the total exergy of fuel (waste heat) entering the HRSG. The higher the exergy of destructive the more inefficient of HRSG of cogeneration plant.

Electrical power output of cogeneration system can be enhanced further by cooling the air intake into the turbine unit through inlet air cooling (IAC) technologies. IAC also is named as combustion turbine inlet air cooling (CTIAC), turbine inlet air cooling (TIAC), combustion turbine air cooling (CTAC) and gas turbine inlet air cooling (GTIAC). Whereby, air intake into the turbine can be cooled through wetted technologies such as fogging, evaporative cooling, vapour compression chiller, vapour absorption chiller and thermal energy storage tank combination as reviewed by Abdulrahman et al.[2]. This technology is suitable use in the hot climates region such as Asian country. Katanobu, K et al. [3] presented a case study by Mitsubishi Heavy Industries, Ltd. (MHI) utilizing Chon Buri city climate in Thailand to design and manufacture large-scale gas turbines for electric power plants at the Takasago Machinery Works facility. Modular electric centrifugal chiller with 7800 RT cooling capacity is needed to cool the inlet air from 35°C to 10°C to the turbine to improve 10% power augmentation. Shahrul N. O. K et al [4] modelled Sitiawan, Perak Malaysia climate for LM6000PD Gas Turbine with power output capacity of 43.5 MW at ISO rating with the aid of Turbine Inlet Air Cooling. Cooling capacity of 1950 RT mechanical chiller is designed to satisfy the 0.4% evaporative design point consists of 98% operating hours in year 2015 that resulted 32.11% power output gained. With reference to year 2014 Malaysia Energy Balance Handbook [5], the total electricity consumption for Malaysia recorded a growth of 4.3 percent from 123,079 GWh in 2013 to 128,330 GWh in 2014. The share of natural gas as energy input in power stations increased from 43.7 percent in 2013 to 43.8 percent in 2014. The oil's share of fuel used for generation continued to decline and in 2014. According to Statistic of Malaysia Energy Commission years 2015 on electricity supply industry in Malaysia [6], there are 27 gas turbine power plant where 7 is running open cycle gas-fired, 1 is open cycle diesel combustion and 19 is running on combined cycle gas-fired. A total 27 gas turbine power plant generated electricity is 11.644GW. Hence, the cogeneration plant or gas fired combined cycle plant in Malaysia that using LM6000PD Gas Turbine is viable to enhance the performance efficiency by using turbine inlet air cooling (TIAC) technology.

Few types of prime movers are adopted in combined heating and power (CHP) plant or combined cooling, heating and Power (CCHP) plant such as reciprocating engines (up to 75 MW energy generating capacity), gas turbines (up to 250 MW), Microturbines (up to 350 kW), Combined cycle gas turbine (up to 100 MW) and as new emerging technologies such as Stirling engines (up to 55 kW) and Fuel cell (up to 2MW), Microturbines (up to 350 kW) and Organic Rankine Cycle (up to 250 MW) [7]. The CCHP prime movers mentioned above integrate with steam absorption or adsorption chiller to produce free chilled water supply. However, in this paper the focus is on the techno economic analysis of the cogeneration plant where gas turbine as the prime mover with 44MW generation power output. An electric compression chiller (ECC) with turbine inlet air cooling (TIAC) and steam absorption chiller (SAC) with turbine inlet air cooling (TIAC) was selected to compare the cogeneration plant without TIAC. The weather condition data of the Sitiawan, Perak for year 2015 is used as input data for simulation as showed in Figure 1. Table 1 shows the design selection and sizing of the cogeneration plant.

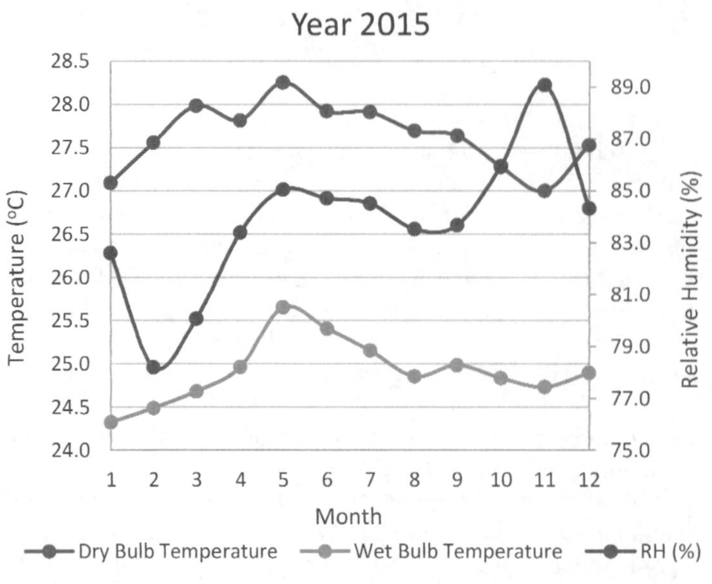

Figure 1. Average 2015 yearly weather data at Sitiawan, Perak

IOP Conf. Series: Earth and Environmental Science **159** (2018) 012040 doi:10.1088/1755-1315/159/1/012040

Table 1. Design criteria for cogeneration plant

	Non TIAC	TIAC with ECC	TIAC with SAC
Ambient temperature (°C)	24.3 – 28.3	24.3 – 28.3	24.3 – 28.3
Relative Humidity (%)	78.2 – 89.1	78.2 – 89.1	78.2 – 89.1
GT inlet temperature set point (°C)	-	12	12
Inlet air cooling daily operational hours	-	24	24
Chiller COP	-	5.4	1.2
Inlet cooling system sizing	-	98% of operating hours in a year	98% of operating hours in a year
Cooling Capacity (tons)	-	1893.2	1893.2

Figure 2 below shows the cogeneration power plant integration where the inlet to the compressor attached with electric compression chiller (ECC) to cool the compressor air intake at 12T(°C) that enhanced the gas turbine generator (GTG) thermal efficiency to 41.32% from 38.94%.

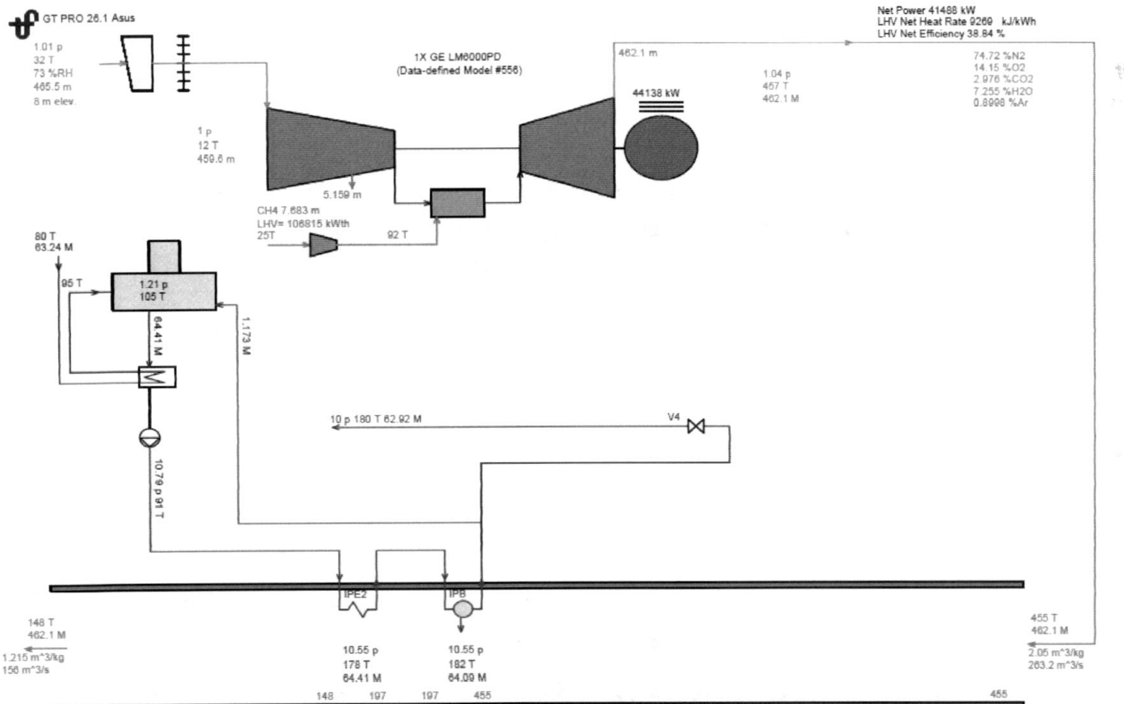

Figure 2. Schematic diagram of cogeneration plant utilizing TIAC with ECC

2. Methodology

Input data of subcategories below has been used for simulation:

a) climatic data
b) user defined economic data assumptions

The local data of dry bulb, wet bulb and relative humidity (RH) for year 2015 from Sitiawan, Perak, Malaysia have been used for this study. The local data were evaluated using the binning method to determine the performance and economic analysis of a cogeneration plant with and without using TIAC technology.

IOP Conf. Series: Earth and Environmental Science **159** (2018) 012040 doi:10.1088/1755-1315/159/1/012040

The data gathered from the meteorology department in the year of 2015 is specified to hourly ambient temperature. Data of wet bulb (WB) temperature, dry bulb (DB) temperature and the relative humidity (RH) were used. Based on these data, a scatter graph of RH versus DB was plotted.
The GT Pro software is used to simulate the annual performance of gas-fired cogeneration plant. Gas turbine models which are GE LM6000PD (aero-derivative GT) operating with and without turbine inlet air cooling, in Malaysian climate is simulated. Sitiawan's weather data is analysed and divided into 27 discrete time periods covering a total of 8756 operating hours. Each time bin indicate the associated operating hours, mean dry bulb and wet bulb temperatures. Electric compression chiller and steam absorption chiller is used to cool the air intake in the gas turbine hence created three comparisons in this paper that is Non TIAC, TIAC with ECC and TIAC with SAC. Public utility rate in Malaysia such as water consumption tariff (RM2.28/m^3), electricity consumption tariff (Average charges RM 0.337/kWhr; Maximum demand charges RM 29.70/kW/mth), natural gas consumption tariff (RM 27.30 GJ) and steam consumption tariff (RM 100/ton) is used in this paper to calculation the project feasibility. While, the maintenance cost of the cogeneration plant is assumed 4% from the capital expenditure cost. The flow chart of methodology showed in Figure 3.

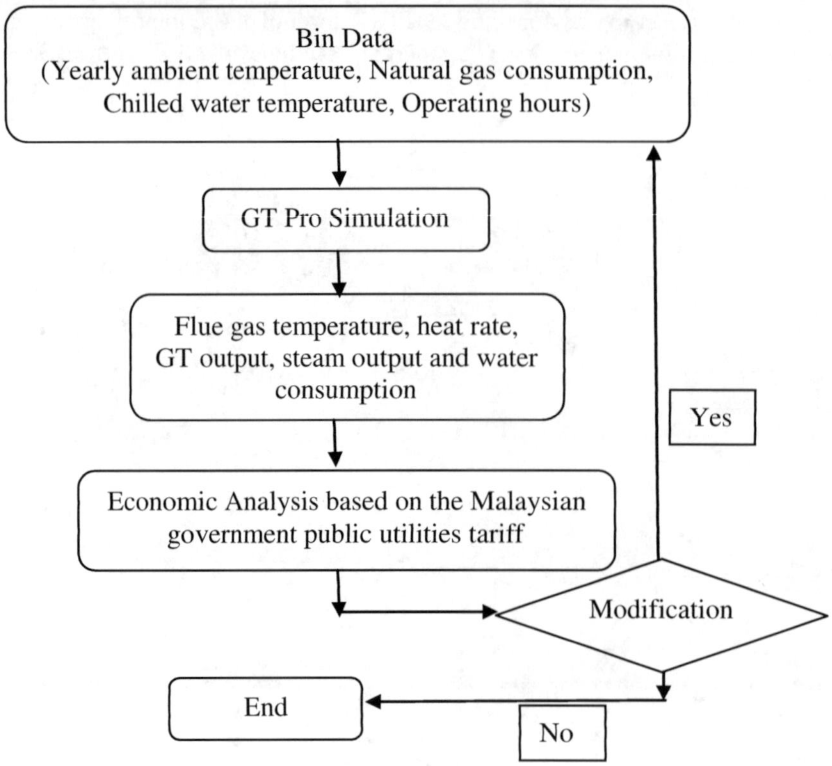

Figure 3. The Methodology flow

3. Results and discussions

Results of the performance analysis, efficiency analysis and economic assessment as tabulated in Table 2 and Table 3 respectively. Calculation of the thermal efficiency of combine heat and power (CHP) plant and combine cooling, heat and power (CCHP) plant according to Mahmet et al. (2009) [8] as using the formula below:

Combined heat and Power Plant (CHP) thermal efficiency, $\eta_{thermal}$

$$\eta_{thermal} = \frac{W_E+\sum Q_{TH}}{Q_{Fuel}} \quad (1)$$

where W_E is total net power output (kW) and $\sum Q_{TH}$ is total net process heat output (kW)

The results on the Table 2 shows that the TIAC technologies has improved the gas turbine generator (GTG) performance by 6.11 %. Hence, by constructing heat recovery steam generator (HRSG), the performance of cogeneration plant is enhanced to 74.74 – 82.66%. Non TIAC cogeneration plant showed high thermal efficiency (82.66%) due to lower fuel consumption of GTG but GTG efficiency unable to improve more than 40%. El-Shazly et al. (2016) [9] explained that GTG higher heat rate (LHV) lead to lower thermal efficiency performance. Hence, TIAC is one of the method to increase the GTG thermal efficiency by reducing the heat rate (LHV). By increasing the thermal efficiency of GTG to 41.32%, GTG fuel consumption increases 18.58% to generate 21.53% more power output compared to Non TIAC. TIAC method too reduced the GTG exhaust flue gas temperature that help to reduce the net plant heat rate by 4%. In the meantime, although the flue gas exhaust temperature decreases but the flow rate output increased by 22.2% that key factor to improve the net process heat output (net steam output) plant production by 8%.

Table 2. Performance of cogeneration plant comparisons

Descriptions	Unit	Non TIAC	TIAC with ECC	TIAC with SAC
GTG Quantity		1	1	1
Power Output per GTG	kW	33564	44138	44138
GTG Heat Rate (LHV)	kJ/kWhr	9246	8712	8712
GTG Efficiency (LHV)	%	38.94	41.32	41.32
Exhaust Flow per GTG	t/hr	378	462	462
GTG Exhaust Temperature	°C	481	457	457
HRSG Quantity		1	1	1
HRSG Efficiency	%	74.86	72.28	72.35
Net Process Heat Output	kW	38693	42652	37244
Plant Gross Heat Rate (LHV)	kJ/kWhr	9246	8712	8712
Plant Nett Heat Rate (LHV)	kJ/kWhr	9530	9269	9030
Plant Gross Power Ouput	kW	33,564.00	44,138.00	44,138.00
Plant Nett Power Output	kW	32,563.00	41,488.00	42,586.00
Plant Thermal Efficiency	%	**82.66**	**78.77**	**74.74**
Nett Electricity Export per Annum	GWhr	301.39	365.98	372.93
Nett Steam Export per Annum	tons	511,111.27	551,018.55	500,530.00
Fuel Consumption per Annum (HHV)	TJ	3,151.05	3,736.46	3,736.46
Water Consumption per Annum	m^3	2,602.83	88,820.42	129,725.73
Capital Cost	RM (Million)	199.21	219.36	224.51

Table 3 shows the economic assessment calculation is based on 25 years revenue return where the first 2 years is the construction period. The capital investment cost varies for Non TIAC (RM 199.21 Million), TIAC with ECC (RM 219.36 Million) and TIAC with SAC (RM 224.51 Million).

IOP Conf. Series: Earth and Environmental Science **159** (2018) 012040 doi:10.1088/1755-1315/159/1/012040

Cogeneration plant integrate TIAC with ECC generated the highest net profit (RM 82.19 Million) compared to TIAC with SAC (RM 79.58 Million) and Non TIAC (RM 70.25 Million). In term of water consumption, TIAC with SAC consumed 14.8 tons/hour of treated water compared to TIAC with ECC that is 10.14 tons/hour. El-Shazly et al. (2016) [9] and Maria et al. (2014) [10] have suggested highest NPV value and fastest payback time (PBT) is the key to choose the best cogeneration plant design.

As conclusion, TIAC with ECC shows highest net present value (NPV) (RM 328.54 Million) and the highest internal rate of return (IRR) (32%) that gave the fastest normalized payback time (PBT) relative to Non TIAC that is 1 years 8 months. Another reason ECC shows the better result is that high coefficient of performance (COP) of ECC than SAC that reduced the electricity consumption and the lower capital cost of investment of TIAC with ECC that generated highest yearly net revenue. Thus, cogeneration plant using TIAC with ECC is selected as the best power plant design amongst the other two design.

Table 3. Economic analysis on cogeneration plant

Descriptions	Unit	Non TIAC	TIAC with ECC	TIAC with SAC
Max Demand Charge	RM/kW/mth		29.60	
Electricity Charges (Average)	RM/kWhr		0.3370	
Water Charges	RM/m^3		2.28	
Fuel Charges	RM/GJ		27.30	
Steam Charges	RM/ton		100.00	
		Cash Inflow		
Maximum Demand Charges	RM/annum (Million)	11.57	14.74	15.13
Electricity Charges	RM/annum (Million)	101.57	123.33	125.68
Steam Charges	RM/annum (Million)	51.11	55.10	50.05
		Cash Outflow		
Fuel Charges	RM/annum (Million)	86.02	102.01	102.01
Maintenance	RM/annum (Million)	7.97	8.77	8.98
Water Charges	RM/annum	5,934.46	202,510.56	295,774.66
Net Revenue	RM/annum (Million)	70.25	82.19	79.58
		Payback Period		
NPV	**RM (Million)**	**270.88**	**328.54**	**307.82**
IRR		**31%**	**32%**	**31%**
PBT	**(Relative to Non TIAC) Years**	**-**	**1.69**	**2.71**

IOP Conf. Series: Earth and Environmental Science **159** (2018) 012040 doi:10.1088/1755-1315/159/1/012040

4. Conclusions

Economic analysis shows that TIAC with ECC cogeneration plant is more favourable due to lower PBT, higher IRR and NPV compare to TIAC with SAC and Non TIAC plant. Although, TIAC with SAC cogeneration plant produced free cooling generated from steam but the initial cost of investment is higher and generate lower net revenue than TIAC with ECC cogeneration plant.

Cogeneration plant can be further improved by reducing the pinch temperature lower than 15°C but will incur higher initial investment cost. TIAC technology produced condensed water at the rate of 5.86 tons/hour that can be source of clean water. In Malaysia scenario, the simulated results proved that natural gas fired turbine power plant performance that use aero-derivate gas turbine can be enhanced by integrating with TIAC technology using SAC or ECC.

5. References

[1] Adzuieen N and Mohd Amin A M 2014 Economic Comparison of Power and Chilled Water Generation by Cogeneration System and Public Utility *MATEC Web of Conferences* **13**(2014)02008

[2] Abdulrahman M A and Abdulhadi V 2010 A Review of Inlet Air-Cooling Technologies for Enhancing the Performance of Combustion Turbines in Saudi Arabia *Applied Thermal Engineering* **30**(2010)1879-88.

[3] Komuro T, Ito E, Sonoda T, Tomita Y, Hidaka K and Shibutani S Power Augmentation for Gas Turbine Combined Power Generation Plant by Inlet Air Cooling under High Ambient Temperature *MHI technical Report* **47** No. 4 (2010)49-54.

[4] Shahrul N O K., Didi A S, Mohd Sharil M F, Danny T H K and Mohd Kamarul Y Y 2017 Feasible Study of Turbine Inlet Air Cooling using Mechanical Chillers in Malaysia Climate *Energy Procedia* **138**(2017) 558-563.

[5] Energy Comission Malaysia National Energy Balance 2014 Putrajaya Malaysia 2016.

[6] Energy Comission Malaysia Performance and Statistical Information on Electricity Supply Industries in Malaysia 2015 Putrajaya Malaysia 2016.

[7] Dusan G, Branka G U, Christoph M Dragan U and Athikom B High Efficiency Cogeneration: CHP and non-CHP Energy *Energy* **135**(2017)269-78.

[8] Mahmet K and Ibrahim D Performance Assessment of Cogeneration Plants *Energy Conversion and Management* **50**(2009)76-81.

[9] El-Shazly A A, Elhelw M, Sorour M M and El-Maghlany W M Gas Turbine Performance Enhancement via Utilizing Different Integrated Turbine Inlet Cooling Techniques *Alexandria Engineering Journal* **55**(2016)1903-14.

[10] Maria U, Jose I L and Eva A Techno-economic Feasibility Assessment of A Biomass Cogeneration Plant Based on Organic Rankine Cycle *Renewable Energy* **66**(2014)707-13.

Acknowledgement

The authors gratefully acknowledge Shinryo (Malaysia) Sdn. Bhd. for providing the financial support to complete the work in this project.

Kinetic of Biogas Production in Co-Digestion of Vegetable Waste, Horse Dung, and Sludge by Batch Reactors

P V Dinh[1], T Fujiwara[1], S T Pham Phu[1], and M G Hoang[2]

1 Okayama University, 3-1-1 Tsushima, Kita Ward, Okayama 700-8530, Japan
2 National University of Civil Engineering, 55 Giai Phong, Ha Noi, Vietnam
E-mail: dinh88.nuce@gmail.com

Abstract. Batch experiments were performed firstly to evaluate co-digestion of vegetable waste (VW), horse dung (HD), and sludge (S). All reactors were set at a temperature of 37oC, pH of 6.7, and total solid 2.5%. Each single-substrate in the mixture played a significant role. In which, VW contributed mainly to the formation of biogas yield, S and HD played nutrient balance role. The biogas yield was in the range of 168-554 Nml/g-TS. Especially, the biogas yield could be estimated from the proportion of the substrates by equation $G1_{Nml/g-TS} = 53.7 + 7.448 \times VW(\%) + 1.922 \times HD(\%)$ or from nutrient ratio (C/N) by equation $G2_{Nml/g-TS} = 1341 - 48.46 \times C/N$. Further, the experimental data was applied to evaluate the kinetic equations of biogas production including the Gompertz (G) and Logistic (L) models. Constants in both models were found out by using the least squares fitting method. Both models showed high potential, in which, G model was completely better than L model. However, both models failed at time t=0 day. Moreover, the constant λ in models did not reflect the right definition itself, it was merely a mathematical constant.

1. Introduction

Anaerobic digestion (AD) is defined as a series of processes including hydrolysis, acidogenesis, acetogenesis, and methanogenesis. In which, microorganisms degrade down biodegradable materials in the condition of the oxygen absence [1]. Generation of solid waste containing a large amount of bio-waste increases rapidly in the recent years causing numerous problems (such as human health, the environmental pollution, the economic burden, ..etc.) that humankind has to deal. Hence, the AD of organic waste materials has been attracted remarkable attention within the scientific community for many years because of bringing two benefits include treating waste and producing biogas as an alternative energy source [2-4].

Inhibition of AD may occur because of losing the nutrient balance including excessiveness of macronutrients (Na, K, etc.) or deficiency of trace elements (Zn, Fe, Mo, etc.) [5, 6]. Particularly, the nutrient balance can be relatively measured by based on C/N ratio. The C/N ratio in the feedstock is too high pointing that the feedstock is not sufficient of nitrogen which needs for the build-up of microbial mass. Low C/N ratio leads to high concentration of generated ammonia, which is harmful to the AD processes [7]. Thus, the combination of the low nutrient material and the high nutrient material is the great ideal to reach nutrient balance. Dai, et al. [8] reported that co-digestion of grass and waste activated sludge enhanced gas production and methane content. Riggio, et al. [9] used a mixture of cow slurry, olive pomace, and apple pulp for the AD, and reported that stable gas generation was obtained with a mixture containing 85% cow slurry, 10% olive pomace, and 5% apple pulp. Wang, et al. [10] performed co-digestion of dairy manure, chicken manure, and wheat straw with the conclusion that co-digestion was better than individual digestion for methane potential. In fact, co-digestion has been studied so much for recent years. However, the combination of vegetable waste (VW), sludge (S), and horse dung (HD) has been still left open-questions.

The final goal in the field of the AD aims to reach high biogas yield. Hence, the accumulation of biogas production is considered as the most important. This process can be described by the biogas production kinetic models. Biogas is generated under the activities of anaerobic microorganism hence many studies used growth kinetics to describe the biogas production [2, 11]. Growth curve often displays a phase in which the specific growth rate begins at a value of zero in a certain period (resulting in a lag time -λ) and then increases to a maximal value (μ_m). Additionally, growth curve contains a final phase in which the growth rate reduces and finally reaches zero so that an asymptote (A) is reached [12]. In which, the modified Gompertz (G) model and logistic (L) model have been received the most attention by good simulation [11, 13, 14]. However, the comparison between the G model and the L model for simulation of cumulative gas production has been not taken care of properly.

This study uses vegetable waste, sludge, and horse dung as feedstock for co-digestion firstly to evaluate different proportions of the mixture. Further, the influence of the C/N ratio to biogas potential is also investigated to contribute information about optimal C/N ratio. The obtained data from co-digestion is used for G and L models to simulate accumulative gas production. In which, advantages and disadvantages of each model are discussed to have an overview in applying the gas production kinetic models.

2. Materials and methods

2.1. Preparation of substrates

Vegetable waste (VW) and horse dung (HD) were collected in Jul 2017 at Okayama University. Raw materials were cut by a household grinder. Sludge (S) was collected from the digestion plant in Kobe City. All materials were stored in a refrigerator below 4°C until use.

2.2. Analytical methods

The 2400 series II CHNS/O analyzer (PerkinElmer) was used to analyze C/N ratio in the raw materials. Total solid (TS) was measured by the standard method 1684-EPA (2001); pH value was determined as following APHA (2000) SM 4500-H+ method by using Laqua-twin pH meter (Horiba).

Gas volume was measured by using the gas bag and a 120ml syringe. The temperature of the gas bag was observed every day for transferring gas volume into standard condition (25°C; P = 1at).

2.3. Reactors

By changing proportion of vegetable waste, horse dung, and sludge in the mixture, there were seven reactors (type 500ml) with different C/N ratios as shown in Table 1. All reactors were set up with a temperature of 37°C, TS of 2.5%, and initial pH of 6.7.

Table 1. Characteristics of the substrate inside reactors

No.	R1	R2	R3	R4	R5	R6	R7
Vegetable (%)	0	50	50	33.3	66.6	16.7	16.7
Horse dung (%)	100	50	0	33.3	16.7	66.6	16.7
Sludge (%)	0	0	50	33.3	16.7	16.7	66.6
C/N	22.5	16.8	18.7	19.8	16.0	21.1	23.9

2.4. Gas production kinetic models

The original Gompertz function is written as:

$$G_t = A \times exp[-\exp(b - ct)] \quad (1)$$

Where, G_t, accumulation gas production (ml/g-TS);
A, biogas production potential (ml/g-TS);

IOP Conf. Series: Earth and Environmental Science **159** (2018) 012041 doi:10.1088/1755-1315/159/1/012041

b, c: constants of model;
t, cumulative time for biogas production (days).
At the flection point, where $t=t_o$, the second derivative is equal to zero, and the first derivative reaches the maximum value (μ_m-ml/g-TS)

$$\begin{cases} \left(\frac{dG_t}{dt}\right)_{to} = A.c.\exp[-\exp(b-c.t_o)].\exp(b-c.t_o) = \lambda_m & (2) \\ \left(\frac{d^2G_t}{dt^2}\right)_{to} = A.c^2.\exp[-\exp(b-c.t_o)].\exp(b-c.t_o).[\exp(b-c.t_o)-1] = 0 & (3) \end{cases}$$

Moreover, the lag time (λ-day) is defined as the t-axis intercept of the tangent through the flection point, then:

$$0 = (\lambda - t_o).\left(\frac{dG_t}{dt}\right)_{to} + G_{t=to} \quad (4)$$

By solving the equations of (2), (3), and (4): $c=\mu_m.e/A$; $b=(\mu_m.\lambda.e+A)/A$. Thus, the equation (1) can be rewritten as (called the modified Gompertz model):

$$G_t = A \times exp\left\{-\exp\left[\frac{\mu_m.e}{A}(\lambda - t) + 1\right]\right\} \quad (5)$$

By the same, the original logistic function $L_t = A \times [1+\exp(b-ct)]^{-1}$ can be rewritten as:

$$L_t = A \times \left\{1 + \exp\left[\frac{4.\mu_m}{A}(\lambda - t) + 2\right]\right\}^{-1} \quad (6)$$

3. Results and discussions

3.1. Biog·as production

Experimental data on biogas yield was shown in **Figure 1**. In which, the whole digestion process had completed for approximately 45 days. Biogas accumulation was obtained in the range of 168-554 Nml/g-TS with the order as following R7<R1<R6<R4<R3<R2<R5. The biogas yield in this study could be compared with others studies about co-digestion. For instance, Heo, et al. [15] performed co-digestion of food waste (FW) and waste activated sludge, reported that biogas potential was from 170 to 443 ml/g-TS. Dong, et al. [16] studied co-digestion of municipal solid waste and obtained biogas yield of 260, 263, 294 ml/g-TS respectively at different total solids of 16%, 13.5%, and 11%. Zhang, et al. [17] focused on co-digestion of FW and cattle manure (CM), reported that biogas production of 484, 441, and 448 ml/g-TS were corresponding to FW/CM of 2, 3, and 4 respectively.

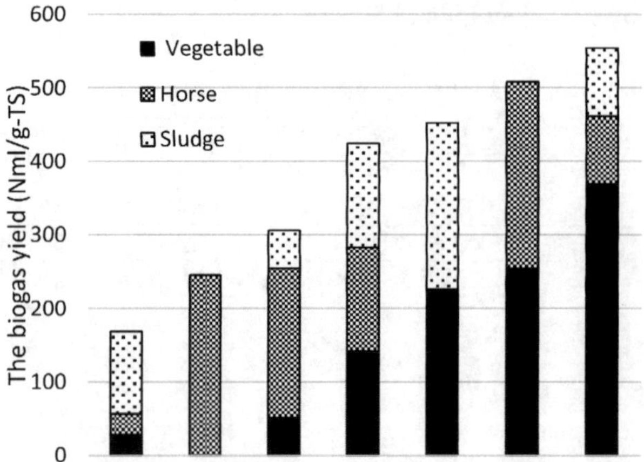

Figure 1. Biogas yield in reactors.

IOP Conf. Series: Earth and Environmental Science **159** (2018) 012041 doi:10.1088/1755-1315/159/1/012041

The lowest values of both the biogas yield and the biogas production rate were found in reactor R7 (168 Nml/g-TS) which contained 66.6% sludge (C/N=29.7). This result was derived from the high C/N ratio leading to deficiency of nitrogen in the feedstock, dragged on the inhibition as reported by Hartmann and Ahring [7]. Except for R7, the biogas yield from co-digestion was significantly higher than that of mono-digestion of horse dung (see Figure 1). The similar results were reported by Smith, et al. [18] and Kalia, et al. [19]. This reinforced the conclusion that co-digestion reduces inhibition and improves biogas production. The second lowest biogas production was in reactor R1 with 100% HD in the feedstock. This result can be explained by HD contains a large amount of cellulose which has been recommended not suitable for the AD [20, 21].

Moreover, except for R7, biogas yield increased accompany with lifting the proportion of VW (see Figure 1). And the highest biogas yield (554 Nml/g-TS) was found in reactor R5 that contained a large amount of VW (66.6%). Hence, a mixture with the proportion of a small amount of sludge, a little amount of HD, and a large amount of VW is recommended for the aim of high biogas yield. In addition, many studies reported that too low C/N in feedstock leads to increase ammonia formation which is toxic to the AD [3, 7]. Thus, a too high proportion of VW (C/N=13.2) leads to occur inhibition instead of enhancing biogas production. Overall, each raw material in this study evenly played a significant role in balancing nutrient of AD process.

Cumulative gas production could be predicted based on the substrate proportion by using the linear regression as equation $G1 = 53.7 + 7.448 \times VW(\%) + 1.922 \times HD(\%)$ (7). **Table 2** showed the significance of all the regression coefficients for equation (7). This result apparently quantized the role of each single-substrate in biogas generation. VW showed a worthier role than HD and S in the contribution of generating biogas. Obviously, from equation (7), the mixtures containing VW in major proportion and HD, S in minor proportion were having synergic effect with biogas yield being the highest. Rao, et al. [22] also found out the relationship between biogas yield and substrate composition in the mixture by a quartic regression model. However, the composition of the substrate in the mixture is merely external expression of chemical components inside.

Table 2. The significance of the regression coefficients

Coefficients	Estimate	T value	Pr (> |t|)
Intercept	53.7	0.933	0.4037
VW	744.8	7.529	0.0017
HD	192.2	2.846	0.0466
R-squared=0.9433; Adjusted R-squared =0.915; p-value=0.00321			

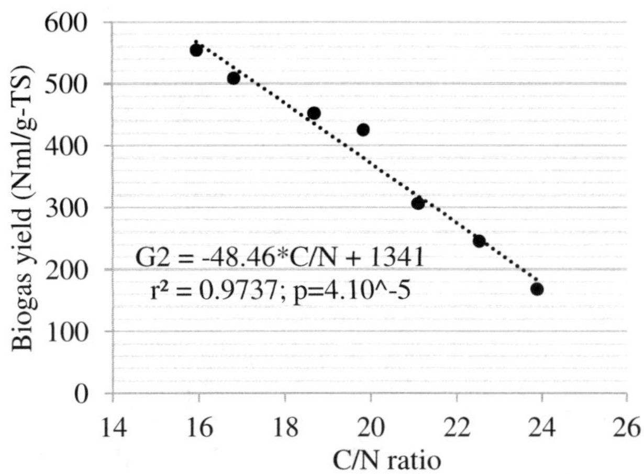

Figure 2. The relationship between biogas yield and C/N ratio.

IOP Conf. Series: Earth and Environmental Science **159** (2018) 012041 doi:10.1088/1755-1315/159/1/012041

3.2. Influence of C/N ratio to biogas yield

In essence, modification the proportion of each substrate in co-digestion aims to change nutrient ratio in the feedstock. In which, the nutrient quality is often evaluated indirectly by C/N ratio [17, 20]. The C/N ratio in the feedstock is too high pointing that the feedstock is not sufficient of nitrogen which needs for the build-up of microbial mass [3, 7]. Low C/N ratio leads to high concentration of generated ammonia, which is harmful to the AD processes [7]. Hence, the relationship between C/N ratio and biogas production was also investigated. Influence of C/N ratio on cumulative gas production was presented in Figure 2.

The high coefficient of determination demonstrated that C/N ratio is one of the key factors affecting the performance of the AD. The C/N ratio in the range of 16-23.9 showed a negative linear effect on biogas yield. The lower the C/N ratio, the higher the produced biogas. And an optimal C/N ratio was 16. This result agreed with the conclusion of some recent studies that low C/N ratio (15-20) was the best nutrient for the AD [8, 17].

3.3. Simulation of biogas accumulation

Both Gompertz (G) and (L) models were originally intended to describe the bacteria growth curve [12, 13]. With the assumptions of the rate of gas production is proportional both to the current microbial mass and to the substrate level, the L model - equation (6) is used to simulate biogas accumulation. The G model – equation (5) is also manipulated to express the biogas generation with the assumption that substrate limitations do not influence growth, and that the growth rate is proportional to the microorganism [13]. The kinetic constants A, μ_m, and λ of two models were determined by using the least squares fitting method (non-linear regression approach) with the aid of the solver function in MS Excel ToolPak. The received kinetic constants were shown in Table 3. By plotting simulation of the two models were also obtained the graph as shown in Figure 3.

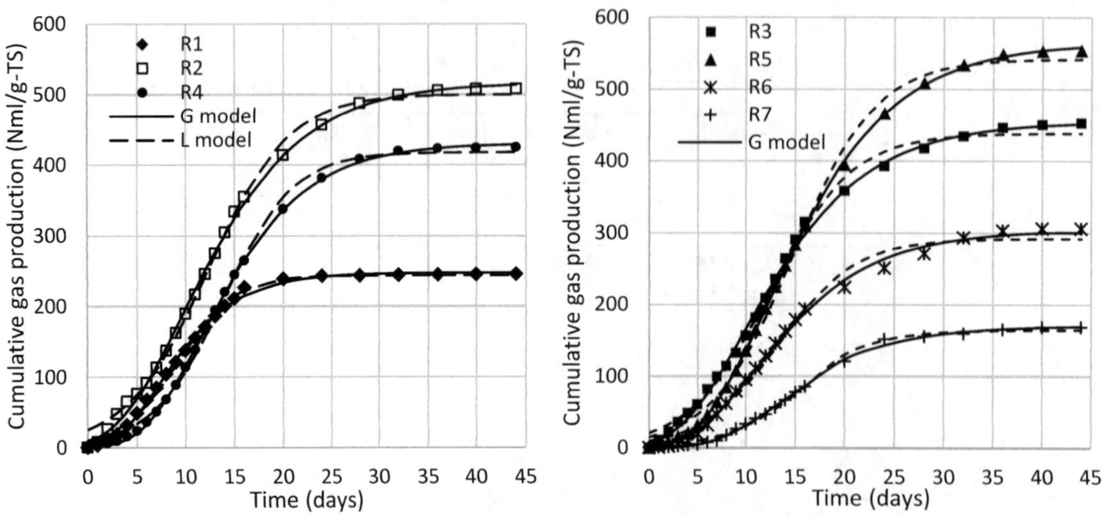

Figure 3. Accumulative gas production from experimental data and kinetic models.

The coefficient of determination was maximum in reactor R5 (r^2=0.9999) for G model and R4 (r^2=0.9979) for L model. However, the best SSE values were in reactor R7 for both models. The minimums of r^2 were 0.9977, 0.9901 in reactor R6 for G model and L model, respectively. Overall, results of the plotting data in Figure 3 and high coefficient of determination in Table 3 proved that both models were capable of simulating well the cumulative biogas production curve. However, the higher coefficient of determination and the lower sum of squared errors (SSE) demonstrated that the G model was better than the L model. This result had the same point of view with published reports from Latinwo and Agarry [11], Lo, et al. [23], and Lay, et al. [24].

IOP Conf. Series: Earth and Environmental Science **159** (2018) 012041 doi:10.1088/1755-1315/159/1/012041

Table 3. Constants of gas production kinetic models

No.	Modified Gompertz model					Modified Logistic model				
	A	μm	λ	R2	SSE	A	μm	λ	r^2	SSE
R1	247	21	2.91	0.9984	307	244	21	3.39	0.9972	637
R2	517	28	2.93	0.9987	997	500	30	3.92	0.9972	2322
R3	455	24	3.13	0.9982	1048	439	26	4.18	0.9963	2324
R4	431	27	5.67	0.9998	145	418	29	6.59	0.9979	3177
R5	565	30	5.44	0.9999	131	542	33	6.51	0.9966	3879
R6	304	17	4.54	0.9977	703	292	19	5.52	0.9901	3177
R7	171	10	6.91	0.9988	118	164	11	7.96	0.9969	339

There was different between the kinetic constants which were obtained from both models. The biogas production potentials (A) in G model were higher 1.5-4.2% than them in the L model. The G model showed the lower 1.7-10% values of μ_m, and also lower 16.6-33.8% values of λ. However, both models had the same expression, the biogas production (A and μ_m) reached maximum and minimum in the reactor R5 and R7, respectively for both models.
At time t=0 day, experimental data showed zero value of cumulative biogas while both models exhibited positive values for all reactors (see Table 4). However, the G model was still better than L model for simulation at zero points. For λ – lag time (days), which is often understanding as the min time taken to appear biogas. Experimental data exhibited that the biogas was generated as soon after t=0 (see Figure 3). While obtained λ -value changed in among 2.9-6.9 days in both models (see Table 3), many studies which used G and L models had displayed the same status [2, 11, 13, 25, 26]. Obviously, obtained λ-values did not reflect right itself definition.

Table 4. Cumulative biogas production at t=0 days

Model	R1	R2	R3	R4	R5	R6	R7
Experiments (Nml/g-TS)	0.0	0.0	0.0	0.0	0.0	0.0	0.0
G (Nml/g-TS)	1.3	7.7	6.1	0.4	1.6	1.1	0.1
L (Nml/g-TS)	9.9	24.7	20.3	8.9	15	8.7	2.9

4. Conclusions

Co-digestion of VW, HD, and S was performed successfully. Experimental results demonstrated that each single-substrate in mixture played a significant role in nutrient balance. And co-digestion of VW, HD, and S improved biogas yield from single-substrates. The high component of sludge in feedstock reduced biogas yield because of high C/N ratio. The high HD proportion in the mixture led to the high cellulose which was not good for the AD. Hence, the mixture contained VW in major proportion had synergic effect with the cumulative gas volume being the highest. The volume of the cumulative biogas (Nml/g-TS) could be estimated from the component ratio of substrates by equation $G1_{Nml/g-TS} = 53.7 + 744.8 \times \text{VW}(\%) + 192.2 \times \text{HD}(\%)$ (r^2=0.9433; p-value=0.00321), or from C/N ratio by equation $G2_{Nml/g-TS} = 1341 - 48.46 \times C/N$ (r^2=0.9737, p-value =0.000). While the equation G1 exhibited external phenomenon then the equation G2 showed internal essence of substrate proportion influences to AD.
The high value of correlation coefficient (r^2=0.9901-0.9999) demonstrated that both the (G) and Logistic (L) models were appropriate to simulate the accumulation of biogas production. However, with higher r^2-values and lower SSE-values in all reactors demonstrated that G model was better than L model. Influences of operating conditions such as temperature, pH, TS to the constants of the models were not investigated, that is a big limitation of this study.

4.1. Limitations of both models.
Experimental data showed both models failed at time t=0 day. The lag time - λ did not reflect the right itself definition, it seemed merely mathematical constant. In addition, a characteristic of biogas production curve $-t_o$ (when μ_m -biogas generation rate reached maximum) does not appear in the G and L models. Hence, it is difficult to make a comparison between experimental data and modeling data. That was the reason previous studies which focused on G and L models did not mention to t_o-value. Overall, for these disadvantages in both models, there should be a complete model for an alternative.

Next study: We would like to try to find out a biogas production kinetic model that can overcome the limitations of conventional models which is mentioned above. As well-known, anaerobic digestion is defined as a series of processes including hydrolysis, acidogenesis, acetogenesis, and methanogenesis. Biogas is mainly including CO_2 and CH_4, in which, methane is generated only from the methanogenic process, and CO_2 can be generated from acidogenic, acetogenic, and also methanogenic processes. Therefore, kinetic of biogas generation should include the two processes respectively.

5. References

[1] Nayono S E 2010 *Anaerobic digestion of organic solid waste for energy production*: KIT scientific Publishing. Vol. 46.

[2] Syaichurrozi I and Sumardiono S 2013 *Biogas production kinetic from vinasse waste in batch mode anaerobic digestion.* World applied sciences journal. **26**(11): p. 1464-1472.

[3] Mao C, Feng Y, Wang X, and Ren G 2015 *Review on research achievements of biogas from anaerobic digestion.* Renewable and Sustainable Energy Reviews. **45**: p. 540-555.

[4] Deng Y, Xu J, Liu Y, and Mancl K 2014 *Biogas as a sustainable energy source in China: regional development strategy application and decision making.* Renewable and Sustainable Energy Reviews. **35**: p. 294-303.

[5] Zhang L, Lee Y-W, and Jahng D 2011 *Anaerobic co-digestion of food waste and piggery wastewater: focusing on the role of trace elements.* Bioresource technology. **102**(8): p. 5048-5059.

[6] Zhang C, Su H, Baeyens J, and Tan T 2014 *Reviewing the anaerobic digestion of food waste for biogas production.* Renewable and Sustainable Energy Reviews. **38**: p. 383-392.

[7] Hartmann H and Ahring B K 2006 *Strategies for the anaerobic digestion of the organic fraction of municipal solid waste: an overview.* Water science and technology. **53**(8): p. 7-22.

[8] Dai X, Li X, Zhang D, Chen Y, and Dai L 2016 *Simultaneous enhancement of methane production and methane content in biogas from waste activated sludge and perennial ryegrass anaerobic co-digestion: The effects of pH and C/N ratio.* Bioresource Technology. **216**: p. 323-330.

[9] Riggio V, Comino E, and Rosso M 2015 *Energy production from anaerobic co-digestion processing of cow slurry, olive pomace and apple pulp.* Renewable Energy. **83**: p. 1043-1049.

[10] Wang X, Yang G, Feng Y, Ren G, and Han X 2012 *Optimizing feeding composition and carbon–nitrogen ratios for improved methane yield during anaerobic co-digestion of dairy, chicken manure and wheat straw.* Bioresource Technology. **120**: p. 78-83.

[11] Latinwo G and Agarry S 2015 *Modelling the kinetics of biogas generation from mesophilic anaerobic co-digestion of sewage sludge with municipal organic waste.* Chem Process Eng Res. **31**: p. 2224-7467.

[12] Zwietering M, Jongenburger I, Rombouts F, and Van't Riet K 1990 *Modeling of the bacterial growth curve.* Applied and environmental microbiology. **56**(6): p. 1875-1881.

[13] Schofield P, Pitt R, and Pell A 1994 *Kinetics of fiber digestion from in vitro gas production.* Journal of animal science. **72**(11): p. 2980-2991.

[14] Teleky B E and Balan M C 2015 *Timeline of gas production under anaerobic conditions.* Journal of Bioprocessing & Biotechniques. **5**(4): p. 1.

[15] Heo N H, Park S C, and Kang H 2004 *Effects of mixture ratio and hydraulic retention time on single-stage anaerobic co-digestion of food waste and waste activated sludge.* Journal of Environmental Science and Health, Part A. **39**(7): p. 1739-1756.

[16] Dong L, Zhenhong Y, and Yongming S 2010 *Semi-dry mesophilic anaerobic digestion of water sorted organic fraction of municipal solid waste (WS-OFMSW).* Bioresource Technology. **101**(8): p. 2722-2728.
[17] Zhang C, Xiao G, Peng L, Su H, and Tan T 2013 *The anaerobic co-digestion of food waste and cattle manure.* Bioresource technology. **129**: p. 170-176.
[18] Smith D B and Almquist C B 2014 *The anaerobic co-digestion of fruit and vegetable waste and horse manure mixtures in a bench-scale, two-phase anaerobic digestion system.* Environmental technology. **35**(7): p. 859-867.
[19] Kalia A K and Singh S 1998 *Horse dung as a partial substitute for cattle dung for operating family-size biogas plants in a hilly region.* Bioresource technology. **64**(1): p. 63-66.
[20] Ostrem K 2004 *Greening waste: Anaerobic digestion for treating the organic fraction of municipal solid wastes.* Earth Engineering Center Columbia University: p. 6-9.
[21] Hills D J and Roberts D W 1981 *Anaerobic digestion of dairy manure and field crop residues.* Agricultural Wastes. **3**(3): p. 179-189.
[22] Rao P V and Baral S S 2011 *Experimental design of mixture for the anaerobic co-digestion of sewage sludge.* Chemical Engineering Journal. **172**(2): p. 977-986.
[23] Lo H, Kurniawan T, Sillanpää M, Pai T, Chiang C, Chao K, Liu M, Chuang S, Banks C, and Wang S 2010 *Modeling biogas production from organic fraction of MSW co-digested with MSWI ashes in anaerobic bioreactors.* Bioresource Technology. **101**(16): p. 6329-6335.
[24] Lay J-J, Li Y-Y, and Noike T 1996 *Effect of moisture content and chemical nature on methane fermentation characteristics of municipal solid wastes.* Doboku Gakkai Ronbunshu. **1996**(552): p. 101-108.
[25] Deepanraj B, Sivasubramanian V, and Jayaraj S 2015 *Experimental and kinetic study on anaerobic digestion of food waste: The effect of total solids and pH.* Journal of Renewable and Sustainable Energy. **7**(6): p. 063104.
[26] Nopharatana A, Pullammanappallil P C, and Clarke W P 2007 *Kinetics and dynamic modelling of batch anaerobic digestion of municipal solid waste in a stirred reactor.* Waste management. **27**(5): p. 595-603.

IOP Conf. Series: Earth and Environmental Science **159** (2018) 012042 doi:10.1088/1755-1315/159/1/012042

Improved Design of Ultrasonic Cleaning Tank Using Harmonic Response Analysis in ANSYS

W Tangsopa[1], T Keawklan, K Kesngam, S Ngaochai and J Thongsri[2]

Computer Simulation in Engineering Research Group, College of Advanced Manufacturing Innovation, King Mongkut's Institute of Technology Ladkrabang, Bangkok, 10520, Thailand
E-mail: [1]worapol.tsp@gmail.com, [2]jatuporn.th@kmitl.ac.th

Abstract. Ultrasonic cleaning tank is a machine that many factories widely used to clean objects. At one factory, a problem occurred in the cleaning process, resulting in the factory not being able to clean objects, but cracks also appeared on some objects. It was anticipated that these were caused by uneven acoustics pressure distribution which resulted in unsuitable cavitation. This directly affected cleaning performance within the tank. In order to improve the tank's efficacy, in this research, we used Harmonic Response Analysis in ANSYS 17.2 to simulate the occurrence of acoustic pressure in the tank to find a suitable type and placing position of the transducer for the tank. We found that transducers made from PZT4 gives greater acoustic pressure than those made from PZT8. Changing the transducer type does not affect the acoustic pressure pattern. PZT4 is therefore the most suitable type of transducer for this tank. Simulation results also suggested that placing the transducer under and beside the tank will result in intense acoustic pressure that is evenly dispersed throughout the tank. Results from this simulation were then passed to the manufacturing factory, and were later accepted that it could truly enhance the tank's efficacy. Information regarding the issue was also used in the development of improved design of ultrasonic cleaning tank to have greater performance levels.

1. Introduction

The ultrasonic (u/s) cleaning tank is a machine used to clean valuable objects that are fragile yet contaminated or dirty to be more cleaned such as lens, jewelry, dental instruments and electronic devices etc. Acoustic pressure occurs from the vigorous vibration of the piezoelectric transducer. Once it has received electric currents, it will send ultrasonic waves into the medium following the transducer's built frequency which is between 20-20,000 kHz [1-2]. Liquid medium will then change pressure phase or lessen following the pulse of the harmonic wave. The rare fraction phase occurs during low or sub-zero pressure while the compression phase occurs during high or positive pressure. When the liquid medium is in the low-pressure phase, many tiny cavitation bubbles occur. When the liquid medium in any area is in the high-pressure state, the bubbles that occur will shrink and absorb energy. The bubbles continuously change according to cycle until they cannot absorb any more energy and will then collapse during the high-pressure phase. When bubbles collapse, they release an intense energy called the liquid jet, with a temperature of 5,500 ^{o}C, pressure of 68 MPa and a speed of approximately 280 m/s [3] that collides with contaminant particles on or nearby an object therefore cleaning the object's surface.

Our research is a collaboration between a u/s tank manufacturer and our research group. The u/s tanks used in the actual cleaning process had a problem of after cleaning, most of objects were cleaned but some objects still had particle contamination, and some were cracked. As for the unclean objects, it required recleaning process until it is clean while the damaged objects must be destroyed. The factory that used these u/s tanks had budget losses from the cleaning process approximately 100 million THB per year, therefore; this problem must be resolved urgently.

From a review, we found that the main cause of this problem leading to the losses is bad acoustics pressure distribution because an abrupt change in acoustic pressure distribution results in the cavitation effect. The main factors that affect acoustics pressure distribution are power, ultrasonic frequency, temperature of solution, sonication time, position of cleaning object and type of transducer. Cavitation intensity increases with increasing power generated into the medium [4]. The cleaning performance depends on the magnitude of acoustic pressure. It is good performance at the location of large acoustic pressure and poor at the location of small acoustic pressure. High frequency is proper for cleaning small objects, but low frequency is in opposite [5]. When the solution's temperature is increased, the cleaning performance decreases. It was found that the suitable temperature of solution for cleaning is in the range of 20 – 40 °C, while the best cleaning temperature is below 20 °C [6, 7]. Since sonication time, a time for cleaning objects using ultrasonic, is increased, corrosion due to cavitation is also increased [8]. The performance of transducers made from PZT4 and PZT8 were investigated by finite element method and experiment [9]. It was reported that PZT4 is better than PZT8. However, this work investigated only the performance of transducer. The u/s tank model and other environmental conditions did not include. Recently, to find ways for improving the performance of ultrasonic cleaning in a simple tank, we simulated the acoustic pressure distribution using the Harmonic Response Analysis in ANSYS based on actual conditions from the factory [10]. The results of simulation were confirmed by the experiment of foil corrosion test. We found that increasing the power applied to the transducer was unaffected the acoustic pressure pattern. Also, increasing ultrasonic frequency gave more uniform acoustic pressure distribution. The position that provided the highest cleaning efficiency was at the middle of the tank.

This article is an extension of the research in references [9-10]. We will simulate the acoustic pressure pattern in the u/s tank by using PZT4 and PZT8 as transducers based on actual operating conditions of the factory to compare the cleaning performance. Moreover, we will determine the optimal placement of transducer providing the highest cleaning performance by using Harmonic Response Analysis in ANSYS 17.2. The simulation results may be used to improve the cleaning process and to develop a prototype of u/s tank with higher efficiency for the manufacturer in near future.

2. Theoretical Background

The cleaning process results from the cavitation effect. It occurs when the solution's pressure changes phase abruptly. However, at present no mathematical equation may directly express the cavitation. Therefore, researchers rely on the wave equation of acoustic pressure in form of Helmholtz equation (1) to explain cavitation's behaviour, since it can illustrate the cavitation bubble occurrence well.

$$\nabla \cdot -\frac{1}{\rho}(\nabla p_t) - \frac{\omega^2 p}{c^2 \rho} = 0 \qquad (1)$$

where acoustic pressure $p = p_0 exp(i\omega t)$. ω represents the frequency of wave. The constants c and ρ are the speed of wave and density of water, respectively.

The temperature and pressure of the bubble collapse are also important factors that affect the cleaning process. When the bubble expands to its maximum size, the area surrounding the bubble's temperature will rise to over 5,500 °C. Because the pressure is greater than 68 MPa [3], the collapse will occur quickly in the micro-seconds unit. The heat will not be able to escape from the bubble in time. This deems that the bubble collapses adiabatically and results in the cleaning process. The equation used to explain the relationship of collapse, temperature and bubble radius can be expressed by the well-known Reyleigh-Plesset equation as [11]:

$$P_{\text{collapse}} = \left(P_0 + \frac{2\sigma}{R_0}\right)\left(\frac{R_0}{R}\right)^{3\gamma} - \frac{2\sigma}{R} - \frac{4\mu}{R}\left(\frac{dR}{dt}\right) \qquad (2)$$

where P_0 is the acoustic pressure in solution. $P_{collapse}$ is the pressure inside the bubble. σ is the solution surface tension. R_0 is the starting radius of the bubble and R is the radius of the bubble while collapsing. γ is the ratio of gas specific heats. μ is the dynamic viscosity. Equation (2) implies that the cavitation bubble radius relates to acoustic pressure. This shows that the increase of acoustic pressure is the increased intensity of the cleaning process.
As for the ANSYS program, we used the Harmonic Response Analysis which is calculated the acoustic pressure based on the finite element method (FEM). The program calculates acoustic pressure in the solution where the cavitation effect occurs and reports as graphic colours. In calculating the acoustic pressure, the finite element formulation is obtained by a testing wave using the Galerkin procedure. The wave equation (1) is multiplied by the testing function (w) and integrated over the volume of the domain with some manipulation to yield [12]:

$$\begin{aligned}&\iiint_{\Omega_F} \frac{1}{\rho_0 c^2} w \frac{\partial^2 p}{\partial t^2} dv + \iiint_{\Omega_F} \nabla w \cdot \left(\frac{4\mu}{3\rho_0^2 c^2} \nabla \frac{\partial p}{\partial t}\right) \mathrm{d}v + \iiint_{\Omega_F} \nabla w \cdot \left(\frac{1}{\rho_0} \nabla p\right) \mathrm{d}v \\ &- \oiint_{\Gamma_F} w\left(\frac{1}{\rho_0} + \frac{4\mu}{3\rho_0^2 c^2} \frac{\partial}{\partial t}\right) \hat{n} \cdot \nabla p ds + \oiint_{\Gamma_F} w \frac{4\mu}{3\rho_0^2} \hat{n} \cdot \nabla Q ds \\ &= \iiint_{\Omega_F} w \frac{1}{\rho_0} \frac{\partial Q}{\partial t} dv + \iiint_{\Omega_F} \nabla w \cdot \left(\frac{4\mu}{3\rho_0^2} \nabla Q\right) \mathrm{d}v\end{aligned} \tag{3}$$

where dv is volume differential of acoustic domain Ω_F, ds is surface differential of acoustic domain boundary Γ_F and $\hat{n}$ is outward normal unit vector to the boundary Γ_F. From equation (3), the water with fluid domain is now adapted to acoustic domain which can be calculated the acoustic pressure by ANSYS. Readers who are interested in more details of derivation and meaning expression of equation (3) can be found in references [12].

3. Methodology

3.1. Ultrasonic cleaning tank

The u/s cleaning tank we used to draw a model is donated from the factory, the model which industrial factories widely purchase the most. The size is 244 mm (wide) × 340 mm (long) × 220 mm (depth). It may contain 18 litres of water. Beneath the tank are 8 PZT4 piezoelectric transducers. The overall power is 400 watts, with a frequency of 28 kHz as shown in figure 1.

Figure 1. Actual u/s cleaning tank.

3.2. Simulation model

To determine a suitable type of transducer for u/s tank, we draw a CAD model as shown in figure 2. This model is similar to the actual u/s tank but using PZT8 as transducer instead of PZT4. To investigate the change after repositioning the transducers and to find the best position for placing transducer. We created a CAD model for improved design of u/s tank by relocating 4 transducers from below to the side. We anticipated that this method will give better distributed acoustics pressure. The model with repositioned transducer is shown in figure 3. All models consist of 4 domains; water (solution), the plate (at the bottom of the ultrasonic tank), foil sheet (located at the middle position of the tank which has the highest cleaning efficiency) and the transducer. The transducers consist of back mass, piezoelectric, front mass and glue.

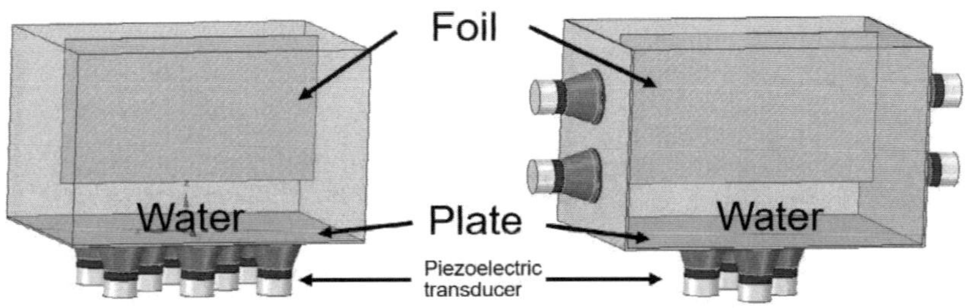

Figure 2. CAD model of actual u/s tank.

Figure 3. CAD model for improved design of u/s tank.

3.3. Mesh model

By simulation using the FEM method, we transferred the CAD models to a mesh model in order to divide large models into elements. The connected points between elements are called nodes. The more number of nodes, the higher accuracy of simulation. Our research simulated pressure occurrence from high frequency waves. Thus, we needed to create a highly detailed mesh. The mesh in each model must have at least 6-12 elements in 1 wave length to ensure that the pattern from our simulation is realistic [10, 12]. The mesh model for the actual and improved design of u/s tanks are shown in figure 4 and 5. We used hexahedral mesh, a mesh that gives more nodes than others but gives more accurate solution than other mesh types. Both models consist of approximately 240,000 elements and 706,000 nodes. They provided high accurate results in a limit of computer performance and computational time usage.

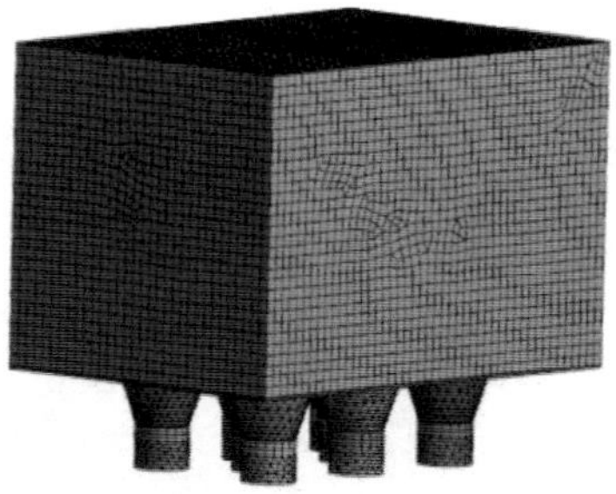

Figure 4. Mesh model of actual u/s tank.

Figure 5. Mesh model of modified u/s tank.

3.4. Harmonic response analysis

Harmonic Response Analysis is a package in ANSYS 17.2 used to determine the steady-state response of a linear structure to loads that vary sinusoidally with time [12]. It was integrated with an additional package of piezoelectric and MEMS [13] to increase the capability for simulating the solution in piezoelectric domain. This integration was ensured that gives the higher accurate solution. Therefore; it was employed to determine the acoustic pressure and intensity of the u/s tank in this research. The

IOP Conf. Series: Earth and Environmental Science **159** (2018) 012042 doi:10.1088/1755-1315/159/1/012042

important values required for simulation are: material properties of water at 45 °C, PZT4 and PZT8, which are shown in Table 1. Other values we used the default setting of the program.

The calculation is separated into 3 domains: in PZT domain, plate domain and water domains. In PZT domain, piezoelectric is coupling between structural and electric fields that occurs in the piezoelectric materials. Applying a voltage to transducer (PZT) creates a displacement, and vibration. The equation governs finite element method in PZT domain is [13]

$$\begin{pmatrix} K_{UU} & K_{UV} \\ K_{UV} & -K_{VV} \end{pmatrix}\begin{Bmatrix} U \\ V \end{Bmatrix}+\begin{pmatrix} C_{UU} & 0 \\ 0 & -C_{VV} \end{pmatrix}\begin{Bmatrix} \dot{U} \\ \dot{V} \end{Bmatrix}+\begin{pmatrix} M_{UU} & 0 \\ 0 & 0 \end{pmatrix}\begin{Bmatrix} \ddot{U} \\ \ddot{V} \end{Bmatrix}=\begin{Bmatrix} F \\ Q \end{Bmatrix} \tag{4}$$

where K_{UU}, K_{VV}, K_{UV} are structural stiffness, dielectric permittivity and piezoelectric coupling element matrices, respectively. C_{UU}, C_{VV} are structural damping and dielectric dissipation, respectively. M_{UU} is mass. U is displacement vector and V is an applied voltage.

After completing the calculation of node displacements and vibration in PZT domain using equation (4), the results will transfer to the plate domain and then to the water domain where the acoustic pressure is numerically calculated using equation (3). Finally, the computer will display the acoustics pressure distribution results into different colors according to the acoustics pressure value in considered area.

Table 1 Material properties required for setting in Harmonic Response Analysis

Domain	Type	Value
Water (45 °C)	Water density	990.15 kg/m^3
	Acoustic velocity	1,533.5m/s
	Dynamic viscosity	5.7977x10^{-4} kg/m·s
Piezoelectric (PZT4)	Density	7,500 kg/m^3
	Permittivity Constant (ε_0)	8.854e-12 F/m
	Stiffness (c^E)	C_{11}=1.39x10^{11}=C_{22}, C_{21}=7.78x10^{10}, C_{31}=7.43x10^{10}=C_{32}, C_{44}=3.06x10^{10}, C_{55}=2.56x10^{10}=C_{66} Pa
	Piezoelectric stress (e)	e_{31}= -5.2 c/m^2, e_{33}= 15.1 c/m^2, e_{15}=12.7
	Relative Permittivity($\frac{\varepsilon_T}{\varepsilon_0}$)	K_{11} = 1,475, K_{33}= 1,300
Piezoelectric (PZT8)	Density	7,600 kg/m^3
	Permittivity Constant (ε_0)	8.854e-12 F/m
	Stiffness (c^E)	C_{11}=1.47x10^{11}=C_{22}, C_{21}=8.11x10^{10}, C_{31}=8.10x10^{10}=C_{32}, C_{44}=3.29x10^{10}, C_{55}=3.13x10^{10}=C_{66} Pa
	Piezoelectric stress (e)	e_{31}= -3.9 c/m^2, e_{33}= 13.9 c/m^2, e_{15}=10.3 c/m^2
	Relative Permittivity($\frac{\varepsilon_T}{\varepsilon_0}$)	K_{11} = 1,290, K_{33}= 1,000

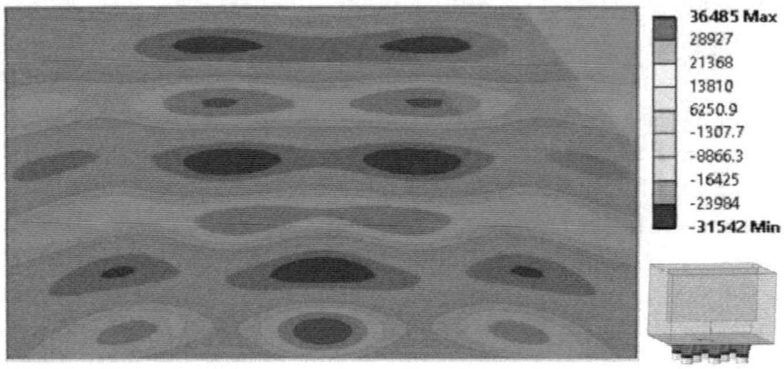

Figure 6. Acoustic pressure distribution of actual u/s tank using PZT4 as transducers [10].

4. Results and Discussion

As mentioned in the introduction that this research will study the results of using the PZT8 as the transducer and the results after repositioning the transducers to the side and its influences towards the acoustic pressure within the tank. We would therefore like to use the acoustic pressure from the simulation in our previous work [10] as reference. The results from reference [10] were confirmed the accuracy by comparing with foil corrosion test. Since our work in this article uses similar methodology, u/s tank model and operating conditions to the work in reference [10], this makes us confident that the methodology used, and simulation results obtained in this research are credible. The simulation results of the acoustic pressure in the middle plane of the tank when we used the PZT4 transducer as shown in figure 6. Once the transducer was changed to type PZT8, acoustic pressure distribution results as shown in figure 7.

From the comparison of figures 6 and 7, once the type of transducer is changed from PZT4 to PZT8, the acoustic pressure distribution pattern did not change. The acoustic pressure distribution in a middle plane, a plane that gave the highest cleaning performance, revealed that using PZT4 transducer has higher intensity of acoustic pressure than PZT8. High intensity of acoustic pressure gives the higher energy of cavitation and better cleaning performance. According to Table 1, we can see that the permittivity constant, stiffness, piezoelectric stress and relative permittivity of PZT8 are lesser than PZT4, therefore; resulting in less cavitation intensity. From the simulation results, we may say that the PZT4 is most suitable material for using as transducer for manufacturing the u/s cleaning tank.

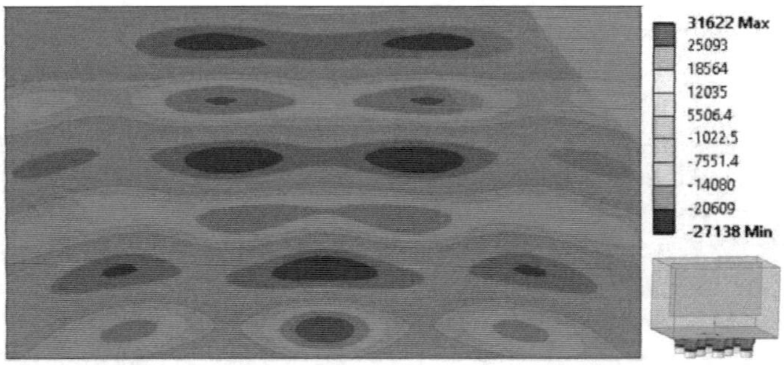

Figure 7. Acoustic pressure distribution of actual u/s tank using PZT8 as transducers.

To check the acoustics pressure distribution results from repositioning 4 transducer heads to the side of the u/s tank so the cleaning process within the tank is evenly done as shown in figure 3 by using the PZT4 as the transducer. Simulation result of the acoustic pressure occurrence is shown in figure 8. Once compared to the simulation results of the original tank without repositioning in figure 6, it was found that, by changing the transducers' position, the acoustic pressure pattern at an overall figure had changed. Better acoustic pressure distributions to all areas in the tank were observed. Other than this, the highest and lowest acoustic pressure are also greatened, resulting in better and more intense cavitation. The better acoustic pressure distribution change is from relocating transducers to the side resulting in acoustic waves from 3 directions that give more constructive and destructive interference points than the original u/s tank which all 8 transducers were installed beneath the u/s tank resulting in acoustic wave from a single direction. All results mentioned above made us confident that by repositioning 4 transducers to the sides will also enhance the performance of cleaning as well. All results of this research were submitted to the tank manufacturer. At the present, they have already used the knowledge to build newer model with better cleaning performance than the original one. This simulation is done at 45 ^{o}C. Cleaning performances depend on the temperature as well, therefore; a further study of the temperature and its effects upon cleaning performances by using simulation is an interesting and challenging topic in the future. In addition, during research, we also found that the transducer has a short life expectancy. If we applied the Harmonic Response Analysis and the Fatigue Analysis in ANSYS to deal with the problem following the work in reference [14], the results could be solved the problem and extended a lifetime usage of the transducers.

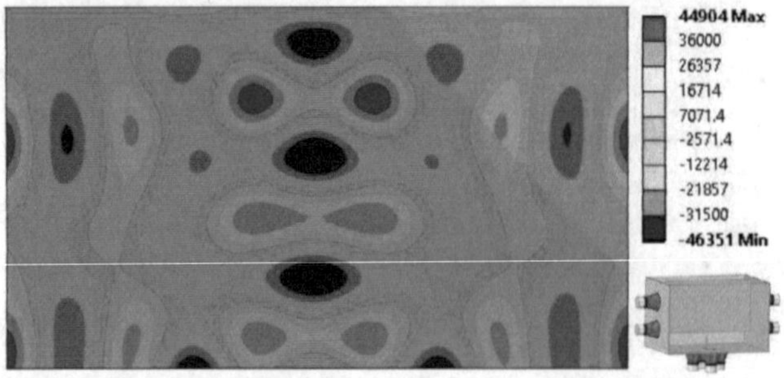

Figure 8. Acoustic pressure distribution of modify u/s tank.

5. Conclusion

This research simulated acoustic pressure distribution occurrence which directly affects the ultrasonic (u/s) tank's object cleaning efficacy. The ultrasonic cleaner tank has a frequency of 28 kHz with power of 400 watt. The tank in this research originated from the manufacturer where a problem occurred. The problem was after cleaning, most of objects were cleaned but some objects still had particle contamination, and some were cracked. This made a low efficiency in cleaning. To solve the problem, we used Harmonic Response Analysis in ANSYS to simulate the acoustic pressure to determine the optimal design of u/s tank with higher efficiency. First, we changed the piezoelectric transducer type from PZT4 which is a conventional material to PZT8. The simulation results showed that the acoustic pressure distribution pattern did not change but using PZT4 as transducer gave the higher intensity of acoustic pressure. To determine suitable positions for placing the transducers, we changed the installation position of the transducer of the tank to the side, it was found that the acoustic pressure distributed more evenly throughout the tank and thus intensified the acoustic pressure within the tank as well. Therefore, we concluded that using PZT4 as transducer and placing the transducers beneath and beside the tank contribute to better cleaning efficacy than the original u/s tank. The finding from this research was proposed to the tank manufacturer. It was used as novel information for developing an improved design of u/s tank with higher efficiency in near future.

6. References

[1] Khmelev V, Shalunov A, Nesterov V, Dorovskikh R and Golykh R 2014 Development of the rectangular ultrasonic radiator of the stair-step form *Proc. Int. Conf. on Micro Nanotechnologies and Electron Devices* (Novosibirsk, Russia) p 187-193.
[2] Khmelev V, Shalunov A, Nesterov V, Dorovskikh R and Golykh R 2015 Ultrasonic radiators for the action on gaseous media at high temperatures *Proc. Int. Conf. of Young Specialists on Micro Nanotechnologies and Electron Devices* (Novosibirsk, Russia) p 224-228.

[3] John F 2002 Ultrasonic Cleaning: Fundamental theory and application p 1-14.
[4] Vetrimurugan R 2012 Optimization of Hard Disk Drive heads cleaning by using ultrasonics and prevention of its damage *Procedia APCBEE* vol 3 p 222-230.
[5] Kim J, Choi S and Kim J. 1999 Vibroacoustic characteristics of ultrasonic *Apply Acoustics* vol 58 p 211-218.
[6] Niemczewski B 2007 Observations of water cavitation intensity under practical ultrasonic cleaning conditions *Ultrasonics Sonochemistry* vol 14 p 13-18.
[7] Overton G, Williams P and Trevena D 1984 The influent of cavitation history and entrained gas on liquid tension strength *J. Phys. D. Apply. Phys.* vol 17 p 979-987.
[8] Azhdast M, Haleh H, Pouladzadeh P, Azhdast A and Soltanabadi M 2009 Theoretical and Experimental Analysis of beating and cavitation phenomenon on erosion in ultrasonic cleaning process *Proc. Int. Conf. on Advanced Intelligent Mechatronics Suntec Convention and Exhibition Center*, (Singapore) p 624-629.
[9] DeAngelis D, Schulze G 2016 Performance of PZT8 versus PZT4 piezoceramic materials in ultrasonic transducers *Physics Procedia* vol 87 p 85-92.
[10] Tangsopa W, Thongsri J. and Busayaporn W 2017 Simulation of ultrasonic uleaning and ways to improve the efficiency *Prof. Int. Conf. Electrical Engineering Congress* (Pattaya, Thailand).
[11] Hilgenfeldt S, Brenner M, Grossmann S and Lohse D 1998 Anslysis of Reyleigh-Plesset dynamics for sonoluminescing bubbles *J. of Fluid Mechanics* vol 365 p 171-204.
[12] ANSYS, Inc 2015 Acoustic ACT×R170: Introduction to Acoustic.
[13] ANSYS, Inc 2015 Piezoelectric & MEMS exposure in ANSYS WorkBench using ACT.
[14] Boonkaew P and Thongsri J 2017 The optimum design of micro gripper for lifetime improvement based on fatigue analysis and six sigma analysis *Int. J. of Applied Engineering Research* vol 12 No 20 p 10233-10241.

Acknowledgement
This work was partially supported by Seagate Technology (Thailand) Ltd., Pasuda Supplies and Services Co., Ltd., College of Advanced Manufacturing Innovation (AMI), and King Mongkut's Institute of Technology Ladkrabang (KMITL).

IOP Conf. Series: Earth and Environmental Science **159** (2018) 012043 doi:10.1088/1755-1315/159/1/012043

The Creation of Modern Electric Vehicles with Additional Source of Energy

K Karpukhin[1], **A Terenchenko**[1], **and A Kolbasov**[1]

1NAMI Russian State Scientific Research Center, 2 Automotornaya st., Moscow, Russia, 125438
E-mail: K.Karpukhin@nami.ru

Abstract. The development of electric vehicles as energy-efficient transport sets the direction of research in the field of renewable energy, such as solar radiation, expressed in the application of photovoltaic converters. NAMI Russian State Scientific Research Center carried out work on the application of the photoelectric Converter in the electric vehicle. Obtained from photoelectric converters of the data give the possibility of compensating for self-discharge and applicable to maintain the climate control, in addition to the low-voltage precharge and high-voltage batteries. The study examined the main climatic zones of the Russian Federation and the level of solar radiation, the possibility of accumulation of electricity using photovoltaic converters in **the** largest regional cities, developed technical solutions for the integration of a module of photoelectric converters in the basic design of the vehicle with a large storage of electrical energy (hybrids, electric vehicles). Results and solutions obtained during the research works, to extend the capabilities of energy electric and hybrid vehicles.

1. Introduction

Personal mobility is a basic need of humans. In 1998, we had a total about 700 million vehicle, but in 2018, we have over 1 billion vehicle in the world. Hybrids and electric vehicles developed differently in the last 30 years, but they had hard way into the market. According to forecasts of the European Commission for transport by 2020 in Europe 7 % of all new passenger and light commercial vehicles will be electric, increasing to a massive 31% by 2030. This measure is a forced connection with the sharp deterioration of the situation on the planet associated with air pollution and global warming. The solution of these problems is the creation of electric vehicles, production of which is established in the leading countries of the world.

2. Main Part

Development and creation of modern electric vehicles for both personal and commercial use, in the last 10 years have become relevant. The transition to vehicles with low emission of harmful substances during operation becomes the primary task in the automotive industry. This is an emerging area of research in the field of electric drive in vehicles for various purposes [1].

Pure Electric vehicle is characterized by storage accumulation and electricity in a high voltage battery and traction motor.

As components of the system of accumulation and storage of electricity (SHI) used a variety of batteries: lead, Nickel-metal hydride, lithium-polymer (Li-Pol), lithium-iron-phosphate (LiFePo4), lithium titanate (LTO), lithium-cobalt (LiNiCo) and others [2,3].

At present time, in total the technical characteristics in the manufacture and development of electric vehicle preference is given to a battery with lithium-ion based. However, they remain expensive and cumbersome to provide adequate reserve of electric vehicles along with the capabilities of a traditional combustion engine.

Often the mileage of the electric vehicle with the storage and accumulation of energy in the battery of lithium-ion batteries the energy capacity of 20 kWh is not more than 160 km [4]. However, it began a trend in which automakers began to equip their electric cars more energy-intensive batteries. Thus, there has been a trend in the increase of the traction battery and not in the discovery and application of new chemical foundations.

Any battery needs charging, that is, the replenishment of energy reserves. It is known that in modern cars the electric cars a full charge time is 4 to 8 hours, and some models have optional on-Board charger that allows, with the appropriate charging station to produce replacement electricity supply to 80% in 30 minutes from an external power source.

Long time stop for charging SHI are one of the main problems of electric vehicles. To reduce the charging time from a stationary source of energy and to increase the mileage we developed a large number of variants of the combined power plant with additional energy sources such as electrochemical fuel cells in the hybrid vehicles on electric (Toyota Mirai), Range Extander composed of a parallel hybrid (BMW i3) and photovoltaic converters (FEP) (prototypes and production models based on Toyota Prius, Ford C-Max, Tesla Model 3, The Fisker Karma) [5,6].

At the moment in the global automotive industry to find a balance between cost and power consumption of traction batteries, as the chemical composition of cells. So the next step is the study and application of ways of reducing charging time and increasing mileage. It should be understood that the additional power Converter and having multiple storage of energy in an electric car puts it in the category of hybrid vehicles. Thus, electric vehicles with a Range Extender classified as "hybrid vehicle" (a vehicle with at least two different energy converters (engines) and two different (side) energy storage systems for the purpose of propelling a vehicle), and is already able to throw out harmful substances into the atmosphere during the process of energy conversion.

Electric vehicles with integrated batteries solar cells is considered not a hybrid, since the energy storage system he structural one. The obvious advantage is that the fuel for solar cells, use of renewable energy – energy solar radiation and in the process of conversion of solar radiation into electrical energy battery solar cells does not produce harmful substances. Thus, the increase of energy efficiency of electric vehicles through the introduction in the design of batteries, solar cells, powered by renewable energy source, is ideologically more correct than vehicle electric vehicle internal combustion engine.

The ideology of solar cells based on the principle of converting optical radiation into an electrical signal. Optical range is a region of the spectrum of electromagnetic radiation from the deep ultraviolet (0.01 μm) to the far infrared (up to 1000 μm). The function of solar cells is the conversion of optical radiation into an electrical signal. This function is performed with various radiation detectors, which mainly belong to two groups — the actual photovoltaic and thermal. Radiation is: a) self radiation of the investigated object; b) reflected or scattered its surface; C) a partially absorbed (with semi-transparent body).

Table 1. Specifications of SUNPOWER solar cells

№	Specifications	
1	Manufacturer	SunPower
2	Country of origin	Taiwan
3	Panel type	Monocrystalline
4	The material of manufacture of the module	Pure silicon
5	Power, W	3.3 - 3.4
6	The voltage in the mode XX,	0.682
7	Power short-circuit current, And	6,25
8	Operating voltage, V	0.574
9	Operating amperage, And	5,8

10	The size of the solar cells, mm	125x125
11	The bending angle of the element is not more degrees	30

NAMI Russian State Scientific Research Center developed a system of additional power source on the solar cells that are integrated into the design of vehicle LADA (Fig. 1,2). The area of coverage of the surface of the car battery photoelectric converters is 1.8 m2 (115 items dimension 125x125 mm). Were applied flexible monocrystalline cells with an efficiency of at least 20% (detailed characteristics in Table 1).

The study examines the main climate zone of the Russian Federation, including those where a high density of road transport with the aim of studying the impact on the environment. Also, the mathematical pilot study in the introduction of photoelectric transducers in the design of electric vehicles was investigated the energy distribution of the sunlight on cities of the Russian Federation. The most effective in this regard were: Astrakhan, Sochi, Vladivostok. These cities represent different climatic zones, but the annual flow of energy is quite high and is approximately at the same level (table 2). These studies indicate that even at 20% efficiency solar cells per square meter the maximum amount of energy will not exceed a total of 274 kWh per year. When connecting to the Smart Grid technology V2G, it is possible that a vehicle could be a mini power station with battery and solar cells to generate electricity for domestic purposes [7]. Table 2 shows town in the Russian Federation with the highest and lowest annual rates of solar radiation energy.

Table 2. Average values of solar radiation for town of Russian Federation

Town	Month low (December)	Month high (June or July)	The total for the year
Astrakhan	95.8 MJ/m^2 (26.6 kWh/m^2)	755.6 MJ/m^2 (209.9 kWh/m^2)	4.94 GJ/m^2 (1371 kWh/m^2)
Sochi	124.9 MJ/m^2 (34.7 kWh/m^2)	744.5 MJ/m^2 (206.8 kWh/m^2)	4.91 GJ/m^2 (1365.1 kWh/m^2)
Vladivostok	208.1 MJ/m^2 (57.8 kWh/m^2)	518.0 MJ/m^2 (143.9 kWh/m^2)	4.64 GJ/m^2 (1289.5 kWh/m^2)
Petrozavodsk	8.6 MJ/m^2 (2.4 кВт·ч / м^2)	601.6 MJ/m^2 (167.1 kWh/m^2)	3.10 GJ/m^2 (860.0 kWh/m^2)
Arkhangelsk	4 MJ/m^2 (1.1 kWh/m^2)	575 MJ/m^2 (159.7 kWh/m^2)	3.06 GJ/m^2 (850 kWh/m^2)
Saint Petersburg	8 MJ/m^2 (2.2 kWh/m^2)	578 MJ/m^2 (160.6 kWh/m^2)	3.02 GJ/m^2 (840 kWh/m^2)

Developed and tested in laboratory conditions, optimization algorithms control the charging process and the generation of solar energy in the modes changes the light and temperature conditions (0…30 ^{o}C). Results and solutions obtained during the research work allow to expand the possibilities of electric transport with renewable energy and improve energy efficiency, reduce losses during transportation of electricity from power plant to consumer.

To perform computational experiments was chosen two basic driving cycle and one additional. The first basic cycle is a mixed cycle New European Driving Cycle (NEDC), which is currently used in the certification procedures, Rules of the United Nations No. 101, in accordance with which will be carried out experimental evaluation of the characteristics of the developed electric vehicle. The second basic cycle is WLTC (Worldwide Harmonized Light Vehicle Test Cycle) – mixed driving cycle, which is designed as a replacement for the outdated NEDC and already introduced in the European certification procedure. An additional cycle is ARTEMIS (Assessment and Reliability of Transport Emission Models and Inventory Systems) will be combined, developed in the 2000s in Europe to assess the environmental properties of vehicles. He is selected as the current model European speed limit of the road.

In the process of developing mathematical models and computational experiment for electric vehicle battery, solar cells were used annual changes in environmental factors in the city of Moscow, as well as the possibility of infrastructure for electric transport with the aim of equipping charging stations with batteries, solar cells [8]. Considered a large number of conditions that impact simulation tests WLTC cycles, NEDC, ARTEMIS change the following conditions:
- move North, South, West and East at different times of day with an interval of 3 hours;
- at different times of the year.

The result was achieved, indicating that extra power source on the basis of solar cells can generate an electric power, which is enough to increase mileage up to 20% in summer daytime in the absence of clouds (NEDC). However, in the winter in January power generation is so small that the values obtained in the calculations does not exceed 1%.

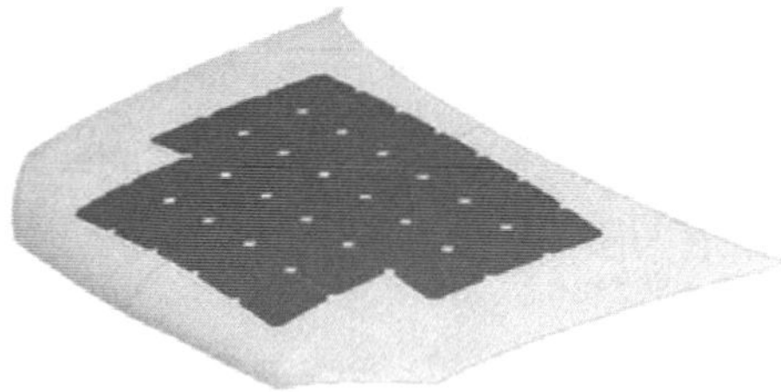

Figure 1. The location of the solar cells on the hood.

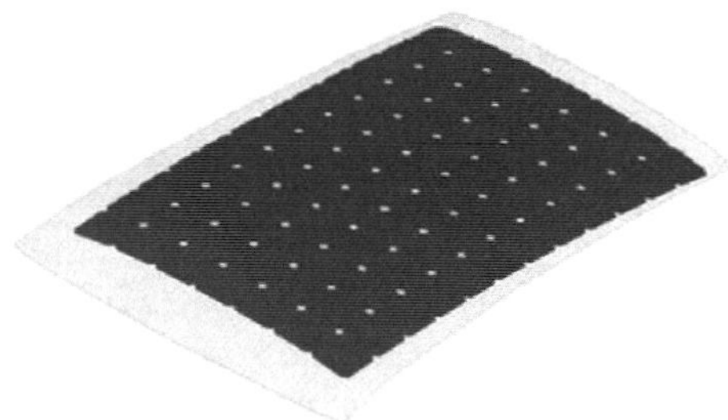

Figure 2. The location of the solar cells on the roof.

3. Results and discussion
The results of the study allow to formulate the main aspects of ensuring the effective functioning of the auxiliary power source to solar cells. Obviously, the best effect is achieved in regions with a warm climate and with minimal shading, for example, Astrakhan. However, it is known that for each degree over 25°C, the efficiency of solar cells be reduced. However, the estimated experiment showed the possibility of increasing mileage up to 20% through the use of a renewable source of energy. Thus, during prolonged idle electric vehicle battery self-discharge and energy costs, in the form of the cabin air, it is possible to compensate for the expense of batteries, solar cells, thus able to avoid CO_2 emissions.

Even with the introduction of Autonomous systems of heating of the battery, for example, by using thermoelectric converters Peltier, daily power consumption at -30°C for at least 1.6 kW [9, 10, 11]. Therefore, long-term Parking lot in the winter periodically possible heating of the battery due to the energy received from the power source based on solar cells.

4. Conclusion
1. Claimed by manufacturers of electric vehicles operating characteristics in the form of a driving range can be increased due to additional power plant for renewable fuels – a battery of solar cells. When the current efficiency of 22%, the use of solar cells as a magnifier mileage already is relevant.
2. The use of additional power sources based on solar cells with a power of 400 watts, will in the course of the year to up to 270 kWh, which corresponds to 132 300 g/CO_2 produced by the energy sector for an equivalent amount of electricity (490 g/kWh).
3. With the introduction of additional power sources based on solar cells in the design of electric vehicles it is possible to achieve increased mileage up to 20% at certain times of the year.
4. The developed system additional power sources based on solar cells for electric vehicles, which is able to increase the mileage of electric vehicle recharging renewable energy.

5. References

[1] John M. Miller, 2010. Propulsion systems for hybrid vehicles 2nd Edition *M. UK: The Institution of Engineering and Technology.*
[2] Gaines L, Cuenca R, Life-Cycle Costs of Lithium-Ion Vehicle Batteries 2000-01-1483 *J. SAE paper*, p.12.
[3] Abdullah AL-Refai, Osamah Rawashdeh, Rami Abousleiman. 2016. An Experimental Survey of Li-Ion Battery Charging Methods. *J. SAE International Journal of Alternative Powertrains*, **5 (1)**.
[4] Terenchenko A, Karpukhin K, Kurmaev R 2015 Features of operation of electromobile transport in the conditions of Russia. *J. Paper of EVS 28 International Electric Vehicle Symposium and Exibition*, KINTEX, Korea.
[5] International Energy Agency. Hybrid and Electric Vehicles *M. The Electric Drive Chauffeurs*, 2017.
[6] John Wiley & Sons 2014 Encyclopedia of Automotive Engineering *M.* ISBN: 978-0-470-97402-5 Online.
[7] Singh S, Mathur A, Das S, Sinha P, et al. 2017 Development of Smart PublicTransport System by Converting the Existing Conventional Vehicles to EV's in Indian Smart Cities. *J. SAE Technical Paper 2017-01-2011.*
[8] Karpukhin K, Terenchenko A, Kolbasov A, Kondrashov V 2017 Using photoelectric converters in road transport in order to improve energy efficiency in Russia. *J. International Journal of Mechanical Engineering and Technology*, **8 (10)** pp. 529-534.
[9] Karpukhin K, Terenchenko A, Shorin A, Bakhmutov S, Kurmaev R 2015 Temperature control of the battery for hybrid or electric vehicle. *J. Biosciences biotechnology research Asia*, **12 (2)** pp. 1297-1301.
[10] Slabosritsky R, Khazhmuradov M, Lukyanova V, Analysis and calculation of battery cooling system 2011 *J. Radioelectronics and informatics*, **3** pp.3-8.
[11] Slabosritsky R, Khazhmuradov M, Lukyanova V, Studying of battery cooling system 2012 *J. Radioelectronics and informatics*, **2** pp.5-8.

Acknowledgments
The paper was prepared under the agreement # 14.624.21.0047 with the Ministry of Education and Science of the Russian Federation (unique project identifier RFMEFI62417X0047).

IOP Conf. Series: Earth and Environmental Science **159** (2018) 012044 doi:10.1088/1755-1315/159/1/012044

The Effect of Urban Development Plans on City's Old Context

S Kalantari and Zh Shan

Department of Architecture and Urban Planning, Huazhong University of Science & Technology, China.
E-mail: S.kalantari_uni@yahoo.com

Abstract. It is about fifty years that the comprehensive and detailed plans have been the directing instrument and documents for Iranian cities spatial development. Meanwhile, the old context of cities not only became adapted with the modern patterns and their urbanism impression, but also this issue impaired their physical-special structure. Isfahan is a city with primary biological cores; and its historical background is indebted to Zayanderoud River. The role of the river and its branches (Madies) has been very important for agriculture and city development. "Madi" is one of Isfahan urban environmental features, which antiquity of this element refers to the ages before Islam. Unfortunately, due to the city development of the city, some part of the city identity, namely Madi, was forgotten. This paper tries to mention the value of such a city environmental element that was very considerable in the past, and evaluate the effects of urban development plans on them.

1. Introduction

The Iranian cities that had evolved innately relying on local innovation and creativity, have changed during the recent decades and have lost most of their unique features. The historic cores of Iranian cities encompass a large number valuable historic landscapes and monuments. These areas are the product of a complex process of interaction between human and physical environment, taking place over many centuries. Until the early twentieth century, cities had been involved in a continuous process of change. Isfahan is a historical city, located in the heart of Iran by a dramatic modernization transformations which started in 1920s. The natural features are the most important factors in Isfahan existential causes and city formation. The special and valuable situation of the *Zayanderoud* river as an identity index of Isfahan and the existence of branching *Madi* (Madi is a local expression means the water stream) and expansion in the urban context accompanying with other historical elements have established an organic context in the city center [1]. The Madis play an important and multifaceted role in Isfahan, and since it is as a field (structural), historical and environmental (natural) element that is considered in the context of Isfahan, rehabilitation and conservation of them has a significant impact on the quality improvement of city.

It lasts over five decades of experience in preparing and implementing the comprehensive urban plans and consequently the detailed plans in Iran, the prepared and implemented plans which aims to solve urban problems and improve the lives of citizens, in some cases not only help to alleviate urban problems, but also make new problems in cities. Regarding the usual methods of urban development plans Isfahan have faced with three ratified comprehensive plans *(kocks 1959, Organic 1969, Revision comprehensive plan 1988)* and three detailed plans *(Organic detailed plan 1973, Revision detailed plan 1993, Review detailed plan 2011)* and a districted plan *(Isfahan districted comprehensive plan 1988)* that each one has affected Isfahan physical order. According to the performed studies on the detailed plans in Isfahan and the use of resources in this field, this paper tries to evaluate the effect of the detailed plans in different times related to the Madis as the environmental prominent elements in Isfahan. Therefore, the status of Madis is reviewed in three detailed plans periods in Isfahan, i.e. pre-revolution (Organic Detailed Plan 1973), post-revolution (Revision Detailed Plan 1993) and contemporary (reviewed plan-detailed plan 2011). In other words, the main purpose of this paper is to analyze the status of Madis in the approved city plans and their effect of them on Madis as indicator elements of the old context of Isfahan. In this regard, the paper seeks to answer the question: "Is the current condition of Madis a feedback of approved urban plans and their instructions in order to protect the Madis?"

Figure 1. Left: Isfahan city– Right: location of Madies pass through the city

2. Materials and methods

The emergence and development of sciences related to urbanism and planning promotes the dissemination of several thoughts and theories of the city, urban development and spatial-physical effects. These theories were often stated during the last 150 years from the well-known researchers such as Eugene Viollet-le-Duc (1814-1879), John Ruskin (1818-1900), Camillo Boito (1836-1914), Camillo Sitte (1843-1903), Luca Beltrami (1854-1933), Ludico Quaroni (1911-1987), Gustavo Giovanni, Patrick Geddes (1854-1932), Le Corbusier (1887-1965), Lewis Mumford (1895), Kevin Lynch (1918-1984), Aldo Rasi (1931-1997), Christopher Alexander (1936), Constantinos Doxiadis (1914-1975), Richard Rogers (1933) who have addressed to recognize and break down the foundations of the old cities in different times and approaches in relation to the urban plans. Preparing the official plan of urban development in Iran has not so long historey and its historical roots are referred to the beginning of a contemporary century. Therefore, the research literature of this subject is influenced by the roots of these new plans related to several current decades. Theoretical and sample fields of the studies that have been mentioned about Iranian cities draw a special pattern of socio-spatial and physical transformation that making the problems and difficulties is an obvious feature of it.

In the physical- spatial structure in Isfahan, Madis are multidimensional and multifaceted elements that many factors influence in their role and function in the urban system and also have a significant impact on the quality of the environmental context. Madis' length in the city is about 200 km and their width is between 15 and sometimes up to 45 meters. Also Madis are scattered across the city linear and can act as the green local and neighborhood axes. Thus, they must have an appearance according to today's life and meeting the needs and desires of people in the city; their rehabilitation is from essential actions for the city life and from those actions that may be done for upgrading the quality of urban environment. According to the socio- spatial and physical factors that interfere in Madis quality, the examination and explanation of Madis status based on all factors not only are impossible because of the wide variety of the required data, but also follow a danger of general speaking and superficial view; in this regard, evaluation and adjustment of qualitative difference of the detailed plans regarding Madis are based on the main components of the detailed plans. The methodology of this study is analytical- applied based on environmental planning. Since the madies network as an urban issue needs precise investigations, field studies and analyses should be used to conduct the research. In order to do data collection, library and field methods were used. In this regard, investigation of approved comprehensive and detailed projects have been used.

3. Urban development plans in Isfahan

Urban planning and providing the related plans have extensively began since facing cities with the phenomenon of rapid growth of urbanization and they are placed in governmental programs. According to Iran Supreme Council of Urbanism and Architecture definition [2], detailed plan is based on the criteria and outlines of a comprehensive plan of the city as a legal document of spatial - physical development of the city and in addition to the planning regulations of urbanization has three main components, including street networks, land use and density of building. However, considering the Madis by urban planners and designers in urban development of Isfahan is tangible, but the current situation of Madis is not a description of rehabilitation and conservation measures of plans. Therefore, in the following to examine more comprehensive research purpose, among what we called the urban development plans mentioned in the physical transformations of Isfahan, we address the approved detailed urban plans.

Organic detailed plan (before Islamic revolution 1973): was approved to follow an organic comprehensive plan in the period of pre-revolution in 1973 [3]. In this plan, by limiting the urban development and by suggesting a population overflow in the satellite towns, constructing the new streets has been avoided in central and historical contexts of city. In organic plan, Madis' routes were seen as a pedestrian network to prevent passing cars and only the limited availability to the small and leading domain to the public parking area allowed. Madis axis is mainly from public service centers with a cultural, religious and educational role in places from neighborhoods' centers, the merchant role is emphasized beside it. In general organic plan, in addition to getting free Madis' route and creating the movement possibility over it, the legal limits of the routes, given hierarchy streams, has been evacuated from the building. In fact, there was a general ideology in Isfahan since the old days, there are two types of seasonal routes in Isfahan, one of winter routes used in cold seasons that include the roof-covered markets and bazars, the vast sunshine passages between the neighborhoods and the second, in the summer and warm routes which the adjacent passages of Madis and streams of water, of which they cover both parts of the city. From the very detailed features of the organic detailed plan is a floating plan; in fact, the management objective of organic detailed plan, with floating quality of land use in the plan is to give more and clearer details to the urban design that is also necessary due to the sensitivities and capacities of the historical context. Concerning height and density of constructions within the historical context, the proposals of plan are based on height and density restrictions.

Revised detailed plan (after Islamic revolution 1993): was prepared by the Naghshe Jahan Consulting Engineers [4]. It was almost a 20-year gap between the time of preparing this plan and organic plan, that due to outbreak of revolutions led to significant evolution in physical field of the city. During this period, there was almost no approved plan and criterion whatever that is an indicator of action sometimes is the municipal decisions and the Commission on Article 5 in the form of case decision mainly political without a comprehensive review. What are first and foremost in this plan are the significant expansion of the automobile transportation network and widening the city's many roads and even constructing the new streets. The enclosure of "Madis" based on the plan throughout its length is released and axially moving on foot has been seen. Passing the roadway in this plan contrary to the previous design alongside Madis is not removed, but this passage is limited to its adjunct home. In getting free the Madis enclosure, the widening has taken place in relation to the existing Madis axis and only valuable historical home remains safe from widening. In this plan, available public uses are reinforced around Madis but in the scale of the neighborhoods, there are no substantial changes unless where Madis are placed in the periphery of the urban street. Residential buildings in both sides of Madis have a special code in the detailed plans. Accordingly, by limiting the height and increasing it along the special angle of Madis on the one hand, the desired breeze can penetrate into the depth of the city and on the other hand by opening the crossing section of Madis, the major part of a green valley is full of trees in the urban landscape. In this plan, the land use are fully fixed and contrary to the organic plan they are not stated in the floating plan.

Revision detailed plan (contemporary era 2011): The review plan ratified in 2011 by five counselor engineers (Atak, Bavand, Saravand, Sharestan, and Shahr va khane) charged as district responsible counselor engineers and Naqshe jahan Pars counselor engineers as the coordinator counselor engineer [5]. The composition of detailed plan revision is based on sustainable development principles and strategies, social development, organizing the population inhabitant, economic development and prospers of activities in the city, communicating net improvement, environment protection, green space expansion, tourism, and reformation and reconstruction of old historical context of Isfahan. Centralize all urban plans under the supervision of municipal urban designing organization such as historical context restoration, cultural places, community and cultural centers, sport places, parks and green spaces geometric reform plan, old contexts exhilarating, different elevation intersections, pedestrians, the third chain of Isfahan traffic, urban furniture, Zayanderoud river boundaries construction, Madis designing and reconstruction in order to optimizing the plans from quality and design value aspect and decrease in designing costs and performance of projects.

4. Finding: analyzing madies in Isfahan's detailed plans

Though the urban plans and particularly in scale of the detailed plan plays an unbearable role at least in shaping the city's body, but because of the variety of the conditions of these plans included, study of all plans is excluded in current paper. Therefore, in the review and analysis of the research process, it has been sought to choose the criteria that have the greatest impact on the form and performance of Madis and selected criteria in the detailed plan in different periods are investigated in association with Madis. Table 1, shows comparison of obligations and criterions of each three plans, concerning the main components of them (land use, buildings density and network) regarding the Madis.

The maps comparison shows that the Madis bound has faced with lots of changes concerning the buildings height and density. The organic detailed plan has considered the height of buildings in the Madi bounds up to two floors (7.5 meters) from the ground. It is not comparable with other two plans from this side; on the other hand the height identification of other two detailed plans indicates the major inequality. The high constructions on the Madis bound followed with the increase in the buildings height, density, the motorist traffic and doesn't indicate the protective approach of the plans (figure 2).

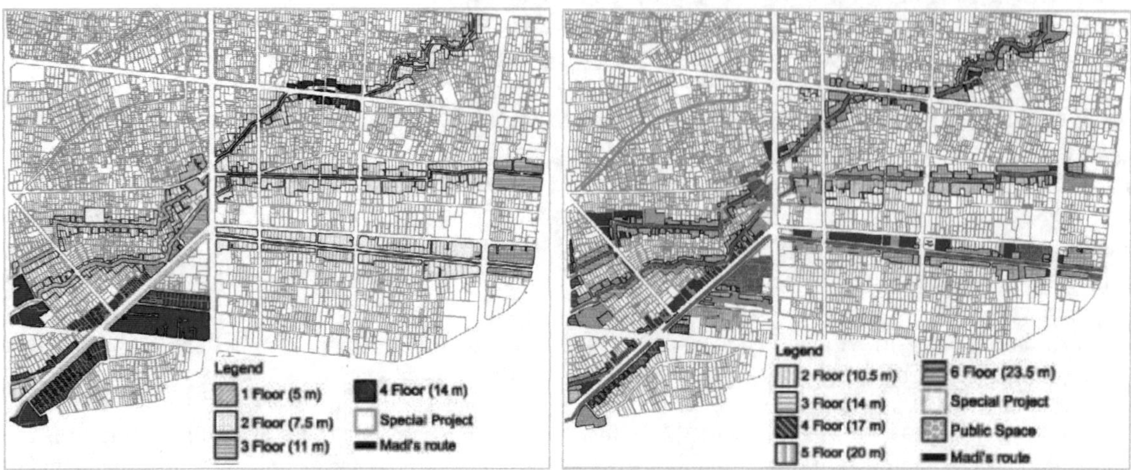

Figure 2. Madi's frontage building density (Left: Revised Detailed Plan, Right review detailed plan)

IOP Conf. Series: Earth and Environmental Science 159 (2018) 012044 doi:10.1088/1755-1315/159/1/012044

Table 1. Comparing of rules and regulations of Isfahan's detailed plans in relation with Madis

	Organic Detailed Plan (1973)	Revision Detailed Plan (1993)	Review Detailed Plan (2011)
Land use	Balanced concentration of public services through the available markets in the vicinity of Madis and distribution of new services to surrounding area; Appropriate activities with historical functions of area; Payment of parking tolls by properties along the pedestrians of Madis, in accordance with a detailed plan to the municipality to create the parking in the properties; The minimum subdivision area for the residential land use 300m^2	The minimum subdivision area for the residential land use 500m^2; Authorized commercial functions of the residential land use with the municipal agreements including permitted use of independent activity or commercial and service activities for quarters (maximum area 30m^2); Dedication of code MS to the Madis bound (special residential land use and public services); Considering the public service uses, educational facilities, green space and religious uses in Madis margin;	Commercial land use (daily market at the neighborhood, region and city level) using fruit and vegetable market and seasonal exhibitions and markets; Public green space as major land use; Commercial neighbor containing historical value, daily market, commercial complex; Authorized commercial usage of residential land use units in frontage of neighboring inter district and sectors main passages under the condition of municipal permission; Specific residential land use only for residential use and authorized items
Density	Considering the average density for residential and commercial land use; The buildings' height up to 7.5m; Not being obligatory to recess from the ally in the pedestrian passages with less than 6m width in the ground floor; Coordination of residential density from the aspect of construction level and the height of the buildings with the existing context;	The maximum floors for the first row of Madi's frontage is two and for the second row is three floors; The maximum number of floors for kindergarten and medical center is one floor and other public services land use is two floors; The permitted height is for the same boarders areas and it can't be more; The maximum number of floors for the same boarders ownerships with the codes: N 1.7, N 1.12, N 1.10 (Jolfa historic area, Takhte Poolad religious cultural complex and historical-cultural areas) is one floor; The maximum height for the public buildings in one floor is 5m and two floors buildings is 7.5m	In the adjacent pavement of Madis with a width of 6 m, the maximum allowable building height of 10.50meters (two storeys); In the adjacent corridor with a width of 6-8 meters of Madis, the maximum allowable building height of 14 meters (three storeys); In the adjacent pavement of Madis with a width of 8-10 m, maximum allowed height of 17 meters (four storeys); In the adjacent corridor with a width of 10-14 meters of Madis, maximum allowed height of 20 meters (five storeys); Recess with at least 2m from Madi, all the floors without any building construction and as the outdoor places under the owner authorization and adding maximum one floor up to the approved height;
Network	Removing the proposed streets with the comprehensive organic plan in this area; Maintaining valuable and important passages; Establishment on motorist access in the low valuable pathways of the context All access restriction limits up to 10 m; Recess only on low valuable passages; Allocating the Madis' margins to main pedestrian crossings and preventing disconnected by motorists;	Building doors with more than 1.5m width in the impending side to the Madis which is regarded for the pedestrian is not permitted; Devotion of around Madis path to the pedestrians and the prohibition of the motorist movement; Limited, non-continuous paths for emergency vehicles with a maximum width of 3 meters in the places that the owners of the plates have the access only through the Madis margin; Widening the Madis passing equally from the existing axis except the existence of valuable buildings from architectural and historical aspect;	Principled separation of the pedestrian and motorist path in a part that is dedicated to the motorist; Prohibition of motorist access from the Madis side in the blocks that have other accesses than the Madis side; Connecting Madis with a network of walking routes between the neighborhoods of surrounding area; The priority of moving Madis around the path with pedestrian crossing and the prohibition of the motorist, except for certain cases; Establishment of especial bikeway with at least 2m width by the differentiation in its flooring along the Madi path and building parking lot for bicycles in every 500m; No obligation to recess in historical buildings or aging trees adjunct to Madis; Establishment of special path with maximum width of 5m for to the emergency vehicles for parts that only have access to Madis axis

IOP Conf. Series: Earth and Environmental Science **159** (2018) 012044 doi:10.1088/1755-1315/159/1/012044

The basic of pathways network of this area, is old axis (Chaharbaq axis and inter sector historical passages), modern urban axis (new streets) and natural axis (Madis). The comparison of aforesaid pathways net map of detailed plans doesn't show significant changes; these changes demonstrate the purification of Madi bounds and antecedence to their boundaries were a few with inconsiderable dimension; the accomplished sitting back from Madis bounds are aimed at facilitating the motorist traffic. Establishment of little squares and urban open spaces in the reconsideration plan for detailed plan are other differences of this plan that are not considered in other detailed plans.
The maps of comprehensive plan that are prepared in a more general scope than the detailed plans shows the dominant land use arena identification as field division. These maps are represented in the form of land use in the detailed plan [6]. In Organic Detailed Plan the land use of the Madis bounds are mostly residential and its changes in this field is completely obvious with other two plans; therefore the major changes of other two detailed plans are shown here (figures 3,4).

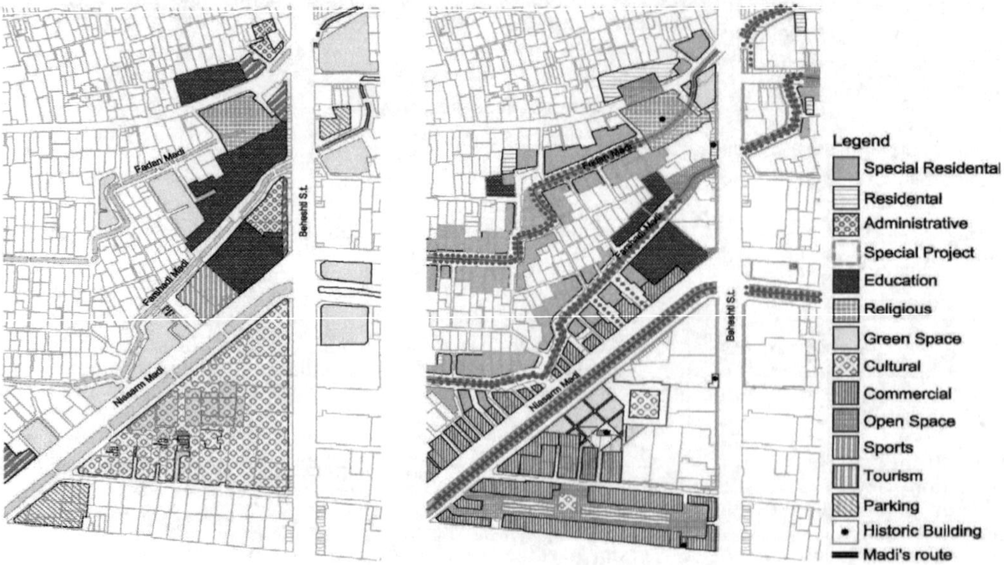

Figure 3. Land use comparison of connection place of Niasarm, Farshadi and Fadan Madis with Beheshti Street (left: Revision Plan 1993 - right: Review Plan 2011)

Figure 4. Land use comparison of Fadan Madi (left: Revision Plan 1993- right: Review Plan 2011)

5. Discussion and conclusion

It has been for five decades from the preparation of urban development plans including directing, comprehensive, and detailed plans aimed at harmonized urban physical development. But they have not been able to achieve their objectives for the lack of scheduling comprehensive system governance on the total country. On the other hand, the origin of these plans shows inconsistencies with Iran land variety and features [7]. The evaluation of current status of Madis indicates that there is considerable distance and difference between the reality (current situation) and the images of Madis derived from the plans. These differences are obvious almost in all determined components -except the main urban pathways- including the buildings height, land use, and Madis boundaries pathways.

Madis are one of the important features of Isfahan that have been able to resistant from the very first day of its formation until now. They have preserved the city life as a coherent generalization in the physical structure and Isfahan space organization. In the current period the socio-economic, union, and ethnic structures in the past provided the causes of sectors' distinction from each other socially and culturally on one hand and the causes of the internal coherence and social and cultural correlation of each sector have been destroyed. Consequently the appropriate physical structures and spaces had been stopped from coordinated changes.

It has been about five decades that Isfahan has systematic detailed plans and all of them have emphasized on Madis values and importance and introduced its bounds as special one and has composed serious criterion and regulations for them. But the Madis problems has become more than before. The high construction issue in their bounds, fast and easy access of the motorists to the neighboring houses and consequently the persistence on the purification of its bounds and over setting back proportionate to their circumstances are from these problems that are investigated in this survey.

Apart from the changes in life style and the evolution of the urban activities, the main reason in weaknesses of local functions of Madis in the current time, can be observed in the urban planning methods and techniques that are unable to produce and establish functional, social, and modern human communications appropriate with the modern requirements in the Madis span. Table 2 [8], evaluates the detailed plans from the view of their objectives and concepts, evaluation and ratification, and their implementation issues and problems.

From the structural examination of the Isfahan Madis and the analytic point of view in the urban development plans role, it is deduced that:

- There is no predetermined and unchangeable destiny for the city and its current identity is caused from the long lasting gradual transformation or rather short time fundamental evolutions that are occurred during the time and will continue in future inevitably. The truth shows the implausibility of some claims regarding the historical identity that seems to be updated resisting in front of any type of dynamic changes.
- Lowering the dignity of historical heritages and flawing the dynamism of past to present and future connection facilitates the way of ignoring the historical cores of the city.
- A stereotyped behavior, regardless of the city incipient movement and the political and social realities brings the defeat by itself. Single-dimensional modernism that is going to establish evolution in the urban structure ignoring the national and traditions heritages and society cultural evolution cannot work in a way but unconsidered and primarily destructive measures.
- Being bounded to the principle of objectivity demonstrates that the composition of objectives must be done through observing society political structure, the real amount people participation in decision making, the existence of the beneficiaries groups, economic situations, and the role of land, work and goods.
- The strong character of the city has had a considerable effect on the designing trend and is elegant despite lots of urban plans offers.
- The urban spaces in Isfahan have not formed from simple arrange of buildings and axis by each other; but each of them are a collection of functional relation and elements that are organized in a body and any flaw to any part of it can ends in its whole eradication. So although the penetration in the context has got historical necessity for entering modern elements of city life, this measure must be taken according to structural –incipient legality.
- Without real participation of people, while there is a possibility of idealistic elements penetration in the plans optimistically, no assurance exists for the plan's objectives and programs success.
- Physical preparations cannot provide required instruments for controlling urban developments. The urban expansion is a multi-dimensional phenomenon that the physical aspect is its

updated observable face; so it is impossible to defeat such a complicated issue successfully with a single-dimensional solution.

- Considering the previous plan ideas in the entire modern ideas codification is a positive point that has brought such endurance in the procedure of programming and has compensated the premature invalidity to some extent.
- The entity of the plans have been needed to the strong and centered interference of the government in many cases; but the responsibility of state institution had been in a way that in many cases people and the city are being impressed by the beneficiaries or victimized by the deliberations. This is the truth is unfortunately not supported by any urban plans and the performance discussion doesn't have any significance there. The separation of scheduling from performance and the custodian organizations for city governing inefficiencies had been always effective in the plan's relative fiascos.

Table 2. The evaluation of detailed plans from the view of their objectives, content, and performance

Content and objectives of the plans	**Evaluation and approval**	**Executive problems**
Lack of connection among the three components of population, activity, and function Lack of appropriate connection between the counselor engineers and urban management Lack of agreement with Madis conditions The assimilation of the detailed plans for the whole city The urbanism offered patterns being stereotyped and the plans' content lack of appropriateness with the sectors needs Separation of Madis from environmental district in effective components on sectors space organization and function The manner of similar interference in all urban contexts types including historical, ancient, developed, and marginal Mere attention to physical subjects and issues and ignorance of the third dimension related matters and the quality dimensions of the environment and socio-economic dimensions Lack of attention to the local conditions and ignorance of people participation Non significance of Isfahan immense politics Lack of dynamism in the comprehensive and detailed plans and being recognized as absolute The criterions being expanded and awkward Lack of connection among the studies and designing	Lack of sufficient attention to the evaluation and composition trend Long distance between the preparation and composition time of comprehensive and detailed plans Being state plan and the priority specification by the governing system The interference of responsible personal elegance Emphasis on the assessment of physical suggestions Lack of executive facilities consideration Absence of composed system in supervision Ignoring the role and participation of people in plans utilization	Lack of people and social institutes participation Absence of any control lever and criterions in the expedite city development Lack of coordination between plans and zonal constructing programs The plan's financial and credit not being predicted Technical employees shortage (human and executive force) Centered decision making system The change in the effective components quality coefficients on time passing derived calculations and predictions

References

[1] Kalantari S 2009 *Spatial & Physical Planning Of Old Fabric Madies of Isfahan as Urban Pedestrian Network.* Master thesis, (Tehran: Islamic Azad University) p 123

[2] Iran supreme council of urbanism and architecture 2009 *urban development and architecture & urban regulation* (Tehran: Iran Development) p 144

[3] Organic Consulting Eng 1973 *Detailed plan of Isfahan master plan*. Isfahan General Directorate of Roads and Urban Development.

[4] Naghshe-E-Jahan Pars Consulting Eng 1993 *Detailed plan of Isfahan: Urbanism construction criteria* (Isfahan General Directorate of Roads and Urban Development)

[5] Isfahan municipality 2011 *Planning and building regulations detailed design plan review* (Tehran: Department of Planning and Architecture)

[6] Behzadfar M 2011 *Design & plan in urban development* (Tehran: Iran University of Science & Technology Press) p 50

[7] Qamami M the concise criticism and evaluation of urban plans 1992 *ABADI* **7** 30 25

[8] Habibi K, Behzadfar M, Meshkini A, Alizadeh H, Mahaki V Evaluation of the effects of urban development plans on qualitative promotion of Iranian old city space 2011 Motaleate-e Shahr-e Irani Islami 4 26

IOP Conf. Series: Earth and Environmental Science **159** (2018) 012045 doi:10.1088/1755-1315/159/1/012045

A Review on the Controlling Factors that Affecting the Stormwater Retention Performance of Green Roof

M F Chow[1], M F Abu Bakar[2] and L M Sidek[1]

1 Center for Sustainable Technology and Environment (CSTEN), Universiti Tenaga Nasional, 43000 Kajang, Selangor, Malaysia.
2 Department of Civil Engineering, Universiti Tenaga Nasional, 43000 Kajang, Selangor, Malaysia.
E-mail: mingfaichow12345@gmail.com

Abstract. Green roofs have become a popular strategy in urban stormwater management in the recent years. However, the hydrological performance of green roofs can vary substantially due to design characteristics and environmental conditions. This study is aimed to review all the controlling factors that affecting the stormwater retention performance of green roofs. Current hydrological research on green roofs involving different plant species, substrate types and environmental conditions are reviewed and evaluated in detail. Better understandings on the controlling factors that affecting the hydrological performance of extensive green roofs will be benefiting to the successful implementation of green roof system in the future.

1. Introduction

Nowadays, urban development is expected to grow with increasing population in the urban area. Green roof has been widely implemented as a sustainable urban stormwater management around the world in the past few decades. According to previous studies, design of green roofs should have high hydraulic conductivity which the infiltration rate exceeds the precipitation rate in most rainfall events that likely to be experienced by that particular site [1]-[3]. Total amount of runoff is usually used as a variable to assess the hydrological response of a green roof. The amount of water that does not flow out from the green roof is considered as water retained. In general, the depth of water retained plus the runoff depth must be equal to the total depth of rainfall. If a rainfall exceeds the maximum water storage capacity of a green roof, runoff may not begin immediately because it also depends on the intensity of rainfall event. Green roofs may detain an additional depth of water more than its storage capacity. The maximum water detention capacity of a green roof is greatly dependents on the growing media composition and the response is usually dynamic to the rainfall intensity [3], [4]. Thus, green roof can delay and reduce the peak flow of runoff during a storm event [5]-[10]. The available storage capacity would allow the rainwater to infiltrate into the growing media. Once the volume of rainwater exceeds the maximum storage capacity of green roof, the rainwater will be drained out from the growing media. The excessive rainwater would percolate into growing media and passing down the filter and drainage layers (bottom layer) before flowing out to the drainage system. Drainage layer often consists of high amount of lateral hydraulic conductivity that ensures all rainwater beyond the storage capacity of growing media is able to flow out [1], [2]. Detained water can continuously to drain for a few hours after a rainfall event until the water level is reduced to the maximum storage capacity of green roof [11]. The whole hydrological processes in green roof system are illustrated in Figure 1. The hydrological performance of green roofs can vary substantially due to design characteristics and environmental conditions. Better understandings of the influencing factors on the hydrological performance of extensive green roofs are important in order to assure the successful implementation of green roofs. Therefore, this study is aimed to review all the controlling factors that affecting the stormwater retention performance of green roofs.

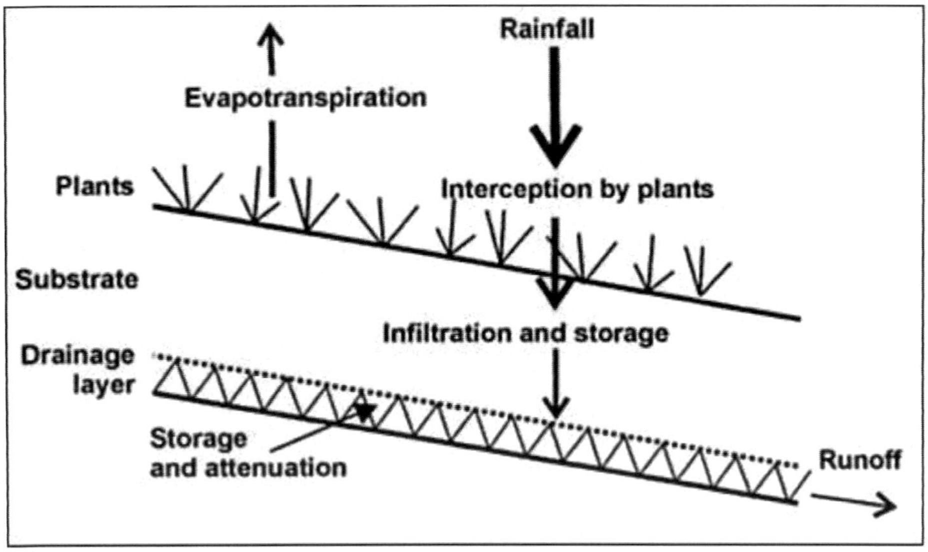

Figure 1. The hydrological processes in green roof system.

2. Factors that affecting retention performance of green roof

2.1. Vegetation

Many studies had agreed that plant species have significant effect on the stormwater retention capacity in green roofs [12]. Green roof vegetation generally retains the rain water through evapotranspiration, interception by leaves, water uptake and retention by root system. Thus, plant selection on a green roof should consider its vegetation coverage, root uptake, root pathways in the substrate, water storage in plant tissue, transpiration, interception, evaporation and shading of media. Generally, plant species with bigger diameter, larger root biomass and taller height would increase the stormwater retention capacity of a green roof [12]. Morgan et al. [13] suggested that the stormwater retention capacity can be increased if the vegetation coverage reaches up to 20-25 %. According to White and Snodgrass [14], the plant species for extensive green roofs are required to show the characteristics of rapid establishment, high ground cover density and easy propagation. Hutchinson et al. [11] also stated that maturity of vegetation was considered as an important factor that leading to the increasing of runoff retention. Nagase and Dunnett [12] studied the differences of plant species types on the stormwater retention performance and found grasses were the most significant vegetation for decreasing runoff volume, followed by forbs and sedum. Succulent plant especially sedum has been identified as the most effective plant in extensive green roofs due to its potential for keeping excessive water, controlling transpiration, drought tolerance and shallow substrate adaptability [15]-[17]. Jarrett et al. [16] stated that sedum has the potential to capture one millimeter of rainfall depth in green roof. Ouldboukhitine et al. [18] stated that evapotranspiration for grass was 60% higher than that of sedum and almost twice fold than the bare substrate. Voyde et al. [19] reported that both Sedum mexicanum and Disphyma australe accounted for approximately 50% of stormwater reduction through evapotranspiration during water abundant conditions. Drought tolerance of sedum species has been well investigated. Lassalle [20] found that S. album L. (white stonecrop) could survive more than 100 days without water. Others also have confirmed that S. album is a drought-hardy species [21], along with S. acre L. (biting stonecrop), S. kamtschati cum ellacombianum Fischer & Meyer (or-ange stonecrop) and S. reflexum L. All of these plant species can survive for 88 day without water [22].

Liesecke [23] in Germany stated that more than five tuberous plant species would be the good selection for extensive green roofs. In Michigan, nine sedum and eighteen native taxa were evaluated for its growing, survival during both establishment and overwintering, and visual appearance in a non-irrigated extensive green roof over 3 years period [24], [25]. The results confirmed that all these nine sedum species were suitable for green roof in Midwestern climatic condition. In United Kingdom, Graceson et al. [26] found that sedum species averagely retained 40% while meadow plant retained 48%, respectively. In Malaysia, Johari et al. [27] had conducted a study on plant selection for green roof. The results identified that Portulaceae, Neoregelia and Carssulaceae are the suitable plant species that can survive in tropical climate. In other study, Ismail et al. [28] indicated that green roof system planted with cardinal creeper (ipomoeahorsfalliae /jasminumsambac), sweet potato (ipomoea batatas), arrowhead plants (syngoniumpodophyllum), and beach morning glory (ipomoea pes-caprae) would led to decreasing of indoor and outdoor temperatures compared to black bare roof and white roof. Raymond and Ahmad [29] found that native fern species outperformed the other non-native sedum species in term of plant coverage and visual appearance during withheld watering test. On the other hand, Chow et al. [30] stated that monoculture of sedum was the most effective single species for retaining water runoff with 55.5%, followed by fern (50.9%), cow grass (34.8%) and manila grass (28.3%). Ayub et al. [31] showed that A. compressus performed the highest runoff reduction percentage in all experiments. Data recorded showed this species managed to attenuate peak flow in the ranges of 30.5% to 67%.

2.2 *Substrate depth and composition*

Substrate composition has a major significant effect on green roof water retention capacity than plan types and cover [22], [32], [33]. The water storage capacity of substrate is always depends on depth, composition and maximum water holding capacity [1], [22], [34], [35]. Many researchers have confirmed that increased substrate depth will improve the stormwater retention performance of green roof [36,33,24,37,38,39]. Mentens et al. [36] concluded that extensive green roofs have capability to retain 45% stormwater runoff (substrate depth: 10 cm) while for intensive green roofs can retain 75% of stormwater runoff (substrate depth: 15 cm), compared to gravel and traditional roof which only retain 25% and 15%, respectively. VanWoert et al. [22] also reported that the substrate depth will affect the water retention capacity of green roof. The effect of growing media depth on the storm water retention performance of green roof was investigated for different depths (2.5, 4.0, and 6.0 cm) and slopes (2% and 6.5%) in their study. For all combined rainfall events, green roof platforms at 2% slope with a 4-cm growing media depth had the greatest mean rainwater retention percentage of 87%.

Mineral contained in substrate will support the growth of vegetation. The selection of substrate composition will be subjected to the availability of materials locally and substrate composition can be formulated for the intended plant selection, anticipated level of maintenance and climatic zone. Vijayaraghavean and Raja [40] have used mixture of expanded perlite, coco-peat (29% moisture content), crushed bricks, exfoliated vermiculite, and sand in the green roof substrate development. They identified that the best soil characteristics of substrate that used in green roof were particle size ranges between 0.25 mm to 4 mm, water holding capacity of 39.4% , bulk density (at maximum water holding capacity) of 912 kg/m3, dry bulk density of 431 kg/m3 and air filled porosity of 19.5%. Recycled wastes also can be used as a substrate in green roof development. Chloe et al. [41] in London, UK have tested four recycled materials which including crushed red brick (the U.K. industry standard substrate base and then used as a control) and three alternative pellets made from: paper ash (from recycled newspaper), clay and sewage sludge (fly ash, sewage sludge and waste clay from excavation), and carbonated limestone (from quarry fines). The outcomes of this study showed the significant interactions between four aggregates and the amount of added organic material, whereby organic material addition did not have the same effect on the plant growth in each aggregate. Eksi et al. [42] in Turkey stated that decomposed material can be used as a green roof substrate. Their result showed that 60% and 80% compost substrate produced the greatest plant growth and fruit yield, respectively. These amounts are much higher than that recommended for standard extensive green roofs planted with succulents or other herbaceous perennials. High compost material will cause negative consequences of substrate shrinkage and nutrient runoff.

Organic content plays an important role in the selection of substrate type in green roof. Amount of organic content will determine the growth of plant. Nagase and Dunnett [43] had carried out a study to examine the relationship between percentage of organic matter in substrate and plant growth in extensive green roof, particularly at the establishment stages. From their results, it can be concluded that addition of 10% organic matter was optimal for all plant species in the study as the plant showed stable growth regardless of the watering regime. Meanwhile, increased organic matter did not result in increased growth in the wet regime of green roof. Nadine et al. [44] had carried out a study to assess whether ECM fungi can enhance P.pinaster growth in the burned soil of green roof. The results showed that inoculation has significantly enhanced the growth of P.pinaster, which R.roseolus exhibited to be the most effective fungi isolate in burned soil with an 8-fold increase in plant fresh weight.

2.3. Rainfall characteristics and seasonal climate

Many green roofs studies showed that the rainfall characteristics have great effects on stormwater runoff retention performance [22], [45], [46]. Carter and Rasmussen [45] stated that an inverse relationship between the retention percentage of stormwater runoff and rainfall depth. Researcher observed 88% rainwater was retained for small storms (<25.4mm), 54% for medium storms (25.4-76.2 mm) and 48% for large storm (>76.2mm). Similarly, Simmons et al. [46] found that rainwater in small storm events (<10mm) were all retained in the green roofs. However, the retention percentage decreased to 8-43% for rain events with depth greater than 28 mm. VanWoert et al. [22] also reported that more than 95 % of total light rainfall (<0.2cm) can be retained by extensive green roof, followed by more than 80% of moderate rainfall event can be retained (0.2-0.6 cm), and only 52% of total heavy rainfall can be retained (>0.6 cm). It was observed that the green roof retention performance also depended on rainfall intensity not just on storm size. Villarreal and Bengtsson [3] mentioned that lower rainfall intensity will result in higher water retention in green roofs. Their results showed that the rainwater retention was 62% for rainfall intensity of 0.4 mm/min but only 21% for rainfall intensity of 1.3 mm/min. Getter et al. [47] also stated that light rainfall event recorded the highest retention percentage of 94% while heavy rainfall recorded the lowest retention percentage of 63%. Lee et al. [48] and Bengtsson et al. [4] both stated similar contrary relationship between rainfall intensity and water retention capacity of green roof. Seasonal conditions may also influence the water retention performance in green roofs as plant evapotranspiration is influenced by climatic aspects such as solar radiation and temperature [36]. The summer season records greater evapotranspiration rate and thus increases the retention capacity of green roof [36,49]. Bengtsson et al. [4] reported that the highest runoff reduction in extensive vegetated roofs was observed for June (88%) while the lowest for February (19%). Meanwhile, Villarreal and Bengtsson [3] found the retention capacities of studied green roof were varied between 34% during September – February and 67% during March-August.

2.4. Slope

Various studies had reported different results on the effect of slope on green roofs runoff retention capacity. Some studies found no correlation between roof slope and runoff retention [6,36,50], while others had observed runoff retention may depend on the slope of green roofs [3,22,47]. Villarreal and Bengtsson [3] found that peak flow and runoff volume increased as slope increased. It was observed that 2% slope has double retention volume recorded compared with 14% slope. Similar results were stated by Getter et al. [47] that runoff volume reduction decreased as roof slope increased. The lowest stormwater retention capacity of 75% was observed for 25% slope while the highest retention capacity of 85% was observed for 2% slope. VanWoert et al. [22] also studied the influences of slope (2% and 6.5%) and substrate depth (2.5, 4.0 and 6.0 cm) on runoff retention capacity of green roofs. In the study, they found the greatest overall retention was 87% for green roof at 2% gradient with 40 cm substrate. Bengtsson [6] examined the influences of slope and length of green roofs on the runoff peak flow retention but found that there are no significant changes on the runoff distributions. Schade [51] in Germany also found no significant difference in runoff retention amounts across different sloped roofs. The contradicting results may be caused by the rainfall patterns in different environments. Many researchers have observed that dry substrate conditions prior to storm event will result in higher stormwater retention compared with initially wet condition in green roof [3,52].

2.5. Age of the green roof

Certain studies claimed that the age of green roofs will affect the rainwater retention efficiency [47,53]. Berndtsson [51] stated that the substrates of green roof experienced several physical and chemical changes over time such as loss of soil particle, washout of dissolvable substances, changes of organic content and changes of soil porosity due to roots development. These changes would influence the hydrology dynamics in the green roof. The water holding capacity increased from 16% to 68% while pore volume and organic content increased twice after 5 years usage of green roofs (from 41% to 82% and from 2% to 4%, respectively) [47]. However, Mentens et al. [36] did not find any significant relationship between the age of green roof and annual runoff volume reduction.

3. Conclusion

This paper has identified all the controlling factors that affecting the hydrological performance of green roofs. The effects of vegetation, substrate composition, physical properties of substrates, climate condition, rainfall pattern, structure and design of green roof have been reviewed thoroughly in this paper. The information provided in this paper are beneficial to the improvement of green roof system in the future.

4. References

[1] Miller C 2003 Proceedings of the 1st Greening Rooftops for Sustainable Communities Conference **1** 177-182.
[2] Beattie D and Berghage R 2004 Proceedings of the 2nd Greening Rooftops for Sustainable Communities Conference 2 411-16.
[3] Villarreal EL and Bengtsson L 2005 Ecological Engineering 25(1) 1-7.
[4] Bengtsson L, Grahn L, Olsson J 2005 Nord. Hydrol. 36(3) 259–268.
[5] Liu K 2003 Proc. of the 18th International Convention of the Roof Consultant Institute. 93-103.
[6] Bengtsson L 2005 Nordic Hydrology 36 269-280.
[7] VanWort ND, Rowe DB, Anderson JA, Rugh CL, Xiao L 2005. HortScience 40 659-664.
[8] Chow, M. F., Abu Bakar, M.F. 2016. International Journal of Architecture and Environmental Engineering. 10(1): 16-20.
[9] Kok, K.H.; Sidek, L.M.; Chow, M.F.; Zainal Abidin, M.R.; Basri, H.; Hayder, G. 2016. Int. J. River Basin Manag. 14, 1–7.
[10] Chow, M.F., Abu Bakar, M.F., Sidek, L.M., Basri, H 2017 *Journal of Engineering and Applied Sciences* 12(21) 5379-5383.
[11] Hutchinson D, Abrams P, Retzlaff R and Liptan T 2003 Proceedings of 1st Greening Rooftops for Sustainable Communities. 372-389.
[12] Nagase A and Dunnett N 2012 *Landscape and Urban Planning* **104(3–4)** 356-363.
[13] Morgan S, Celik S and Retzlaff W 2012 *J. Environ. Eng.* **139** 471–8.
[14] White JW, Snodgrass E 2003 Proc. 1st Intl. Green Roof Conf.: Greening rooftops for sustainable communities 1166–176.
[15] Dunnett NP and Kingsbury N 2004 Planting Green Roofs and Living Walls. Portland (OR): Timber Press.
[16] Jarrett AR, Hunt WF and Berghage RD 2006 Proceedings of the American Society of Agricultural and Biological Engineers Annual International Meeting, Portland OR. Paper # 062310.
[17] Durhman A, Rowe DB and Rugh CL 2007 *HortScience* **42 (3)** 588-95.
[18] Ouldboukhitine S, Belarbi R, Jaffal I, Trabelsi A 2012 *Building and Environment* **46(12)** 2624–2631 ·
[19] Voyde E, Fassman E, Simcock R 2010 *Journal of Hydrology* **394(3-4)** 384-395.
[20] Lassalle, F. 1998. Stadt und Gru¨n 47:437–443.
[21] Kirschstein C 1997 *Stadt und Grun* **46** 252–256.

[22] VanWoert ND, Rowe DB, Andresen JA, Rugh CL, Fernandez RT, Xia L 2005 *Journal Environment Quality* **34** 1036-1044.
[23] Liesecke HJ 2001 *Stadt und Grun* **50(2)**133–139.
[24] Rowe BM, Monterusso and Rugh C 2005 Proc. of 3rd North American Green Roof Conference: Greening rooftops for sustainable communities, Washington, DC. pp. 469–481.
[25] Monterusso MA, Rowe DB and Rugh CL 2005 *HortScience* **40** 391-396.
[26] Graceson A, Martin H, Nige H, Monaghan J 2014 *Biosystems Engineering* **124** 1–7.
[27] Johari J, Rasidi MH, Said I 2011 5th South East Asian Technical University Consortium Symposium. Hanoi.
[28] Ismail A, Samad MHA, Rahman AMA 2011 World Academy of Science, Engineering and Technology 76.
[29] Raymond K and Ahmad H 2014 International Journal of High-Rise Buildings **3(4)** 1-6.
[30] Chow MF, Abu Bakar MF, Roslan MAA, Fadzailah FA, Idrus MFZ, Ismail NF, Sidek LM, Basri H 2015 *ARPN Journal of Engineering and Applied Sciences* **10(15)** 1819-6608.
[31] Ayub KR, Ghani AA, Zakaria NA 2015 Proceedings of the 1st Young Scientist International Conference of Water Resources Development and Environmental Protection, 5–7.
[32] Dunnett N, Nagase A, Booth R, Grime P 2008 *Urban ecosystems* **11** 385–398.
[33] Monterusso MA, Rowe DB, Rugh CL and Russell DK 2004 *Acta Hort* **639** 369-376.
[34] DeNardo JC, Jarrett AR, Manbeck HB, Beattie J and Berghage RD 2004 *Transactions of ASAE* **48(4)** 1491- 1496.
[35] Raes D, Geerts S, Kipkorir E, Wellens J and Sahli A 2006 *Agricultural water management* **81(3)** 335-357.
[36] Mentens J, Raes D and Hermy M 2006 *Landscape and Urban Planning* **77** 217–226.
[37] Nardini A, Andri S, Crasso M 2012 *Urban Ecosyst.* **15** 697–708.
[38] Schmidt M 2006 Proceedings of the RIO6 World Climate and Energy Event.
[39] Carson TB, Marasco DE, Culligan PJ, McGillis WR 2013 *Environ. Res. Lett.*, **8** p. 024036.
[40] Vijayaraghavan K and Raja FD 2014 *Water Resouces* **2(63)** 94-101.
[41] Chloe JM, Charles HF and Alan CG 2009 *Ecology* **35(10)**1507–1513.
[42] Eksi M, Rowe DB, Fernández-Cañero R and Cregg BM 2015 *Urban Forestry & Urban Greening* **14(2)** 315-322.
[43] Nagase A and Dunnett N 2011 *Landscape and Urban Planning* **103(2)** 230-236.
[44] Nadine RS, Albina RF, Miguel AR, Paula MLC and Rui SO 2011 *Soil Biology and Biochemistry* **43(10)** 2115-2120.
[45] Carter TL and Rasmussen TC 2006 Journal of the American Water Resources Association **42(5)** 1261-1274.
[46] Simmons MT, Gardiner B, Windhager S, Tinsley J 2008 *Urban Ecosystems* **11** 339–348.
[47] Getter KL, Rowe DB and Andresen JA 2007 *Ecological Engineering* **31** 225-231.
[48] Lee JY, Moon HJ, Kim TI, Kim HW and Han MY 2013 *Environmental Pollution* **181** 257–261.
[49] Villarreal EL 2007 *Nordic Hydrol* **38(1)** 99–105.
[50] Liesecke HJ 1998 *Stadt und Gru¨n* **47** 46–53.
[51] Schade C 2000 *City and Green* **49** 95-100.
[52] Connelly M and Liu K 2005 Proc. of 3rd North American Green Roof Conference: Greening rooftops for sustainable communities, Washington, DC: 4–6.
[53] Berndtsson JC 2010 *Ecol Eng.* **36** 351-60.

Acknowledgments
The authors would like to thank Ministry of Higher Education, Malaysia (MOHE) for providing the research grant (Vot no: 20140130FRGS). We would like to acknowledge the BOLD Research Grant provided by Universiti Tenaga Nasional (Project No. 10289176/B/9/2017/50) and TNB Seeding Grant provided by Tenaga Nasional Berhad (Project No: U-TG-RD-17-06).

IOP Conf. Series: Earth and Environmental Science **159** (2018) 012046 doi:10.1088/1755-1315/159/1/012046

Waste Recycling System for a Tourism City in Vietnam: Situation and Sustainable Strategy Approach – Case Study in Hoi An City, Vietnam

S T Pham Phu[1,2], T Fujiwara[1], P V Dinh[1] and Hoa K T[2]

1 Okayama University, 3-1-1 Tsushima, Kita Ward, Okayama 700-8530, Japan
2 University of Technology and Education, The University of Danang, 48 Cao Thang, Danang, Vietnam
E-mail: ppstoan@gmail.com

Abstract. Recycling waste brings many benefits for environment and society so that the establishment of sustainable recycling system is inevitable. This study aims to describe the situation of the recycling system in Hoi An City (HAC), through that establish sustainable recycling strategy for the future. The results show that HAC has a high potential for improving recycling. Whereby, recyclable waste accounts for three-fourths of total waste, and the rate of sorting waste is high. Besides, the proportion of recycling practice is slight high with 62% for residents, 39% for hotel sectors, and 56% for restaurants. However, the recycling system in HAC is simple with the rudimentary facility, in which informal sectors play important roles. Thus, the recycling effectiveness was low (24%), mainly papers, plastics, and metals. The concept of sustainable recycling strategy is established by the combination of the government, community, business sectors toward the balance between economic, environment and society.

1. Introduction

Recently, tourism has rapidly developed and become an essential industry in the world [1]. Also, the arrivals to Vietnam increased by 130% in the last ten years and has brought many benefits for society. However, the dark side of tourism development is the negative influences on the environment in which solid waste generation (SWG) is one of the most significant impacts [2, 3]. For developing countries, solid waste management (SWM) is still a big problem, and the green SWM for sustainable tourism development is a significant challenge.

Waste recycling and resource recovery bring many benefits to SWM, economy, environment, and society. In developing countries, both formal and informal sectors play significant roles in recycling system [4] but have several risks of conflict [5]. Thus, combining and promoting the roles of both sectors towards the establishment of sustainable recycling system (SRS) means significant challenges. Salau et al., reported that in developing countries (Lagos State, Nigeria) the informal recycling activities contribute to recycling rate more than the formal sectors [6]. Also, in Thailand (a case study of Phitsanulok City), informal sectors, which played a primary role in SRS were suggested to be supported technique, and finance for improvement [4]. Likewise, Wilson et al., revealed that informal recycling system brings many benefits to SWM in developing countries [5]. Furthermore, the establishment of new formal recovery system to substituted for the informal system is very expensive. Thus, strategy planning of sustainable SWM (SSWM) should be considered based on existing informal collection and recycling systems. Establishment of SRS inevitably approaches sustainability of SWM. Planning SRS strategy for a developing country should focus on three dimensions of sustainability: waste collection and separation, municipal SWM plan, and local recycled-material market. Also, the SRS goes towards the balance between environmental impacts, economic benefits and social acceptance [7].

Recycling system in Vietnam was mentioned as an example in the publication of Troschinetz and Mihelcic with the recovery rate by 13-20%, are mainly paper, plastic, glass, and metal [7]. The situation and strategy of SRS have not studied more. Thus, this study aims to provide the current status of the recycling system in a tourism city, so that contributes to clarify a comprehensive view of the recycling system in Vietnam. Also, the strategy of SRS for this tourism city will be proposed as a specific case which should be considered to apply to other cities in Vietnam.

2. Study Sites

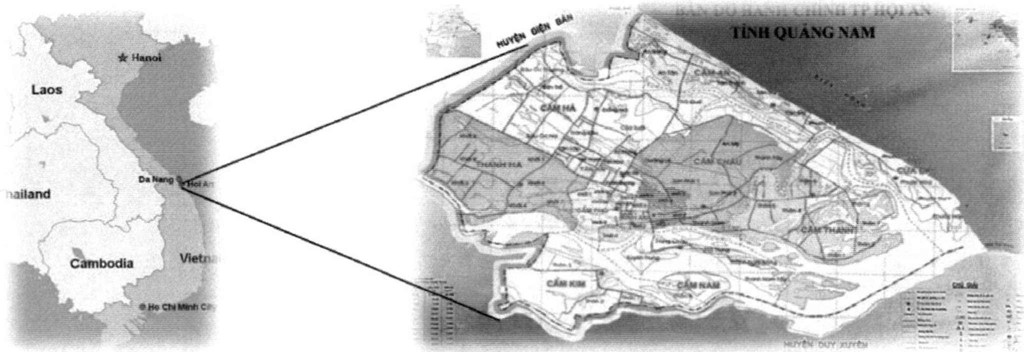

Figure 1. Hoi An City, Vietnam.

Hoi An City (HAC) is a famous tourism city located at the center of Vietnam (Figure 1) with the population of 94,331 people in 2016. Tourism is the primary industry in HAC. In the last decade, the growth rate of arrivals in HAC was 243% and reached 2,517,217 tourists in 2016, led the increase of a number of hotels and restaurants by 3.3 times. Notably, 446 accommodations with 7,795 rooms and 558 restaurants are mainly located in the center areas and tourism destinations [8]. On the other hand, the municipal solid waste (MSW) of HAC climbed from 19,282 tons in 2009 to 30,131 tons in 2016, while almost no difference in population from 2009 (93,808 people) to 2016 (94,331 people) [9]. Consequently, the tourism industry is a significant source of waste in HAC.

The situation of SWM of HAC has analyzed in detail. Giang et al. identified solid waste composition and generation of MSW, and developed a model of SWG from households [10, 11]. Whereby, domestic solid waste was accounted for about 21% of total MSW. Hotel waste characterization and management practices were also analyzed clearly [9]. These studies presented that recyclable waste accounted 78.7% for domestic waste and 84.3% for the hospitality industry. However, the rate of recycling practice in HAC is low and justified by the weak driving force of residents, and many barriers against recycling practice for commercial sectors [9, 12]. Therefore, in this study, the situation of recycling system of HAC is clarified to design the SRS strategy for HAC.

A survey was conducted to collect data of recycling situation, and the face-to-face interviews were implemented to gather information of not only waste management but also recycling practice from recycling sectors. About 300 households, 120 hoteliers, 50 managers of restaurants and 17 owners, of recycling shops, were collected in this survey.

3. Results and discussions

3.1. The recycling paths in HAC

Figure 2 shows the path of recycling activity in HAC from generation to manufactures. In the structure of recycling system in HAC, waste segregation is the first step which plays an essential role in recycling strategy. Waste separation at source was implemented in HAC since 2012. Whereby, waste is sorted into the organic and inorganic waste, and collected separately by the public work company. Although no regulation of recycling waste classification, recyclable waste as mainly papers, metals

IOP Conf. Series: Earth and Environmental Science **159** (2018) 012046 doi:10.1088/1755-1315/159/1/012046

and plastics are also sorted by householders and commercial sectors (Figure 3-a). These recyclable materials are sold to itinerant buyers to get income. A part of recycling waste is picked out of trucks by collection crews. After transferring to disposal areas, mainly PET and metals are picked up again by scavengers. The collected recycling materials are sold to junk shops and moved to manufactures. The rest of recyclable waste goes to the land fill sites.

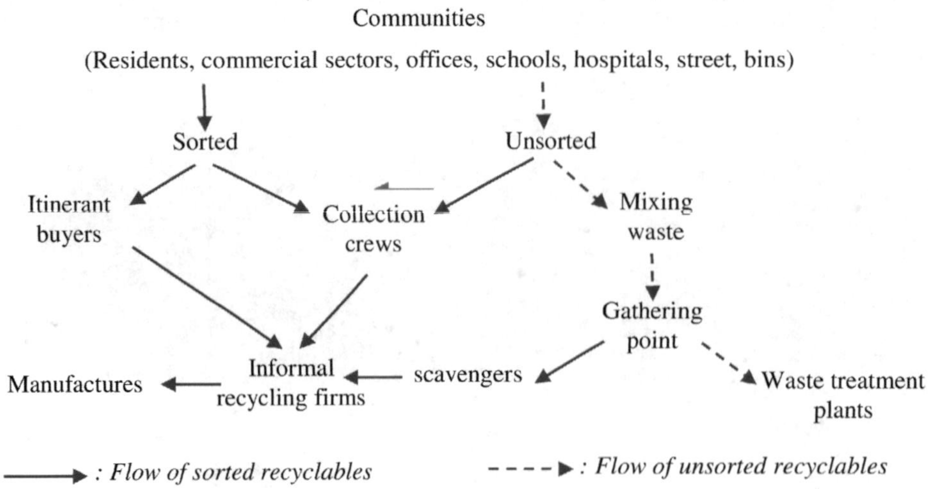

Figure 2. Waste recycling paths in Hoi An City.

The recycling system in HAC consists informal sectors and formal firms. The informal sectors, who are itinerant buyers, street picking, collection crews, scavengers and junk shops at treatment sites collect directly recycling material around the city. The formal sector is the public work company which operating waste collection and treatment in HAC. Until now, all of the recyclable materials were bought to recycling manufactures in Dien Ban industry zone and Danang cities.

3.1.1. Informal recycling waste collectors in HAC

An itinerant waste buyer is a recycling collector who buys sorted dry recyclable waste from householders and commercial sectors, and transfers to the junk shops for selling. In HAC, there are about 100 itinerant buyers. They buy recycling materials door to door by bicycles, tricyclics or bikes (Figure 3). At the public points, the recyclable materials from the mixed waste in the garbage bins or on the street are collected by street waste pickers. Both itinerant buyers and street waste pickers are poor and marginalized social groups. Trading recyclable materials bring them the main income.

(a) Itinerant buyers *(b) By bicycle* *(c) By the burden* *(d) By bike* *(e) Tricyclics*

Figure 3. Itinerant waste buyers in HAC.

IOP Conf. Series: Earth and Environmental Science **159** (2018) 012046 doi:10.1088/1755-1315/159/1/012046

Figure 4. Recycling collection crews **Figure 5.** Scavengers at treatment plants

For non-recycling practice at the source, the recyclable materials in the mixed waste are recovered by municipal waste collection crews when collecting by trucks or sweeping on the street (Figure 4). Sometimes, recycling waste is also given to collection crews by householders and commercial sectors for free. The rest of recyclable materials in the mixed waste is transferred to the treatment area. Here, around 20 scavengers (figure 5) pick recyclable waste again before treatment by composting, incinerator or dumping at the open landfill site.

3.1.2. Informal junk shops

Recyclable materials are sold to the junk shops which are informal commercial sectors. HAC has 17 informal junk shops, in which the recycling materials in 30% of shops are disassembled, separated and washed before compressing and packing for selling to manufactures (figure 6). Whereas, the rest of junk shops try to minimize the volume of recyclables by hand and pack in the form of sacks (figure 6-a).

(a) *(b)*

Figure 6. Junk shops

3.1.3.. Formal firms – waste treatment plants

In HAC, The Public Work Join Stock Company, which is the formal firm response for collecting and treating municipal waste. Composting and incineration are two main treatment methods. However, compost product has low quality with many impurities of plastic, metals, and glass, so farmers do not use that. Whereas, the new open incinerator is working with a capacity of 30 tons per day (30% of design capacity) so that it cannot handle the daily waste of the city. The rest of waste is dumped at the open landfill sites. HAC has not any recycling factory for recovering recyclable materials.

3.2. Recycling material flows

3.2.1. Proportion of recyclable waste of municipal solid waste

IOP Conf. Series: Earth and Environmental Science **159** (2018) 012046 doi:10.1088/1755-1315/159/1/012046

Table 1. The proportions of recyclable waste of MSW in HAC [9, 10].

		Household	Hotels	Restaurants	Street
Quantity		94,331 people	6,096 guests	558 restaurants	-
SWG rate/day		0.223 kg/cap	2.28 kg/guest	26.18 kg/res	6.99 kg/100m
Recyclables	Biodegradable waste (%)	57	58.5	45.9	40.8
	Plastics (%)	14	13.7	8.4	13.8
	Papers (%)	6	8.9	11.5	7.7
	Metals (%)	0.7	1.2	2.2	0.4
	Glass (%)	1	2	9.4	4.5

In HAC, plastics, paper, metals, nylon bags, biodegradable waste and glass are the raw materials which are considered to recover. However, the minority of the plastics, papers, and metals are sorted at the source for recycling by households and commercial sectors. The rest are usually mixed into the garbage. Table 1 presents the SWG rate and the proportion of recyclable materials from household waste, hospitality waste, and street waste in HAC. In general, recycled waste accounted for more than three-fourths of total municipal waste, in which the inorganic recycling materials was about 25% of total waste. Whereas, the rate of dry recycled waste of the hotels and restaurants are higher than such of households with 33.4% and 31.5%, respectively. Furthermore, solid waste from commercial sectors accounted for 76% of the total waste. Therefore, the tourism industry is a significant source of recyclable materials in HAC.

3.2.2. The flows of recycling material in HAC

Figure 7 shows that the paths of sorted dry recycling materials from sources to manufactures. It is quite simple by commercial transactions through intermediaries as recycling collectors and shops. Also, the price of recyclable materials increases follows the trading flow. Whereby, the value of plastics, papers, and metals increase to 7 times for plastics and two times for papers and metals from waste sources to manufactures. Whereas, the government must pay for separation glass and treatment cost ($ US 267/ton). Hence, the financial benefit of glass in value flow from generation sources to manufactures is negative. Besides, the negative economic value of unsorted waste is also one of the burdens of the government by the fee of transportation, segregation, and treatment.

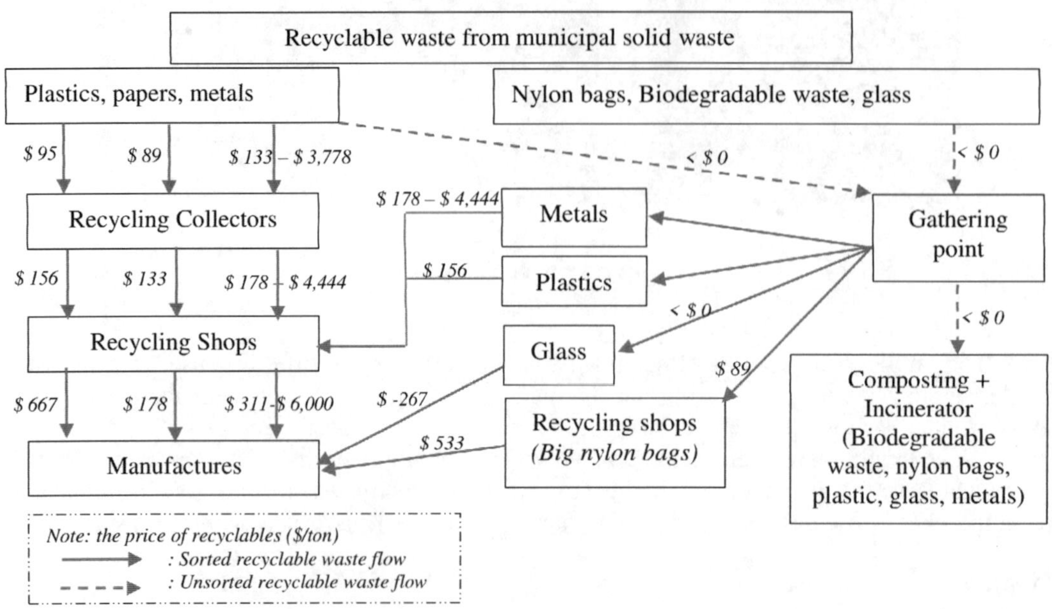

Figure 7. The recycling waste flow in HAC.

3.3. Separation and recycling practice at source in HAC

This study conducted surveys of separation at source and recycling practices on two main waste sources of HAC: households and hospitality sectors. Waste separation at source has been implemented since 2012 and strictly checked by the government. Thus, waste segregation rate was high. However, the majority of low scale hotels (LSH), homestay (HOM), and restaurants (RES) have not sorted waste. This is the weakness of recycling system of HAC which should be improved. Recycling practice depends on many factors, in which economic motivation and condition of storing space are the main influencing factors. Figure 8 shows that waste segregation at source was homologous, recycling practice in various areas in HAC was significantly different. Especially, the rate of recycling practice in commercial sectors was lower than such of residents. Waste from commercial sectors accounted for two-thirds of total municipal waste. Therefore, recycling practice of commercial sectors plays a significant role in recycling system of HAC.

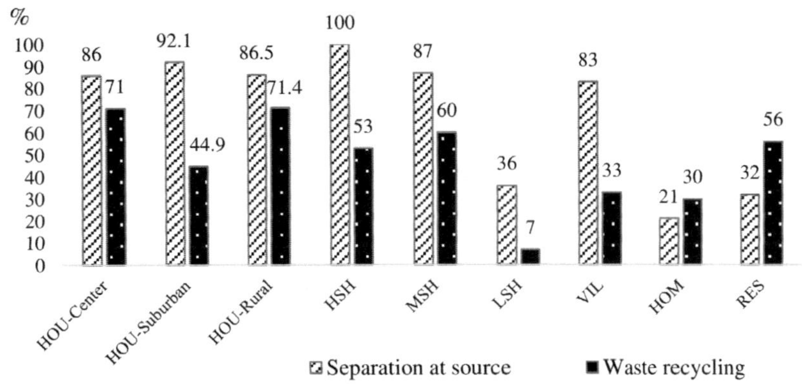

Figure 8. Waste separation rate at source and recycling practices in HAC.

3.4. Recycling practice rate and the recovery effectiveness in HAC

The recovery rate of recycled waste in shown in Figure 9. In general, about 17.5 tons of dry recyclable waste (accounted for 25%) were generated per day, in which 24% of recyclable waste was recovered by junk shops (17%) and scavengers (1%). Glass was collected one ton/day which accounted for 6% of total recyclable waste. 76% of recycling waste was mixed into the garbage and have not recovered.

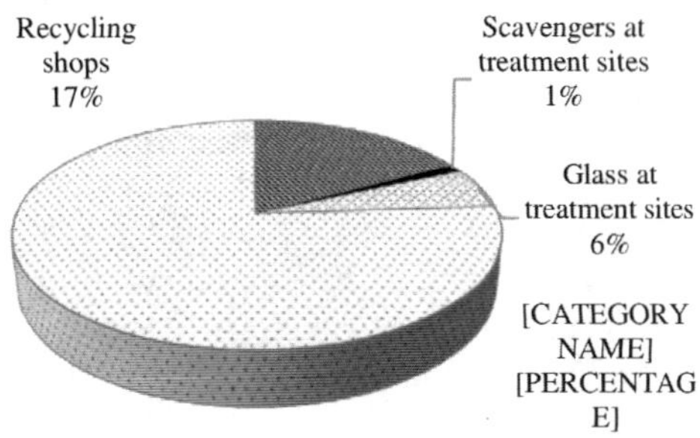

Figure 9. The recovery rate of recycled waste in HAC.

3.5. Sustainable recycling strategy for HAC

The informal recycling system that is the common feature of recycling activity in developing countries. There are many factors influencing a recycling system such as environmental education, waste management collection and separation, government finances and policy, and waste management planning as well [7]. Furthermore, Salau *et al.*, reported that in developing countries the informal sectors play a more important role in recycling system than the formal sectors [6]. Also, the establishment of new formal recovery systems to replace the informal one is very expensive [5]. Thus, sustainable recycling strategy for HAC should be established a base on the gradual improvement the role of informal sectors combines with enhancing waste management practice, government regulation, and supports of stakeholders (figure 10). The core value of sustainable development is the balance between environment (ENV), economy (ECO) and society (SOC) [13]. Figure 10 presents the concept of a sustainable recycling system for HAC. Whereby, the government (GOV), business sectors (BUS) and communities (COM) are the three main subjects that directly perform this strategy. The BUS is as hotels, restaurants, tourism commercial sectors which produce a significant amount of waste of the city. Whereas, COM is the important factor influencing the effectiveness of recycling system of the city [7]. Furthermore, GOV plays a necessary role in making a decision, promulgating regulations, calling for support from stakeholders (STA), managing and investing recovery facility for SRS. Three of them should cooperate, monitor and manage with each other to gradually implement sustainable recycling strategy.

Moreover, the implement process approach SRS is described in figure 10. Whereby, three primary factors as solid waste practices (SWPs) at source, socialization of recycling system and supports of GOV and STA are the foundations to establish SRS. Troschinetz and Mihelcic proved that waste collection and segregation is one of three most significant barriers to recycling [7]. Furthermore, Loan et al., reported that residents in HAC did not sort waste unless they understand its positive effect [12]. Hence, to enhance SWPs, waste separation at source should be improved by a combination of regulation and training. Also, composting should be considered to deploy at the big hotels and the rural areas after segregation. The better practices of SWM is, the less SWG is, the smaller cost of transportation and disposal, the higher economic benefit of sectors and society as well as the higher effectiveness of SRS.

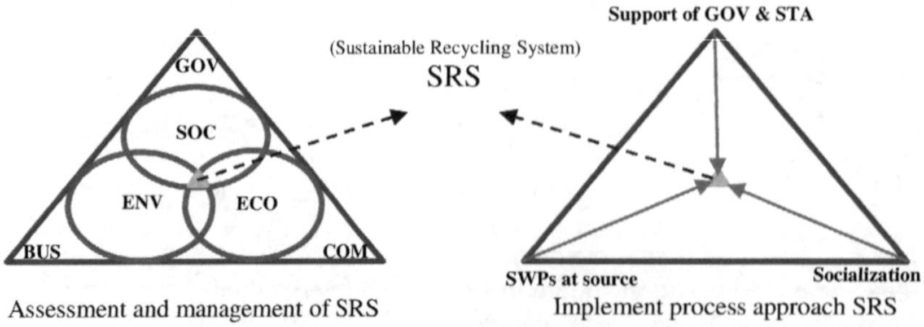

Figure 10. Sustainable recycling strategy for HAC.

Additionally, socialization in recycling system should be considered to expand and develop. From 2010 to 2012, a socialized model of SWM in HAC conducted by Women's Union of HAC and funded by Global Environment Facility Investing in our planet. This project contributed to enhancing the effectiveness of waste recycling, improving awareness of local people and developing the number of itinerant buyers [14]. However, the stability of the project is not maintained due to lack of organization and management of the government. Moreover, socialization in recycling system should be synchronous. Whereby, collection system, recycling shops, and manufacturers should be encouraged to invest in private sectors. The economic benefit from recycling may be higher if recycling materials are purer. In term of recycling management, the role of government is irreplaceable. The recycling training guideline and regulation should be promulgated, supports and funds should be provided, and the coherence between SRS and STA should be enhanced by the GOV. All in all, the synchronized implementation of three factors above will gradually establish and develop SRS for HAC.

4. Conclusion

Sustainable recycling system is the necessity toward sustainability of the municipal SWM in Vietnam. This study contributes to an issue on current concerns of waste management in Vietnam, especially in tourism cities. The situation of the recycling system in HAC is provided as below.

(1) Solid waste characterizations were the favorable condition for developing recycling activities in HAC. Whereby, the recyclable waste was 78.7% for household waste, 84.3% for hotel waste, 77.4% for restaurant waste and 67.2% for street waste. In which, organic waste accounted for two-thirds.

(2) Recycling practice in HAC achieved some initial successes. Notably, the sorting rate was about 88% for residents, 76% for hoteliers, and 32% for restaurants. Also, the recycling rate was 62% for residents, 39% for hotel sectors, and 56% for restaurants.
(3) The recycling system in HAC was simple with rudimentary facilities, which mainly consists separation and collection of recyclable materials by informal sectors, who played a significant role in the recycling system.
(4) Recovery effectiveness of recyclable waste in HAC was 24%, mainly metals, plastics, and papers. While recycling of plastics, metals, and papers bring positive financial benefits for informal sectors, recycling glass has negative economic value.
(5) The sustainable recycling strategy for HAC was suggested by combining the role of government and the collaboration of business sectors and communities based on the balance of social acceptance, economic benefit, and environmental impact. Besides, the implementing process of sustainable recycling strategy was suggested by improving waste management practice, the socialization in recycling activities, supports of the NGOs, and the management of the government.

References

[1] Shamshiry E, Nadi B, Mokhtar M B, Komoo I, Hashim H S, and Yahaya N 2011 Integrated Models for Solid Waste Management in Tourism Regions: Langkawi Island, Malaysia. *J. Environ. Public Health* 2011 P.1-5
[2] Ezio R, Stefano A, Irina A I, and Tiberiu A 2014 Municipal solid Waste Management in Italian and Romanian Tourist areas *UPB Sci Bull*, 76(2): p. 277–288.
[3] Mateu-Sbert J, Ricci-Cabello I, Villalonga-Olives E, and Cabeza-Irigoyen E 2013 The impact of tourism on municipal solid waste generation: The case of Menorca Island (Spain) *Waste Manag.* 33: pp. 2589–2593.
[4] Suchada P, Cholada K, Trankler J, and Scholl W 2003 The role of formal and informal sectors in solid waste management of developing countries (Italy: International Waste Management and Landfill Symposium).
[5] Wilson D C, Velis C, and Cheeseman C 2006 Role of informal sector recycling in waste management in developing countries *Habitat Int.* 30 p. 797–808.
[6] Salau O, Sen L, Osho S, and Adejonwo-Osho O 2016 Empirical Investigation of Formal and Informal Sectors in Waste Recycling of the Municipal Waste Management System of Developing Countries: The Case Study of Lagos State *J. Environ. Ecol.* 7 p. 21-33.
[7] Troschinetz A M and Mihelcic J R 2009 Sustainable recycling of municipal solid waste in developing countries *Waste Manag.* 29 p. 915–923.
[8] Dept of Commercial and Tourism, 'Interview data', 2017.
[9] Song Toan P P, Giang H M and Takeshi F 2017 Analyzing solid waste management practices for the hotel industry *Glob. J. Environ. Sci. Manag.* 4 p. 19–30.
[10] Giang H M, F. Takeshi, and Song Toan P P 2017 Municipal Waste Generation and Composition In A Tourist City - Hoi An, Vietnam *J. JSCE* 5, p. 123–132.
[11] Giang H M, Takeshi F, Song Toan P P and Kim Thai N T 2017 Predicting waste generation using Bayesian model averaging *Glob. J. Environ. Sci. Manag.* 3 p. 385–402.
[12] Loan L T T, Nomura H, Takahashi Y and Yabe M 2017 Psychological driving forces behind households' behaviors toward municipal organic waste separation at source in Vietnam: a structural equation modeling approach *J. Mater. Cycles Waste Manag.* 19 p. 1052–1060.
[13] Chang N B and Pires A 2015 Sustainable solid waste management: a systems engineering approach. *Hoboken, New Jersey: John Wiley & Sons Inc.*
[14] GEFSGP 2012 Building a socialized model of domestic waste management in Hoi An *Global Environment Facility Investing in Our Planet* (VN/SGP/0P4/Y3/CORE/10/07).

Acknowledgments

We are thankful to The People Committee, Environmental and Resources Division, Commercial and Tourism Division of Hoi An City for their support. Also, the authors indebted to Okayama University for the financial assistance.

Sustainable Design of Curved Wood Rice Barrel with Healthy Considerations

F L Chao[1], C K Wu[1], H S Hsieh[1]

1 Department of Industrial Design, Chaoyang University of Technology, Taichung 41349, Taiwan R.O.C.
E-mail: jkwu@cyut.edu.tw

Abstract: The impacts of plasticizers on the human become a controversial topic, avoiding the effects of plasticization is one of the demands of wood barrel design. Through the green design and the bentwood forming skill, we consider the environmental factors in the design stage. By collaborating with bentwood manufacturers, we emphasizing on humidity and insect prevention and design bentwood rice barrel for the small household. The proposed rice barrels have the following features: using of green materials, reducing the environmental impact of processing, providing options for appropriate size by modular design. During humidity test, a curled wood and an equivalent plastic rice barrel were chosen and checked their difference on humidity absorption. High relative humidity (65%) was observed within the plastic barrel. The absorption performance of bulk wood is better than that of plywood. Based on measurements, we will refine design by combining the advantages of materials in future.

1. Introduction

Excessive unsustainable consumption of resources may cause considerable damage to the environment. Sustainable design is a design process [1] in which the designer, in response to possible environmental cost to reduce the impact of a product in the design phase by considering into the material, transportation, processing, and use of the product. The green design enables designers to look detail to adopt the proper concept of reducing the impact of the environment. "The green design reflects the reflection of the environment and ecological damage caused by modern science and technology culture, as well as the return of the designer's moral and social responsibility." Based on this view, we incorporate bentwood forming technology [2, 3] in the design process so that both environment and health concerns can be met. The three objectives of this study are:

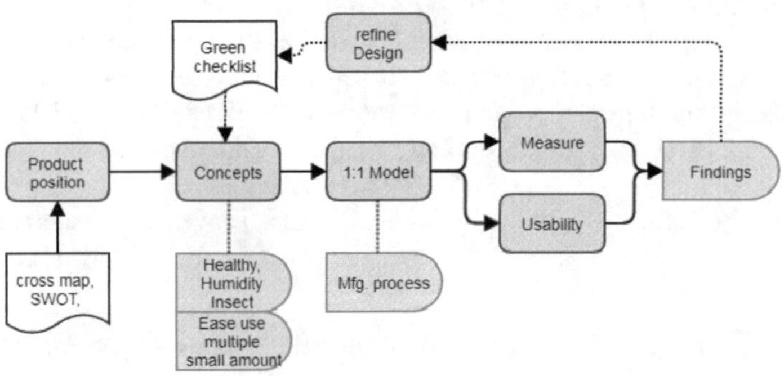

Figure 1. The research flow.

1. To design a rice barrel using bentwood based on scenarios for a small, healthy household. The emphasis of the design is on humidity prevention and insect prevention, and a proposition fit the target household's need;
2. Measure humidity absorption performance of rice barrel prototype model. We select an equivalent plastic rice barrel, and check their difference on humidity absorption, and evaluate humidity absorption performance of bulk wood and plywood;
3. Based on measurements, we refine design concept by combining the advantages of materials.

Figure 1 indicates the research flow of this study.
The intention is to involve scientific data in the design process. Designer can test the system and implement the required improvements. In re-design phase, we acquire new idea to address the finding during the test. After that, some details regarding the usability are provided. The notion of design-based research has been introduced by Collins [4], who described design experiments as a way to advance design and practice. Based on collaboration among designers and practitioners in real-world settings, we adopted this iterative evaluation-driven design-based research approach.

2. Literature

2.1. Green design
Green design is also known as eco-design, environmentally friendly design. Although the argument behind each name is different, the content is consistent. The basic idea is: environmental factors and pollution prevention measures are taken into account in the design stage; the objective of performance parameters are set as the product design goals and the starting point and strive to decrease the product's environmental impact. In [5], authors pointed out that the green design has the following characteristics: safety, reducing materials or using renewable materials, and avoiding improper ecological pollution. Also, Rebecca Proctor [1] pointed out that the sustainable environmental design has the following characteristics: biodegradable, fair trade, local materials, low energy/ waste, and no toxins.

2.2. Bentwood technology
Composite wood assembled with smaller pieces of wood glued together. Wood bending is a specialized technology [6]. After an extended period of development, varieties of shapes successfully developed with anti-moisture, anti-torsion ability. In bentwood process, two kinds of sliced wood are bent and stack; place into the steam tank, softening it by using a hydraulic bending machine, and proceed with compress. The wood bending stereotypes usually divided into dry and high-temperature treatment types, such as high-frequency drying and microwave drying [6]. Stacking bending and high-frequency drying are utilized in the present study.

2.3. Wood and humidity
Wood is a hygroscopic material; it naturally takes on and gives off water to balance out with its surrounding. Wood can absorb large quantities of water before reaching moisture content levels [7]. Moisture content (MC) is a measure of how much water is in a piece of wood relative to the wood in oven dry. Dry wood's moisture content is less than 19%. Fiber saturation is the level of moisture content where the cell walls are holding as much water as they can. Decay can get started if the moisture content of the wood is above fiber saturation (28%). Below this, the water content in the cell cavity will not affect the dimension of the wood [7]. Wood will reach equilibrium moisture content (EMC) that yields no vapor pressure difference between the wood and the surrounding air. The timber used indoors will eventually stabilize at 8-14% moisture content which allows the wood to function as a natural humidity controller.

2.4. Rice storage

IOP Conf. Series: Earth and Environmental Science **159** (2018) 012047 doi:10.1088/1755-1315/159/1/012047

Relative humidity and temperature of the storage environment are the most critical factors affecting maintenance of seed quality during the storage period. Of these two factors, relative humidity is most important because of its direct relation to seed moisture content [8]. The airtight storage means that the store is sealed to prevent air exchange with the ambient atmosphere. For large amount storage, container or storeroom was popular for a long period and less frequently access. In rural areas, hermetic storage is a cost-efficient and chemical-free insect control [9, 10]. Although plastic package can achieve hermetically; it cannot maintain airtight once open the seal.

Maintenance of relative humidity in a storeroom substantially lower than the ambient condition successfully enhanced with a moisture vapor barrier in walls, floor, and ceiling. Moisture vapor diffuse caused by a difference in pressure or relative humidity. The small size of H_2O molecular makes it difficult to stop the water vapor diffusion. The possible approach is O-ring combine with pressure on the cap.

In [11], an airtight storage of brown rice with low moisture content was tested and evaluated as an alternative to refrigerated storage. Brown rice samples (500 g each) at 16.2- 11.0% moisture contents (m.c.w.b.) were stored in plastic bottles for 6 months at 25° C. It was found that germination rate with 16.2% m.c.w.b. at 25° C declined from 97% to 27%. Low-moisture-content storage could be as effective as low-temperature storage. In [12], the storage's oxygen concentration dropped to 2.7% within 30 days, and carbon dioxide rose to 9.1%. The change in gas composition was caused by insect respiration. Hence, insect prevention is also important.

3. Design of Rice Barrel

Many creative techniques have been practiced in concept generation stage. At the beginning of the design, the divergent thinking was used to identify the possible concepts. After the initial direction was found, the convergence of thinking can be used in order to avoid too much information to interfere with the design.

3.1. Positioning and affinity analysis

Positioning analysis help designers to determine the relationship between existing products correctly, and find own opportunities according to the material, functional structure, shape, etc. Figure 2 shows the location of wood barrel by positioning analysis. The existing commercial barrels have a large volume; some are not easy to classify or stack. In order to meet the habits of modern life, the bentwood barrels with healthy considerations will be our target domain.

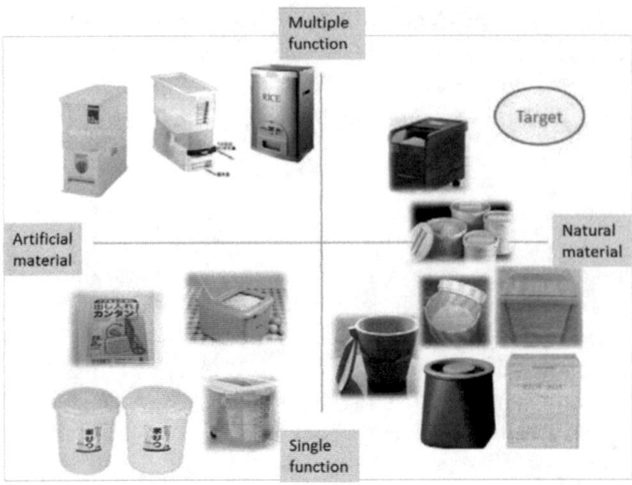

Figure 2. Cross positioning analysis of rice barrel.

SWOT analysis helps reveal of Strengths, Weaknesses, Opportunities, and Threats in competitive markets which identified by the external and internal environment. We will pay attention to the modern people focus on health trends, natural materials to development of small capacity barrel with flexible storage. High-quality design can increase the product market penetration and improve consumer acceptance. Bentwood products need to invest in high manpower, resulting in higher production costs. On the Weaknesses hand, if consumers are unfamiliar with the new product, the product receives less attention. The existing low-cost plastic products (lower-left of figure 2), we cannot rely on low price to get competitiveness.

KJ method is known as Affinity Diagram. It is a way to sort out ideas from complex factors and organized to find a solution to the problem. From the KJ analysis in figure 3, there are three tasks to fulfill namely, healthy, elegant design, and sustainability. The main design issues are the use, the cost of manufacturing, and the environmental assessment considerations. In-depth understanding of the production and promote benefit of natural material are important.

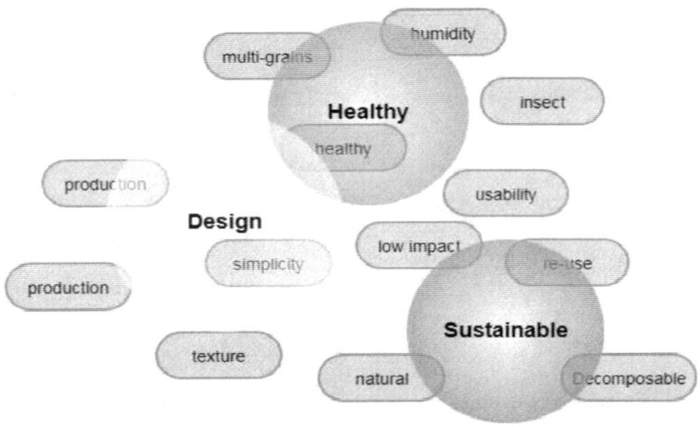

Figure 3. KJ analysis of bentwood barrel.

3.2. Design implementation

After confirming the design direction, we carry out the design steps such as concept development, sketching, and model making. Eating habits shifted to less amount, as well as more variety of grains mixed. We also considered the small packaging capacity with several choices. To reduce processing steps, we chosen simplicity elements (square, rectangular, round shape). The modular and stackable design reduces space required for transportation. The transparent window can be included to identify the internal contents quickly. The module has the same cross-sectional area, and the height of bentwood is changed to allow different capacity.

Prototype model (shown in figure 4) was made in the Canstar a local expertise on bentwood [3]. The barrel bottom uses cypress for pest control and enhances the aesthetic quality. Choosing faster growth wood (beech or birch wood) causes less environmental burden. It can be single barrel or a three-tier structure that allows different rice or grain storage. If bentwood processing can be standardized, the modular design can enable mass production. The consumers are free from plastic barrels, which may have plasticizer ingestion concerns.

Figure 4. Rice barrel during utilization.

3.3. Manufacturing consideration

Prototyping stage was carried out in Canstar. To increase the strength of the curved joints and manufacturing yield, engineer arranged the length of the wooden layers. The joining surface progressively terminated from inside to outer layers (10-degree skew). In figure 5, we found laminated wood joint traces run smoothly on the inner and outer walls (as shown in the tags). Through the thickness compensation, they built a constant thickness wall after the hot pressing process.

(a) (b)

Figure 5. The junction regions of wood barrel (a) inner surface, (b) outer surface. Notice the offset between the junctions of two sides (indicated by 2 tags).

4. Humidity Test

4.1. Compare plastic and wood barrel

We adjusted the moisture content of the white rice to the same level, by placing in the base floor for two days. It is summer time; the average temperature was 32°C. Then we put rice evenly into plastic and wood barrel. During usage, the cover lid is not fixed. For comparing with the same situation, the caps of a plastic barrel are placed without clamping two sidebars (figure 6).

Firstly, we placed both samples in the living room with proper ventilation. After one day, we moved both samples to underfloor without ventilation. The relative humidity (RH) is recorded using four Onset data loggers. The sampling rate is 20 min each. For comparison, four data loggers were placed (figure 6-left) within two rice sample barrels, the rice surface, and the ambient.

Relative humidity was measured according to the method as specified. Figure 7 is moisture change during three days. Researchers observed high relative humidity (65%) within the plastic barrel. In the wood barrel, it is 62%. According to the environment, relative humidity is high (above 67%) in the evening and low (50%) during the day. There is several hours' response delay within the barrel. It suggested that low-moisture storage could be as effective as low-temperature storage [13]; hence the wood performs better.

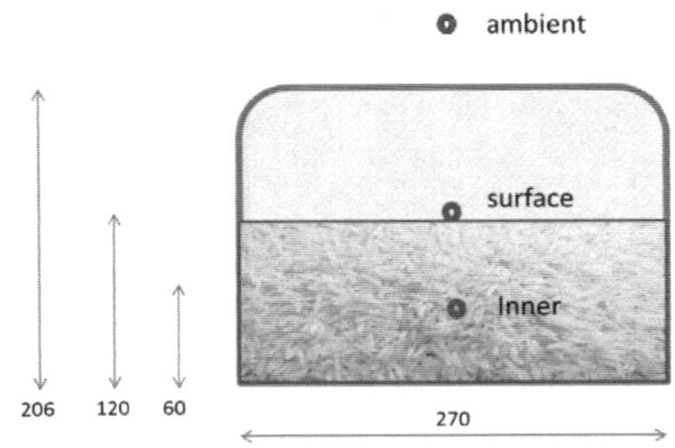

Figure 6. Relative humidity: (a) measurement setup, (b) comparing plastic and wood barrel.

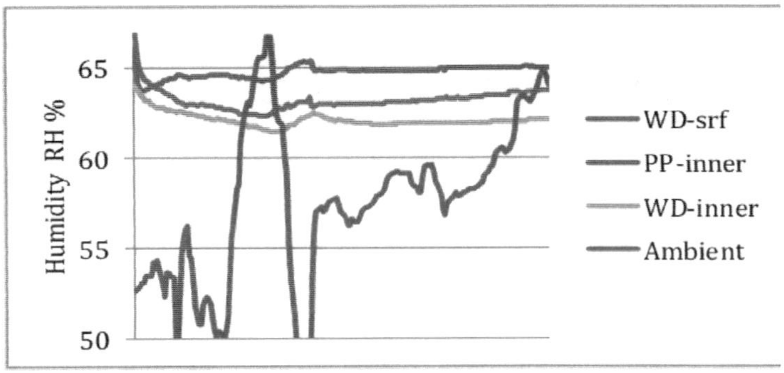

Figure 7. Relative humidity measured results.

4.2. Compare buck wood and plywood

What is the difference of absorption between different wood materials? The curved wooden barrel is made up of two parts, the cylinder is bentwood, and barrel's bottom and lid are solid wood. Are there any differences in the water absorption characteristics of the bulk and the plywood?

Experiment 2 compares the water absorption characteristics of these two materials. Because the plastic bucket has a sealed feature, so we put rice into the plastic bucket and then insert two pieces wood into the rice. After researcher closed the lid, the wood will absorb the moisture in this confined space. We placed the bucket on the basement floor, the solid wood and plywood were placed individually (shown schematically in Figure 8). The period for measurement is 30 minutes, and sampling interval is 20 seconds. Figure 9 shows the change in the moisture content of barrel during 30 min. We can see the gradual decline in the humidity for the solid barrel (with 6% difference). While we put the plywood into the barrel, the humidity decreased slowly (with only 1.5% difference). We can imagine that when the stack of the laminated process, viscous glue is used between different layers with pressure, so the adhesive decreased the permeability of wood, resulting in moisture absorption ability to drop.

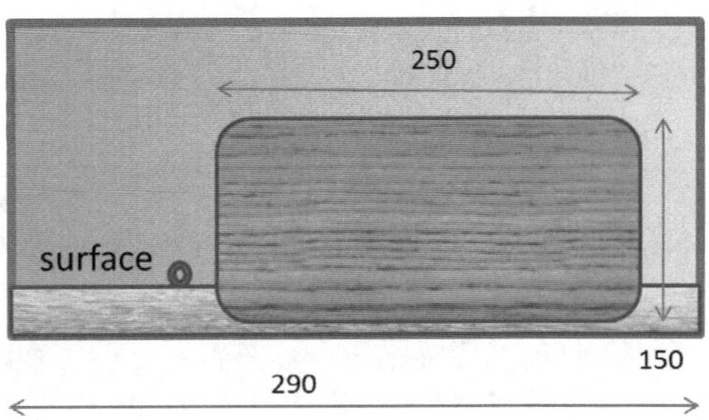

Figure 8. Schematic arrange of water absorption test of two materials

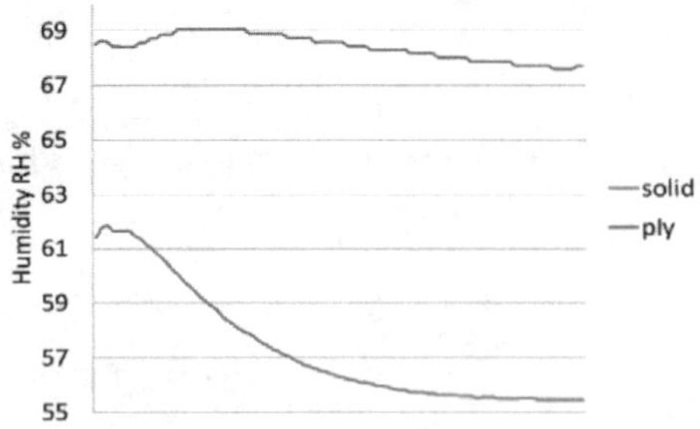

Figure 9. The moisture adsorption curves for bulk wood and plywood.

4.3. Design notice based on findings

How can we apply the findings of the experiment to the design? Since the wood and the wood are different in water absorption, so we can use their advantages. Curl wood can be used for the structural element; solid wood can be used in the internal support bar, the base, and the cover. We also need to pay attention; solid wood cannot be waterproof paint, to provide the original intention. For example, in the re-design phase, one can divide the rice bucket into three areas (white rice, brown rice, barley); then the partition board can be made with solid wood to reveal its advantages.

5. Conclusions

Health issue and the trends of diet lead habits of organic food and natural materials. Rice barrel of multiple usages was developed with curl wood. The design methods used in this study are parameter test and verification, KJ method, and SWOT. Through the design method and the mastering of bentwood forming technology, we considered the environmental factors and prevented the infiltration of pollution. We apply the green design checklists to evaluate its effectiveness. The possibility of technology application, design, and health can coexist.

The humidity test of moisture change during three days, high relative humidity (65%) was observed within the plastic barrel. There is several hours' humidity response delay within the barrel. For buck wood and plywood, there was 6% decline in the humidity for the solid wood and only 1.5% for the plywood. During the laminated process, the viscous glue used between different layers of wood decreased its permeability.

It suggested that low-moisture storage could be as efficient as low-temperature storage. Hence the wood barrel performs better than the plastic one. According to the findings of the experiments, curl wood can be used in curved structural parts; solid timber applied in the internal support bar, the base, and the cover. The future design will implement it to reveal its advantages.

6. References

[1] Rebecca Proctor 2013 Ecological design classic 1000- new eco designs and where to find them, Facebook group

[2] Jiang Haibo, Qiao Ying, Sun Hongbin 2008, Wood bending process research, forestry machinery and woodworking equipment, 2008, No. 36

[3] Dakun Quwood Furniture Co., Ltd. website, song introduction, website: http://www.kenstarwood.tw/html/index.html

[4] A. Collins, Towards a design science of education, New Directions in Educational Technology, E. Scanlon and T. O'Shea, eds., pp. 15-22, Springer, 1992.

[5] William Lidwell; Kritina Holden; Jill Butler Original, 2011 Design rule: 125 key laws affecting cognition, increasing beauty, making design better

[6] Zheng Shunfu 1989 Furniture Woodworking Ability Training Materials - Use of Special Press, Vocational Training Research and Development Center of China (PCF-CBM1103), Taipei.

[7] T. Genkawaa, T. Uchinob 2008 Development of a low-moisture-content storage system for brown

rice: Storability at decreased moisture contents, Biosystems engineering pp 515-522
[8] Al-yahya S A 2001 Effect of storage conditions on germination in wheat. Journal of Agronomy and Crop Science, 186(4), pp 273–279
[9] Moreno-Martinez E; Rivera A; Badillo M V 1998 Effect of fungiand fungicides on the preservation of wheat seed stored with high and low moisture content. Journal of Stored Products Research, 34, pp 231–236
[10] Thilakarathna B. Adhikarinayakea, 2006 Quality change and mass loss of paddy during airtight storage in a ferro-cement bin in Sri Lanka, Journal of Stored Products Research 42 pp 377–390
[11] De Lima, C.P.F. 1990 Airtight Storage: Principle and Practice, Food Preservation by Modified Atmospheres. CRC Press Inc, Boca Raton, pp 9–19
[12] Varnava, A., Navarro, S., Donahaye, E.J. 1995. Long term hermetic storage of barley in PVC-covered concrete platforms under Mediterranean conditions. Postharvest Biology and Technology 6, pp 177–186.
[13] Martha. Y. Quezada, 2006 Hermetic storage system preventing the proliferation of Prostephanus truncatus Horn and storage fungi in maize with different moisture contents, Postharvest Biology and Technology 39 pp 321–326.

CAE Supported Sustainable Product Development - Case Study of Compact Flow Guiding with Flame Lift Mechanism

W L CHEN[1] and F L CHAO[1]

[1] Department of Industrial Design, Chaoyang University of Technology, Taichung, 436, Taiwan, R.O.C
E-mail: flin@cyut.edu.tw

Abstract. The flame device design process and lift mechanism concerns were investgated. Combine qualitative and CAE simulation were proposed to fulfill its functions. Redesign structures allow the user to add fuel and cleaning base parts safely. For enhancing usability, airflow guidance structure cannot tolerate interference with the fuel intake. Within space constrain, we changed guidance channel to produce an ornamental flame with acceptable flame height. The cover is movably engaged with the base and selectively closes the opening. The lift mechanism comprises a supporting member and bearing allow moving to the predetermined height. Rigid body linkage and flow field simulated enables fine tune of geometric design. Functional test and measurement are carried out to confirm the product features. The measured average temperature of the metal chassis is 37.25 °C and that of glass is 141°C. The result shows successful simplicity mechanism and low surface temperature design.

1. Introduction

For enhancing usability, airflow guidance structure needs producing an ornamental flame with acceptable flame height; and lift container vertically within space constrain. Sustainable design are assessed based on functional, economic, and ecological criteria to achieve sustainability [1], [2]. The checklist [3,4] support for the analysis of a product's impact on the environment to avoid missing significant features. The designers work out systematically of redesign tasks on parts, function and product levels. The main strategies are (1) selection of low-impact materials, (2) reduction of material usage, and (3) modification of production techniques, and (4) reduction of impact during use.

1.1. CAE in Sustainable design

The burning and heat transfer mechanisms caused the high temperature of body and parts. The primary target of flame module redesign is providing new structures that allow user can quickly perform the task without potential safety issues. Computational fluid dynamics (CFD) techniques are the foundation to predict airflow. Finite volume method is a conventional approach for large problems and source term dominated flows. Detailed calculation steps by CAE can support engineer to get right parameters. The abilities of prediction and insight of problem make CAE simulation important in sustainable design.

1.2. Qualitative guidelines

General design guidelines enable the designer to understand trend and major factors. The previous flow guidance design required large space and more material. To maintain production cost and reduce raw material used, the size and shape of the airflow guiding channel need redesign to preserve ventilation path with minimum turbulence. The small size of guiding channel means lower natural convection flow caused by negative pressure.

High surface temperature causes user risk. From non-quantitative design guidelines, we can not only transfer the successful cases but also prevent possible pitfall that may happen. Those cases collected in different domains such as computer chassis, avionic facilities. Careful considering cases and guidelines will reduce risky design. The examples of the flow guidelines are as follows [5,6]:

- Air flow reversal may happen if inlet velocity is too high.
- Do not place the inlet of a blower directly downstream of the exhaust of another blower.
- Do not obstruct coolant flow, guide flow using splitters and turning vanes.
- Use gradual turns, enlargements, and contractions in flow passages.
- Install cooling fins vertically. Do not place parts directly above high dissipating elements.

Although some design guide prepared for electronic cooling, the fluid characteristic is the same with our combustion situation. For example, the effect of "Air flow reversal may happen, if inlet velocity is high" can be found in the real-world situation. As the velocity is increased, the static pressure will decrease. A high air velocity can result in such a low static pressure at the inlet [6]. When a flow reversal happens, the hot air will then be drawn in from the exit plenum. In Figure 1(d), a tapper inlet channel is used so that the cross-section area at any point is higher than that of downstream. Without qualitative guidelines, usually, the CAE simulation gets no direction for selecting proper geometric parameters.

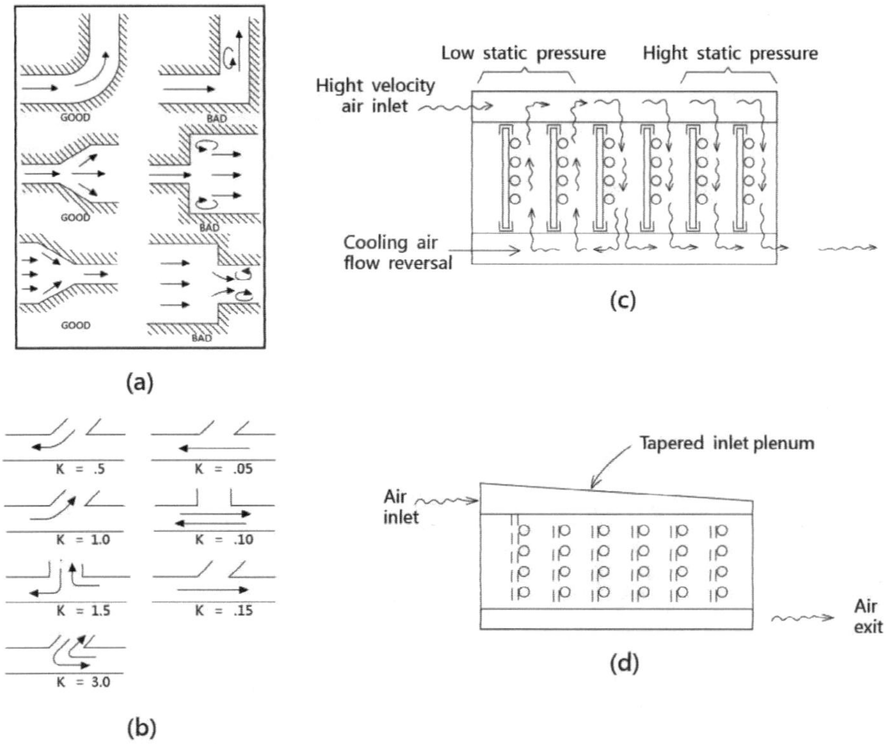

Figure 1. (a) Use gradual turns, enlargements, and contractions in flow passages; (b) Turning loss coefficients for fluids [5]; (c) Air flow reversal may happen if inlet velocity is high; (d) Improvement in the plenum design by tapper channel [6].

1.3. Lift Mechanism

Redesign structures must allow the user to add fuel and cleaning base efficiently and safely. Mechanical CAE allows three-dimensional simulation of moving parts operating in conjunction with one another. Multibody dynamics get right parameters and prevent ill mechanism before prototyping.
In [7], CAE was utilized to motivate student's creativity in product design. Indications for evaluation provided first, and then students analyzes and classifies the proposed mechanisms. Authors found the importance of pre-defined evaluation criteria for benchmarking a high number of potential mechanisms. Students put a sketch of their solution in a common folder and shared directory concept with group members. Considering functional blocks, we can derive proper approach efficiently [7].
In [8], a scissor lift having upper and lower platforms and actuated by a cable reeved to pull at least one set of the ends of the scissor together at its extended position. It reeved a cam roller interposed between the arms and working toward the pivot connection. In [9], a lift system incorporates a transverse bar which carries a patient lift system for movement between two laterally extending bars. The transverse bar is mounted at the same vertical height as the laterally extending bars and carries a motor for lifting. In our design, evaluation criteria were set by the available benchmark. Fuel adding structures are chosen by designer. The CAE is then utilized to determine suitable parameters that fit pre-defined compact space.

2. Case study of flame redesign

For enhancing usability, airflow guidance structure cannot tolerate interference with the fuel intake areas. The design goals are within space constrain,

- changing guidance channel that producing an ornamental flame with acceptable flame height.
- choosing lift mechanism that can move vertically without skew to ensure liquid fuel will not spill out.

2.1. Design procedure

Biofuel which is low oxygen consumption is the fuel to producing an ornamental flame. First, the digital model were built using CATIA solid modeling package. Depending on the type of simulation (flow or mechanism), digital model was discretized into elements or mesh for setting up the system equation with boundary conditions. Static solutions of field distribution or time depended on the motion obtained by solving the linear matrix equations.
The prototype model was built with the actual size and material so that engineer can measure corresponding physical parameters. The measured data are utilized to adjust the simulation parameters iteratively till results meet each other. When the simulation parameters are confirmed, digital model confidently presents a real system which reduces project developing cycle time.

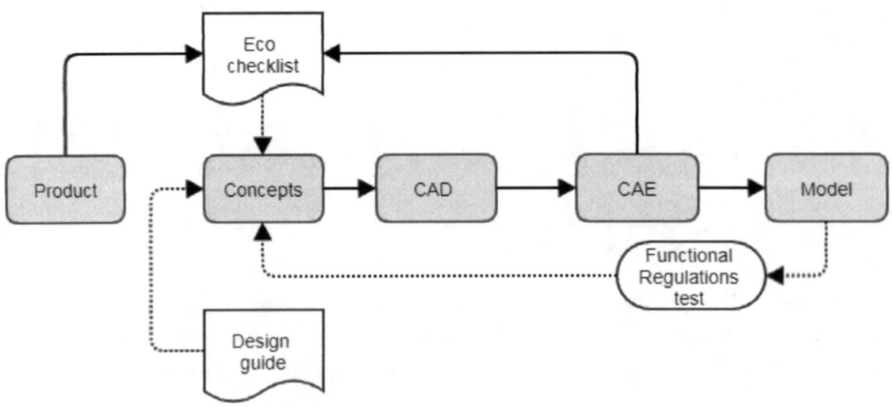

Figure 2. Design process

IOP Conf. Series: Earth and Environmental Science **159** (2018) 012048 doi:10.1088/1755-1315/159/1/012048

As shown in figure 2, the design guideline help prevent mistake arrangements. CAD/ CAE are used to realize the idea quickly. The flow field or mechanism simulation enables fine-tuning before a series of measurements. After the design process, the checklist was reviewed to confirm the design.

2.2. Flow simulation results

In the case of the flame design, the complicated phenomena make CAE simulation much harder. Flow channel analysis was performed using non-premixed combustion mode. With the alcohol fuel combustion equation to make sure the amount of oxygen supply is sufficient.

$$C_2H_5OH + 3O_2 \rightarrow 2CO_2 + 3H_2O \quad (1)$$

In non-premixed combustion, fuel and oxidizer enter the reaction zone in distinct streams [10]. Under certain assumptions, the thermochemistry can be reduced to a single parameter: the mixture fraction. The mixture fraction is the mass fraction that originated from the fuel stream. The profiles of scalar properties provide an alternative thermochemical relationship to that of chemical equilibrium [9]. Velocity vector of XZ section (figure 3) indicates the oxygen supply is sufficient. The upwardly convection is on the outer edge of the cylinder. The overall flow line velocity diagram is helically raised. Table 1 suggests parameters based on simulation and measurement.

Table 1. Analysis results of exhaust port velocity

Quantization value	entry value	Outlet value
Amount of airflow	3.1995(g/sec)	3.278(g/sec)
Oxygen content ratio	0.232	0.1948
Average speed	0.120(m/sec)	1.273(m/sec)

Figure 3. Velocity of XZ section, Aluminium castings and iron plate metal, metal contact temperature is not higher than 60 degrees. (transparent parts: heat-resistant glass)

IOP Conf. Series: Earth and Environmental Science **159** (2018) 012048 doi:10.1088/1755-1315/159/1/012048

2.3. Mechanism simulation results

A flame structure can lift a flame to a predetermined height includes a base, a cover, and a lift mechanism. The base consists of a chamber with an opening; fuel is adapted to contain within the fuel container (figure 4). Flame produced after igniting the fuel. The cover is movably engaged with the base and selectively closes the opening. The lift mechanism includes a supporting member adapted for sliding. This supporting member being moved down from the predetermined height to the different height allow user to ignite the fuel safely and to facilitate the removal of the fuel container.

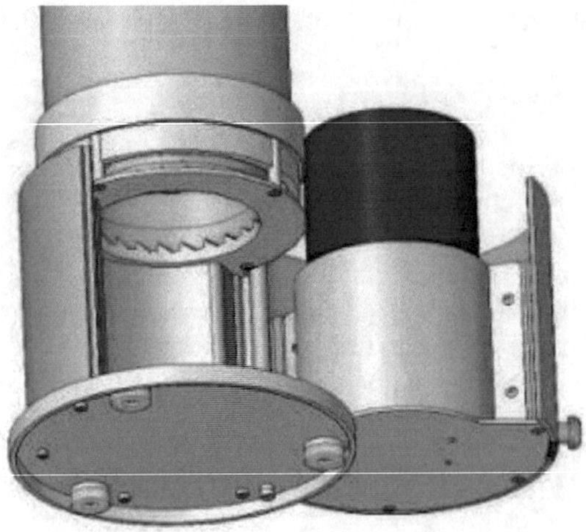

Figure 4. Perspective view of a flame device including a lift mechanism for moving a flame to a predetermined height where the cylinder is fuel tank.

Figure 5. Cross-sectional view of lift mechanism during moving a flame to a predetermined height

By lifting the alcohol tank onto the blade ring, making the product more convenient to use. As the use of lifting structure is very often, different travel range were analyzed (figure 5). Alcohol tank placed under the plane pressure of 50 Newton (alcohol tank five times the weight), because the fixed point set in the bottom of the four screw holes, the material is aluminum, were analyzed stress and deformation. Upward travel 50%, the local stress is mainly concentrated on the pin connecting area of the hitch and the push rod. The nail in the middle has a small amount of stress concentration, the maximum stress is 43.335 and the safety factor is 6.461. Note that the container keeps in horizontal posture. Hence fuel can stay inside safely.

2.4. Prototyping measurement

The redesign of the flame height is lower than the previous. Although the low diversion angle (below 40 degrees) caused outlet oxygen ratio slightly increased (complete flammability), the benefits are small. Prototyping models (figure 6) are used in the test to find better parameters. The main bodies of the ring and spiral material are steel. According to the regulations, we surveyed the most stringent CSA specification for the stove. There is specification about the glass part of the open fire, but there is no high-temperature limit. Manufacture attach warning mark on the glass. Thermal measurement Figure 6 showed average temperature of the metal chassis is 37.25°C (figure 7) which are touchable with finger; the glass temperature is 114°C which is also lower than related products. It revealed the effectiveness of the CAE flow simulation.

Figure 6. Prototyping model with fuel tank lifting mechanism shown detail of the guide vane (where central of gravity locate: X= 0mm，Y= 0mm，Z= 132.405mm)

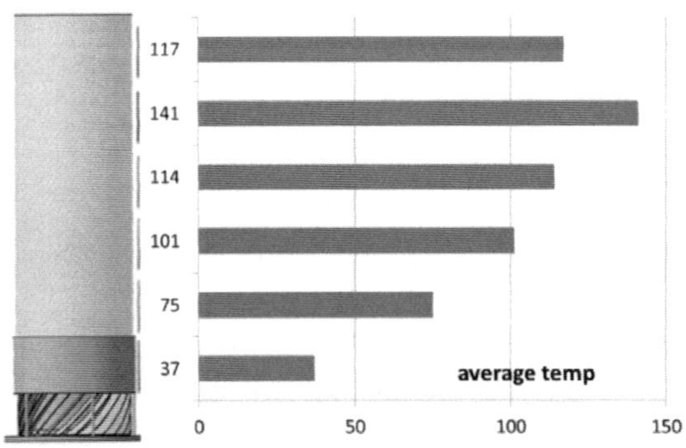

Figure 7. Measurement of surface temperature uses the prototyping model. Measured average temperature of the metal chassis is 37.25°C which are touchable with finger; and that of the glass: the highest part (upper section) is 141° and that of central section is 114°

3. LCA

Industry trend promoted low-carbon production. We selected life cycle assessment (Eco-indicator 99) [10] using carbon footprint as an indicator, category from cradle to the door. The module of case study contains the raw materials (mining and manufacturing), transportation, processing stage till the module is completed [11,12]. Figure 8 explodes perspective view of principal parts. The fuel is ethanol based. The cover is movably engaged with the base and selectively closing the opening. The cover (#2) includes a peripheral wall including a receptacle (#25) extended therein. The first fixing edge is

pivotally engaged with the base, with the first fixing edge pivotally engaged with the first engaging edge (#112) of the lateral wall (#11) of the base.
Carbon footprints of the manufacturing stage are summarized with Eco-it in table 2. Flame module's total carbon footprint is 9.34 $KgCO_2$ and that of processing stages 4.6 Kg which is 40% less than the previous model. Miniature of the flame size and guidance structure, although the mechanism added to provide new functions and keep the Carbon footprints in lower level.

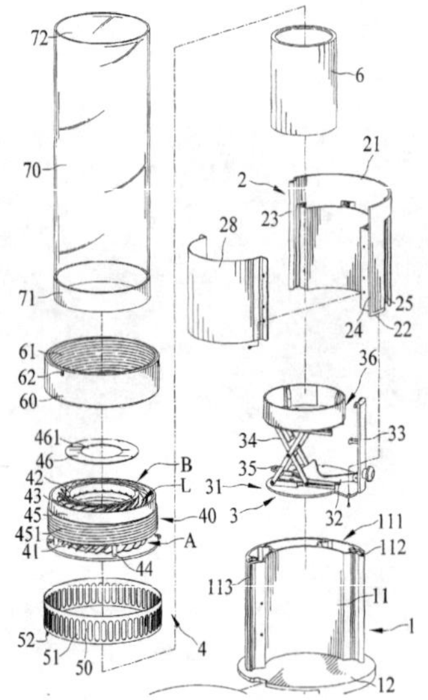

Figure 8. Parts of flamer (exploded perspective view) [13], where (#42) is flow guidance rib and channel, (L) indicate skew angle, (#40) is module, (#12) is base, and (#34,35) are the linkage bar.

Table 2. Raw material carbon footprint: 9.34 ($KgCO_2$)

Part No.	Name	Quantity	Weight (Kg)	Material	Footprint
1	Glass	1	0.921	Glass	0.51
2	Core	1	0.783	cast iron	1.16
3	Disc	1	0.402	White iron	1.9
4	Shell	1	2.082	cast iron	3.1
5	Wick	4	0.064	Stainless steel	0.16
6	Light body	1	0.108	White iron	0.51
7	Base	1	1.351	cast iron	2.0

4. Conclusions

It becomes increasingly essential for company in addressing eco-performance to satisfy global environmental regulations. We use combine qualitative and CAE design process for flame device including a lift mechanism. The case study shows successful simplicity mechanism and low surface temperature. Design process helps understand natural convection and four bar linkage phenomenon.

Through the model test, modify and verify the rationality of the simulation parameters, the usability and safety features are greatly enhanced.

5. References

[1] Ramani K, Ramanujan D, Bernstein W, Zhao F, Sutherland J, Handwerker C, Choi J, Kim H, 2010 Integrated sustainable life cycle design: a review ASME J Mech Design p 132
[2] Arnette A., Brewer B, Choal T 2014 Design for sustainability (DFS): the intersection of supply chain and environment. J Clean Prod 83 p 374-390
[3] Spangenberg J, Fuad-Luke A, Blincoe K 2010 Design for Sustainability (DfS): the interface of sustainable production and consumption J Clean Product 18 p 1483-1491
[4] Kara S Honke I Kaebernick H 2005 An Integrated Framework for Implementing Sustainable Product Development, Environmentally Conscious Design and Inverse Manufacturing 2005. Eco Design Fourth International Symposium p 684-691
[5] Morrison G N 1983 Thermal guide for reliability engineer, Hughes.
[6] Dave S. Steinberg, Cooling Techniques for electronic equipment, John Wiley & Sons, 1980
[7] Fauroux, J. C., Bouzgarrou, B. C., & Gogu, G. 2004. Innovative mechanism design with CAE software. In DS 33: Proceedings of E&PDE 2004, the 7th International Conference on Engineering and Product Design Education, Delft, the Netherlands, 02.-03.09. 2004.
[8] Capaldi, G., & Sinreich, M. G. (1998). U.S. Patent No. 5,809,591. Washington, DC: U.S. Patent and Trademark Office.
[9] Selyutin, L., & Zhao, J. (1999). U.S. Patent No. 5,951,776. Washington, DC: U.S. Patent and Trademark Office.
[10] Liew S K, Bray K N, Moss J B 1981 A flamelet model of turbulent non-premixed combustion. Combustion Science and Technology, 27(1-2), p 69-73
[11] PRé Consultants, Goedkoop M, Oele, M, Schryver A, Vieira M, Hegger S 2010 SimaPro Database Manual Methods library vol.2.4, p 16-17
[12] Huang Shijie, Chen Jiahao 2015 Research on Carbon Footprint Evaluation of Mechanical Equipment Products, Int. Conference on sustainability development p B-1
[13] Chen W L 2013, US Patent Pub. No. 2013/0011800 A1.